W9-BBO-893

Combinatorics of Experimental Design

Combinatorics of Experimental Design

ANNE PENFOLD STREET
Department of Mathematics, University of Queensland
and
DEBORAH J. STREET
Waite Agricultural Research Institute, The University of Adelaide

CLARENDON PRESS · OXFORD
1987

Oxford University Press, Walton Street, Oxford OX2 6DP
Oxford New York Toronto
Delhi Bombay Calcutta Madras Karachi
Petaling Jaya Singapore Hong Kong Tokyo
Nairobi Dar es Salaam Cape Town
Melbourne Auckland
and associated companies in
Beirut Berlin Ibadan Nicosia

Oxford is a trade mark of Oxford University Press

Published in the United States
by Oxford University Press, New York

British Library Cataloguing in Publication Data
Street, Anne Penfold
Combinatorics of experimental design.
1. Combinatorial analysis
I. Title II. Street, Deborah J.
511'.6 QA164
ISBN 0-19-853256-3
ISBN 0-19-853255-5 Pbk

Library of Congress Cataloging in Publication Data
Street, Anne Penfold.
Combinatorics of experimental design.
Bibliography: p.
Includes index.
1. Experimental design. 2. Combinatorial designs and configurations. I. Street, Deborah J. II. Title.
QA279.S77 1987 519.5 86-12587
ISBN 0-19-853256-3
ISBN 0-19-853255-5 (pbk.)

Set by Macmillan India Ltd., Bangalore 25.
Printed and bound in Great Britain by
Biddles Ltd, Guildford and Kings Lynn

Preface

The study of designs is a large and rapidly growing area of mathematics, and it is not possible to cover all its aspects in a book of this size. There is an obvious dichotomy in the literature of designs; they are considered as incidence structures by combinatorialists, and as experimental plans or layouts by statisticians, and members of each of these groups are sometimes unaware of related developments and problems arising in the other area. We aim to bridge this gap by providing the background necessary to make the combinatorial aspects of statistical literature more easily accessible to combinatorialists, and vice versa.

The areas of design theory which we have not covered include: efficiency-balance, variance-balance, and consequently all considerations of optimality; analysis of experiments; t-designs and their automorphism groups; packing and covering designs, row and column designs, n-ary designs, Doehlert–Klee designs, Howell designs, Room squares, magic squares, Youden squares, F-squares, (r, λ) arrays, and large sets of pairwise disjoint designs. We have considered the relationship between designs and graphs only in so far as we need it for the representation of designs, and we have not dealt with the relationship between designs and error-correcting codes.

We have concentrated solely on designs of interest in both statistics and combinatorics, with pairwise balance as the unifying concept, and we deal with constructions and existence results for such designs, and with their properties.

Chapter 1 outlines the early history of designs, their relation to linear models, and their use as sampling schemes. Chapters 2, 3, and 4 cover basic properties of balanced incomplete block designs, standard constructions using difference sets, the structure of designs as described by their automorphism groups, and the question of which designs are the irreducible building-blocks from which all others can be constructed.

Chapters 5, 6, and 7 deal with Latin squares and families of Latin squares, and the relation between such squares and block designs. Chapter 8 takes up the topic of resolvability and in particular of affine geometries and their extension to projective geometries. Chapters 9 and 10 deal with factorial designs: first the symmetrical case, where the design is based on the flats of a finite geometry; then some constructions for the single replicate case, both symmetrical and asymmetrical.

Chapter 11 concerns partially balanced designs and their relation to planes and biplanes. Chapters 12 and 13 cover some of the existence results for balanced incomplete block designs: first, the symmetrical designs; next, designs with index one and given block size.

Finally, Chapters 14 and 15 deal with designs in which the arrangement of the treatments relative to each other, within each block, or within an array, is important.

Since we are aiming at an audience with a wide range of backgrounds, including third- and fourth-year honours students in statistics or combinatorics or both, we have assumed a reasonable knowledge of linear algebra but very little else. Number theory, finite field theory, and some essential statistical concepts are developed as needed. The accompanying diagram shows the logical dependence of the chapters.

To keep the list of references to a reasonable length, we have included only items of particular historical interest, or closely related to our treatment of the subject, and survey articles from which more of the literature can be followed up.

Many colleagues and students have made helpful comments on this manuscript and space does not allow us to thank them all. Those to whom we are especially grateful include the following: Kaye Basford, Norman Biggs, Elizabeth Billington, Derrick Breach, Clive Davis, Roger Duke, Ken Gray, Rudi Mathon, Sheila Oates-Williams, Peter Robinson, Christopher Rodger, Jennifer Seberry, Ralph Stanton, Bill Venables, Peter Wild, and Therese Wilson. Obviously all responsibility for errors and omissions rests with the authors.

During the writing of this book, one or other of us received support or hospitality from Auburn University, the University of Manitoba, the University of Queensland, the University of Adelaide, Imperial College (London), and the Australian Research Grants Scheme; we thank them all. We also thank Bill Wilson and Stan Eckert for drawing the excellent diagrams, and Carolyn Davey for a beautiful job of typing the original manuscript. Finally we thank our husbands, Norm Street and Bill Wilson, who endured.

St Lucia and Glen Osmond A.P.S.
1986 D.J.S.

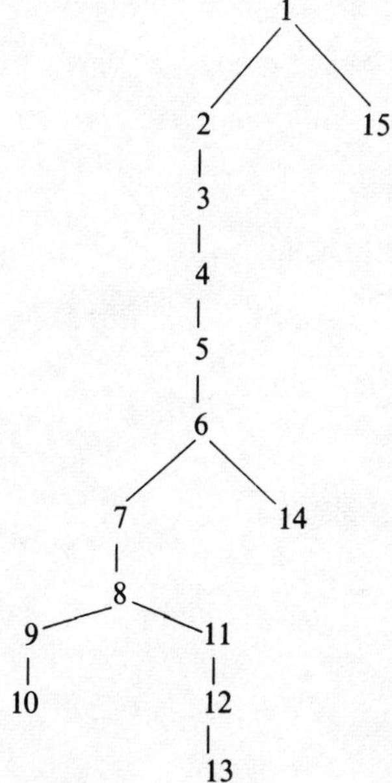

Logical dependence of chapters

Contents

Notation, terminology, and numbering

$\lvert X\rvert$	Cardinality (number of elements) of the set X.
$X \setminus Y$	Set of elements which belong to X but not to Y.
$X+y$ yX	If $X=\{x_1,\ldots,x_n\}$, then $\begin{cases} X+y=\{x_1+y,\ldots,x_n+y\}, \\ yX=\{yx_1,\ldots,yx_n\}. \end{cases}$
$GF[p^n]$	Galois field of order p^n.
Primitive element of $GF[p^n]$	Generator of multiplicative group of $GF[p^n]$.
$\binom{a}{b}$, read 'a choose b'.	Number of ways of choosing b objects from a set of a distinct objects, without repetitions.
$a \mid b$	The integer a exactly divides the integer b.
$lcm[a, b]$	Least common multiple of the integers a and b.
(a, b) or $gcd\,(a, b)$	Greatest common divisor of the integers a and b
Relatively prime integers.	Those with greatest common divisor equal to 1.
$[x]$	Integer part of real number x.
$(a_1 a_2 \ldots a_m)$	Cyclic permutation mapping a_1 to a_2, a_2 to $a_3, \ldots, a_m$ to a_1.
$(a_1 \ldots a_m)(b_1 \ldots b_n)$	Product of two cyclic permutations.
I, I_n	Identity matrix (subscript indicates size).
$J, J_{m\times n}, J_n$	Matrix with each entry 1.
j, j_n	Column vector with each entry 1.
$M \times N$	$[m_{ij}N]$, the Kronecker product of matrices M and N.
$\mathop{\times}\limits_{i=1}^{n} M_i$	Kronecker product $M_1 \times \ldots \times M_n$.
$\mathrm{diag}(x_1,\ldots,x_n)$	Matrix with ith diagonal entry x_i, off-diagonal entries 0.

Equations are numbered in sequence straight through each chapter. Theorems, Lemmas, Examples, and Constructions are also numbered in one sequence throughout each chapter. A Corollary of any Theorem or Lemma is given a number referring to that Theorem or Lemma. Thus, in Chapter 2, equation (2.8) falls in Section 2.2, and is referred to as (2.8) within Chapter 2, and as equation (2.2.8) within any other chapter. In the same chapter, Lemma 23 (following Example 21 and Lemma 22) in Section 2.6, is referred to as Lemma 23 within Chapter 2, and as Lemma 2.6.23 within any other chapter. Its Corollaries are numbered 23.1 and 23.2 respectively.

1 Introduction

1.1. Background

Combinatorics is the branch of mathematics which deals with the problems of selecting and arranging objects in accordance with certain specified rules. In particular, a *combinatorial design* is a way of choosing, from a given finite set, a collection of subsets with particular properties. For example, the sizes of the subsets and of intersections of pairs of subsets may be restricted, and so may the numbers of subsets to which individual elements or pairs of elements belong. Sometimes a design is best represented in two dimensions, by being written out as an array.

Probably the earliest systematic study of designs was that published by Euler in 1782, inspired by the problem of the 36 officers. The officers were chosen, six from each of six different regiments, so that the selection from each regiment included one officer from each of six ranks (the same six ranks in every case). The problem was whether it was possible for the officers to parade in a six-by-six formation, such that each row and each column contained one member of each rank and one member of each regiment.

Let us first consider just the regiment to which each officer belongs. An arrangement of the officers such that every row and every column contains one member of each regiment can be represented as a 6×6 array whose (i, j) entry is the name of the regiment to which the officer belongs who stands in row i and column j. If we name the regiments 1, 2, 3, 4, 5, 6, then such an array is shown in Table 1.1; many other solutions are of course possible. Notice that each row and each column contains the whole set 1, 2, 3, 4, 5, 6 in some order or, equivalently, no row or column can contain any of these numbers more than once.

Table 1.1

1	2	3	4	5	6
3	1	2	6	4	5
2	3	1	5	6	4
4	5	6	1	2	3
6	4	5	3	1	2
5	6	4	2	3	1

If we now consider just the rank of each officer, and if we name the ranks similarly 1, 2, 3, 4, 5, 6, then again we obtain a 6×6 array in which each row and column contains each number precisely once. But to solve the whole problem more is needed. For if the entry in row i and column j is x in the first array and y in the second, then the officer who stands in that position belongs to regiment x and rank y, and to solve the problem it is necessary that in the six positions where the first array has a 1, the second array has 1, 2, 3, 4, 5, 6 once each; and similarly for the positions where the first array has a 2, a 3, and so on. In other words, if we regard the two arrays as being superimposed on each other and thus defining a 6×6 array of ordered pairs, then each of the 36 ordered pairs (1, 1), (1, 2), . . . , (1, 6), (2, 1), . . . , (6, 6) occurs precisely once.

Euler, who wrote the two arrays in terms of Greek and Roman letters respectively, called the pair of superimposed arrays a *Graeco-Latin square.* Nearly one and a half centuries later, Fisher called each of the individual arrays with no repeated element in any row or column a *Latin square*, the name in current use. A pair of Latin squares of the same size which when superimposed contain every possible ordered pair of symbols are said to be *orthogonal.* Euler conjectured (and in 1901 Tarry proved) that no pair of orthogonal Latin squares of order 6 could exist. He conjectured further that no pair of orthogonal Latin squares of order n could exist for any $n \equiv 2 \pmod 4$, but in fact for $n \geq 10$, this conjecture was shown to be false in 1960 by Bose, Shrikhande, and Parker (Euler's spoilers).

The earliest mention of Latin squares seems to have been made in 1624 in connection with puzzles. By 1788, de Palluel had used a 4×4 Latin square to define an experimental layout. The sixteen plots of the experiment were sixteen sheep. There were four sheep of each of four breeds (the rows of the square) and one sheep of each breed was fed on one of four diets (the columns of the square). At four different times (the symbols of the square) four sheep were killed and the amount of saleable produce recorded for each. The four sheep killed at a given time included one sheep of each breed and one sheep on each diet. The actual layout is given in Table 1.2. (The experiment, in fact, showed that sheep could be fattened on a diet of root vegetables, a new idea at the time.)

Table 1.2. The Latin square used by de Palluel

		Diet			
		Potatoes	Turnips	Beet	Corn
Breed	Isle de France	A	B	C	D
	Beauce	D	A	B	C
	Champagne	C	D	A	B
	Picardy	B	C	D	A

Date killed: A = 20 Feb, B = 20 Mar, C = 20 Apr, D = 20 May

But it was Fisher who, in 1926, pointed out that Latin squares and families of mutually orthogonal Latin squares could be used systematically as experimental designs which allow for more than one source of variation in an experiment.

The property of being a Latin square can in fact be defined in terms of orthogonality, where we say that two $n \times n$ arrays, each containing elements from an n-set X, are *orthogonal* if the two arrays, when superimposed, contain every possible ordered pair of elements. Thus for the case of 3×3 arrays, Table 1.3 shows R where each element of row i is i, C where each element of column j is j, and two further arrays L_1 and L_2. To say that L_1 and L_2 are Latin squares is, in fact, equivalent to saying that they are simultaneously orthogonal to R and to C. They are also mutually orthogonal.

Table 1.3

1	1	1	1	2	3	1	3	2	1	2	3	1	2	3
2	2	2	1	2	3	2	1	3	2	3	1	4	5	6
3	3	3	1	2	3	3	2	1	3	1	2	7	8	9
	R			C			L_1			L_2			S	

What is more, they are closely related to another kind of design. Suppose that we take the last array, S, of Table 1.3 and consider it superimposed on R, C, L_1, L_2 in turn. We make a list of the three elements of S that occur with 1 when it is superimposed on R, namely 1, 2, 3, those that occur with 2, namely 4, 5, 6, and so on, giving altogether twelve lists, three from each of the four arrays R, C, L_1, L_2. These lists are shown as rows in Table 1.4.

Table 1.4. The design (D1)

1	2	3	1	4	7	1	5	9	1	6	8
4	5	6	2	5	8	3	4	8	2	4	9
7	8	9	3	6	9	2	6	7	3	5	7
from R			from C			from L_1			from L_2		

We shall disregard the ordering of elements in each list, and simply regard the twelve lists as twelve 3-subsets of the set $\{1, 2, \ldots, 9\}$. Not only does each subset have the same cardinality, but also each element occurs in the same number of subsets, namely four, and each unordered pair of elements occurs in precisely one of the subsets. This is our first example of a *block design*, which we denote by (D1). The subsets listed are called *blocks* and (D1) is, in particular, a *triple system*; that is, a design with three elements per block.

This particular triple system was described by Plücker in 1835, though he studied it in quite a different context, in connection with properties of plane curves of the third order. By 1839, Plucker was aware that if a triple system exists on v elements, where each unordered pair of elements occurs in precisely one block, then $v \equiv 1$ or 3 (mod 6). The same designs were studied by Woolhouse (1844), Kirkman (1847, 1850a) and Steiner (1853); they are now generally known as *Steiner* triple systems.

The design (D1) is also an affine plane of order 3. Let the field GF[3] consist of the integers 0, 1, 2 with addition and multiplication modulo 3. Then the points of such a plane can be represented by the nine ordered pairs

$$00,\ 10,\ 20,\ 01,\ 11,\ 21,\ 02,\ 12,\ 22$$

where we have abbreviated (x, y) to xy. Just as the lines of the real Euclidean plane may be described by the points satisfying equations of the form

$$ax+by=c$$

for various real constants a, b, c, so may the lines in the plane over GF[3] be described by the twelve equations

$$\begin{aligned} x&=c, & c&=0, 1, 2;\\ y&=c, & c&=0, 1, 2;\\ x+y&=c, & c&=0, 1, 2;\\ x+2y&=c, & c&=0, 1, 2. \end{aligned}$$

These lead to lines, each containing precisely three points, as shown in the rows of Table 1.5.

Table 1.5

$c=0$	00, 01, 02	00, 10, 20	00, 11, 22	00, 12, 21
$c=1$	10, 11, 12	01, 11, 21	02, 10, 21	01, 10, 22
$c=2$	20, 21, 22	02, 12, 22	01, 12, 20	02, 11, 20
	$x=c$	$y=c$	$x+2y=c$	$x+y=c$

This shows our design (D1) in its guise as an affine plane, where the fact that two points determine a unique line corresponds to the previous statement that each unordered pair of elements occurs in precisely one of the subsets. In each of the four classes, there are three lines. Two lines belonging to distinct classes intersect in precisely one point. Two lines from the same class never intersect and the lines in a class form a set of *parallels*.

Finite planes over fields of order n were first studied by von Staudt (1856) and more extensively by Veblen and Bussey (1906) and Veblen and Wedderburn (1907). The particular case we considered has three points per line and is thus a triple system. In general, an affine plane will have n points per line, but it is still an example of a block design. To see how such a design might be used in experimental work, we consider the following example.

Suppose that we have v varieties of wheat available, and that we need to know which one (or ones) will be the highest yielding in some particular agricultural region. To help decide this question, we choose n plots of land in the area, planning to sow each plot with one of the v varieties and then to treat all the plots in the same way. How should we allocate the v varieties to the n plots?

Suppose that we can reasonably make three assumptions:

(A1) the n plots are homogeneous, that is, they are essentially the same and can be expected to give the same yields, except for random fluctuation, if sown with the same variety and managed in the same way;

(A2) the inherent variability of all the plots is the same and is independent of the variety sown on the plot;

(A3) the yield of any plot is independent of the yield of any other plot.

(We note in passing that the second and third of these assumptions have been found reasonable in practice, and are generally made in such experimental work. The first assumption can only be made if it is consistent with information available from previous sowings.)

We let r_i plots be sown with variety i, so that $\sum_{i=1}^{v} r_i = n$. To avoid any source of bias, we should decide in some random manner which particular r_i plots receive variety i, perhaps by drawing numbered discs out of a bag or by using a random number generator. This is called a *completely randomized design*. But how should we choose the value for each r_i, $i = 1, 2, \ldots, v$?

In Section 1.3 we will show that to compare the yields of the v varieties it is reasonable to compare the v average yields, where the average yield for variety i is the mean of the yields of the r_i plots sown with variety i. Given the three assumptions above we will also show that the variability of these average yields decreases as r_i increases. Thus r_i should be as large as possible for each i. Hence we would like to have $n = vr$ and to set $r_i = r = n/v$, $i = 1, 2, \ldots, v$. (But see Section 1.7 and Exercise 1.8.)

Provided we can find a sufficiently large set of homogeneous plots, a completely randomized design gives the simplest method of comparing v varieties.

But suppose that a sufficiently large set of homogeneous plots is not available. What should we do? The usual approach, known as *blocking*, is to carry out the experiment using a family of smaller homogeneous sets of plots, the *blocks*. If each block contains v plots, then we can allocate each variety to one plot of each block, which is sensible if we have reason to believe that plots in different blocks have different properties. Such an arrangement is called a *randomized complete block design*, since the complete set of varieties appears in each block. But if, as is more often the case, the blocks contain k plots where $k < v$ then the method of allocation of varieties to blocks becomes a matter of some interest. A block is said to be *incomplete* if it does not contain all the v varieties. From now on we assume that each plot has one variety allocated to it, that no variety is repeated in a block and that a block is specified by its list of varieties. Note that some blocks may be repeated.

The designs now known as 'balanced incomplete block designs' (BIBDs) were introduced by Yates in 1935 in a paper read to the Royal Statistical Society. At that time he proposed the name 'symmetrical incomplete randomized blocks', and a year later the name 'symmetrical incomplete randomized block arrangement'. The current term was introduced by Bose in 1939. The term 'symmetrical BIBD' now refers to a BIBD in which the number of varieties equals the number of blocks.

One form of experiment fairly common in the 1930s was the comparison of v varieties in b blocks of size 2. One variety, called the *control*, occurred in every one of the b blocks; the remaining $v-1$ varieties occurred in $r = b/(v-1)$ blocks each, always together with the control. The difference between the control and any other variety, A, was then estimated directly, by taking the difference between the respective means in the blocks containing A; the difference between two non-control varieties A and B had to be estimated indirectly, by considering the differences between the control and A, and between the control and B.

This led to the comparison of A with B being made with less precision than the comparison of either with the control.

Yates pointed out that if the same number of blocks of size 2 were used, but laid out so that each of the $v(v-1)/2$ pairs of varieties appeared equally often, then every comparison could be made equally precisely. Thus, this 'symmetrical' arrangement of varieties was preferable.

In both his 1935 and 1936 papers, he discussed optimal choice of block size and showed that the desirable properties of the designs still held with constant block size larger than 2, provided only that:

(i) no variety appeared more than once in a block;
(ii) every variety appeared equally often;
(iii) every (unordered) pair of varieties appeared in the same number of blocks.

(These three properties, with the assumption of constant block size, define a BIBD. Indeed, assumption (ii) follows from the others.) The 1936 paper also discussed construction of some small BIBDs; Fisher and Yates in 1938 gave a table of small designs and Bose in 1939 the first systematic methods for their construction. Bose in particular exploited the properties of finite geometries and their applications as designs.

Before looking at designs in general, we note one especially important property of the design (D1). The blocks, as written in Table 1.4 and Table 1.5, are arranged in four classes of three blocks each, and the three blocks of any one class contain between them all the points of the design precisely once. This is similar to the situation in real Euclidean geometry where all the lines with given gradient contain between them all points of the plane precisely once. In other words, one point of a line, together with its gradient, determines the line uniquely. This property is known as *resolvability*. It is important from a statistical point of view, because the union of the blocks of a parallel class (or resolution class) of such a design can be regarded as though it were one block of a randomized complete block design, thus simplifying the analysis of the results. It is important also from the combinatorial point of view, because it allows the design to be extended, just as the real affine plane can be extended to the projective plane.

For instance, starting from the representation of (D1) given in Table 1.5, we shall adjoin to the design four new points, namely 0, 1, 2, ∞, each of which labels a class. We shall adjoin to each block one extra point, namely its class label. Finally, we adjoin one new block, consisting of the four new points. This gives the design (D2) shown in Table 1.6, which is a projective plane. Each pair of elements occurs in precisely one block; each pair of blocks intersects in precisely one element.

Table 1.6. The design (D2)

00, 01, 02, ∞	00, 10, 20, 0	00, 11, 22, 1	00, 12, 21, 2
10, 11, 12, ∞	01, 11, 21, 0	02, 10, 21, 1	01, 10, 22, 2
20, 21, 22, ∞	02, 12, 22, 0	01, 12, 20, 1	02, 11, 20, 2

0, 1, 2, ∞

1.2. Designs

We now look more closely at what is meant by the term 'design'.

Let X be a finite set of points and let $\mathscr{B} = \{B_i | i \in I\}$ be a family of subsets, B_i, of X. The subsets are called *blocks* and the pair $(X, \mathscr{B})$ is called a *design based on the set* X. The *order* of a design $(X, \mathscr{B})$ is $|X|$, the cardinality of X, and the set $\{|B_i| | B_i \in \mathscr{B}\}$ is the set of *block sizes* of the design. Notice that not all the blocks need be distinct; in fact we shall look later at applications of designs with repeated blocks. A design is said to be *incomplete* if at least one of its blocks is a proper subset of X. Thus, the varieties in an experiment correspond to the points of X, and the randomized complete block design to r blocks, each equal to X. The two examples given above are incomplete: (D1) has order 9, (D2) has order 13, and their block sizes are 3 and 4 respectively.

A *pairwise balanced design* (PBD) is a design in which each pair of points occurs in λ blocks, for some constant λ called the *index* of the design. (We shall simply refer to this property as 'balance' but other forms of balance, such as variance balance and efficiency balance are also important; see Section 1.7.) An example of a PBD of order 7 is given by $(X, \mathscr{B})$ where $X = Z_7$, the integers modulo 7, and

$$\mathscr{B} = \{123456, 01, 02, 03, 04, 05, 06\}. \tag{D3}$$

In this case $\lambda = 1$, and the set of block-sizes is $\{2, 6\}$. Another PBD based on the same set, with $\lambda = 1$, is given by

$$\mathscr{B} = \{124, 235, 346, 450, 561, 602, 013\} \tag{D4}$$

which has all blocks of size 3. The designs (D1) and (D2) are also PBDs with $\lambda = 1$.

A *linked design* is one in which each pair of blocks intersect in μ points for some constant μ. Designs (D2), (D3), and (D4) are all linked, with $\mu = 1$ in each case; design (D1) is not linked, since any two blocks in the same class are disjoint but any two blocks in different classes have one common element.

Although (D3) and (D4) have several features in common, being incomplete, balanced and linked and having the same numbers of blocks, they also differ in several ways: in (D4) each block contains three varieties and each variety occurs in three blocks, but in (D3) one block has size 6 and six blocks have size 2, and, similarly, one variety belongs to six blocks and six varieties belong to two blocks.

A design in which all the blocks contain the same number of varieties, and all the varieties occur in the same number of blocks, is called a *block design.* Usually in dealing with such a design, we use the symbols v, b, r, k to denote the number of varieties, the number of blocks, the *replication number* (that is, the number of occurrences) of each variety and the size of each block respectively. If such a design is also balanced, and if $1 < k < v$, we call it a *balanced incomplete block design* (BIBD) with parameters (v, b, r, k, λ). Thus (D1), (D2), and (D4) are BIBDs with parameters (9, 12, 4, 3, 1), (13, 13, 4, 4, 1), and (7, 7, 3, 3, 1) respectively, but (D3) is not a block design. Again, if $X = Z_7$, the design with blocks

$$123, 234, 345, 456, 560, 601, 012 \qquad \text{(D5)}$$

is a block design with $v = b = 7$, $r = k = 3$, but is not balanced since pairs of elements which differ by ± 1 (such as 01, 12, and so on) occur in two blocks each, those differing by ± 2 (02, 13, and so on) in one block each, and those differing by ± 3 (03, 14, and so on) never occur together.

Like (D4), the design based on $X = Z_7$ with blocks

$$123, 145, 160, 246, 250, 340, 356 \qquad \text{(D6)}$$

is also a BIBD with parameters (7, 7, 3, 3, 1), but we can go even further. Suppose we apply to each block of (D6) the permutation (3456). This means that each 3 in the design (D6) becomes a 4, each 4 a 5, each 5 a 6, each 6 a 3, and each 1, 2, and 0 is left unchanged. Thus the block 123 becomes 124, 145 becomes 156, and so on. Altogether (3456) applied to the design (D6) produces the design (D4), with its blocks reordered as shown in Table 1.7. These designs are therefore *isomorphic* – that is, one can be obtained from the other by relabelling blocks and varieties, but they have essentially the same structure. (Sometimes a non-trivial permutation applied to each block of a design will give the same design – the permutation is then called an *automorphism* of the design.) On the other hand, since the elements 1 and 4 never occur together in any block of (D5), no permutation of the elements of (D5) can produce a balanced design. Thus (D5) is not isomorphic to (D4) and (D6).

Table 1.7

123	145	160	246	250	340	356		(D6)
↓	↓	↓	↓	↓	↓	↓	(3456)	↓
124	156	130	253	260	450	463		(D4)

In many applications we are interested only in distinguishing between designs, up to isomorphism, and in classifying designs with given parameters into their isomorphism classes.

A variety and a block are said to be *incident* if the variety belongs to the block. One convenient way to represent a design is by means of an *incidence matrix*. For a design $(X, \mathscr{B})$ with v varieties and b blocks, the incidence matrix is a $v \times b$ matrix, $A = [a_{ij}]$, such that

$$a_{ij} = \begin{cases} 1 & \text{if variety } i \text{ belongs to block } j, \\ 0 & \text{otherwise.} \end{cases}$$

The incidence matrices of designs (D1), (D3), (D4), and (D6) are shown in Table 1.8. For (D3), (D4), and (D6) we will write 7 for the zero of Z_7.

Table 1.8. Incidence matrices of some designs; only ones are printed

(D1)

	1	2	3	4	5	6	7	8	9	10	11	12
1	1			1			1			1		
2	1				1				1		1	
3	1					1		1				1
4		1		1				1			1	
5		1			1		1					1
6		1				1			1	1		
7			1	1					1			1
8			1		1			1		1		
9			1			1	1				1	

(D3)

	1	2	3	4	5	6	7
1	1	1					
2	1		1				
3	1			1			
4	1				1		
5	1					1	
6	1						1
7		1	1	1	1	1	1

(D4)

	1	2	3	4	5	6	7
1	1				1		1
2	1	1				1	
3		1	1				1
4	1		1	1			
5		1		1	1		
6			1		1	1	
7				1		1	1

(D6)

	1	2	3	4	5	6	7
1	1	1	1				
2	1			1	1		
3	1					1	1
4		1		1		1	
5		1			1		1
6			1	1			1
7			1		1	1	

Notice that in the incidence matrix of a block design, each row contains r 1s, corresponding to the r blocks containing that variety, and each column contains k 1s, corresponding to the k varieties belonging to that block. Thus, if J_n denotes an $n \times n$ matrix, and $J_{m \times n}$ an $m \times n$ matrix, with all entries equal to 1, we have

$$AJ_b = rJ_{v \times b} \tag{1.1}$$

and

$$J_v A = kJ_{v \times b} \tag{1.2}$$

for any block design with incidence matrix A.

Table 1.9 illustrates the relationship between incidence matrices of isomorphic block designs, in this case (D4) and (D6). Starting from the incidence matrix of (D6), we permute rows (relabelling the varieties) as shown, thus row 3 of the matrix of (D6) is moved to row 4, and so on. Then we permute columns (rearranging the blocks) of the intermediate matrix as shown, so that column 4 is moved to column 2, and so on. This gives the incidence matrix of (D4).

Table 1.9. Isomorphism of the designs (D6) and (D4)

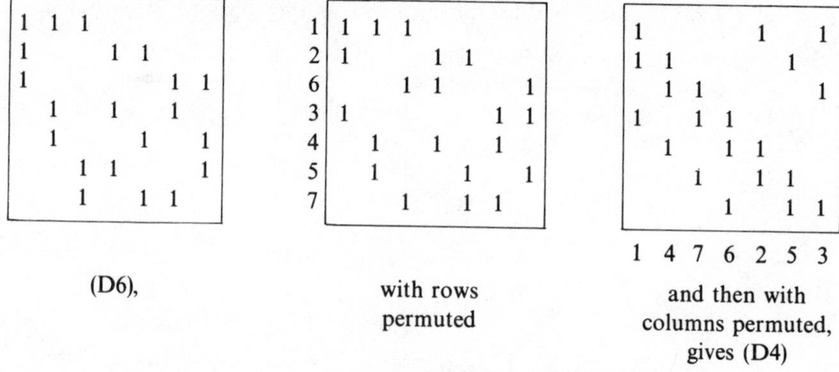

(D6), with rows permuted and then with columns permuted, gives (D4)

In general, permutation of rows and columns of the incidence matrix of one design simply gives the incidence matrix of an isomorphic design, and two designs are isomorphic precisely when their incidence matrices are related in this way. In other words, if A_1 and A_2 are incidence matrices of designs D_1 and D_2 respectively, then D_1 and D_2 are isomorphic if and only if there exist permutation matrices P and Q such that

$$PA_1Q = A_2.$$

Theorem 1. *In any block design, we have*

(i) $vr = bk$. (1.3)

If the block design is balanced, we have also

(ii) $\lambda(v-1) = r(k-1)$. (1.4)

Proof. (i) Consider the sum, s, of all the entries of the incidence matrix of the design. Each row contains r 1s, and there are v rows, so $s = vr$. Each column contains k 1s and there are b columns, so $s = bk$. Hence (1.3) follows.

(ii) We rearrange the blocks of the design so that blocks 1 through r are those containing variety 1, giving the incidence matrix shown in Table 1.10. Here we find the sum, again denoted by s, of the entries in the $(v-1)\times r$ submatrix consisting of rows 2 through v, columns 1 through r. Since each variety occurs with the first in precisely λ blocks, each row of the submatrix contains λ 1s, and there are $v-1$ rows, so $s = \lambda(v-1)$. Since each block contains $k-1$ varieties other than the first, each column of the submatrix contains $k-1$ 1s, and there are r columns, so $s = r(k-1)$. Hence (4) holds, no matter which variety was labelled as 1. □

Table 1.10

	← r →	← $b-r$ →
1	1 1 1	0 0 0
2 . . . v		

Equations (1.3) and (1.4) show that given any three of the five parameters of a BIBD, we can determine the other two. In view of this, we shall usually refer to a $B[k, \lambda; v]$ from now on, rather than to a (v, b, r, k, λ) BIBD, listing all the parameters only when we specifically want to draw attention to them.

Theorem 2. *If A is the incidence matrix of a $B[k, \lambda; v]$ design, then*

$$AA^T = (r-\lambda)I_v + \lambda J_v \tag{1.5}$$

and

$$J_v A = kJ_{v\times b}. \tag{1.2}$$

Conversely, if A is a $v\times b$ (0, 1) matrix which satisfies (1.5) *and* (1.2), *then*

$$v = \frac{r(k-1)}{\lambda} + 1,$$

$$b = \frac{vr}{k},$$

and, provided that $k < v$, A is the incidence matrix of a $B[k, \lambda; v]$ design.

Proof. (i) Suppose that A is the incidence matrix of a $B[k, \lambda; v]$ design. Equation (1.2) holds, since the design has k varieties per block, and hence A has k 1s per column (and all other entries are zeros).

The (i, j) entry of AA^T is the dot product of row i of A with column j of A^T, that is, of rows i and j of A. Hence, it must be

$$\sum_{n=1}^{b} a_{in} a_{jn}.$$

But

$$a_{in} a_{jn} = \begin{cases} 1 & \text{if } a_{in} = 1 = a_{jn}, \\ 0 & \text{otherwise,} \end{cases}$$

so that we get exactly r 1s if $i = j$ (counting the r blocks to which variety i belongs) and exactly λ 1s if $i \neq j$ (counting the λ blocks to which varieties i and j both belong). Hence the diagonal entries of AA^T are all equal to r, and the off-diagonal entries to λ, proving eqn (1.5).

(ii) Now suppose that A is a $v \times b$ (0, 1) matrix which satisfies eqns (1.5) and (1.2). Define a block design D with varieties $c_1, c_2, \ldots, c_v$ and blocks $B_1, B_2, \ldots, B_b$, where

$$c_i \in B_j \text{ if and only if } a_{ij} = 1.$$

A straightforward check shows that D is a BIBD with the required parameters. □

1.3. The completely randomized design and the linear model

We begin by stating assumptions (A1), (A2), and (A3) formally. To do this we must introduce some more notation.

Let Y_{ij} be the yield of the jth plot sown with variety i. Suppose that the *expected yield* for a plot sown with variety i is τ_i and that Y_{ij} differs from τ_i by some random error term only. Thus,

$$Y_{ij} = \tau_i + E_{ij},$$

and the expected value of the error term, $\mathscr{E}(E_{ij})$ is zero. The E_{ij} are the terms representing the random fluctuation mentioned in (A1). We also know that the expected value of Y_{ij} is τ_i, written

$$\mathscr{E}(Y_{ij}) = \mathscr{E}(\tau_i + E_{ij}) = \mathscr{E}(\tau_i) + \mathscr{E}(E_{ij}) = \tau_i.$$

This follows because the *expectation operator*, $\mathscr{E}$, is a linear operator and because the expected value of a constant is that constant. Y_{ij} and E_{ij} are called *random variables.*

We are interested in a measure of the magnitude of the difference between the random variable and its expected value. Clearly,

$$\begin{aligned} \mathscr{E}(Y_{ij} - \mathscr{E}(Y_{ij})) &= \mathscr{E}(\tau_i + E_{ij} - \tau_i) \\ &= \mathscr{E}(E_{ij}) \\ &= 0 \end{aligned}$$

so using the difference itself gives no information. Hence the *variance* is defined as the expected value of the square of the difference between the random variable and its expected value. Thus,

$$\begin{aligned}\mathrm{Var}(Y_{ij}) &= \mathscr{E}((Y_{ij} - \mathscr{E}(Y_{ij}))^2)\\ &= \mathscr{E}((\tau_i + E_{ij} - \tau_i)^2)\\ &= \mathscr{E}(E_{ij}^2)\\ &= \mathscr{E}((E_{ij} - \mathscr{E}(E_{ij}))^2)\\ &= \mathrm{Var}(E_{ij}),\end{aligned}$$

and (A2) says that this is constant for all i and j. Thus

$$\mathrm{Var}(Y_{ij}) = \mathrm{Var}(E_{ij}) = \sigma^2.$$

Finally, (A3) says that the yield of any plot is independent of the yield of any other plot. Formally,

$$\mathscr{E}(Y_{ij}Y_{st}) = \mathscr{E}(Y_{ij})\,\mathscr{E}(Y_{st}) \text{ for all } (i, j) \neq (s, t).$$

Often we say the random variables Y_{ij}, and hence the E_{ij}, are *uncorrelated.*

For the remainder of the discussion it is convenient to list the plots in some fixed, but arbitrary, order and to let $\boldsymbol{Y}$ and $\boldsymbol{E}$ be vectors containing Y_{ij} and E_{ij}, respectively, in the same order. We choose to list the r_1 plots sown with variety 1, followed by the r_2 plots sown with variety 2, and so on. Then $\boldsymbol{Y}$ is a random vector in $\mathrm{R}^n = \{(y_1, \ldots, y_n)^T | y_i \in \mathrm{R}\}$ with vector of expected values (or *mean vector*) $\boldsymbol{\eta}$, where $\eta_{ij} = \tau_i$. Thus

$$\boldsymbol{Y} = \boldsymbol{\eta} + \boldsymbol{E},$$

and $\boldsymbol{\eta}$ is a fixed, but unknown, vector in R^n. What we want to do is to use the yields in $\boldsymbol{Y}$ to estimate the unknown quantities in $\boldsymbol{\eta}$.

Let $\boldsymbol{\tau} = (\tau_1, \tau_2, \ldots, \tau_v)^T$. Then all the unknown quantities in $\boldsymbol{\eta}$ may be expressed as linear combinations of the unknown elements of $\boldsymbol{\tau}$, by using the plot–variety incidence matrix, X, an $n \times v$ (0, 1) matrix. Then

$$\boldsymbol{\eta} = X\boldsymbol{\tau}$$

and hence

$$\boldsymbol{Y} = X\boldsymbol{\tau} + \boldsymbol{E}.$$

Such a formulation, where X is a known matrix, $\boldsymbol{\tau}$ is a vector of unknown parameters and $\boldsymbol{E}$ is an error vector with mean vector $\boldsymbol{0}$, defines a class of statistical models known as *linear models.* In this case, X is of full column rank, and the model is known as a *full rank* linear model.

Example 3. Suppose we have $v = 3, r_1 = 3, r_2 = 2, r_3 = 3$. Then

$$\begin{aligned}\boldsymbol{Y} &= (Y_{11}, Y_{12}, Y_{13}, Y_{21}, Y_{22}, Y_{31}, Y_{32}, Y_{33})^T,\\ \boldsymbol{E} &= (E_{11}, E_{12}, E_{13}, E_{21}, E_{22}, E_{31}, E_{32}, E_{33})^T,\\ \boldsymbol{\eta} &= (\tau_1, \tau_1, \tau_1, \tau_2, \tau_2, \tau_3, \tau_3, \tau_3)^T\end{aligned}$$

and the matrix X is given in Table 1.11. □

Table 1.11

$$X = \begin{bmatrix} 1 & 0 & 0 \\ 1 & 0 & 0 \\ 1 & 0 & 0 \\ 0 & 1 & 0 \\ 0 & 1 & 0 \\ 0 & 0 & 1 \\ 0 & 0 & 1 \\ 0 & 0 & 1 \end{bmatrix}$$

Geometrically, we have a situation in which the columns of X define a hyperplane in $\mathbf{R}^n$, through the origin, known as the *model subspace*, or as the *range* of X, $\mathscr{R}(X) = \{X\mathbf{c} \mid \mathbf{c} \in \mathbf{R}^v\}$. Clearly $\boldsymbol{\eta} = X\boldsymbol{\tau} \in \mathscr{R}(X)$. The *kernel* of X is $\mathscr{K}(X) = \{\mathbf{w} \mid X\mathbf{w} = \mathbf{0}\}$, and $\mathscr{R}(X)^{\perp} = \{\mathbf{d} \mid \mathbf{d}^T X\mathbf{c} = 0 \text{ for every } X\mathbf{c} \in \mathscr{R}(X)\}$ is called the *orthogonal complement* of $\mathscr{R}(X)$ in $\mathbf{R}^n$.

To estimate $\boldsymbol{\eta}$ we have a data vector $\boldsymbol{Y}$ which differs from $\boldsymbol{\eta}$ by $\boldsymbol{E}$. (Note that we are using $\boldsymbol{Y}$ to denote both the random vector and the vector of actual yields (the data vector).) By (A2) and (A3) we know $\boldsymbol{E}$ has no preferred direction in $\mathbf{R}^n$. Hence the most natural estimate of $\boldsymbol{\eta}$ is that vector in the model subspace which is closest to $\boldsymbol{Y}$ in the usual Euclidean sense. Thus we want to choose an estimate, $\hat{\boldsymbol{\eta}}$, of $\boldsymbol{\eta}$ such that the function

$$(\boldsymbol{Y} - \boldsymbol{\eta})^T (\boldsymbol{Y} - \boldsymbol{\eta})$$

is a minimum at $\boldsymbol{\eta} = \hat{\boldsymbol{\eta}}$.

Choose $\boldsymbol{b}$ such that

$$(\boldsymbol{Y} - X\boldsymbol{b})^T X = 0,$$

or, equivalently, so that

$$X^T X\boldsymbol{b} = X^T \boldsymbol{Y}. \tag{1.6}$$

The second form gives the so-called *normal equations* of the design.

Now $X\boldsymbol{b} \in \mathscr{R}(X)$, $\boldsymbol{Y} - X\boldsymbol{b} \in \mathscr{R}(X)^{\perp}$ so $X\boldsymbol{b}$ and $\boldsymbol{Y} - X\boldsymbol{b}$ are the components of $\boldsymbol{Y}$ in $\mathscr{R}(X)$ and $\mathscr{R}(X)^{\perp}$ respectively. Thus

$$\begin{aligned}(\boldsymbol{Y} - \boldsymbol{\eta})^T (\boldsymbol{Y} - \boldsymbol{\eta}) &= (\boldsymbol{Y} - X\boldsymbol{b} + X\boldsymbol{b} - \boldsymbol{\eta})^T (\boldsymbol{Y} - X\boldsymbol{b} + X\boldsymbol{b} - \boldsymbol{\eta}) \\ &= (\boldsymbol{Y} - X\boldsymbol{b})^T (\boldsymbol{Y} - X\boldsymbol{b}) + (X\boldsymbol{b} - \boldsymbol{\eta})^T (X\boldsymbol{b} - \boldsymbol{\eta}),\end{aligned}$$

since $\boldsymbol{Y} - X\boldsymbol{b} \in \mathscr{R}(X)^{\perp}$, $X\boldsymbol{b} \in \mathscr{R}(X)$, and $\boldsymbol{\eta} \in \mathscr{R}(X)$. As $(\boldsymbol{Y} - X\boldsymbol{b})^T (\boldsymbol{Y} - X\boldsymbol{b})$ is constant, we see that $(\boldsymbol{Y} - \boldsymbol{\eta})^T (\boldsymbol{Y} - \boldsymbol{\eta})$ is minimized if $\hat{\boldsymbol{\eta}} = X\boldsymbol{b}$.

If we wish to estimate $\boldsymbol{\tau}$, where $\boldsymbol{\eta} = X\boldsymbol{\tau}$, then any vector $\hat{\boldsymbol{\tau}}$ satisfying

$$X\hat{\boldsymbol{\tau}} = \hat{\boldsymbol{\eta}} = X\boldsymbol{b}$$

produces the same vector $\hat{\boldsymbol{\eta}}$ and is called a *least squares estimator* of $\boldsymbol{\tau}$. Now

$$X(\hat{\boldsymbol{\tau}} - \boldsymbol{b}) = \mathbf{0}$$

so $\hat{\boldsymbol{\tau}} = \boldsymbol{b} + \boldsymbol{w}$, where $\boldsymbol{w} \in \mathscr{K}(X)$. Thus $\hat{\boldsymbol{\tau}}$ is uniquely defined if $\mathscr{K}(X) = \{\mathbf{0}\}$; that is, if X is of full column rank.

In the case of the completely randomized design, since X is of full column rank, we have

$$\hat{\tau} = (X^T X)^{-1} X^T Y$$

and so

$$\hat{\tau}_i = \frac{1}{r_i} \sum_{j=1}^{r_i} Y_{ij} = \bar{Y}_i,$$

the mean of the plots sown with variety i. What is the variance of $\hat{\tau}_i$? First we calculate the expected value of $\hat{\tau}_i$.

$$\begin{aligned}\mathscr{E}(\hat{\tau}_i) &= \mathscr{E}\left[\frac{1}{r_i}\sum_j Y_{ij}\right] \\ &= \frac{1}{r_i}\sum_j \mathscr{E}(Y_{ij}) = \tau_i.\end{aligned}$$

Now

$$\begin{aligned}\operatorname{Var}(\hat{\tau}_i) &= \mathscr{E}((\hat{\tau}_i - \tau_i)^2) \\ &= \mathscr{E}\left[\left(\left(\frac{1}{r_i}\sum_j Y_{ij}\right) - \tau_i\right)^2\right] \\ &= \mathscr{E}\left[\left(\frac{1}{r_i}\sum_j (Y_{ij} - \tau_i)\right)^2\right] \\ &= \left(\frac{1}{r_i}\right)^2 \mathscr{E}\left[\left(\sum_j (Y_{ij} - \tau_i)\right)^2\right] \\ &= \left(\frac{1}{r_i}\right)^2 \mathscr{E}\left(\sum_j (Y_{ij} - \tau_i)^2 + 2\sum_j \sum_{j'>j} (Y_{ij} - \tau_i)(Y_{ij'} - \tau_i)\right) \\ &= \left(\frac{1}{r_i}\right)^2 \sum_j \operatorname{Var}(Y_{ij}) + 2\sum_j \sum_{j'>j} \mathscr{E}((Y_{ij} - \tau_i)(Y_{ij'} - \tau_i)) \\ &= \sigma^2/r_i + 0\end{aligned}$$

since (A3) says that $\mathscr{E}(Y_{ij} Y_{ij'}) = \mathscr{E}(Y_{ij})\mathscr{E}(Y_{ij'})$. Thus, to have the variance of $\hat{\tau}_i$ as small as possible we should try to make r_i as large as possible, in accordance with intuition. As this is true for each of $r_1, r_2, \ldots, r_v$, we would like to have $n = vr$, and to set $r_i = r = n/v$, for $i = 1, 2, \ldots, v$.

1.4. Block designs and the linear model

Now consider a block design, laid out on $n = vr = bk$ plots, where the design is complete if $k = v$ and incomplete if $k < v$. Let Y_{ij} denote the yield of the plot in block j sown with variety i. (Of course, in a BIBD there may be no such plot.) Then

$$Y_{ij} = \tau_i + \beta_j + E_{ij},$$

where τ_i is the effect of variety i, β_j is the effect of block j and E_{ij} is an error term with the same properties as before. Again, we can write this in vector notation as

$$Y = \eta + E.$$

Now η contains $v+b$ unknown quantities. Let

$$\tau = (\tau_1, \tau_2, \ldots, \tau_v)^T \text{ and } \beta = (\beta_1, \beta_2, \ldots, \beta_b)^T.$$

Then all the unknown elements in η may be expressed as linear combinations of the unknown elements of τ and β, using the plot–variety incidence matrix, T, and the plot–block incidence matrix, B. Here T and B are respectively $n \times v$ and $n \times b$ (0, 1) matrices. Then

$$\eta = T\tau + B\beta.$$

Now, putting $X = [T\,B]$ and $\gamma = \begin{bmatrix} \tau \\ \beta \end{bmatrix}$, we have

$$\begin{aligned} Y &= T\tau + B\beta + E \\ &= X\gamma + E. \end{aligned}$$

Example 4. Suppose $v = 3$ and $b = 2$. Then

$$\begin{aligned} Y &= (Y_{11}, Y_{12}, Y_{21}, Y_{22}, Y_{31}, Y_{32})^T, \\ E &= (E_{11}, E_{12}, E_{21}, E_{22}, E_{31}, E_{32})^T, \\ \eta &= (\tau_1+\beta_1, \tau_1+\beta_2, \tau_2+\beta_1, \tau_2+\beta_2, \tau_3+\beta_1, \tau_3+\beta_2)^T, \\ \beta &= (\beta_1, \beta_2)^T, \tau = (\tau_1, \tau_2, \tau_3)^T \end{aligned}$$

and the matrices B and T are given in Table 1.12. □

Table 1.12

$$B = \begin{bmatrix} 1 & \\ & 1 \\ 1 & \\ & 1 \\ 1 & \\ & 1 \end{bmatrix} \qquad T = \begin{bmatrix} 1 & & \\ 1 & & \\ & 1 & \\ & 1 & \\ & & 1 \\ & & 1 \end{bmatrix}$$

Proceeding as before we have

$$X\hat{\gamma} = \hat{\eta} = X\mathbf{b}$$

where

$$X^T X\mathbf{b} = X^T Y \tag{1.6}$$

so

$$\hat{\gamma} - \mathbf{b} \in \mathcal{K}(X).$$

In a randomized complete block design or in a BIBD, X is not of full rank, so $\mathscr{K}(X) \neq \{\mathbf{0}\}$ and $\hat{\gamma}$ is any vector in the coset $\boldsymbol{b}+\mathscr{K}(X)$.

Are there any linear functions of the elements of γ which have the same value for all the vectors in the coset $\gamma+\mathscr{K}(X)$? Let $\mathbf{w} \in \mathscr{K}(X)$. Then a linear function

$$\theta = \mathbf{v}^T(\gamma+\mathbf{w})$$

which has the same value for all the vectors in $\gamma+\mathscr{K}(X)$ is said to be *estimable*. Clearly

$$\mathbf{v}^T\gamma = \mathbf{v}^T(\gamma+\mathbf{w})$$

so

$$\mathbf{v}^T\mathbf{w} = 0 \text{ for all } \mathbf{w} \in \mathscr{K}(X).$$

Let $\mathbf{v}^T = \mathbf{r}^T X$. Then $\mathbf{v}^T\mathbf{w} = 0$ so $\mathbf{r}^T X\gamma$ is estimable and, indeed, the vector of coefficients of any estimable function may be represented as a linear combination of the rows of X. For any estimable function, θ,

$$\hat{\theta} = \mathbf{v}^T\boldsymbol{b},$$

is called the *least squares estimator* of θ.

We now evaluate $\boldsymbol{b}$ for a randomized complete block design and for a BIBD in turn.

(i) For a randomized complete block design we have

$$X^TX = \begin{bmatrix} T^TT & T^TB \\ B^TT & B^TB \end{bmatrix} = \begin{bmatrix} bI_v & J_{v\times b} \\ J_{b\times v} & vI_b \end{bmatrix}.$$

Let $\mathbf{w} = [\mathbf{j}_v^T, -\mathbf{j}_b^T]^T$, where $\mathbf{j}_v = J_{v\times 1}$. Then $X^TX\mathbf{w} = \mathbf{0}$ so the rank of X^TX is at most $v+b-1$. The normal equations (1.6) may be written as

$$bI_v\hat{\tau} + J_{v\times b}\hat{\beta} = T^TY,$$

and

$$J_{b\times v}\hat{\tau} + vI_b\hat{\beta} = B^TY,$$

where $\boldsymbol{b} = [\hat{\tau}^T, \hat{\beta}^T]^T$ and $\hat{\tau}^T = [\hat{\tau}_1, \hat{\tau}_2, \ldots, \hat{\tau}_v]$ and $\hat{\beta}^T = [\hat{\beta}_1, \hat{\beta}_2, \ldots, \hat{\beta}_b]$. From these equations we get

$$vbI_v\hat{\tau} + vJ_{v\times b}\hat{\beta} = vT^TY$$

and

$$J_{v\times b}J_{b\times v}\hat{\tau} + vJ_{v\times b}\hat{\beta} = J_{v\times b}B^TY$$

whence

$$(vbI_v - bJ_v)\hat{\tau} = (vT^T - J_{v\times b}B^T)Y.$$

The rank of $(vbI_v - bJ_v)$ is $v-1$ (see Exercise 1.5), and we solve the equation subject to one constraint: we choose to have $\mathbf{j}_v^T\hat{\tau} = 0$. Thus

$$vb\hat{\tau} = (vT^T - J_{v\times b}B^T)Y = vT^TY - J_{v\times bv}Y.$$

Now $T^TY = [T_1, T_2, \ldots, T_v]^T$ where $T_i = \sum_j Y_{ij}$ and $J_{v\times bv}Y = Gj_v$ where $G = \sum_i \sum_j Y_{ij}$. Hence

$$\hat{\tau}_i = \frac{vT_i}{bv} - \frac{G}{bv} = \frac{T_i}{b} - \frac{G}{bv};$$

that is, the difference between the average yield of the plots sown with variety i and the average yield of all the plots. Note that τ_i is not an estimable function, so the estimate of τ_i depends on the constraint imposed (see Exercise 1.6). Observe also that

$$\hat{\beta}_j = B_j/v,$$

where $B_j = \sum_i Y_{ij}$, so that the rank of X^TX is exactly $v+b-1$, since we can solve for all the unknowns after imposing one constraint.

(ii) For a (v, b, r, k, λ) BIBD with incidence matrix A we have

$$X^TX = \begin{bmatrix} T^TT & T^TB \\ B^TT & B^TB \end{bmatrix} = \begin{bmatrix} rI_v & A \\ A^T & kI_b \end{bmatrix}.$$

Again $X^TX\mathbf{w} = \mathbf{0}$. The normal equations may be written as

$$rI_v\hat{\tau} + A\hat{\beta} = T^TY,$$

and

$$A^T\hat{\tau} + kI_b\hat{\beta} = B^TY,$$

which give

$$krI_v\hat{\tau} + kA\hat{\beta} = kT^TY$$

and

$$AA^T\hat{\tau} + kA\hat{\beta} = AB^TY.$$

Hence

$$(krI_v - AA^T)\hat{\tau} = (kT^T - AB^T)Y.$$

But by Theorem 2

$$krI_v - AA^T = krI_v - (r-\lambda)I_v - \lambda J_v.$$

Now

$$kr - (r-\lambda) = kr - r + \lambda = r(k-1) + \lambda = \lambda(v-1) + \lambda = \lambda v,$$

so

$$\lambda(vI_v - J_v)\hat{\tau} = (kT^T - AB^T)Y.$$

Again we find the solution $\hat{\tau}$ for which $j_v^T\hat{\tau} = 0$. Thus

$$\lambda v\hat{\tau}_i = kT_i - \sum_{\{j \mid i \in \text{block } j\}} B_j,$$

and note that τ_i is not an estimable function. Observe also that

$$k\hat{\beta} = B^TY - A^T\hat{\tau}$$

so that again X^TX is of rank $v+b-1$.

Example 5. Consider the design (D4) where we again replace the variety 0 by 7. Then

$$Y = (Y_{11}, Y_{15}, Y_{17}, Y_{21}, Y_{22}, Y_{26}, Y_{32}, Y_{33}, Y_{37}, \ldots, Y_{74}, Y_{76}, Y_{77})^T.$$

The matrices T and B are given in Table 1.13; A appears in Table 1.8. $X = [T\ B]$. Thus, for instance,

$$\lambda v\hat{\tau}_1 = 7\hat{\tau}_1 = 3(Y_{11} + Y_{15} + Y_{17}) \\ - (Y_{11} + Y_{21} + Y_{41} + Y_{55} + Y_{65} + Y_{15} + Y_{77} + Y_{17} + Y_{37}). \quad \square$$

Table 1.13

$$B = \begin{bmatrix}
1 & & & & & & \\
 & & & & 1 & & \\
 & & & & & & 1 \\
1 & & & & & & \\
 & 1 & & & & & \\
 & & & & & 1 & \\
 & 1 & & & & & \\
 & & 1 & & & & \\
 & & & & & & 1 \\
1 & & & & & & \\
 & & 1 & & & & \\
 & & & 1 & & & \\
 & 1 & & & & & \\
 & & & 1 & & & \\
 & & & & 1 & & \\
 & & 1 & & & & \\
 & & & & 1 & & \\
 & & & & & 1 & \\
 & & & 1 & & & \\
 & & & & & 1 & \\
 & & & & & & 1
\end{bmatrix}
\qquad
T = \begin{bmatrix}
1 & & & & & & \\
1 & & & & & & \\
1 & & & & & & \\
 & 1 & & & & & \\
 & 1 & & & & & \\
 & 1 & & & & & \\
 & & 1 & & & & \\
 & & 1 & & & & \\
 & & 1 & & & & \\
 & & & 1 & & & \\
 & & & 1 & & & \\
 & & & 1 & & & \\
 & & & & 1 & & \\
 & & & & 1 & & \\
 & & & & 1 & & \\
 & & & & & 1 & \\
 & & & & & 1 & \\
 & & & & & 1 & \\
 & & & & & & 1 \\
 & & & & & & 1 \\
 & & & & & & 1
\end{bmatrix}$$

1.5. Latin squares and the linear model

Let Y_{ij} be the yield of the plot in row i and column j and let $k(i, j)$ be a function whose value is the variety sown in the plot in row i and column j. Thus

$$Y_{ij} = \rho_i + \kappa_j + \tau_{k(i,j)} + E_{ij},$$

where ρ_i is the effect of row i, κ_j is the effect of column j, $\tau_{k(i,j)}$ is the effect of variety $k(i, j)$ and E_{ij} is an error term with the same properties as before. Let

$$Y = (Y_{11}, Y_{12}, \ldots, Y_{1n}, Y_{21}, \ldots, Y_{2n}, \ldots, Y_{nn})^T.$$

Let R, C, and T be $n^2 \times n$ (0, 1) matrices, where R is the plot–row incidence matrix, C the plot–column incidence matrix and T the plot–variety incidence matrix. Thus if $R=[r_{su}]$, $C=[c_{su}]$, and $T=[t_{su}]$ then, for example,

$$r_{su}=\begin{cases}1 & \text{if the } s\text{th element of } Y \text{ is in row } u,\\ 0 & \text{otherwise.}\end{cases}$$

Let $\tau=(\tau_1,\ldots,\tau_n)^T$, $\boldsymbol{\rho}=(\rho_1,\ldots,\rho_n)^T$, $\boldsymbol{\kappa}=(\kappa_1,\ldots,\kappa_n)^T$, and $\gamma=\begin{bmatrix}\rho\\ \kappa\\ \tau\end{bmatrix}$.

Then

$$\begin{aligned} \boldsymbol{Y} &= R\boldsymbol{\rho}+C\boldsymbol{\kappa}+T\boldsymbol{\tau}+\boldsymbol{E}\\ &=[R\;C\;T]\boldsymbol{\gamma}+\boldsymbol{E}\\ &=X\boldsymbol{\gamma}+\boldsymbol{E}\\ &=\boldsymbol{\eta}+\boldsymbol{E}.\end{aligned}$$

Example 6. We use the Latin square given in Table 1.2. The matrices R, C and T are given in Table 1.14. □

For a Latin-square design we have

$$X^TX=\begin{bmatrix} R^TR & R^TC & R^TT\\ C^TR & C^TC & C^TT\\ T^TR & T^TC & T^TT\end{bmatrix}=\begin{bmatrix} nI_n & J_n & J_n\\ J_n & nI_n & J_n\\ J_n & J_n & nI_n\end{bmatrix}.$$

Let $\mathbf{w}_1=[\,j_n^T,\mathbf{0},-j_n^T]^T$ and $\mathbf{w}_2=[\,j_n^T,-j_n^T,\mathbf{0}]^T$. Then $X^TX\mathbf{w}_1=X^TX\mathbf{w}_2=\mathbf{0}$ so the rank of X^TX is at most $3n-2$. The normal equations can be written as

$$\begin{aligned} nI_n\hat{\rho}+J_n\hat{\kappa}+J_n\hat{\tau} &= R^TY,\\ J_n\hat{\rho}+nI_n\hat{\kappa}+J_n\hat{\tau} &= C^TY,\end{aligned}$$

Table 1.14

$$R=\begin{bmatrix}1&&&\\1&&&\\1&&&\\1&&&\\&1&&\\&1&&\\&1&&\\&1&&\\&&1&\\&&1&\\&&1&\\&&1&\\&&&1\\&&&1\\&&&1\\&&&1\end{bmatrix}\qquad C=\begin{bmatrix}1&&&\\&1&&\\&&1&\\&&&1\\1&&&\\&1&&\\&&1&\\&&&1\\1&&&\\&1&&\\&&1&\\&&&1\\1&&&\\&1&&\\&&1&\\&&&1\end{bmatrix}\qquad T=\begin{bmatrix}1&&&\\&1&&\\&&1&\\&&&1\\&&&1\\1&&&\\&1&&\\&&1&\\&&1&\\&&&1\\1&&&\\&1&&\\&1&&\\&&1&\\&&&1\\1&&&\end{bmatrix}$$

and

$$J_n\hat{\rho}+J_n\hat{\kappa}+nI_n\hat{\tau}=T^T Y.$$

We find the solution for which $j_n^T\,\hat{\rho}=j_n^T\,\hat{\kappa}=0$. Let $R_i=\sum_j Y_{ij}$, $C_j=\sum_i Y_{ij}$, $T_k=\sum_{\{(i,j)|k(i,j)=k\}} Y_{ij}$ and $G=\sum_i\sum_j Y_{ij}$. Then

$$nI_n\hat{\tau}=T^T Y \qquad \text{so } \hat{\tau}_i=\frac{T_i}{n},$$

$$nI_n\hat{\kappa}=C^T Y-J_n\hat{\tau} \text{ so } \hat{\kappa}_i=\frac{C_i}{n}-\frac{G}{n^2},$$

and

$$nI_n\hat{\rho}=R^T Y-J_n\hat{\tau} \text{ so } \hat{\rho}_i=\frac{R_i}{n}-\frac{G}{n^2}.$$

Again, none of the parameters τ_i, ρ_i and κ_i is an estimable function.

1.6. Designs as sampling schemes

Consider a population of v individuals (or elements). For this population we are interested in the mean value, μ, say of some attribute (such as height or weight) which can be measured for each individual.

If we can measure the attribute for each of the v individuals then we can determine μ exactly. Indeed,

$$\mu=\frac{1}{v}\sum_{i=1}^{v} Y_i=\overline{Y}$$

and the variance, σ^2, is given by

$$\begin{aligned}
\sigma^2&=\frac{1}{v-1}\sum_i (Y_i-\overline{Y})^2\\
&=\frac{1}{v-1}\left\{\sum_i Y_i^2-v\overline{Y}^2\right\}\\
&=\frac{1}{v-1}\left\{\sum_i Y_i^2-v\cdot\frac{1}{v^2}\left(\sum_i Y_i\right)^2\right\}\\
&=\frac{1}{v-1}\left\{\sum_i Y_i^2-\frac{1}{v}\left(\sum_i Y_i^2+2\sum_i\sum_{j>i} Y_iY_j\right)\right\}\\
&=\frac{1}{v-1}\left\{\frac{v-1}{v}\sum_i Y_i^2-\frac{2}{v}\sum_i\sum_{j>i} Y_iY_j\right\}\\
&=\frac{1}{v}\left\{\sum_i Y_i^2-\frac{2}{v-1}\sum_i\sum_{j>i} Y_iY_j\right\}.
\end{aligned}$$

But if measuring the attribute for each of the v individuals is not a realistic option, say because it is prohibitively expensive, then a sample of k distinct individuals is chosen (sampling without replacement) and the measurement recorded for these k individuals. Let these measurements be $Y_{i_1}, Y_{i_2}, \ldots, Y_{i_k}$ and let

$$\bar{S} = \frac{1}{k} \sum_{j=1}^{k} Y_{i_j}.$$

We now evaluate the mean and variance of $\bar{S}$. Let $t_i = 1$ if the ith unit is included in the random sample of size k and let $t_i = 0$ otherwise. Then let $Pr(t_i = 1)$ denote the probability that the ith unit is included in the sample and put

$$Pr(t_i = 1) = \pi_i;\ Pr(t_i = 1, t_j = 1) = \pi_{ij}, i \neq j;\ Pr(t_i = 1, t_i = 1) = \pi_{ii} = \pi_i.$$

Now

$$\mathscr{E}(t_i) = 0 \,.\, Pr(t_i = 0) + 1 \,.\, Pr(t_i = 1) = \pi_i$$

and

$$\begin{aligned} \operatorname{Var}(t_i) &= \mathscr{E}[(t_i - \mathscr{E}(t_i))^2] = \mathscr{E}[(t_i - \pi_i)^2] \\ &= \mathscr{E}(t_i^2 - 2t_i\pi_i + \pi_i^2) \\ &= \mathscr{E}(t_i^2) - \pi_i^2 \\ &= \pi_i - \pi_i^2 = \pi_i(1 - \pi_i). \end{aligned}$$

Define the *covariance* of two random variables, $\operatorname{Cov}(t_i, t_j)$, by

$$\operatorname{Cov}(t_i, t_j) = \mathscr{E}[(t_i - \mathscr{E}(t_i))(t_j - \mathscr{E}(t_j))];$$

in particular,

$$\operatorname{Cov}(t_i, t_i) = \operatorname{Var}(t_i).$$

Then here we get

$$\begin{aligned} \operatorname{Cov}(t_i, t_j) &= \mathscr{E}[(t_i - \pi_i)(t_j - \pi_j)] \\ &= \mathscr{E}(t_i t_j) - \pi_i \pi_j \\ &= \pi_{ij} - \pi_i \pi_j \end{aligned}$$

since

$$\begin{aligned} \mathscr{E}(t_i t_j) &= 0.0 \,.\, Pr(t_i = 0, t_j = 0) + 0.1 \,.\, Pr(t_i = 0, t_j = 1) \\ &\quad + 1.0 \,.\, Pr(t_i = 1, t_j = 0) + 1.1 \,.\, Pr(t_i = 1, t_j = 1) \\ &= Pr(t_i = 1, t_j = 1). \end{aligned}$$

Now we can write

$$\bar{S} = \frac{1}{k} \sum_{i=1}^{v} t_i Y_i.$$

Then, looking at the distribution of t_i (and so viewing Y_i as a constant), we get

$$\mathscr{E}(\bar{S}) = \frac{1}{k} \sum_{i=1}^{v} Y_i \,.\, \mathscr{E}(t_i) = \frac{1}{k} \sum_{i=1}^{v} \pi_i Y_i$$

and

$$\begin{aligned}
\operatorname{Var}(\bar{S}) &= \operatorname{Var}\left(\frac{1}{k}\sum_i t_i Y_i\right)\\
&= \mathscr{E}\left[\left(\frac{1}{k}\sum_i t_i Y_i - \mathscr{E}\left(\frac{1}{k}\sum_i t_i Y_i\right)\right)^2\right]\\
&= \frac{1}{k^2}\mathscr{E}\left[\left(\sum_i Y_i(t_i-\pi_i)\right)^2\right]\\
&= \frac{1}{k^2}\sum_i\sum_j Y_i Y_j \mathscr{E}[(t_i-\pi_i)(t_j-\pi_j)]\\
&= \frac{1}{k^2}\sum_i\sum_j Y_i Y_j(\pi_{ij}-\pi_i\pi_j)\\
&= \frac{1}{k^2}\left\{\sum_i Y_i^2\,\pi_i(1-\pi_i)+2\sum_i\sum_{j>i} Y_i Y_j(\pi_{ij}-\pi_i\pi_j)\right\}.
\end{aligned}$$

Suppose that each of the $\binom{v}{k}$ samples of k distinct individuals is equally likely. Then

$$\pi_i = \binom{v-1}{k-1}\bigg/\binom{v}{k} = k/v$$

and

$$\pi_{ij} = \binom{v-2}{k-2}\bigg/\binom{v}{k} = k(k-1)/v(v-1), \text{ for } i \neq j, \text{ and } \pi_{ii} = \pi_i.$$

Thus

$$\mathscr{E}(\bar{S}) = \frac{k}{v}\cdot\frac{1}{k}\sum_{i=1}^{v} Y_i = \bar{Y} = \mu$$

and

$$\begin{aligned}
\operatorname{Var}(\bar{S}) &= \frac{1}{k^2}\left\{\sum_i Y_i^2\frac{k}{v}\left(1-\frac{k}{v}\right)+2\sum_i\sum_{j>i} Y_i Y_j\left(\frac{k(k-1)}{v(v-1)}-\frac{k^2}{v^2}\right)\right\}\\
&= \frac{1}{k}\frac{1}{v}\left\{\left(\frac{v-k}{v}\right)\sum_i Y_i^2+2\left(\frac{k-1}{v-1}-\frac{k}{v}\right)\sum_i\sum_{j>i} Y_i Y_j\right\}\\
&= \frac{1}{k}\frac{1}{v}\frac{v-k}{v}\left\{\sum_i Y_i^2-\frac{2}{v-1}\sum_i\sum_{j>i} Y_i Y_j\right\}\\
&= \frac{1}{k}\frac{v-k}{v}\sigma^2\\
&= \frac{1}{k}\left(1-\frac{k}{v}\right)\sigma^2.
\end{aligned}$$

Now suppose that the blocks of a BIBD with parameters (v, b, r, k, λ) are used as the possible samples of k distinct individuals. Then

$$\pi_i = r/b = k/v$$

and

$$\pi_{ij} = \lambda/b = \lambda(v-1)/b(v-1) = r(k-1)/b(v-1) = k(k-1)/v(v-1).$$

Hence, if the possible samples of k individuals are reduced from all $\binom{v}{k}$ to the b blocks of a (v, b, r, k, λ) BIBD then the sample mean $\overline{S}$ has the same mean and variance and so is just as good an estimator.

The blocks of the design are chosen to exclude samples which are particularly expensive to collect (for instance, because they are geographically widespread) and may well include repeated blocks, corresponding to samples particularly cheap to collect.

1.7. References and comments

The references already mentioned in this chapter include some that are of historical importance. For further information on the history of designs, the following are recommended. Dudeney (1917) attributes a card puzzle (Problem 304) equivalent to a Latin square to Claude Gasper Bachet de Méziriac who published it in the 1624 edition of *Problèmes plaisans et délectables*. Preece (1983) refers to both this and Jacques Ozanam's 1723 publication of solutions to a similar puzzle in *Récreations mathématiques et physiques*. Dénes and Keedwell (1974) give a very detailed treatment of Latin squares. Fisher (1935–66) and Pearce (1983) include much material on the development of experimental design. Doyen and Rosa (1980) give a full bibliography on Steiner systems, and the paper by de Vries (1984) has additional information on early development of the triple systems. See Evans (1969) for an account of the relationship between universal algebra and Euler's officer problem. See also Hughes and Piper (1985).

Linear models are covered in detail in several textbooks, including Graybill (1961) and Searle (1971). The approach used here is that of Venables (1985). The results on sampling can be found in Chakrabarti (1964), Raj (1971), and Foody and Hedayat (1977). Hedayat and Li (1979, 1980) have constructed designs with certain parameters and various numbers of repeated blocks. In some experiments, not all comparisons are of equal interest. In these cases unequal allocations of treatments to plots may be appropriate; see Exercise 1.8. D. G. Hoffman, Schellenberg and Vanstone (1976) discuss (r, λ) systems; see Exercise 1.3.

A block design is said to be *variance-balanced* if every elementary contrast (a term defined in Exercise 1.7) of treatment effects is estimated with the same variance. The *efficiency* of a contrast $\sum_i \lambda_i \tau_i$ is defined to be the ratio of the variance of $\sum_i \lambda_i \hat{\tau}_i$ in a randomized complete block design on vr plots to the variance of $\sum_i \lambda_i \hat{\tau}_i$ in the design under consideration. A design in which all elementary contrasts are estimated with the same efficiency is said to be *efficiency-balanced*. BIBDs are pairwise, variance-, and efficiency-balanced. Results about the relationships between the three types of balance appear in Puri and Nigam (1977), as do several constructions and references to earlier work.

Another way to describe block designs is in terms of the optimality properties that they satisfy. Let V be a $v \times v$ matrix whose (i, j)th position contains Cov $(\hat{\tau}_i, \hat{\tau}_j)$. Consider the set of all designs for v treatments and with blocks of size k. Then a design in the set is said to be *D-optimal* if the determinant of V is a minimum for the designs in the set; to be *A-optimal* if the trace of V is a minimum for the designs in the set; and to be *E-optimal* if the maximum eigenvalue of V is a minimum for designs in the set. For a given experiment we can, using the data, construct an ellipsoid, called the *confidence ellipsoid*, which contains the true value of τ with probability equal to p (often p is 0.95 or 0.99). In the case where the error terms follow a normal distribution, D-optimality corresponds to minimizing the square of the volume of the confidence ellipsoid, A-optimality corresponds to minimizing the sum of squares of the lengths of the axes of the confidence ellipsoid and E-optimality corresponds to minimizing the square of the maximum diameter. Kiefer (1980, 1981) looks at optimal design theory and combinatorics, Silvey (1980) describes the mathematical background and John and E. R. Williams (1982) look at the current state of a number of conjectures about optimal designs.

Steinberg and Hunter (1984) review a number of different topics in designs including optimal designs. Federer (1980) gives a bibliography of papers in experimental design (including references to earlier bibliographies).

Exercises

1.1. Show that, in the design (D4), each pair of distinct blocks intersect in precisely one variety. Use this property to give an estimate of β_j.

1.2. Let $X = \{1, 2, 3, 4, 5, 6\}$ and let designs be defined by the sets of blocks

$$123, 234, 345, 456, 561, 612 \qquad \text{(D7)}$$

and

$$123, 145, 126, 246, 345, 356 \qquad \text{(D8)}$$

respectively. Show that (D7) and (D8) both have parameters $v = b = 6$, $r = k = 3$, are not balanced and are not isomorphic to each other.

(Hint: to show non-isomorphism, consider the $\binom{6}{2} = 15$ unordered pairs of elements of X and how many times each such pair replicates in each design.)

1.3. An (r, λ)-*system* is a PBD of index λ in which every variety is replicated r times. Show that

$$123,\ 456,\ 14,\ 15,\ 16,\ 24,\ 25,\ 26,\ 34,\ 35,\ 36$$

is a (4, 1)-system on six elements. Is (D3) an (r, λ)-system?

1.4. Consider the design (D1) as given in Table 1.4, and the array S given in Table 1.3. To each block of (D1), adjoin the new element 0, giving 12 blocks with four elements each. Now take each element of S in turn, and corresponding to it construct two blocks of size 4 in the following way: if the element occupies position (i, j) in S, then one block consists of the other elements of row i and column j, and the other block consists of the elements in neither row i nor column j. Thus, corresponding to the element 2, we obtain the blocks 1358 and 4679. This gives altogether 18 more blocks of size 4, for a total of 30 blocks, chosen from the set $\{0, 1, \ldots, 9\}$. Show that each element occurs 12 times among these blocks, each unordered pair four times and each unordered triple once. (This is an example of an *inversive* plane.)

1.5. Show that $vbI_v - bJ_v$ has rank $v - 1$.

1.6. Consider a randomized complete block design and instead of finding the $\boldsymbol{b}$ for which $j_v^T \hat{\tau} = 0$ find the $\boldsymbol{b}$ for which $\hat{\beta}_1 = 0$. In this case we have

$$\begin{bmatrix} \hat{\tau}_1 \\ \hat{\tau}_2 \\ \vdots \\ \hat{\tau}_v \\ \hat{\beta}_2 \\ \vdots \\ \hat{\beta}_b \end{bmatrix} = \begin{bmatrix} bI_v & J_{v \times (b-1)} \\ J_{(b-1) \times v} & vI_{b-1} \end{bmatrix}^{-1} \begin{bmatrix} T_1 \\ T_2 \\ \vdots \\ T_v \\ B_2 \\ \vdots \\ B_b \end{bmatrix}.$$

Verify that

$$\begin{bmatrix} bI_v & J_{v \times (b-1)} \\ J_{(b-1) \times v} & vI_{b-1} \end{bmatrix}^{-1} = \begin{bmatrix} \frac{1}{b} I_v + \frac{b-1}{bv} J_v & -\frac{1}{v} J_{v \times (b-1)} \\ -\frac{1}{v} J_{(b-1) \times v} & \frac{1}{v} I_{b-1} + \frac{1}{v} J_{b-1} \end{bmatrix}.$$

Hence find $\hat{\tau}_i, \hat{\beta}_j$. Show that $\tau_i - \tau_m$ is an estimable function. Verify that the estimates of this function obtained from the method given in the text and the method of this problem are the same.

1.7. We define a *contrast* to be a linear parametric function $\sum_{i=1}^{v} \lambda_i \tau_i$, where $\sum_i \lambda_i = 0$. A contrast is said to be *elementary* if, for some $i_1 \neq i_2$, we have

$$\lambda_{i_1} = 1,\ \lambda_{i_2} = -1, \text{ and } \lambda_i = 0 \text{ for } i \neq i_1, i \neq i_2.$$

Show that in a randomized complete block design and in a BIBD all elementary contrasts are estimable.

1.8. Suppose an experiment is to be designed to compare t new treatments with a control treatment. A completely randomized design is to be used; there are to be r_0 observations on the control and r observations on each of the new treatments. Let N be the total number of plots available for the experiment. Show that for any given value of N, the variance of the comparisons between the control and the new treatments is minimized if $r_0 = r\sqrt{t}$.

2 Balanced incomplete block designs

2.1. Elementary properties of BIBDs

Several simple relations between the parameters of a design can be conveniently proved using incidence matrices. Here we let x_i denote the number of blocks which intersect some particular block of our design in precisely i elements, for $i = 0, 1, \ldots, k$. (In general, the values of x_i may vary from one block to another in the same design.)

Theorem 1. *In any block design, we have:*

$$\text{(i)} \quad \sum_{i=0}^{k} x_i = b - 1; \tag{2.1}$$

$$\text{(ii)} \quad \sum_{i=0}^{k} ix_i = k(r-1). \tag{2.2}$$

Proof. (i) Suppose that we consider one particular block of the design and that, by rearranging blocks if necessary, we can call it B_1. Now of the remaining $b-1$ blocks, x_0 are disjoint from B_1, x_1 intersect B_1 in one element each, and so on, till finally x_k blocks intersect B_1 in k elements each (so they are copies of B_1). This accounts for all the other blocks of the design so (2.1) holds.

(ii) We now rearrange the rows of the incidence matrix or, in other words, relabel the varieties, so that

$$B_1 = 123 \ldots k,$$

and rearrange the columns so that the x_0 blocks disjoint from B_1 correspond to columns $2, \ldots, x_0+1$, the x_1 blocks intersecting B_1 in one variety each correspond to columns $x_0+2, \ldots, x_0+x_1+1$, and so on. The incidence matrix is shown in Table 2.1. Now we sum the entries in the $k \times (b-1)$ submatrix consisting of rows 1 through k, and columns 2 through b. Since each of these varieties occurs in $r-1$ blocks other than B_1, each row of this submatrix contains $(r-1)$ 1s, and there are k rows. Since there are no 1s in x_0 columns, one 1 in each of x_1 columns, and so on, this means that

$$0.x_0 + 1.x_1 + \ldots + k.x_k = k(r-1),$$

proving (2.2). □

Table 2.1

	B_1	$\leftarrow x_0 \rightarrow$	$\leftarrow x_1 \rightarrow$	$\ldots$	$\leftarrow x_i \rightarrow$	$\ldots$	$\leftarrow x_k \rightarrow$
$\uparrow$ k $\downarrow$	1 1 1	all 0s	one 1 per column		i 1s per column		all 1s
$\uparrow$ $v-k$ $\downarrow$							

If the block design is balanced, we can say more.

Theorem 2. *In any balanced incomplete block design, we have*:

(i) $\sum_{i=0}^{k} \binom{i}{2} x_i = \binom{k}{2}(\lambda - 1);$ (2.3)

(ii) $v \leq b$ *(Fisher's inequality).* (2.4)

Proof. (i) To prove this, consider again the incidence matrix permuted as in Table 2.1. We count the number of pairs of varieties ab, where both a and b belong to the first block. There are $\binom{k}{2}$ such pairs, each of which occurs $\lambda - 1$ times in blocks other than the first; that is, $\binom{k}{2}(\lambda - 1)$ pairs occur altogether outside the first block. No such pairs could occur in the $x_0 + x_1$ blocks which intersect the first in zero or one variety, $\binom{2}{2} = 1$ such pair occurs in each of the x_2 blocks which intersect the first in two varieties, $\binom{3}{2}$ such pairs in each of the x_3 blocks which intersect the first in three varieties, and so on, giving altogether

$$\sum_{i=0}^{k} \binom{i}{2} x_i = \binom{k}{2}(\lambda - 1)$$

such pairs, and verifying (2.3).

(ii) For Fisher's inequality, we give a proof based on considering the mean and variance of the intersection size of one fixed block B with all the others. Let i represent the size of the intersection of B with some other block, and let μ be the mean value of i, that is,

$$\mu = \left(\sum_{i=0}^{k} ix_i\right) \Big/ \left(\sum_{i=0}^{k} x_i\right) = k(r-1)/(b-1) \text{ by eqns (2.1) and (2.2).} \tag{2.5}$$

Now consider the sum

$$\begin{aligned}
\sigma^2 &= \sum_{i=0}^{k} (i-\mu)^2 x_i \\
&= \sum_i (i^2 - 2i\mu + \mu^2)x_i \\
&= \sum_i (i(i-1) + i - 2i\mu + \mu^2)x_i \\
&= \sum_i \left(2\binom{i}{2} + i - 2i\mu + \mu^2\right)x_i \\
&= k(k-1)(\lambda-1) + k(r-1) - (k^2(r-1)^2)/(b-1)
\end{aligned}$$

by eqns (2.1), (2.2), (2.3), and (2.5).
Thus

$$(b-1)\sigma^2 = (b-1)[k(k-1)(\lambda-1) + k(r-1)] - k^2(r-1)^2.$$

By eqns (1.2.3) and (1.2.4), we have

$$b - 1 = \frac{vr}{k} - 1 = \frac{r^2(k-1)}{\lambda k} + \frac{r}{k} - 1.$$

Thus, in terms of the independent variables r, k and λ, $(b-1)\sigma^2$ is a cubic in r, and the coefficient of r^3 is

$$\frac{k(k-1)}{\lambda k} = \frac{k-1}{\lambda}. \tag{2.6}$$

If $r = k$, then $b = v$ and $k(k-1) = \lambda(b-1)$, in which case $(b-1)\sigma^2 = 0$. This means that $(r-k)$ is a factor of the cubic. Again, if $r = \lambda$, then $k = v$ and $b = \lambda$, in which case $(b-1)\sigma^2 = 0$, so that $(r-\lambda)$ is also a factor of the cubic. Finally, $\dfrac{d}{dr}[(b-1)\sigma^2] = 0$ when $r = \lambda$, and thus $(r-\lambda)^2$ is a factor.

Hence $(b-1)\sigma^2 = \alpha(r-\lambda)^2(r-k)$ for some constant α. Comparing with (2.6) shows that

$$\alpha = (k-1)/\lambda$$

and

$$(b-1)\sigma^2 = (k-1)(r-\lambda)^2(r-k)/\lambda. \tag{2.7}$$

Now

$$(b-1)\sigma^2 \geq 0, \text{ by definition of } \sigma^2,$$
$$k-1 \geq 0,\ (r-\lambda)^2 \geq 0,\ \lambda \geq 0,$$

and thus

$$r - k \geq 0.$$

But now, by eqn (1.2.3), we must have $v \leq b$. (Note that σ^2 is independent of the choice of the block B.) □

In our examples so far, we have always had $v \leq b$. Since (D5), (D7), and (D8) are not balanced, we can see that $v \leq b$ is a necessary but not a sufficient condition for an incomplete block design to be balanced.

2.2. Dual designs and symmetric designs

If A is the incidence matrix of a block design D with parameters v, b, r, k, then, as we have already pointed out in Section 1.2,

$$AJ_b = rJ_{v \times b}, \tag{2.8}$$

$$J_v A = kJ_{v \times b}. \tag{2.9}$$

What about A^T? It is a $b \times v$ matrix, with k 1s per row and r 1s per column, so that

$$J_b A^T = rJ_{b \times v} \quad \text{and} \quad A^T J_v = kJ_{b \times v};$$

that is, A^T is the incidence matrix of a block design with parameters b, v, k, r, constructed in the following way. If the original design has varieties $a_1, a_2, \ldots, a_v$, and blocks $B_1, B_2, \ldots, B_b$, then the new design has varieties $c_1, c_2, \ldots, c_b$ and blocks $C_1, C_2, \ldots, C_v$, where c_i belongs to C_j if and only if a_j belongs to B_i. This is called the *dual design*, D', of D.

For example, consider the design based on $X = \{1, 2, 3, 4, 5, 6\}$ with blocks

$$123,\ 124,\ 135,\ 146,\ 156,\ 236,\ 245,\ 256,\ 345,\ 346. \tag{D9}$$

The incidence matrix, A, of (D9) is given in Table 2.2, together with A^T, the incidence matrix of (D9)'. Here variety 1 occurs in blocks 1, 2, 3, 4, 5, of (D9), so block 1 of (D9)' consists of varieties 1, 2, 3, 4, 5, and so on. (D9)' has parameters 10, 6, 3, 5 and is certainly a block design but it is certainly not balanced, since $10 > 6$. (D9) itself is a $B[3, 2; 6]$ design.

For a BIBD, Fisher's inequality, $v \leq b$, must hold. In the dual design, blocks and varieties interchange roles, so in general Fisher's inequality cannot hold for both a BIBD and its dual. Consequently, the dual is not a balanced design. The only exception occurs when $v = b$, in which case the design is said to be a *symmetric* BIBD (SBIBD); if this property is particularly important we denote it by $SB[k, \lambda; v]$. To show that the dual of an SBIBD is also an SBIBD, and to obtain an alternative proof of Fisher's inequality, we consider again the incidence matrix.

Table 2.2

1	1	1	1	1	0	0	0	0	0
1	1	0	0	0	1	1	1	0	0
1	0	1	0	0	1	0	0	1	1
0	1	0	1	0	0	1	0	1	1
0	0	1	0	1	0	1	1	1	0
0	0	0	1	1	1	0	1	0	1

Incidence matrix, A, of (D9)

1	1	1	0	0	0
1	1	0	1	0	0
1	0	1	0	1	0
1	0	0	1	0	1
1	0	0	0	1	1
0	1	1	0	0	1
0	1	0	1	1	0
0	1	0	0	1	1
0	0	1	1	1	0
0	0	1	1	0	1

Incidence matrix, A^T, of the dual, (D9)′

Theorem 3. *If A is the incidence matrix of a $B[k, \lambda; v]$ design, then*:

(i) $\det(AA^T) = (r-\lambda)^{v-1}[r+(v-1)\lambda]$;

(ii) $v \le b$. (2.4)

Proof. (i)

$$AA^T = \begin{bmatrix} r & \lambda & \lambda & . & . & . & \lambda \\ \lambda & r & \lambda & . & . & . & \lambda \\ \lambda & \lambda & r & . & . & . & \lambda \\ . & . & . & . & . & . & . \\ \lambda & \lambda & \lambda & . & . & . & r \end{bmatrix}$$

so its determinant equals that of

$$M = \begin{bmatrix} r & \lambda-r & \lambda-r & . & . & . & \lambda-r \\ \lambda & r-\lambda & 0 & . & . & . & 0 \\ \lambda & 0 & r-\lambda & . & . & . & 0 \\ . & . & . & . & . & . & . \\ \lambda & 0 & 0 & . & . & . & r-\lambda \end{bmatrix}$$

(which was found by subtracting the first column of AA^T from every other column). Now, adding rows 2 through v to row 1 of M, we find

$$\det(AA^T) = \det M = \det \begin{bmatrix} r+(v-1)\lambda & 0 & 0 & . & . & . & 0 \\ \lambda & r-\lambda & 0 & . & . & . & 0 \\ \lambda & 0 & r-\lambda & . & . & . & 0 \\ . & . & . & . & . & . & . \\ . & . & . & . & . & . & . \\ \lambda & 0 & 0 & & & & r-\lambda \end{bmatrix}$$

$= (r-\lambda)^{v-1}[r+(v-1)\lambda]$, proving (i).

(ii) Since $\lambda(v-1) = r(k-1)$, we see that $r = \lambda$ implies $v = k$, which is not possible for a BIBD. Also,

$$\det(AA^T) = (r-\lambda)^{v-1}[r + r(k-1)] = (r-\lambda)^{v-1}rk \neq 0.$$

Hence, $v = \text{rank}\,(AA^T) = \text{rank}\,A \leq \min(v, b)$, proving eqn (2.4) by a different approach. □

Now we can get some more information concerning the dual design.

Theorem 4. *The dual of a $B[k, \lambda; v]$ design is a balanced incomplete block design if and only if $v = b$, that is, the design is symmetric.*

Proof. (i) If $b > v$, then the dual design has more treatments than blocks, and cannot be balanced by Theorem 2(ii).

(ii) If $v = b$, we consider the incidence matrix A of the design. Since $k = r$, Theorem 1.2.2 shows that

$$AA^T = (k-\lambda)I_v + \lambda J_v \tag{2.10}$$

and

$$J_v A = kJ_v. \tag{2.11}$$

We prove that

$$A^TA = (k-\lambda)I_v + \lambda J_v \tag{2.12}$$

and

$$J_v A^T = kJ_v; \tag{2.13}$$

that is, that we can replace A by A^T in (2.10) and (2.11), with $r = k$. If these equations hold, then by Theorem 1.2.2, we know that A^T is the incidence matrix of a BIBD, in fact one with the same parameters as the design we started from.

We already know that

$$AJ_v = kJ_v,$$

since $v = b$ and $r = k$. Transposing, we have

$$J_v^T A^T = kJ_v^T,$$

but $J_v^T = J_v$, so we have (2.13).

We know by Theorem 3 that $\det(AA^T) \neq 0$. But since $v = b$, the matrix A is square. Hence, $\det(AA^T) = (\det A)\det A^T = (\det A)^2 \neq 0$. (2.14)
Thus, $\det A \neq 0$ and A^{-1} must exist. Since

$$kJ = AJ,$$

we have

$$kA^{-1}J = A^{-1}AJ = J$$

so that

$$A^{-1}J = k^{-1}J.$$

Now from (2.10),

$$A^{-1}(AA^T)A = (k-\lambda)A^{-1}I_vA + \lambda A^{-1}(J_vA),$$

so

$$\begin{aligned} A^T A &= (k-\lambda)I_v + \lambda A^{-1}.kJ_v \\ &= (k-\lambda)I_v + \lambda k.k^{-1}J_v \\ &= (k-\lambda)I_v + \lambda J_v, \end{aligned}$$

proving (2.12). □

Corollary 4.1. *A symmetric balanced incomplete block design is linked, with* $\mu = \lambda$.

Proof. Let the design have incidence matrix A. The number of varieties common to blocks i and j equals the dot product of columns i and j of A. But this is the (i,j) entry of $A^T A$, and thus always equals λ, for $i \neq j$, by Theorem 4. □

The converse of Corollary 4.1 is also true so that in fact a BIBD is linked if and only if it is symmetric, but we can say even more.

Theorem 5. *If there is one block in a BIBD such that every other block intersects it in a fixed number of points, then the design is symmetric.*

Proof. By definition,

$$\sigma^2 = \sum_{i=0}^{k} (i-\mu)^2 x_i^2$$

so if there is a block B which intersects every other block in precisely i points, then $\mu = i$ and $x_j = 0$ for $j \neq i$. Hence $\sigma^2 = 0$ for that block and so for all blocks. Thus, by eqn (2.7),

$$(k-1)(r-\lambda)^2(r-k) = 0.$$

But $k > 1$ and $r > \lambda$ so $r = k$.

Hence the design is symmetric. □

Theorem 5 tells us that the presence of just one linked block in a BIBD ensures that the whole design is balanced and symmetric.

2.3. Representing designs by graphs

Now we look at some small examples of designs, many with interesting properties which are discussed more generally in later chapters.

Example 6. Suppose that from a set of v varieties, we choose every 2-set to be a block, as Yates did; see Chapter 1. Then $b = \binom{v}{2} = v(v-1)/2$ and $k = 2$.

Further, every variety occurs with each of the $v-1$ other varieties exactly once, so that $r = v-1$ and $\lambda = 1$. Thus, we have a

$$(v, v(v-1)/2, v-1, 2, 1,) \text{ BIBD}.$$

Also, if we write down each of the pairs λ times, for any positive integer λ, we form a design with parameters

$$(v, \lambda v(v-1)/2, \lambda(v-1), 2, \lambda)$$

and this is the only design with these parameters. □

In the particular case where $v = 6$, $\lambda = 1$, we have blocks

12, 13, 14, 15, 16, 23, 24, 25, 26, 34, 35, 36, 45, 46, 56. (D10)

This design has the following interesting property. Suppose we arrange its fifteen blocks into five classes of three blocks each as shown in the rows of Table 2.3.

Table 2.3

12	34	56
13	25	46
14	26	35
15	24	36
16	23	45

Then each class of three blocks contains each variety precisely once. Certainly if v were odd, we could not split up the $v(v-1)/2$ blocks of the design in this way – each class would need a half-block – but what if v is even, $v \neq 6$? In fact, such a partition always exists. To describe it, we introduce some terminology.

A *resolution class* in a BIBD is a set of blocks which together contain each variety of the design precisely once. A *resolvable* BIBD is one whose blocks can be partitioned into mutually disjoint resolution classes; this partition can in some cases be carried out in several different ways (Exercise 2.16).

The design (D10) is the second example of a resolvable design that we have seen: the $B[3, 1; 9]$ design (D1) was also resolvable. The property of resolvability is not restricted to balanced designs alone. For instance, if $X = \{1, 2, \ldots, 8\}$, the design with blocks

1234, 5678, 1256, 3478 (D11)

is not balanced but is resolvable into two classes.

Theorem 7. *For every positive integer n, the BIBD with parameters* $(2n, n(2n-1), 2n-1, 2, 1)$ *is resolvable.*

Proof. Let the underlying set of varieties S be

$$S = \{\infty, 0, 1, \ldots, 2n-2\},$$

and define addition on the set S to be carried out modulo $2n-1$, except that

$$\infty + i = \infty \text{ for all } i \in S.$$

Now define the initial resolution class to be

$$\begin{aligned} R_0: &\{\infty, 0\}, \{1, 2n-2\}, \{2, 2n-3\}, \ldots, \{n-1, n\} \\ &= \{\infty, 0\}, \{1, -1\}, \{2, -2\}, \ldots, \{n-1, -(n-1)\}, \end{aligned}$$

and the ith resolution class for $i = 1, 2, \ldots, 2n-2$ to be

$$\begin{aligned} R_i: &\{\infty, i\}, \{1+i, 2n-2+i\}, \{2+i, 2n-3+i\}, \ldots, \{n-1+i, n+i\} \\ &= \{\infty, i\}, \{i+1, i-1\}, \{i+2, i-2\}, \ldots, \{i+(n-1), i-(n-1)\}. \end{aligned}$$

Then $\{R_0, R_1, \ldots, R_{2n-2}\}$ is a resolution of the BIBD.

To check this, we see that each R_i consists of n blocks, that altogether they cover $n(2n-1)$ blocks, and that each variety occurs precisely once in each resolution class. Now suppose that $\{x, y\} \subseteq S$. Then if $x = \infty$, $\{\infty, y\} \in R_y$. Otherwise

$$x + y = \alpha \quad \text{and} \quad x - y = \beta,$$

and we can choose a, b, such that

$$2a \equiv \alpha \quad \text{and} \quad 2b \equiv \beta \pmod{2n-1},$$

and $0 \leq a, b \leq 2n-2$. In this case $\{x, y\} \in R_a$, where it appears in the form $\{a+b, a-b\}$. □

This construction is most easily understood in terms of a diagram. Choose $2n$ vertices (or points) and join all the $\binom{2n}{2} = n(2n-1)$ pairs of distinct points, with one edge each. (This structure is K_{2n}, the *complete graph* on $2n$ vertices.) Each of these edges corresponds to one block in our design, and Theorem 7 thus states that the edges in K_{2n} can be partitioned into $2n-1$ sets, where each set contains precisely one edge incident with each vertex. (Such a set is called a *one-factor* of K_{2n}, and the set of $2n-1$ edge-disjoint one-factors is called a *one-factorization.*)

To illustrate the construction of Theorem 7, choose one vertex, labelled ∞, as the centre of a clockface, and arrange the vertices $0, 1, \ldots, 2n-2$, cyclically around it. Each resolution class (or one-factor) is obtained by rotating the previous one through $(360/(2n-1))^\circ$, as shown for $n = 3$ in Fig. 2.1.

What conditions are necessary for resolvability? Certainly, we must have $k|v$ if the varieties can be partitioned into blocks of k varieties each, but this condition is not sufficient for $k > 2$. For example, in the design (D8), which has $v = 6$, $k = 3$, no two blocks are disjoint: they intersect in one or two elements. But a resolution class would have to consist of two disjoint blocks and thus cannot exist. We consider some generalizations of Theorem 7 in Chapter 8; see

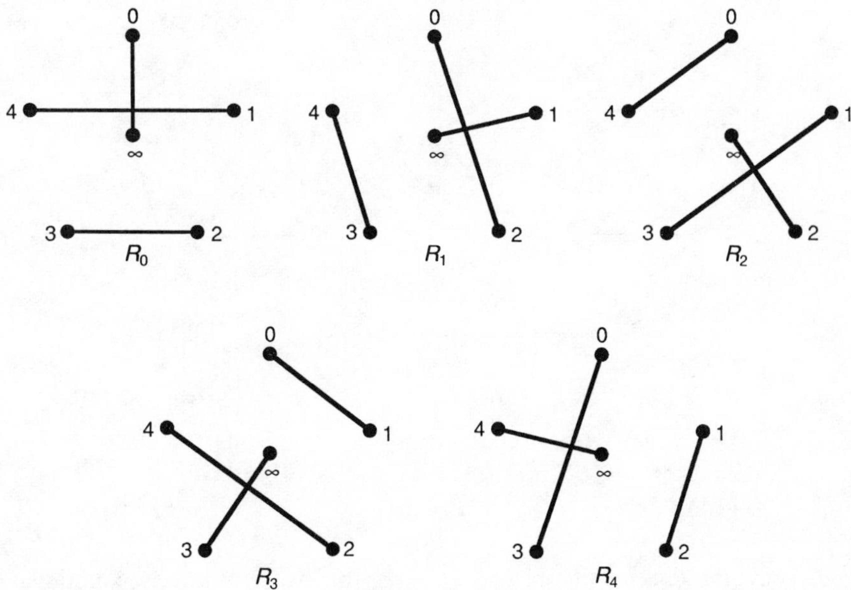

Fig. 2.1

also Section 2.7. For the moment we look at some other examples of applications of graph theory to designs.

If a block design with $k = 3$ is balanced with $\lambda = 1$, then it is called a *Steiner* triple system, denoted also by $B[3, 1; v]$. An *adjacency graph* for one particular variety, x, in a triple system $B[3, \lambda; v]$ is a graph with $v - 1$ vertices, labelled with the varieties other than x, such that the edge yz occurs in the graph precisely when xyz is a block of the design. If the block xyz occurs twice in the design, then the edge yz occurs twice in the adjacency graph (so, strictly speaking, it is a multigraph).

Example 8. Consider a $B[3, 2; 6]$. We have already encountered such a design, namely (D9); now we see how it might have been constructed.

Consider the adjacency graph of variety 1. The *valency* of a vertex in a graph is the number of edges incident with it; this graph has five vertices, each with valency $\lambda = 2$. Thus it can take either of the forms shown in Fig. 2.2. The two graphs shown correspond to the five blocks of the design which must contain variety 1. In case (a), varieties 2 and 3 have occurred twice each, hence they must each occur three more times. But they have occurred together twice already, so we must have six additional blocks in the design. This is impossible, and the design in (a) cannot be completed. Thus, we can see that each variety

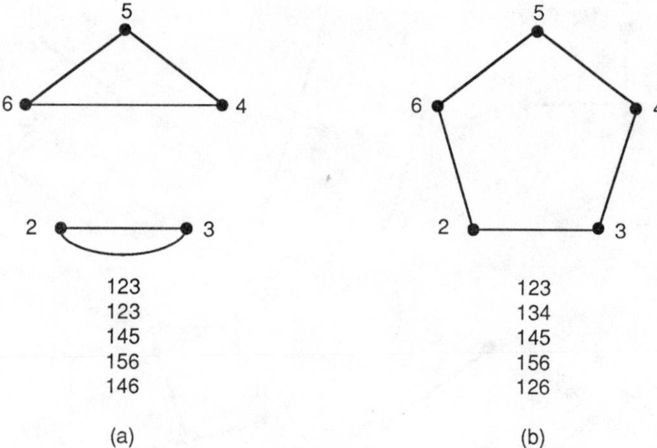

Fig. 2.2

must have the adjacency graph shown in case (b). In case (b), the five additional blocks must be of the form:

$$23*,\ 2**,\ 2**,\ 3**,\ 3**,$$

where the stars indicate spaces to be filled in by three replications each of 4, 5, and 6. We could fill in the remaining places by trial and error but it is easier to consider the adjacency graphs of varieties 2 and 3. Each must be a cycle, as in Fig. 2.2(b), and each can be partly labelled from the existing blocks; see Fig. 2.3.

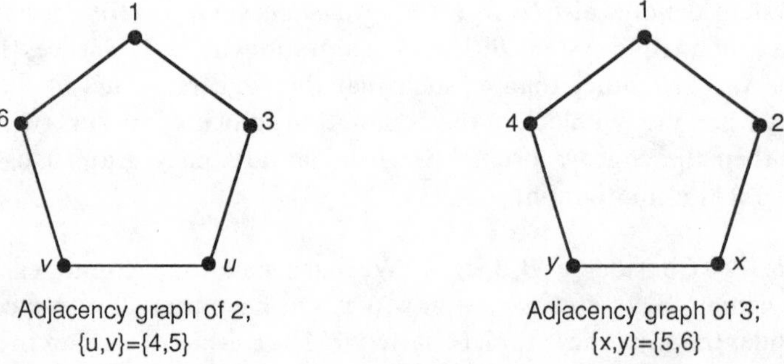

Fig. 2.3

If we let $u = 4$, then we have the block 234 in the design; this would mean that the edge 24 would have to occur in the adjacency graph of 3 which is impossible. Hence $u = 5$, $v = 4$, giving blocks 235, 245, 246. Now $x = 5$, $y = 6$, giving the remaining blocks 346, 356. Hence up to isomorphism, the only

BIBD with the given parameters has blocks

$$123, 134, 145, 156, 126, 235, 245, 246, 346, 356. \qquad \text{(D12)}$$

Notice that if we apply the permutation (23) (56) to the blocks of (D9) we obtain (D12), another example of isomorphism. □

If we choose all $\binom{v}{k}$ of the k-subsets of a v-set (as in Example 6 where $k = 2$), then certainly we have a design: $b = \binom{v}{k}$, $r = \binom{v-1}{k-1}$ (since each variety occurs in the k-subsets with $k - 1$ varieties chosen from among the other $v - 1$), and similarly $\lambda = \binom{v-2}{k-2}$. This design always exists but is not especially interesting. But the designs (D9) and (D12) show that if $v = 6, k = 3$, so that we have $\binom{6}{3} = 20$ blocks, then we can choose 10 of them to form a $B[3,2;6]$ design. Hence the remaining 10 blocks form a design with the same parameters, in fact an isomorphic design (since those parameters lead to a unique design). This is an illustration of the *garbage heap principle*: whatever has not been used in making a design is just as interesting as the design itself.

Suppose, instead of the 3-subsets chosen from a 6-set, we consider the $\binom{7}{3} = 35$ 3-subsets chosen from a 7-set.

Example 9. Let $X = Z_7$ be the set on which the design is based. The possible blocks are listed in five cyclic sets in Table 2.4.

Table 2.4

I	II	III	IV	V
013	023	012	024	036
124	134	123	135	140
235	245	234	246	251
346	356	345	350	362
450	460	456	461	403
561	501	560	502	514
602	612	601	613	625

The blocks in set I form the $B[3, 1; 7]$ design (D4). Similarly, the blocks in set II form a BIBD isomorphic to (D4), where the permutation (16) (25) (34) takes set I to set II. Since the complete set of 35 blocks forms a balanced design, so must sets III, IV, and V together, namely a $B[3, 3; 7]$ design. But does this design contain a subdesign with just seven blocks; that is with the same parameters as sets I and II?

By Corollary 4.1, a $B[3, 1; 7]$ design must be linked and, since $\lambda = 1$, any pair of its blocks must intersect in precisely one variety. Hence from set III, we could choose at most two blocks of such a design, say 012, 234 (without loss of generality). Similarly, from each of sets IV and V we can also choose at most two blocks. (To see this, it may be easier to rewrite these cyclic sets as 024, 246, . . . , 502 and 036, 362, . . . , 403 respectively). Hence these 21 blocks do not contain any subdesign – that is, they form an *irreducible design*: although the parameters are such that a $B[3, 3; 7]$ design might be a union of a $B[3, 1; 7]$ design and a $B[3, 2; 7]$ design, or of three $B[3, 1; 7]$ designs, this particular design is not.

This immediately raises the question of whether we could have chosen designs differently to start with, so that we could in fact have formed five pairwise disjoint designs from the 35 blocks, or even three $B[3, 1; 7]$ designs and one $B[3, 2; 7]$ design.

In considering this problem, we need first to know the structure of a $B[3, 2; 7]$ design with no repeated blocks. (It is perfectly legitimate and in some cases very useful for a design to have repeated blocks, but at present we are trying to choose designs from a set with no repeated blocks available.) The possible adjacency graphs for variety 0 are shown in Fig. 2.4; no repeated edges are allowed.

In case (a), the design has six blocks 012, 023, 013, 045, 056, 046, and the remaining eight blocks must contain four replications each of 1, 2, 3, 4, 5, 6. Thus, seven blocks must be 12*, 1**, 1**, 1**, 2**, 2**, 2**, where the stars denote varieties yet to be found. If 3 belongs to the first of these blocks, then the pairs 12, 23, 13 have occurred twice each and three additional blocks will be needed to accommodate the remaining three occurrences of 3. This gives altogether $6 + 7 + 3 > 14$ blocks, which is impossible. Hence, 3 must belong to, say, the second and fifth of these blocks, giving

12*, 13*, 1**, 1**, 23, 2**, 2**,

and two additional blocks containing 3 will be needed. This still gives $6 + 7 + 2 > 14$ blocks, so case (a) is impossible.

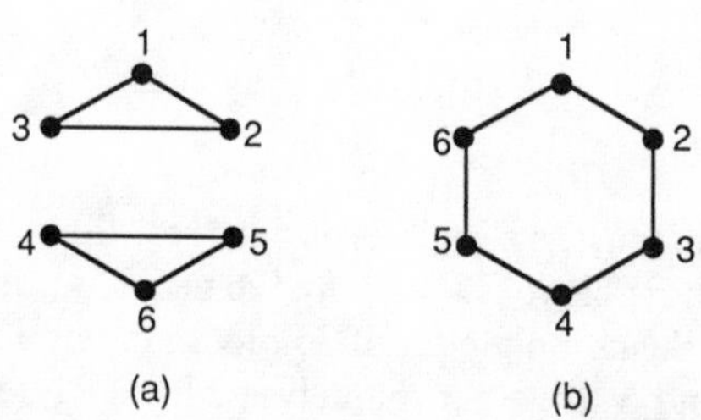

Fig. 2.4

Thus, 0 has the adjacency graph shown in (b), and every other variety must have an isomorphic adjacency graph. We start with the six blocks:

$$012,\ 023,\ 034,\ 045,\ 056,\ 061$$

and have to determine eight remaining blocks. An argument similar to that for case (a) shows that these must be:

$$12*,\ 13*,\ 13*,\ 1**,\ 23*,\ 2**,\ 2**,\ 3**,$$

with twelve blanks to be filled in with four occurrences each of 4, 5, 6. The adjacency graphs of 1, 2, 3 are shown in Fig. 2.5. Now $\{x, y, z\} = \{3, 4, 5\}$, $\{u, v, w\} = \{4, 5, 6\}$ and $\{a, b, c\} = \{1, 5, 6\}$, but we can say more than this. We must have a block $12x$, and also a block $12u$, but since $\lambda = 2$ and the block 012 is already present, this forces $x = u \in \{4, 5\}$. Similarly, $a = w \in \{5, 6\}$.

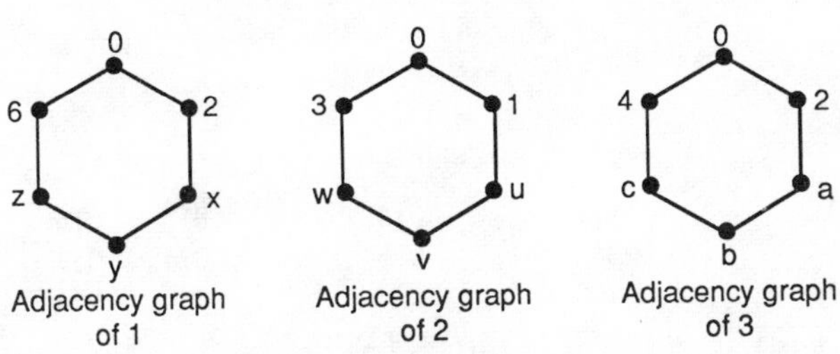

Fig. 2.5

Suppose that we let $x = u = 4$, $y = 5$, $z = 3$. This leads to blocks 124, 145, 135, 136. The blocks 135, 136 imply that $b = 1$, and $\{a, c\} = \{5, 6\}$. If we choose $a = 5$, so that $w = 5$, $c = 6$, $v = 6$, the remaining blocks must be 235, 246, 256, 346. Other choices are possible, but all of them lead to designs isomorphic to this one. Hence, we have the design:

$$012,\ 034,\ 056,\ 145,\ 136,\ 235,\ 246;\ 023,\ 045,\ 061,\ 124,\ 135,\ 256,\ 346$$

where blocks 1 to 7 form a $B[3, 1; 7]$ design, and so do blocks 8 to 14. Hence, any $B[3, 2; 7]$ design with distinct blocks is reducible. (It turns out that, in fact, *any* $B[3, 2; 7]$ design is reducible.)

Thus, if we are going to choose more than two $B[3, 1; 7]$ designs from the 35 3-sets of a 7-set, we must in fact partition those 3-sets into five subdesigns. We now show that this is impossible.

Consider first the $\binom{6}{2} = 15$ blocks containing 0. These must be partitioned into five sets, of three blocks each, just as the blocks of (D10) were partitioned in Example 6. In other words, we need a one-factorization of K_6, where the vertices of K_6 are labelled 1, 2, 3, 4, 5, 6. Without loss of generality, we take our first one-factor to be 12, 34, 56. This gives two choices for the next one-factor, but after that the one-factorization is determined as shown in Table 2.5. These two one-factorizations are isomorphic since the permutation (12) (34) (56) transforms A to B, and B to A. Hence, we may assume that the blocks containing 0 are partitioned as shown in Table 2.6, and that the remaining blocks in each of the five subdesigns are, for example, 135, 146, 236, 245 in case (i) or 136, 145, 235, 246 in case (ii) of the first subdesign.

Table 2.5

	13	14	15	16	
	→25	26	24	23	A
12 ↗	46	35	36	45	
34					
56 ↘	13	14	15	16	
	→26	25	23	24	B
	45	36	46	35	

Table 2.6

012			013			014			015	016
034			025			026			024	023
056			046			035			036	045
1	35	36	1	26	24	1	23	25	1	1
1	46	45	1	45	56	1	56	36	1	1
2	36	35	3	24	26	4	25	23	5	6
2	45	46	3	56	45	4	36	56	5	6
	(i)	(ii)		(i)	(ii)		(iii)	(iv)		

If we choose the first subdesign to have blocks as in case (i), then the second subdesign can have blocks (i) but not (ii), and similarly if the first subdesign has blocks (ii), so must the second. But now for the third subdesign, (iii) contains 245, repeating a block in (i), and (iii) also contains 156, repeating a block in (ii). Thus, (iii) cannot be chosen for the third subdesign. Similarly, (iv) would repeat blocks 234 of (i) and 136 of (ii). Hence, only two subdesigns are possible. □

If all the $\binom{v}{k}$ k-sets chosen from a v-set can be partitioned to form $B[k, 1; v]$ designs, these designs are said to form a *large set*. For all but six of the possible values of v, large sets of Steiner triple systems have been constructed; see Section 2.7.

2.4. Designs with larger blocks

Now we consider some parameters for designs with $k > 3$, namely (11, 11, 5, 5, 2) and (22, 22, 7, 7, 2).

Example 10. The $B[5, 2; 11]$ design must, if it exists, be a symmetric design and hence a linked design, by Corollary 4.1. Thus, any pair of its blocks intersect in exactly two varieties, and they must take the form shown in Table 2.7(a) where the six varieties 5, 6, 7, 8, 9, A have still to be allocated to ten blocks. (Here A denotes 10.)

Since $\lambda = 2$, each of the remaining six varieties will occur with a series of blocks containing (ab), (bc), (cd), (de), (ea), where

$$\{a, b, c, d, e\} = \{0, 1, 2, 3, 4\},$$

but the order in which varieties are arranged has yet to be determined. In shortened form, this sequence of blocks is called the *λ-chain* $(abcde)$.

We may assume without loss of generality that 5 occurs with the λ-chain (01234), and that 6 and 7 occur with 0, 1 to complete that block; this is shown in Table 2.7(b). The other possible λ-chains containing 0, 1 are (01243), (01324), (01342), (01423), (01432). Suppose that we attempt to place variety 6 in accord with the first of these, (01243). Then we would have blocks 01567 (already complete) and 1256*, which intersect in at least three varieties. Hence, this λ-chain cannot be used. Similarly, we can rule out the second and the fifth, so that we can only assign 6 and 7 to the chains (01342) and (01423), in either order. This is shown in Table 2.7(c).

Now the block 026** needs two additional varieties, say 8 and 9, which then occur in a block with 6. Since 6 was assigned to the λ-chain (01342) = (02431), 8 and 9 must be assigned to chains consistent with it, namely (02143) and (02314). This leaves only one way to place A, the last variety. Hence, the completed design is that in Table 2.7(d).

Our argument shows that this design is unique up to isomorphism. □

Table 2.7

(a)	(b)	(c)	(d)
0 1 2 3 4	0 1 2 3 4	0 1 2 3 4	0 1 2 3 4
0 1 * * *	0 1 5 6 7	0 1 5 6 7	0 1 5 6 7
0 2 * * *	0 2 * * *	0 2 6 * *	0 2 6 8 9
0 3 * * *	0 3 * * *	0 3 7 * *	0 3 7 8 A
0 4 * * *	0 4 5 * *	0 4 5 * *	0 4 5 9 A
1 2 * * *	1 2 5 * *	1 2 5 * *	1 2 5 8 A
1 3 * * *	1 3 * * *	1 3 6 * *	1 3 6 9 A
1 4 * * *	1 4 * * *	1 4 7 * *	1 4 7 8 9
2 3 * * *	2 3 5 * *	2 3 5 7 *	2 3 5 7 9
2 4 * * *	2 4 * * *	2 4 6 7 *	2 4 6 7 A
3 4 * * *	3 4 5 * *	3 4 5 6 *	3 4 5 6 8

Example 11. Consider the design with parameters (22, 22, 7, 7, 2). Let *A*

denote its incidence matrix and apply eqn (2.14). We have

$$(\det A)^2 = \det(AA^T)$$
$$= (7-2)^{22-1}(7)7 \qquad \text{by Theorem 3(i).}$$

But now det A must be an integer, since all the entries in A are integers, and hence $\sqrt{(5^{21}(49))} = 7\sqrt{(5^{21})}$ must be an integer. This is obviously impossible, so no design with these parameters can exist. □

Notice in the example above that the parameters satisfy the conditions we had proved earlier for balanced incomplete block designs, and hence that these conditions are necessary *but not sufficient* for the existence of the design.

Example 12. In some experiments, there may be more sources of variability than can be controlled by ordinary blocking. For example, if the experimental units are the half-leaves of plants, there may be more treatments (varieties) than there are half-leaves per plant. As there will be variation between plants, between leaves within plants, and between half-leaves within leaves, there are two natural blocking systems, the plants and the leaves, with one system nested inside the other. By using this natural blocking we can, when analysing the results, remove components due to differences between plants and due to differences between leaves within plants. By randomly assigning treatments to half-leaves within plants, we can estimate the uncontrolled variation due to differences between half-leaves. Hence we can accurately compare treatment effects.

We define a *nested* balanced incomplete block design with parameters $(v; b_1, b_2; r; k_1, k_2; \lambda_1, \lambda_2; m)$ to be a design on v varieties, each replicated r times, with two systems of blocks such that:

(i) each block of the first system (subsequently called a 'block') contains exactly m blocks of the second system ('sub-blocks');
(ii) ignoring the second system leaves a BIBD with b_1 blocks of size k_1 each, and index of balance λ_1:
(iii) ignoring the first system leaves a BIBD with b_2 blocks of size k_2 each, and index of balance λ_2.

Thus $vr = b_1k_1$ by (ii) and eqn (1.2.3)
$= b_1mk_2$ by (i)
$= b_2k_2$ by (iii),

$\lambda_1(v-1) = r(k_1-1)$ and $\lambda_2(v-1) = r(k_2-1)$ by eqn (1.2.4), and hence

$$(v-1)(\lambda_1 - m\lambda_2) = (m-1)r.$$

For instance, the (8, 14, 7, 4, 3) design shown in Table 2.8 may be nested in several ways, to give a design with parameters (8; 14, 28; 7; 4, 2; 3, 1; 2). The blocks have four varieties each, and are listed in two cyclic sets of seven blocks each. The sub-blocks may be chosen to be the rows, columns or diagonals of the blocks as listed, thus the sub-blocks of the block 1235 may be 13 and 25, or 12 and 35, or 15 and 23. Once chosen, the nesting arrangement must be continued through the cycles, thus sub-blocks 13 and 25 must accompany 24 and 36, 23 and 5∞, and so on. □

Table 2.8

13 25	23 5∞
24 36	34 6∞
35 40	45 0∞
46 51	56 1∞
50 62	60 2∞
61 03	01 3∞
02 14	12 4∞

2.5. Complementary, multiple, derived, and residual designs

Given a balanced incomplete block design, we can often construct new designs from it. For example, if the design is based on a set S, and has blocks $B_1, B_2, \ldots, B_b$, then the collection of blocks $S \backslash B_1, S \backslash B_2, \ldots, S \backslash B_b$ is also a balanced incomplete block design, called the *complement* or *complementary design* of the original.

Theorem 13. *The complement of a* $B[k, \lambda; v]$ *design is a* $B[v-k, b-2r+\lambda; v]$ *design, provided that* $b-2r+\lambda > 0$.

Proof. Certainly there are v varieties and b blocks in both designs. Any variety belongs to r blocks in the original design, and hence to the other $b-r$ blocks in the complement. Since $B_i \subset S, |S| = v, |B_i| = k$, we must have $|S \setminus B_i| = v-k$, for $i = 1, 2, \ldots, b$.

Finally, consider varieties x and y in the original design. In λ blocks, both x and y occur; in $r-\lambda$ blocks, x occurs without y, in $r-\lambda$ blocks, y occurs without x; in the remaining $b-2(r-\lambda)-\lambda = b-2r+\lambda$ blocks, neither x nor y occurs. Hence, x and y occur together in the complements of these blocks. Since the number is independent of the choice of x and y, the design is balanced providing $b-2r+\lambda > 0$. □

See Exercise 2.15 for the conditions under which $b-2r+\lambda = 0$.

Example 14. The $B[3,1;7]$ design

013, 124, 235, 346, 450, 561, 602

has the complementary $B[4,2;7]$ design

2456, 3560, 4601, 5012, 6123, 0234, 1345. □

We note in passing that since the set of all $\binom{v}{k}$ k-sets of a given v-set is a BIBD with parameters

$$\left(v, \binom{v}{k}, \binom{v-1}{k-1}, k, \binom{v-2}{k-2}\right),$$

the collection of all k-sets *not* occurring as blocks of a given (v, b, r, k, λ) design must form a BIBD with parameters

$$\left(v, \binom{v}{k} - b, \binom{v-1}{k-1} - r, k, \binom{v-2}{k-2} - \lambda\right).$$

Thus, for example, the set of all 3-subsets of a 7-set forms a BIBD with parameters (7, 35, 15, 3, 5) as in Example 9, and if we delete the seven blocks listed as cycle 1, we are left with a (7, 28, 12, 3, 4) BIBD. (The garbage heap principle strikes again.) Such a design might also reasonably be described as the complement of the original design, but in fact the term complement is used only for the design constructed as in Theorem 13.

Another easy construction is that of a *multiple design*. If we have a $B[k, \lambda; v]$ design, and we write down m copies of each block, we have a $B[k, m\lambda; v]$ design, which is a multiple of the original design. On the other hand, a $B[k, m\lambda; v]$ design is an *m-quasi-multiple* design which may or may not be a multiple design. In Example 9, for instance, we saw a 5-quasi-multiple of a $B[3, 1; 7]$ design which turned out to be a union of three pairwise disjoint subdesigns, namely two copies of the $B[3, 1; 7]$ design and one irreducible 3-quasi-multiple of this design. Further, we saw that a 2-quasi-multiple, that is, a $B[3, 2; 7]$ design, with no repeated blocks must be reducible.

If we start from a symmetric BIBD, with parameters (v, v, k, k, λ), we have two additional constructions.

Theorem 15. *If the blocks of an* $SB[k, \lambda; v]$ *are* $B_1, B_2, \ldots, B_v$, *then:*

(i) *the sets* $B_1 \cap B_v, B_2 \cap B_v, \ldots, B_{v-1} \cap B_v$ *are the blocks of a* $(k, v-1, k-1, \lambda, \lambda-1)$ *BIBD called the* derived *design of the original, with respect to the block* B_v;

(ii) *the sets* $B_1 \setminus B_v, B_2 \setminus B_v, \ldots, B_{v-1} \setminus B_v$ *are the blocks of a* $(v-k, v-1, k, k-\lambda, \lambda)$ *BIBD called the* residual *design of the original, with respect to the block* B_v.

Proof. The block B_i contains k varieties, of which λ belong to B_v and hence to $B_i \cap B_v$, and $k-\lambda$ belong to $B_i \setminus B_v$. Hence, the derived design is based on the set of k varieties belonging to B_v, and the residual on the set of $v-k$ varieties outside B_v, and each has $v-1$ blocks by definition. Any variety in the block B_v belongs to $k-1$ blocks other than B_v in the original design, and hence to $k-1$ blocks of the derived design. Similarly, any variety not appearing in B_v belongs to k blocks of the residual design.

Finally, we have to check that both these designs are balanced. Any two varieties, both belonging to B_v, must also both belong to $\lambda-1$ other blocks of the original design, and hence to $\lambda-1$ blocks of the derived design. Any two varieties, neither of which belongs to B_v, must both belong to λ other blocks of the original design, and hence to λ blocks of the residual. □

Example 16. Consider the $SB[5,2;11]$ design of Example 10, shown in Table 2.7(d). For convenience, we take derived and residual designs with respect to the block 01234, so that these designs are based on the sets {5, 6, 7, 8, 9, A} and {0, 1, 2, 3, 4} respectively. Table 2.9 shows these designs: (a) is the original, (b) the derived and (c) the residual. The parameters are (5, 10, 4, 2, 1) and (6, 10, 5, 3, 2) for (b) and (c) respectively. □

Table 2.9

	(a)	(b)	(c)
B_0:	0 1 2 3 4		
	0 1 5 6 7	0 1	5 6 7
	0 2 6 8 9	0 2	6 8 9
	0 3 7 8 A	0 3	7 8 A
	0 4 5 9 A	0 4	5 9 A
	1 2 5 8 A	1 2	5 8 A
	1 3 6 9 A	1 3	6 9 A
	1 4 7 8 9	1 4	7 8 9
	2 3 5 7 9	2 3	5 7 9
	2 4 6 7 A	2 4	6 7 A
	3 4 5 6 8	3 4	5 6 8
	$SB[5, 2; 11]$ with B_0 as distinguished block	$B_i \cap B_0$, $i = 1, \ldots, 10$	$B_i \setminus B_0$, $i = 1, \ldots, 10$

2.6. Hadamard designs

The last topic we consider in this chapter is the family of $SB[2h-1, h-1; 4h-1]$ designs, called *Hadamard* designs. If $h = 2$, we have an $SB[3,1;7]$ design; if $h = 3$, an $SB[5,2;11]$ design, and so on. These designs are frequently studied in terms of matrices related to their incidence matrices.

An *Hadamard matrix* H of side n is an $n \times n$ matrix with each element either 1 or -1, which satisfies

$$HH^T = nI.$$

In other words, if $H = [h_{ij}]$, then

$$\sum_{r=1}^{n} h_{ir} h_{jr} = \begin{cases} n & \text{if } i = j, \\ 0 & \text{otherwise.} \end{cases}$$

Notice that if H is an Hadamard matrix, then so is H^T.

Example 17. Certainly [1] is an Hadamard matrix. Less trivially, so, are $\begin{bmatrix} 1 & 1 \\ 1 & -1 \end{bmatrix}$, and the matrices of Table 2.10. (We often write '$-$' for '-1'.) □

If we take an Hadamard matrix, H, and negate some of its rows or columns, or permute its rows, or permute its columns, then what we obtain is still an Hadamard matrix. Thus the matrix H_1 of Table 2.10 can be transformed by negating its second, third, and fourth rows, to give H_2, or by interchanging its first and third columns, to give H_3, and then negating the fourth row of H_3 to give H_4. Each of these matrices is Hadamard. An Hadamard matrix which has every entry in its first row and first column equal to $+1$ is said to be *standardized*; H_2 and H_4 of Table 2.10 are examples of such matrices.

Table 2.10. Hadamard matrices of side 4

$$\underset{H_1}{\begin{bmatrix} 1 & 1 & 1 & 1 \\ - & 1 & 1 & - \\ - & - & 1 & 1 \\ - & 1 & - & 1 \end{bmatrix}} \underset{H_2}{\begin{bmatrix} 1 & 1 & 1 & 1 \\ 1 & - & - & 1 \\ 1 & 1 & - & - \\ 1 & - & 1 & - \end{bmatrix}} \underset{H_3}{\begin{bmatrix} 1 & 1 & 1 & 1 \\ 1 & 1 & - & - \\ 1 & - & - & 1 \\ - & 1 & - & 1 \end{bmatrix}} \underset{H_4}{\begin{bmatrix} 1 & 1 & 1 & 1 \\ 1 & 1 & - & - \\ 1 & - & - & 1 \\ 1 & - & 1 & - \end{bmatrix}}$$

Theorem 18. *There is an Hadamard matrix of side* 4m *if and only if there is an* $SB[2m-1, m-1; 4m-1]$ *design.*

Proof. (i) Suppose that H is an Hadamard matrix of side $4m$. Without loss of generality, we may assume that it is standardized, so that $h_{1j} = 1 = h_{i1}$, for all $i, j = 1, 2, \ldots, 4m$. Since $HH^T = 4mI_{4m}$, we know that

$$\sum_{j=1}^{4m} h_{1j} h_{rj} = 0 \quad \text{for } r \neq 1,$$

so that

$$\sum_{j=1}^{4m} h_{rj} = 0, \; r \neq 1,$$

and similarly

$$\sum_{i=1}^{4m} h_{ir} = 0, r \neq 1.$$

In other words, the entries in any row or column, except the first, sum to zero.

Let M be the square matrix of side $4m-1$ formed by deleting the first row and column from H; that is,

$$H = \begin{bmatrix} 1 & j^T \\ j & M \end{bmatrix}.$$

Then each row and column of M sums to -1, or in other words

$$MJ = -J = JM.$$

Also

$$HH^T = \begin{bmatrix} 1 & j^T \\ j & M \end{bmatrix}\begin{bmatrix} 1 & j^T \\ j & M^T \end{bmatrix} = 4mI_{4m},$$

so

$$MM^T = 4mI_{4m-1} - J_{4m-1}.$$

Now let $A = \frac{1}{2}(M+J)$. Then

$$AJ = \tfrac{1}{2}(MJ+JJ) = (2m-1)J = JA. \tag{2.15}$$

Also

$$\begin{aligned} AA^T &= \tfrac{1}{4}(M+J)(M^T+J) \\ &= \tfrac{1}{4}(MM^T+JM^T+MJ+JJ) \\ &= \tfrac{1}{4}\{(4mI-J)-J-J+(4m-1)J\} \\ &= mI+(m-1)J. \end{aligned} \tag{2.16}$$

Now eqns (2.15) and (2.16), together with Theorem 1.2.2, show that A is the incidence matrix of an $SB[2m-1, m-1; 4m-1]$ design.

(ii) The proof can be reversed. If A is the incidence matrix of an $SB[2m-1, m-1; 4m-1]$ design, and we let $M = 2A-J$, then

$$H = \begin{bmatrix} 1 & j^T \\ j & M \end{bmatrix}$$

is an Hadamard matrix of side $4m$. □

Example 19. Let A be the incidence matrix of the $SB[3,1;7]$ design (D4) and let $M = 2A-J$. In other words M is the matrix formed from A by replacing every zero by -1. Then

$$H = \begin{bmatrix} 1 & j^T \\ j & M \end{bmatrix}$$

is an 8×8 Hadamard matrix; see Table 2.11. □

Table 2.11

$$
\underset{A}{\begin{bmatrix}
1\,0\,0\,0\,1\,0\,1 \\
1\,1\,0\,0\,0\,1\,0 \\
0\,1\,1\,0\,0\,0\,1 \\
1\,0\,1\,1\,0\,0\,0 \\
0\,1\,0\,1\,1\,0\,0 \\
0\,0\,1\,0\,1\,1\,0 \\
0\,0\,0\,1\,0\,1\,1
\end{bmatrix}}
\underset{M}{\begin{bmatrix}
1 & - & - & - & 1 & - & 1 \\
1 & 1 & - & - & - & 1 & - \\
- & 1 & 1 & - & - & - & 1 \\
1 & - & 1 & 1 & - & - & - \\
- & 1 & - & 1 & 1 & - & - \\
- & - & 1 & - & 1 & 1 & - \\
- & - & - & 1 & - & 1 & 1
\end{bmatrix}}
\underset{H}{\begin{bmatrix}
1 & 1 & 1 & 1 & 1 & 1 & 1 & 1 \\
1 & 1 & - & - & - & 1 & - & 1 \\
1 & 1 & 1 & - & - & - & 1 & - \\
1 & - & 1 & 1 & - & - & - & 1 \\
1 & 1 & - & 1 & 1 & - & - & - \\
1 & - & 1 & - & 1 & 1 & - & - \\
1 & - & - & 1 & - & 1 & 1 & - \\
1 & - & - & - & 1 & - & 1 & 1
\end{bmatrix}}
$$

We have seen Hadamard matrices of sides 1, 2, and 4; suppose that an Hadamard matrix has side n, for $n > 2$. What values might n take?

Lemma 20. *If there is an Hadamard matrix H of side n, for $n > 2$, then $n \equiv 0$* (mod 4).

Proof. We may assume that the first row of H has all its entries 1, so that each column of H begins with entries.

$$1, 1, 1 \text{ or } 1, 1, -1 \text{ or } 1, -1, 1 \text{ or } 1, -1, -1.$$

By permuting columns we can ensure that the first three rows of H are arranged as shown in Table 2.12, where there are respectively a, b, c, d columns of each type.

Table 2.12

$$
\begin{array}{cccc}
1\;1\ldots1 & 1\;1\ldots1 & 1\;1\ldots1 & 1\;1\ldots1 \\
1\;1\ldots1 & 1\;1\ldots1 & -\,-\ldots- & -\,-\ldots- \\
1\;1\ldots1 & -\,-\ldots- & 1\;1\ldots1 & -\,-\ldots- \\
\leftarrow a \rightarrow & \leftarrow b \rightarrow & \leftarrow c \rightarrow & \leftarrow d \rightarrow
\end{array}
$$

Certainly

$$a + b + c + d = n.$$

By considering the (1, 2), (1, 3), and (2, 3) entries in $HH^T = nI_n$, we find that

$$a + b - c - d = 0,$$
$$a - b + c - d = 0,$$
$$a - b - c + d = 0,$$

and hence that $a = b = c = d$, and $n = 4a$. □

No other restrictions on the side of an Hadamard matrix are known, and the *Hadamard conjecture* states that for every $n \equiv 0 \pmod 4$ there is an Hadamard matrix of side n.

Recall that if M and N are $a \times b$ and $c \times d$ matrices respectively, then their *Kronecker product* $M \times N$ is the $ac \times bd$ matrix defined by

$$M \times N = [m_{ij}N].$$

Example 21. If

$$M = \begin{bmatrix} 1 & 1 \\ 1 & -1 \end{bmatrix}, \quad N = \begin{bmatrix} x & y \\ y & -x \end{bmatrix},$$

then

$$M \times N = \begin{bmatrix} x & y & x & y \\ y & -x & y & -x \\ x & y & -x & -y \\ y & -x & -y & x \end{bmatrix}. \qquad \square$$

Lemma 22. *Let $M = [m_{ij}]$ be an $a \times b$ matrix and let $N = [n_{ij}]$ be a $c \times d$ matrix. Their Kronecker product $M \times N$ is an $ac \times bd$ matrix with the following properties:*

(i) $\alpha(M \times N) = (\alpha M) \times N = M \times (\alpha N)$ *for any scalar* α;
(ii) $(M_1 + M_2) \times N = (M_1 \times N) + (M_2 \times N)$ *and* $M \times (N_1 + N_2) = (M \times N_1) + (M \times N_2)$;
(iii) $(M_1 \times N_1)(M_2 \times N_2) = M_1 M_2 \times N_1 N_2$;
(iv) $(M \times N)^T = M^T \times N^T$;
(v) $(L \times M) \times N = L \times (M \times N)$ *for any matrix* L. $\square$

The proof is left as an exercise.

From the properties of the Kronecker product, we have immediately

Lemma 23. *If H_1 and H_2 are Hadamard matrices of sides h_1 and h_2 respectively, then $H_1 \times H_2$ is an Hadamard matrix of side $h_1 h_2$.* $\square$

Corollary 23.1. *Since there is an Hadamard matrix of side 2, namely* $\begin{bmatrix} 1 & 1 \\ 1 & - \end{bmatrix}$, *there are Hadamard matrices of side 2^s for every positive integer s.* $\square$

Corollary 23.2. *If there is an Hadamard matrix of side h, for any positive integer h, then there is an Hadamard matrix of side $2^s h$ for every positive integer s.* $\square$

2.7. References and comments

The proofs given for Fisher's inequality are the original one of Fisher (1940) and that of Bose (1949). Ryser (1963), Hall (1967), and Raghavarao (1971) give much material on the use of incidence matrices. Many of the numerical relations between parameters of a design were originally obtained by Sprott and Stanton (1964). For the factorization of complete graphs see, for example, König (1950). Theorem 7 is the simplest case of a result of Baranyai (1975) that the family of all t-subsets of an n-set X is resolvable if and only if $t \mid n$; see also Cameron (1976). For the construction of large sets of triple systems, see Lu (1983, 1984).

For some applications of graph theory to designs, see Stanton and Goulden (1981), Morgan (1977), and Mathon and Rosa (1977) who point out that the use of adjacency graphs dates back at least some sixty years. An application to efficiency-balance was given by Paterson (1983). The idea of a λ-chain was used by Husain (1945). Nested designs are discussed by Preece (1967). Systematic methods for the construction of balanced incomplete block designs are discussed in later chapters. For Hadamard designs and their generalizations see, for instance, W. D. Wallis, A. P. Street, and J. S. Wallis (1972), and Geramita and Seberry (1979).

For a comprehensive survey of designs, see Beth, Jungnickel, and Lenz (1985). The most recent table of BIBDs is due to Mathon and Rosa (1985).

Exercises

2.1. Find a balanced, linked, incomplete design with $v = 6$, $\mu = \lambda = 1$. Is it a block design? Is each variety replicated equally often? How many such designs exist (up to isomorphism)?

2.2. Show that there are exactly three non-isomorphic block designs with $v = 4$, $b = 6$, $r = 3$, $k = 2$. Is any of them balanced? Is any of them linked?

2.3. Consider a BIBD with parameters (v, v, k, k, λ) where v is even. Can λ be odd? Can $k - \lambda$ be odd?

2.4. In a BIBD with parameters (8, 14, 7, 4, 3), what are the possible sizes of the intersections of one particular block with each of the other thirteen blocks? If the design has $v = 8$, $b = 14$, $r = 7$, $k = 4$, but is not necessarily balanced, what are the possible intersection sizes? What are the intersection sizes of blocks in the dual design of an (8, 14, 7, 4, 3) BIBD?

2.5. If A is the incidence matrix of an (8, 14, 7, 4, 3) BIBD, what is the value of $\det(AA^T)$?

2.6. Does there exist a BIBD with parameters (46, 46, 10, 10, 2)?

2.7. Without using Theorem 5, show that the dual of a balanced incomplete block design is a linked block design.

2.8. Show that a balanced, linked, incomplete block design must be symmetric.

2.9. Consider the BIBDs with parameters (7, 14, 6, 3, 2). If repeated blocks are allowed, how many non-isomorphic designs exist with these parameters? Show that they are all reducible.

2.10. Find the complementary design of an $SB[5, 2; 11]$. Find the derived and residual designs of this complementary design.

2.11. Consider the array A of Table 2.13. Define 16 blocks in the following way: the ith block B_i consists of the other three varieties that appear in the same row of the array as i, together with the other three varieties that appear in the same column as i, so that, for instance, $B_5 = 19D467$. Show that B_0, $B_1, \ldots, B_F$ form a (16, 16, 6, 6, 2) BIBD, S. Find the λ-chains associated with each variety of the design S, relative to B_0. Find the derived and residual designs of S.

2.12. Define a block design with parameters $v = 8$, $b = 14$, $r = 7$, $k = 4$ as follows: blocks 1 through 7 are those of an $SB[3, 1; 7]$, each with the extra variety ∞ adjoined; blocks 8 through 14 are those of the $SB[4, 2; 7]$ complementary to the original design. Show that this design is balanced, with $\lambda = 3$. How many times does each of the $\binom{8}{3}$ possible 3-sets appear in the design?

2.13. Let D be an SBIBD based on a set X. Let B be one of its blocks, and let $C = X \setminus B$. Prove that the complement with respect to C of the residual design of D with respect to B is identical to the derived design with respect to C of the complement of D.

2.14. Consider the array B_0 of Table 2.14, where the elements of the array are integers modulo 19. Let B_0 be the block of nine varieties listed in the array and let B_i be the block $B_0 + i$, with addition modulo 19, so that, for instance, B_2 is as shown in Table 2.14. Show that, using either the rows or the columns of the array B_0 as sub-blocks, we have a nested BIBD with parameters (19; 19, 57; 9; 9, 3; 4, 1; 3).

Table 2.13

0	1	2	3
4	5	6	7
8	9	A	B
C	D	E	F

A

Table 2.14

2	5	17
8	10	18
12	7	6

B_0

4	7	0
10	12	1
14	9	8

B_2

2.15. Show that $b - 2r + \lambda = 0$ for a $B[k, \lambda; v]$ design if and only if $k = v$ or $v - 1$. (Hint: show that $b - 2r + \lambda = \lambda(v - k)(v - k - 1)/k(k - 1)$.)

2.16. Can the resolvable design (D10) be broken up into resolution classes in any other way?

2.17. Prove Lemma 22.

3 Difference set constructions

3.1. Some small examples

Consider the $B[3, 1; 7]$ and $B[4, 2; 7]$ designs that we encountered earlier, where $X = \{1, 2, 3, 4, 5, 6, 0\}$ and the blocks are

$$\mathscr{B}_1 = \{013, 124, 235, 346, 450, 561, 602\}$$

and

$$\mathscr{B}_2 = \{2456, 3560, 4601, 5012, 6123, 0234, 1345\}$$

respectively. If we regard the seven elements of X as integers modulo 7, then each block of each design is obtained from the one before it by just adding 1 modulo 7. Thus

$$\{3, 4, 6\} + 1 = \{3 + 1, 4 + 1, 6 + 1\} = \{4, 5, 0\} \pmod 7.$$

Alternatively, we can think of forming the design by starting from any one block and adding to it each integer modulo 7 in turn. Thus

$$\{0, 1, 3\} + 1 = \{1, 2, 4\},\ \{0, 1, 3\} + 2 = \{2, 3, 5\}$$

and so on.

Suppose that we now adjoin to X a new element denoted by ∞, and make the convention that

$$x + \infty = \infty + x = \infty$$

for any $x \in X$. If we adjoin to each block in $\mathscr{B}_1$ the symbol ∞, and take the union of the family of these extended blocks and the blocks of $\mathscr{B}_2$, we have a $B[4, 3; 8]$ design based on the set $X' = X \cup \{\infty\}$ with the set of blocks

$$\mathscr{B}_3 = \begin{Bmatrix} \infty 013, & \infty 124, & \infty 235, & \infty 346, & \infty 450, & \infty 561, & \infty 602, \\ 2456, & 3560, & 4601, & 5012, & 6123, & 0234, & 1345 \end{Bmatrix}.$$

Why is it possible to develop a block design from the set $\{0, 1, 3\}$, or $\{2, 4, 5, 6\}$, or from the pair of sets $\{\infty, 0, 1, 3\}$, $\{2, 4, 5, 6\}$? The key to this question is found in looking at the differences between the elements of the set. For example, consider the block $\{0, 1, 3\}$. The difference between the elements 1 and 3 is $3 - 1 = 2$. When 1 is added to each element of the block, these numbers become 4 and 2, but their difference remains the same. There are seven ordered pairs of integers modulo 7 with difference 2: 1 and 3; 2 and 4; 3 and 5; 4 and 6; 5 and 0; 6 and 1; 0 and 2. The original block contained 1 and 3, the next block 2 and 4, the one after that 3 and 5, and so on: the seven blocks between them contain each of the pairs precisely once. In the same way, we see that each ordered pair with difference 1, 3, 4, 5, or 6 appears just once. But every pair of different numbers modulo 7 must have difference 1, 2, 3, 4, 5, or 6. So every possible pair of elements will occur precisely once in the design. Thus the design is balanced, with $\lambda = 1$. Similarly, the set $\{2, 3, 5, 6\}$ contains precisely two ordered pairs having each non-zero difference and yields a balanced design, with $\lambda = 2$. (Remember here that each unordered pair $\{i, j\}$ is associated with two ordered pairs, and hence with differences $i-j$ and $j-i$.)

Suppose that B is a set of k integers modulo v. We define the design *developed from B* (modulo v) to be the set of v k-sets formed by first taking B, then adding 1 to each element of B to give the set $B+1$, and so on: thus the design consists of the blocks

$$B, B+1, B+2, \ldots, B+v-1,$$

formed by adding the different integers (modulo v) to B once each, and B is called the *starter block* of the design. In fact, any block in such a design may be taken as a starter block. Notice that such a design may contain repeated blocks; that is, two or more blocks which contain exactly the same elements.

Lemma 1. (i) *The design developed from B is a block design with $b = v$ and $r = k$.*

(ii) *If λ_i of the $k(k-1)$ ordered pairs of elements of B have difference i, and if $x - y = i$, then $\{x, y\}$ is a subset of precisely λ_i blocks.*

Proof. (i) Since B has k elements, so has $B + x$, and the design has constant block size.

Suppose that $x \in B$. Then $x+1 \in B+1$, $x+2 \in B+2$, and so on, ending with $x+v-1 = x-1 \in B+v-1$. Since x is fixed, the elements $x, x+1, \ldots, x+v-1$ run through the whole of Z_v, that is,

$$\{x, x+1, \ldots, x+v-1\} = \{0, 1, \ldots, v-1\}.$$

The same is true for any element of B. Thus every element of Z_v occurs in the design once for each element of B. Hence each element has replication number $r = k$.

(ii) Now suppose that the ordered pairs of elements of B with difference i are

$$(b_1, c_1), (b_2, c_2), \ldots, (b_{\lambda_i}, c_{\lambda_i}),$$

where $c_j - b_j = i$ for $j = 1, 2, \ldots, \lambda_i$. Consider any x and y such that $x - y = i$. Then the block $B + (x - c_1)$ must contain both

$$b_1 + x - c_1 = x - i = y$$

and

$$c_1 + x - c_1 = x.$$

Similarly, $x, y \in B+(x-c_2), \ldots, B+(x-c_{\lambda_i})$, so that $\{x, y\}$ is a subset of at least λ_i blocks.

Finally, suppose that $x, y \in B+d$. Then $(x-d)-(y-d)=i$, and $x-d$, $y-d \in B$. Thus $x-d=c_j$, $y-d=b_j$ for some j, and $B+d=B+(x-c_j)$, one of the blocks we have already listed. Therefore $\{x, y\}$ is a subset of precisely λ_i blocks. □

Corollary 1.1. *If $\lambda_i=\lambda$ for $i=1, 2, \ldots, v-1$, then the design developed from B is pairwise balanced, and, provided $k<v$, it is an $SB[k, \lambda; v]$ design.* □

Thus we make a definition: a (v, k, λ) *cyclic difference set* is a k-set B of integers modulo v with the property that, given any non-zero integer d modulo v, there are precisely λ ordered pairs of elements of B whose difference is d. Notice here that if $x-y=i$, then $y-x=v-i$, so that $\lambda_i=\lambda_{v-i}$, whether B is a difference set or not. If v is even, the integer $v/2$ will sometimes play a special role in designs developed from B.

This idea may be generalized in several ways.

First, we may choose a difference set with k elements from an abelian group of order v, not necessarily cyclic, and develop a design by adding each element of the group in turn. (We could generalize further by choosing elements from any finite group, but we shall not need the non-Abelian case.) For example, consider $Z_4 \times Z_4$, where the elements of the group are ordered pairs of integers modulo 4, and the group operation is addition modulo 4 in each component. Thus $12+31=03$, and so on. The set $B=\{00, 02, 10, 30, 11, 33\}$ is a starter block which can be developed to give an $SB[6, 2; 16]$ design as shown in Table 3.1.

Table 3.1. An $SB[6, 2; 16]$ design based on $Z_4 \times Z_4$

B:	00, 02, 10, 30, 11, 33	$B+20$:	20, 22, 30, 10, 31, 13
$B+01$:	01, 03, 11, 31, 12, 30	$B+21$:	21, 23, 31, 11, 32, 10
$B+02$:	02, 00, 12, 32, 13, 31	$B+22$:	22, 20, 32, 12, 33, 11
$B+03$:	03, 01, 13, 33, 10, 32	$B+23$:	23, 21, 33, 13, 30, 12
$B+10$:	10, 12, 20, 00, 21, 03	$B+30$:	30, 32, 00, 20, 01, 23
$B+11$:	11, 13, 21, 01, 22, 00	$B+31$:	31, 33, 01, 21, 02, 20
$B+12$:	12, 10, 22, 02, 23, 01	$B+32$:	32, 30, 02, 22, 03, 21
$B+13$:	13, 11, 23, 03, 20, 02	$B+33$:	33, 31, 03, 23, 00, 22

Secondly, we could develop two or more starter blocks and consider the collection of all the blocks obtained from them as the design. If we have h starter blocks, each of size k, chosen from the integers modulo v, then the resulting design will have

$$b=hv, \; r=hk,$$

since v blocks will arise from each starter block. If these starter blocks are denoted by $B_1, B_2, \ldots, B_h$, we let λ_{ij} be the number of ordered pairs of elements of B_j with difference i. As in the proof of Lemma 1, it follows that if $x-y=i$, then $\{x, y\}$ is a subset of λ_{ij} of the blocks developed from B_j, and hence of $\sum_{j=1}^{h} \lambda_{ij}$ blocks of the design. Thus if

$$\sum_{j=1}^{h} \lambda_{ij} = \lambda, \quad \text{for } i = 1, 2, \ldots, v-1,$$

we have a balanced design, and in this case $B_1, B_2, \ldots, B_h$ are also called *h-supplementary difference sets.*

If $v \equiv 0 \pmod{k}$, then we may have a starter block

$$\{0, v/k, 2v/k, \ldots, (k-1)v/k\}$$

from which only v/k distinct blocks are obtained. If we include only one copy of each of these distinct blocks in the design, then this is said to be an *extra* (or *short*) starter block, as opposed to the *full* starter blocks which give rise to v distinct blocks each. For example in Z_{15}, the two full starter blocks $\{0, 1, 4\}$ and $\{0, 2, 8\}$, together with the short starter block $\{0, 5, 10\}$, give rise to the $B[3, 1; 15]$ design shown in Table 3.2. Here we use hexadecimal notation, where 10, 11, 12, 13, 14, 15 are represented by A, B, C, D, E, F respectively. (In this particular case, we have no occasion to write 15 or F, since we use numbers 0 to 14.)

Table 3.2. A $B[3, 1; 15]$ design based on Z_{15}

B_1:				B_2:				B_3:		
0	1	4		0	2	8		0	5	A
1	2	5		1	3	9		1	6	B
2	3	6		2	4	A		2	7	C
3	4	7		3	5	B		3	8	D
4	5	8		4	6	C		4	9	E
5	6	9		5	7	D				
6	7	A		6	8	E				
7	8	B		7	9	0				
8	9	C		8	A	1				
9	A	D		9	B	2				
A	B	E		A	C	3				
B	C	0		B	D	4				
C	D	1		C	E	5				
D	E	2		D	0	6				
E	0	3		E	1	7				

Finally, we may adjoin the fixed element ∞ to our original set, as in our example of $(X', \mathscr{B}_3)$. In that example the differences modulo 7 all occur equally often, namely $\lambda = 3$ times. The new element ∞ occurs with a total of λ elements in the starter blocks, and thus becomes a member of $v\lambda$ pairs in all, with each distinct pair appearing λ times.

3.2. The method of differences

The fundamental results relating to difference sets were proved by Bose in 1939. Let G be an Abelian group of order m, written additively, and let there be n symbols corresponding to each group element, so that from the original element $x \in G$, we now have symbols $x_1, x_2, \ldots, x_n$. Then altogether there are mn symbols corresponding to G, and we shall say that symbols with the same subscript, say j, belong to the jth class.

Suppose that we choose a k-subset, S, from the mn symbols, and suppose that p_i symbols of S belong to the ith class. Then

$$\sum_{i=1}^{n} p_i = k.$$

We denote by $x_i^{(1)}, x_i^{(2)}, \ldots, x_i^{(p_i)}$ the symbols of class i belonging to the set S, and similarly by $y_j^{(1)}, y_j^{(2)}, \ldots, y_j^{(p_j)}$ those of class j, where $x^{(1)}, x^{(2)}, \ldots, x^{(p_i)}$, $y^{(1)}, y^{(2)}, \ldots, y^{(p_j)}$ are elements of G.

A difference $x_i^{(\alpha)} - x_i^{(\beta)}$ between two distinct symbols of class i is said to be a *pure difference* of type (i, i) arising from S. Similarly, a difference $x_i^{(\alpha)} - y_j^{(\gamma)}$ is said to be a *mixed difference* of type (i, j) arising from S. Since $\alpha \neq \beta$, and $1 \leq \alpha, \beta \leq p_i$, we have $p_i(p_i - 1)$ pure differences of type (i, i) arising from S; since $1 \leq \gamma \leq p_j$, we also have $p_i p_j$ mixed differences of type (i, j) arising from S. Altogether there are n types of pure differences and $n(n-1)$ of mixed differences possible.

Consider a family of sets, $S_1, S_2, \ldots, S_s$, of symbols, where p_{il} denotes the number of symbols of class i belonging to S_l. Suppose that the sets satisfy the following conditions.

(1) Among the $\sum_{l=1}^{s} p_{il}(p_{il} - 1)$ pure differences of type (i, i) arising from $S_1, S_2, \ldots, S_s$, every non-identity element of G is repeated exactly λ times, for each i.

(2) Among the $\sum_{l=1}^{s} p_{il} p_{jl}$ mixed differences of type (i, j) arising from $S_1, S_2, \ldots, S_s$, every element of G is repeated exactly λ times, for each (i, j).

Then we say that the differences are *symmetrically repeated*, each occurring λ times, in the sets $S_1, S_2, \ldots, S_s$.

Example 2. Let $G = Z_5$, so that $m = 5$, and let $n = 2$. This gives altogether the 10 symbols $0_1, 0_2, 1_1, 1_2, 2_1, 2_2, 3_1, 3_2, 4_1, 4_2$. Let $s = 6$ and consider the sets

$$S_1 = \{0_1, 3_1, 1_2\}, S_2 = \{1_1, 2_1, 1_2\}, S_3 = \{1_2, 4_2, 4_1\},$$
$$S_4 = \{0_2, 4_2, 1_2\}, S_5 = \{0_1, 3_1, 4_2\}, S_6 = \{1_1, 2_1, 4_2\}.$$

The non-zero differences of type (1, 1) occur twice each:

$$1 = 2-1 \text{ and } 4 = 1-2 \text{ in both } S_2 \text{ and } S_6;$$
$$2 = 0-3 \text{ and } 3 = 3-0 \text{ in both } S_1 \text{ and } S_5.$$

Similarly, the non-zero differences of type (2, 2) occur twice each:

$$1 = 1-0 = 0-4 \text{ and } 4 = 0-1 = 4-0 \text{ in } S_4;$$
$$2 = 1-4 \text{ and } 3 = 4-1 \text{ in both } S_3 \text{ and } S_4.$$

The differences of type (1, 2) occur twice each:

$$0 = 1-1 = 4-4 \text{ in } S_2 \text{ and } S_3 \text{ respectively;}$$
$$1 = 2-1 = 0-4 \text{ in } S_2 \text{ and } S_5 \text{ respectively;}$$
$$2 = 3-1 = 1-4 \text{ in } S_1 \text{ and } S_6 \text{ respectively;}$$
$$3 = 4-1 = 2-4 \text{ in } S_3 \text{ and } S_6 \text{ respectively;}$$
$$4 = 0-1 = 3-4 \text{ in } S_1 \text{ and } S_5 \text{ respectively.}$$

Finally, the differences of type (2, 1) are those of type (1, 2) with the signs changed, which thus occur twice each.

Hence the differences are symmetrically repeated in this family of sets. □

Example 3. Let $G = Z_{13}$, $m = 13$, $n = 1$. Thus there are just thirteen symbols altogether. Let $s = 2$ and consider the sets

$$S_1 = \{0, 1, 4\},\ S_2 = \{0, 2, 7\}.$$

Each difference occurs once; ± 1, ± 3, ± 4 in S_1; ± 2, ± 5, ± 6 in S_2. Thus differences are symmetrically repeated, and $\lambda = 1$. □

We make the convention that if $x + y = z$ in G, then $x_i + y = z_i$ in our set, X, of mn symbols. In other words adding a group element to a symbol does not change its class.

Theorem 4. *Let G be an Abelian group of order m, and for each element $x \in G$, let there be defined n symbols $x_1, x_2, \ldots, x_n$, giving a set X of mn symbols altogether. Suppose that s k-subsets $S_1, S_2, \ldots, S_s$ of X are chosen so that:*

(i) *among the ks symbols occurring in the s subsets, exactly r' symbols belong to each of the n classes;*

(ii) *the differences from the s subsets are symmetrically repeated, each occurring λ times.*

Then by adding each element of G in turn to each of the sets $S_1, \ldots, S_s$, we develop a balanced incomplete block design with parameters

$$v = mn;\ b = ms;\ r = r';\ k;\ \lambda.$$

(Note that if $n = 1$, $s = 1$, then we have a block, S_1, which is a difference set in G.)

Proof. The parameters v, b, k need no explanation, and since property (i) implies that

$$ks = nr',$$

the parameters satisfy the usual equations, as in Section 1.2. We need only check replication and balance.

Corresponding to the symbol $x_i^{(\alpha)} \in S_l$, we have a symbol $(x^{(\alpha)} + y)_i \in S_l + y$. As y runs through all the elements of G, the symbols corresponding to $x_i^{(\alpha)}$ run through all the symbols of the ith class exactly once. Thus by (i), every symbol is replicated exactly r' times in our design, so that $r = r'$.

Now consider a pair of symbols in the ith class, say $x_i^{(\alpha)}$ and $x_i^{(\beta)}$. These symbols occur together in the same block of the design if and only if there are symbols $x_i^{(\gamma)}$ and $x_i^{(\delta)}$ occurring together in one of the sets S_l such that

$$x^{(\alpha)} - x^{(\beta)} = x^{(\gamma)} - x^{(\delta)};$$

the argument is now exactly that of Lemma 1, using condition (ii). A similar argument applies to a pair of symbols from different classes. Thus the design is balanced. □

Example 2. (continued). The blocks of the design developed from the sets $S_1, \ldots, S_6$ are shown in Table 3.3. They form a BIBD with parameters

$$v = 5 \times 2,\ b = 5 \times 6,\ r = 9,\ k = 3,\ \lambda = 2. \qquad \square$$

Table 3.3. $B[3, 2; 10]$ design from Theorem 4

$0_1\ 3_1\ 1_2$	$1_1\ 2_1\ 1_2$	$1_2\ 4_2\ 4_1$	$0_2\ 4_2\ 1_2$	$0_1\ 3_1\ 4_2$	$1_1\ 2_1\ 4_2$
$1_1\ 4_1\ 2_2$	$2_1\ 3_1\ 2_2$	$2_2\ 0_2\ 0_1$	$1_2\ 0_2\ 2_2$	$1_1\ 4_1\ 0_2$	$2_1\ 3_1\ 0_2$
$2_1\ 0_1\ 3_2$	$3_1\ 4_1\ 3_2$	$3_2\ 1_2\ 1_1$	$2_2\ 1_2\ 3_2$	$2_1\ 0_1\ 1_2$	$3_1\ 4_1\ 1_2$
$3_1\ 1_1\ 4_2$	$4_1\ 0_1\ 4_2$	$4_2\ 2_2\ 2_1$	$3_2\ 2_2\ 4_2$	$3_1\ 1_1\ 2_2$	$4_1\ 0_1\ 2_2$
$4_1\ 2_1\ 0_2$	$0_1\ 1_1\ 0_2$	$0_2\ 3_2\ 3_1$	$4_2\ 3_2\ 0_2$	$4_1\ 2_1\ 3_2$	$0_1\ 1_1\ 3_2$

Example 3 (continued). The blocks of the design developed from the sets S_1, S_2, form a $B[3, 1; 13]$ design, as in Theorem 4. The blocks are as follows:

014, 125, 236, 347, 458, 569, 67A, 78B, 89C, 9A0, AB1, BC2, C03;
027, 138, 249, 35A, 46B, 57C, 680, 791, 8A2, 9B3, AC4, B05, C16. □

We consider a related result.

Theorem 5. *Let the set X of mn symbols be as defined in Theorem* 4, *and adjoin to it the new symbol ∞. Suppose that $(s+t)$ k-subsets $S_1, S_2, \ldots, S_s$, $T_1, T_2, \ldots, T_t$* of $X \cup \{\infty\}$ *are chosen so that:*

(i) *each of $S_1, \ldots, S_s$ contains only symbols from X, and each of $T_1, \ldots, T_t$ contains ∞, together with symbols from X;*

(ii) *among the ks symbols of X occurring in* $S_1, \ldots, S_s$, *exactly* $mt-\lambda$ *belong to each of the n classes; among the* $(k-1)t$ *symbols of X occurring in* $T_1, \ldots, T_t$, *exactly* λ *belong to each of the n classes;*

(iii) *the differences from the symbols of X in the* $s+t$ *subsets are symmetrically repeated, each occurring* λ *times.*

Then by adding each element of G in turn to each of the sets $S_1, \ldots, S_s$, $T_1, \ldots, T_t$, *with the convention that* $\infty + x = \infty$, *we develop a balanced incomplete block design with parameters*

$$v = mn+1;\ b = m(s+t);\ r = mt;\ k;\ \lambda.$$

Proof. We note that (ii) implies that $ks = n(mt-\lambda)$ and $(k-1)t = n\lambda$. The proof follows that of Theorem 4; see also Exercise 3.3. □

Example 6. Let $G = Z_3 \times Z_3$, that is the set of ordered pairs of integers modulo 3, so that $m = 9$. Choose $n = 3$, $s = 6$, $t = 1$. Then we may choose seven starter blocks which lead to a balanced incomplete block design with $v = 28$, $b = 63$, $r = 9$, $k = 4$, $\lambda = 1$, as follows. Here xy_i denotes the symbol $(x, y)_i$ corresponding to the ordered pair (x, y) in G.

$$S_1 = \{01_1, 02_1, 10_2, 20_2\}, \qquad S_2 = \{01_2, 02_2, 10_3, 20_3\},$$
$$S_3 = \{01_3, 02_3, 10_1, 20_1\}, \qquad S_4 = \{21_1, 12_1, 22_2, 11_2\},$$
$$S_5 = \{21_2, 12_2, 22_3, 11_3\}, \qquad S_6 = \{21_3, 12_3, 22_1, 11_1\},$$
$$T_1 = \{\infty, 00_1, 00_2, 00_3\}.$$

Developing S_1 gives nine blocks as follows:

$$\begin{aligned}
S_1 \quad &= \{01_1, 02_1, 10_2, 20_2\},\\
S_1+01 &= \{02_1, 00_1, 11_2, 21_2\},\\
S_1+02 &= \{00_1, 01_1, 12_2, 22_2\},\\
S_1+10 &= \{11_1, 12_1, 20_2, 00_2\},\\
S_1+11 &= \{12_1, 10_1, 21_2, 01_2\},\\
S_1+12 &= \{10_1, 11_1, 22_2, 02_2\},\\
S_1+20 &= \{21_1, 22_1, 00_2, 10_2\},\\
S_1+21 &= \{22_1, 20_1, 01_2, 11_2\},\\
S_1+22 &= \{20_1, 21_1, 02_2, 12_2\}.
\end{aligned}$$

Other blocks are developed similarly, except that, for instance,

$$T_1+01 = \{\infty, 01_1, 01_2, 01_3\},$$

since $\infty + xy = \infty$. □

In order to make this method work in practice, we must be able to construct the k-subsets S_i and T_j. Frequently we need to use finite fields, and this is the next topic we discuss.

3.3. A first look at finite fields

Roughly speaking, a *field* is an algebraic system designed to reflect the main properties common to the rational, real, and complex number systems. More precisely, a field $\{F, +, \cdot\}$ is a set, F, closed under two operations, addition (denoted by $+$) and multiplication $(\cdot)$, both of which are associative and commutative and for both of which there are identity elements (denoted by 0 for addition, and 1 for multiplication, where $0 \neq 1$). Further, every element $a \in F$ has an additive inverse (denoted by $-a$), every non-zero element $a \in F$ has a multiplicative inverse (a^{-1}) and multiplication distributes over addition. Since $0 \neq 1$, the smallest possible field has two elements. Such a field in fact exists: it consists of the integers modulo 2 with addition and multiplication modulo 2 as shown in Table 3.4.

Table 3.4. *GF*[2]

+	0	1
0	0	1
1	1	0

·	0	1
0	0	0
1	0	1

It is easily checked that for any prime p, the integers modulo p form a field, which we denote either by Z_p or by $GF[p]$ if we want to emphasize the field properties. (Here '*GF*' stands for 'Galois field' after the French mathematician Evariste Galois (1811–32).)

If p is prime and if i is any integer such that $0 < i < p$, then (i, p), the greatest common divisor of i and p, must equal 1. Hence by the Euclidean algorithm there exist integers a and b such that

$$ai + bp = 1.$$

But now

$$ai \equiv 1 \pmod{p}$$

so $a = i^{-1}$ in $GF[p]$. However, if n is composite, then $i^{-1} \pmod{n}$ may not exist: for instance, 2^{-1} does not exist modulo 4, since $2.1 \equiv 2 \equiv 2.3 \pmod{4}$ and $2.2 \equiv 0 \pmod{4}$. Nevertheless, a field of order 4, also denoted by $GF[4]$, may be constructed from $GF[2]$ by using an appropriate polynomial.

To find such a polynomial, consider the quadratics over $GF[2]$, namely x^2, $x^2+x = x(x+1)$, $x^2+1 = (x+1)^2$, and x^2+x+1. The first three quadratics factorize over $GF[2]$, but $1^2+1+1 = 0^2+0+1 = 1$ in $GF[2]$ and hence the polynomial x^2+x+1 does not factorize over $GF[2]$ and is said to be *irreducible* over $GF[2]$. We now try to embed $GF[2]$ in a larger field in which x^2+x+1 will factor.

Suppose we choose α to be a solution of $x^2+x+1 = 0$. Since we are working modulo 2, we have

$$(\alpha+1)^2+(\alpha+1)+1 = (\alpha^2+1)+(\alpha+1)+1$$
$$= \alpha^2+\alpha+1$$

which is also 0, so $\alpha+1$ is the other solution of our equation. Also

$$\alpha^2 = \alpha+1.$$

Thus we may complete the addition and multiplication tables of the field as shown in Table 3.5.

Table 3.5. $GF[4]$

$+$	0	1	α	$\alpha+1$
0	0	1	α	$\alpha+1$
1	1	0	$\alpha+1$	α
α	α	$\alpha+1$	0	1
$\alpha+1$	$\alpha+1$	α	1	0

$\cdot$	0	1	α	$\alpha+1$
0	0	0	0	0
1	0	1	α	$\alpha+1$
α	0	α	$\alpha+1$	1
$\alpha+1$	0	$\alpha+1$	1	α

We are in fact taking the ring of polynomials over $GF[2]$, and working with them modulo x^2+x+1 to give the field $GF[4]$. The integers modulo n form a field if and only if n is prime. In the same way if we start from Z_p, for some prime p, and consider the ring of polynomials, $Z_p[x]$, over Z_p, modulo a polynomial $f(x)$, this forms a field if and only if $f(x)$ is irreducible over Z_p.

Notice that in this case the set of non-zero elements $\{1, \alpha, \alpha+1\}$ forms a cyclic multiplicative group: we can choose α or $\alpha+1$ as a generator since the powers of α, namely

$$\alpha, \alpha^2 = \alpha+1, \alpha^3 = \alpha(\alpha+1) = 1,$$

and the powers of $\alpha+1$, namely

$$\alpha+1, (\alpha+1)^2 = \alpha^2+1 = \alpha, (\alpha+1)^3 = \alpha(\alpha+1) = 1,$$

give all the group elements. Similarly, if we choose θ to be a solution of $g(x) = 0$, where $g(x) = x^4+x+1$, an irreducible quartic over $GF[2]$, then we generate $GF[2^4]$, the field with 16 elements shown in Table 3.6. Here we let the quadruple $abcd$ denote $a\theta^3+b\theta^2+c\theta+d$, where $a, b, c, d \in \{0, 1\}$. Then the elements are $\theta = 0010$, $\theta^2 = 0100$, $\theta^3 = 1000$, $\theta^4 = \theta+1 = 0011$, $\theta^5 = \theta(\theta+1) = \theta^2+\theta = 0110$, and so on. Since

$$(a\theta^3+b\theta^2+c\theta+d)+(A\theta^3+B\theta^2+C\theta+D)$$
$$= (a+A)\theta^3+(b+B)\theta^2+(c+C)\theta+(d+D),$$

where each sum of coefficients is evaluated modulo 2, the field addition is carried out componentwise modulo 2. Thus

$$0110+1100 = 1010.$$

The multiplicative group is again cyclic, consisting of the powers of θ, and has order 15. Notice that

$$\{0, \theta^5 = 0110, \theta^{10} = 0111, \theta^{15} = 1 = 0001\}$$

form a subfield isomorphic to $GF[4]$.

Table 3.6. The elements of $GF[16]$

θ	0010	θ^5	0110	θ^9	1010	θ^{13}	1101
θ^2	0100	θ^6	1100	θ^{10}	0111	θ^{14}	1001
θ^3	1000	θ^7	1011	θ^{11}	1110	θ^{15}	0001
θ^4	0011	θ^8	0101	θ^{12}	1111	0	0000

In this book, we make no attempt at a full discussion of finite fields but simply list their properties as we need to use them, and give some small examples.

The number of elements in a finite field is p^n, where p is prime and n is a positive integer. For every prime power p^n, there is a finite field $GF[p^n]$, unique up to isomorphism, which may be represented as the set of residue classes of polynomials over $GF[p]$, modulo $f(x)$, where $f(x)$ is a polynomial of degree n, irreducible over $GF[p]$. The standard construction for a finite field uses a *primitive* irreducible polynomial of degree n; that is, one which divides $x^m - 1$ for $m = p^n - 1$ but for no smaller m. (The same definition applies to polynomials over $GF[p^l]$, but now $m = (p^l)^n - 1$.)

The $p^n - 1$ non-zero elements of $GF[p^n]$ form a cyclic group under multiplication and any generator of this group is called a *primitive element* of the field. Thus any non-zero element of $GF[p^n]$ satisfies the equation

$$h(x) = x^{p^n - 1} - 1 = 0, \tag{3.1}$$

and any element of $GF[p^n]$ the equation

$$x^{p^n} - x = 0.$$

Hence $xh(x)$ is the product of all the irreducible monic polynomials over $GF[p]$ of degree dividing n.

In $GF[16]$, as we have seen, every non-zero element satisfies the equation

$$x^{15} = 1,$$

a particular case of (3.1), and

$$x^{15} - 1 = (x-1)(x^2+x+1)(x^4+x+1)(x^4+x^3+1)(x^4+x^3+x^2+x+1),$$

the product of all the irreducible (monic) polynomials over $GF[2]$ of degree dividing 4. Since

$$x^5 - 1 = (x-1)(x^4+x^3+x^2+x+1),$$

we have only two primitive irreducible polynomials of degree 4 available to generate $GF[16]$. Also θ^5 satisfies the irreducible polynomial x^2+x+1 of degree 4/2 over $GF[2^2]$. Similarly, so does θ^{10}.

We conclude this section with a brief look at the field of order 9. There are nine monic polynomials of degree 2 over $GF[3]$, namely

$$x^2, x^2+x, x^2+2x, x^2+1, x^2+2, x^2+x+1, x^2+x+2, x^2+2x+1, x^2+2x+2,$$

and of these only

$$p(x) = x^2+1,\ q(x) = x^2+x+2,\ r(x) = x^2+2x+2$$

are irreducible. Note here that

$$x^8-1 = (x-1)(x+1)(x^2+1)(x^4+1)$$

and that, over $GF[3]$,

$$(x^2+x+2)(x^2+2x+2) = x^4+1$$

so that $x(x^8-1)$ is indeed the product of all the irreducible polynomials of degree dividing 2 over $GF[3]$.

We let a, b, c respectively be solutions of $p(x)=0$, $q(x)=0$, $r(x)=0$, and attempt to generate $GF[9]$. Since $a^2+1=0$, we have $a^2=2$, $a^3=2a$, $a^4=1$. Hence a is not a primitive element, an example of the difficulties that arise because $p(x)|(x^4-1)$ and is therefore not a primitive polynomial. However, with the primitive polynomials $q(x)$ and $r(x)$, there is no such problem. The equation $q(b)=0$ leads to the elements

$$b,\ b^2 = 2b+1,\ b^3 = 2b+2,\ b^4 = 2,\ b^5 = 2b,\ b^6 = b+2,\ b^7 = b+1,\ b^8 = 1,$$

and the equation $r(c)=0$ to the elements

$$c,\ c^2 = c+1,\ c^3 = 2c+1,\ c^4 = 2,\ c^5 = 2c,\ c^6 = 2c+2,\ c^7 = c+2,\ c^8 = 1.$$

The correspondence between b and $2c$ is an isomorphism between these two representations of $GF[9]$: we see that

$$q(2c) = r(c) = 0.$$

Table 3.7. The field $GF[9]$

+	00	01	02	10	11	12	20	21	22
00	00	01	02	10	11	12	20	21	22
01	01	02	00	11	12	10	21	22	20
02	02	00	01	12	10	11	22	20	21
10	10	11	12	20	21	22	00	01	02
11	11	12	10	21	22	20	01	02	00
12	12	10	11	22	20	21	02	00	01
20	20	21	22	00	01	02	10	11	12
21	21	22	20	01	02	00	11	12	10
22	22	20	21	02	00	01	12	10	11

	00	01	02	10	11	12	20	21	22
00	00	00	00	00	00	00	00	00	00
01	00	01	02	10	11	12	20	21	22
02	00	02	01	20	22	21	10	12	11
10	00	10	20	21	01	11	12	22	02
11	00	11	22	01	12	20	02	10	21
12	00	12	21	11	20	02	22	01	10
20	00	20	10	12	02	22	21	11	01
21	00	21	12	22	10	01	11	02	20
22	00	22	11	02	21	10	01	20	12

The multiplication and addition tables of $GF[9]$ are shown in Table 3.7, in terms of b, where the field elements are represented as ordered pairs of integers modulo 3, thus xy denotes $xb+y$, and addition is defined componentwise as before.

3.4. Finite fields and difference sets

As we have just seen, a field of order p^n exists for every prime p and every positive integer n; we denote it by $GF[p^n]$. Its multiplicative group is cyclic and any generator of this group is said to be a primitive element of the field.

Suppose that p is odd. Then the multiplicative group $GF[p^n]\backslash\{0\}$, which we shall denote by F^*, has order p^n-1, which is even, and consists of

$$\theta, \theta^2, \theta^3, \ldots, \theta^{p^n-2}, \theta^{p^n-1}=\theta^0=1,$$

where θ is a primitive element. The elements θ^{2a}, $a=0, 1, \ldots, (p^n-3)/2$ (that is, the even powers of θ) are called the *quadratic residues* of the field since each of them is the square of a field element. Notice that since

$$1=\theta^{p^n-1},$$

and since $(-1)^2=1$, we have

$$-1=\theta^{(p^n-1)/2}.$$

Thus -1 is itself a quadratic residue if $p^n\equiv 1 \pmod 4$, but not if $p^n\equiv 3 \pmod 4$. Also if neither x nor y is a quadratic residue, so that

$$x=\theta^{2h+1}, y=\theta^{2k+1},$$

then $xy=\theta^{2(h+k+1)}$, a quadratic residue. Similarly, if x is a quadratic residue but y is not, then xy is not a quadratic residue.

Theorem 7. *Let $v=p^n=4t-1$, where p is prime. Then there is a $(4t-1, 2t-1, t-1)$ difference set.*

Proof. Let θ be a primitive element of the field $GF[v]$. Then the set of non-zero quadratic residues

$$Q=\{\theta^{2a} | a=1, 2, \ldots, 2t-1\}$$

is the required difference set.

Certainly it contains $2t-1$ elements. Since $v\equiv 3 \pmod 4$, $-1\notin Q$. Hence for any non-zero element, x, of the field, we have $x\in Q$ if and only if $-x\notin Q$, or in other words, precisely one of x and $-x$ is a quadratic residue for each x. The odd powers of θ form the set $-Q$.

If $q_1, q_2\in Q$, then precisely one of q_1-q_2 and q_2-q_1 belongs to Q. Suppose that $q_1-q_2=q\in Q$. If q' is any other element of Q, then $q'q_1, q'q_2, q'q\in Q$. So every equation

$$q_1-q_2=q$$

corresponds to the equation

$$q'q_1 - q'q_2 = q'q,$$

and vice versa. Hence every quadratic residue occurs equally often, say λ times, as a difference of elements of Q.

If we now multiply each of these equations by -1, then

$$q_2 - q_1 = -q$$

and so on; this gives every element of $-Q$ λ times also, showing that Q is a difference set.

Since $v = 4t - 1$ and $k = 2t - 1$, we must have $\lambda = t - 1$ by Lemma 1, Corollary 1.1, and Theorem 1.2.1. □

Theorem 7 is an example of the construction in Theorem 4, with $n = s = 1$. We have an analogous result with $n = 1$, $s = 2$, as follows.

Theorem 8. *Let $v = p^n = 4t + 1$, where p is prime. Then there are two supplementary difference sets which can be developed to give a BIBD with parameters* $(4t + 1, 2(4t + 1), 4t, 2t, 2t - 1)$.

Proof. Let θ be a primitive element of the field $GF[v]$. Then the required sets are

$$Q = \{\theta^{2a} | a = 1, 2, \ldots, 2t\},$$

the set of non-zero quadratic residues, and

$$R = \{\theta^{2a+1} | a = 1, 2, \ldots, 2t\} = \theta Q,$$

the set of the remaining non-zero residues.

Certainly each of them has $2t$ elements. Since $v \equiv 1 \pmod 4$, $-1 \in Q$. Hence $Q \equiv -Q$ and $R \equiv -R$. This means that if $q_1, q_2 \in Q$, then both or neither of $q_1 - q_2$ and $q_2 - q_1$ belong to Q. Suppose that $q_1 - q_2 = q \in Q$, and that q' is any element of Q. Then to each equation

$$q_1 - q_2 = q$$

there corresponds the equation

$$q'q_1 - q'q_2 = q'q$$

and vice versa. Hence every element of Q occurs equally often, say λ_1 times, as a difference of elements of Q, and similarly every element of R occurs λ_2 times.

Since $R = \theta Q$, to each difference $q_1 - q_2$ of elements of Q, there corresponds the difference $\theta q_1 - \theta q_2$ of elements of R. Hence, among the differences of elements of R, every element of Q occurs λ_2 times, and every element of R occurs λ_1 times. Counting up differences of elements of Q, and differences of elements of R, we find each non-zero element of $GF[v]$ occurs $\lambda_1 + \lambda_2 = \lambda$ times. Since there are $4t$ such elements, and $4t(2t - 1)$ differences, we have $\lambda = 2t - 1$. □

Example 9. Let $v = 13 = 4t + 1$ where $t = 3$. In $GF[13]$, we may choose 2 as a primitive element, so the non-zero field elements are

$$2, 4, 8, 3, 6, 12, 11, 9, 5, 10, 7, 1$$

with $Q = \{4, 3, 12, 9, 10, 1\}$, $R = \{2, 8, 6, 11, 5, 7\}$. Each element of Q occurs $\lambda_1 = 2$ times as a difference of elements of Q, thus

$$1 = 4 - 3 = 10 - 9,\ 3 = 4 - 1 = 12 - 9 \text{ and so on,}$$

and $\lambda_2 = 3$ times as a difference of elements of R, thus

$$1 = 6 - 5 = 7 - 6 = 8 - 7,\ 3 = 5 - 2 = 8 - 5 = 11 - 8 \text{ and so on.}$$

Thus $\lambda = \lambda_1 + \lambda_2 = 5 = 2t - 1$, and we can construct a $B[6, 5; 13]$ design, as required. □

Other residue difference sets exist, besides the quadratic residues. For example, if $p^n = 1 + 4t^2$, with t odd, then the quartic residues in $GF[p^n]$ form a difference set; the smallest example is $p^n = 37$, $t = 3$. Similarly, if

$$p^n = 1 + 8t^2 = 9 + 64y^2, \quad t \text{ odd, } y \text{ odd,}$$

then the octic residues in $GF[p^n]$ form a difference set; the smallest example is $p^n = 73$, $t = 3$, $y = 1$.

A residue class, together with 0, may form a difference set. Examples of this kind include the cubic residues in $GF[16]$, and the quartic residues in $GF[p^n]$ where

$$p^n = 9 + 4t^2, \quad t \text{ odd,}$$

the smallest example here being $p^n = 13$, $t = 1$.

The following property of $GF[p^n]$, where $p^n = 4t + 1$, allows construction of another difference set.

Lemma 10. *Let θ be a primitive element of $GF[p^n]$, where p is prime and $p^n = 4t + 1$. Then there exists a pair of odd integers c, d such that*

$$(\theta^c + 1)/(\theta^c - 1) = \theta^d.$$

Proof. As θ is a primitive element, $\theta^{2t} = -1$. Thus if

$$c \in \{1, 2, \ldots, 4t - 1\} \backslash \{2t\},$$

then

$$\theta^c + 1 \neq 0, \quad \theta^c - 1 \neq 0$$

and

$$(\theta^c + 1)/(\theta^c - 1) = \theta^d,$$

for some integer d.

If $\theta^d = 1$, then $\theta^c + 1 = \theta^c - 1$, which implies that $p = 2$. If $\theta^d = -1$, then $\theta^c + 1 = 1 - \theta^c$, again implying that $p = 2$. Since p^n is odd, neither is possible; so

$$d \in \{1, 2, \ldots, 4t - 1\} \backslash \{2t\}$$

also. Further, $\theta^c = (\theta^d + 1)/(\theta^d - 1)$ and c and d are uniquely determined by each other.

Of these $4t - 2$ possible values of d and c, $2t - 2$ are even and $2t$ are odd. Hence for at least one choice of c, both c and d are odd. □

Theorem 11. *In the notation of Lemma* 10, *the starter blocks*

$$\left.\begin{aligned}&\{\theta_1^{2i}, \theta_1^{2t+2i}, \theta_2^{2i+c}, \theta_2^{2t+2i+c}\},\\ &\{\theta_2^{2i}, \theta_2^{2t+2i}, \theta_3^{2i+c}, \theta_3^{2t+2i+c}\},\\ &\{\theta_3^{2i}, \theta_3^{2t+2i}, \theta_1^{2i+c}, \theta_1^{2t+2i+c}\},\end{aligned}\right\}\quad i = 0, 1, \ldots, t-1,$$

$$\{\infty, 0_1, 0_2, 0_3\}$$

give a $B[4, 1; 12t+4]$ *design, when developed in* $GF[p^n]$.

Proof. We use Theorem 5. Certainly there are $3(4t+1)+1 = v$ varieties, $(3t+1)(4t+1) = b$ blocks each of size $k = 4$, and each variety belongs to $r = 4t+1$ blocks. It remains only to check that $\lambda = 1$.

From the last block, we get all the occurrences of ∞ with elements of the three classes, and of the mixed difference zero for the three possible types.

Since $\theta^{2t} = -1$, the pure differences of type (1, 1) are

$$\pm 2\theta^{2i} \text{ and } \pm 2\theta^{2i+c},\ i = 0, 1, \ldots, t-1.$$

But c is odd, so this gives all non-zero differences of type (1, 1), and similarly of types (2, 2) and (3, 3).

In the same way, the mixed differences of type (2, 1) are

$$\theta^{2i}(\theta^c - 1),\ \theta^{2i}(\theta^c + 1),\ \theta^{2i+2t}(\theta^c - 1),\ \theta^{2i+2t}(\theta^c + 1),\ i = 0, 1, \ldots, t-1.$$

Now

$$\theta^c + 1 = \theta^d(\theta^c - 1)$$

so these become

$$(\theta^c - 1)\theta^{2i},\ (\theta^c - 1)\theta^{2i+d},\ -(\theta^c - 1)\theta^{2i},\ -(\theta^c - 1)\theta^{2i+d},\ i = 0, 1, \ldots, t-1,$$

and since d is odd, these differences run through all the non-zero elements of $GF[p^n]$. Hence all mixed differences of type (2, 1) occur, and similarly of types (1, 2), (1, 3), (2, 3), (3, 1), and (3, 2).

Since all possible differences occur, they can occur only once each. Hence $\lambda = 1$. □

Example 12. Let $t = 7$, and consider $GF[29]$. We apply Theorem 11, choosing the primitive element $\theta = 2$. The powers of 2, modulo 29, are 2, 4, 8, 16, 3, 6, 12, 24, 19, 9, 18, 7, 14, 28, 27, 25, 21, 13, 26, 23, 17, 5, 10, 20, 11, 22, 15, 1.

If we try $c = 1$, then

$$\frac{\theta^c + 1}{\theta^c - 1} = \frac{2+1}{2-1} = 3 = 2^5 \pmod{29};$$

so $d = 5$ is a feasible choice. Our starter blocks become

$$\{\infty, 0_1, 0_2, 0_3\},$$

together with twenty-one other blocks as shown in the first column of Table 3.8 since h takes values 1, 2, 3. Substituting for the powers of 2, we have starter blocks as shown in the second column of Table 3.8; differences are checked in the remaining columns.

These starter blocks give a $B[4, 1; 88]$ design. □

Similar arguments give us further families of starter blocks.

Table 3.8. $h, h+1 \in \{1, 2, 3\}$ with addition modulo 3

Starter blocks		Differences		
as powers of 2 (mod 29)	as elements of Z_{29}	Type (h, h)	Type $(h+1, h+1)$	Type $(h+1, h)$
$2^0_h, 2^{14}_h, 2^1_{h+1}, 2^{15}_{h+1}$	$1_h, 28_h, 2_{h+1}, 27_{h+1}$	± 2	± 4	$\pm 1, \pm 3$
$2^2_h, 2^{16}_h, 2^3_{h+1}, 2^{17}_{h+1}$	$4_h, 25_h, 8_{h+1}, 21_{h+1}$	± 8	± 13	$\pm 4, \pm 12$
$2^4_h, 2^{18}_h, 2^5_{h+1}, 2^{19}_{h+1}$	$16_h, 13_h, 3_{h+1}, 26_{h+1}$	± 3	± 6	$\pm 13, \pm 10$
$2^6_h, 2^{20}_h, 2^7_{h+1}, 2^{21}_{h+1}$	$6_h, 23_h, 12_{h+1}, 17_{h+1}$	± 12	± 5	$\pm 6, \pm 11$
$2^8_h, 2^{22}_h, 2^9_{h+1}, 2^{23}_{h+1}$	$24_h, 5_h, 19_{h+1}, 10_{h+1}$	± 10	± 9	$\pm 5, \pm 14$
$2^{10}_h, 2^{24}_h, 2^{11}_{h+1}, 2^{25}_{h+1}$	$9_h, 20_h, 18_{h+1}, 11_{h+1}$	± 11	± 7	$\pm 9, \pm 2$
$2^{12}_h, 2^{26}_h, 2^{13}_{h+1}, 2^{27}_{h+1}$	$7_h, 22_h, 14_{h+1}, 15_{h+1}$	± 14	± 1	$\pm 7, \pm 8$

Theorem 13. *Let $v = 6t + 1 = p^n$ for some prime p, and let θ be a primitive element of $GF[p^n]$. Then the starter blocks*

$$\{\theta^i, \theta^{2t+i}, \theta^{4t+i}\}, \; i = 0, 1, \ldots, t-1,$$

can be developed to give a $B[3, 1; v]$ design.

Proof. Since $\theta \in GF[p^n]$, we have $\theta^{6t} = 1$. Thus

$$(\theta^{3t} - 1)\,(\theta^{3t} + 1) = 0.$$

Since θ is a primitive element, $\theta^{3t} \neq 1$, so we must have

$$\theta^{3t} + 1 = 0.$$

Also $\theta^{2t} \neq 1$, and we define s by requiring

$$\theta^{2t} - 1 = \theta^s.$$

Then the differences arising from our starter block with $i = 0$ are

$$\pm(\theta^{2t} - 1), \; \pm(\theta^{4t} - 1), \; \pm(\theta^{4t} - \theta^{2t}).$$

Now

$$\begin{aligned}\theta^{2t}-1&=\theta^{s},\\ -(\theta^{2t}-1)&=\theta^{s+3t},\\ \theta^{4t}-1&=\theta^{4t}-\theta^{6t}=\theta^{4t}(1-\theta^{2t})=\theta^{t}(\theta^{2t}-1)=\theta^{s+t},\\ -(\theta^{4t}-1)&=\theta^{s+4t},\\ \theta^{4t}-\theta^{2t}&=\theta^{2t}(\theta^{2t}-1)=\theta^{s+2t},\\ -(\theta^{4t}-\theta^{2t})&=\theta^{s+5t}.\end{aligned}$$

Thus the differences are

$$\theta^{s}, \theta^{s+t}, \theta^{s+2t}, \theta^{s+3t}, \theta^{s+4t}, \theta^{s+5t},$$

and for each starter block, these differences are multiplied by θ^{i}, $i=0, 1, \ldots, t-1$. Hence altogether all θ^{j}, $j=0, 1, \ldots, 6t-1$, occur as differences. Thus each occurs just once, proving the theorem. □

Example 14. Let $v=19$, $t=3$. Since 2 is a primitive element of $GF[19]$, we have starter blocks

$$\{2^{i}, 2^{6+i}, 2^{12+i}\}, \quad i=0, 1, 2.$$

The powers of 2 (modulo 19) are

2, 4, 8, 16, 13, 7, 14, 9, 18, 17, 15, 11, 3, 6, 12, 5, 10, 1,

so the starter blocks are

$$\{1, 7, 11\}, \{2, 14, 3\}, \{4, 9, 6\} \pmod{19}$$

which give a $B[3, 1; 19]$ design. □

Theorem 15. *Let $v=6t+1=p^{n}$, for some prime p, and let θ be a primitive element of $GF[p^{n}]$. Then the starter blocks*

$$\{0, \theta^{i}, \theta^{2t+i}, \theta^{4t+i}\}, \quad i=0, 1, \ldots, t-1,$$

can be developed to give a $B[4, 2; v]$ design. □

Theorem 16. *Let $v=4t+1=p^{n}$, for some prime p, and let θ be a primitive element of $GF[p^{n}]$. Then the starter blocks*

$$\{\theta^{i}, \theta^{t+i}, \theta^{2t+i}, \theta^{3t+i}\}, \quad i=0, 1, \ldots, t-1,$$

can be developed to give a $B[4, 3; v]$ design. □

The proofs of Theorems 15 and 16 are left as exercises. Notice that the design of Theorem 13 is nested in the design of Theorem 15.

3.5. Subfields and Singer's theorem

We have so far constructed the field $GF[p^n]$ by starting with $GF[p]$, and using a primitive irreducible polynomial of degree n over that field to extend it. But if n is composite, there is another construction which leads of course to the same field, possibly relabelled.

Example 17. Consider $GF[2^2]$, constructed in Table 3.5. There are four monic linear polynomials over this field, namely

$$x, x+1, x+\alpha, x+\alpha+1.$$

If we multiply any two of these linear factors together, we obtain altogether ten distinct reducible monic quadratics. But over this field there are altogether sixteen monic quadratics, six of which must therefore be irreducible. They are

$$x^2+\alpha x+1,\ x^2+(\alpha+1)x+1,\ x^2+x+\alpha,\ x^2+x+(\alpha+1),\ x^2+\alpha x+\alpha,$$
$$x^2+(\alpha+1)\,x+(\alpha+1).$$

But now the quartic, x^4+x+1, which is irreducible over $GF[2]$ and was used to generate $GF[2^4]$ in Section 3.3, can be factored over $GF[2^2]$ as

$$x^4+x+1=(x^2+x+\alpha)\,(x^2+x+(\alpha+1)).$$

The primitive element, θ, which we took in Section 3.3 as the generator of the multiplicative group of $GF[2^4]$, now satisfies the irreducible quadratic $x^2+x+\alpha$, over $GF[2^2]$, since we may take $\alpha=\theta^5$. □

In general, the field $GF[p^n]$ has a unique subfield $GF[p^l]$ if and only if l is a positive integer dividing n. If β is a primitive element of $GF[p^n]$, then β is a root of an irreducible monic polynomial over $GF[p^l]$ of degree n/l and the non-zero elements of the subfield are

$$\beta^{bt},\ b=1,2,\ldots,p^l-1, \tag{3.1}$$

where $t=(p^n-1)/(p^l-1)$. Thus in Section 3.3, $\beta=\theta$, and $t=15/3=5$.

We apply these ideas to prove the simplest case of Singer's theorem. (The general case will be dealt with, from a geometrical point of view, in Chapter 8.)

Theorem 18. *Let $m=p^l$ for some prime p. Then there exists a difference set with parameters $v=m^2+m+1$, $k=m+1$, $\lambda=1$.*

Proof. Let $F=GF[p^l]$, $G=GF[p^{3l}]$, and let F^*, G^* be their respective multiplicative groups. Then any primitive element, β, of G is a solution of a primitive irreducible monic cubic equation over F, and any element $g\in G^*$ can be written as

$$\begin{aligned} g &= a\beta^2+b\beta+c,\quad a,\ b,\ c\in F,\ \text{not all zero,}\\ &= \beta^d,\quad 0\le d\le p^{3l}-2. \end{aligned}$$

Applying (3.1) above, we see that

$$\beta^d \in F^*$$

if and only if $d = bt$ for some $b = 1, 2, \ldots, p^l - 1$, and $t = (p^{3l} - 1)/(p^l - 1) = (m^3 - 1)/(m - 1) = v$. Also, if $g, h \in G^*$, where

$$g = \beta^d, h = \beta^e,$$

then g and h are linearly dependent over F if and only if $d \equiv e \pmod{v}$.

Let $F = \{c_1, c_2, \ldots, c_m\}$ and consider the set $\{1\} \cup (\beta + F)$. Its elements are $g_0 = 1 = \beta^0, g_i = \beta + c_i = \beta^{d_i}, i = 1, 2, \ldots, m$, each of which corresponds either to the constant polynomial

$$P_0(\beta) = 1$$

or to a linear polynomial

$$P_i(\beta) = \beta + c_i, \ i = 1, 2, \ldots, m,$$

but in any case to a monic polynomial.

Now let D be the set

$$D = \{d_0 = 0, d_1, \ldots, d_m\},$$

where each element of D is taken modulo v. Thus $D \subseteq Z_v$ and we show that it is a difference set. For suppose that

$$d_{i_1} - d_{j_1} \equiv d_{i_2} - d_{j_2} \pmod{v}.$$

Then

$$d_{i_1} + d_{j_2} \equiv d_{i_2} + d_{j_1} \pmod{v}$$

so that

$$P_{i_1}(\beta) . P_{j_2}(\beta) \text{ and } P_{i_2}(\beta) . P_{j_1}(\beta)$$

are linearly dependent over F. But monic polynomials are linearly dependent if and only if they are equal. Since the factorization of any polynomial into monic linear factors is unique we have

$$\{d_{i_1}, d_{j_2}\} = \{d_{i_2}, d_{j_1}\};$$

the elements of D are distinct, so this forces $d_{i_1} = d_{i_2}, d_{j_1} = d_{j_2}$ and no two distinct ordered pairs of elements have the same difference. Since there are $m(m+1)$ non-zero differences of elements of D, this completes the proof. □

Example 19. Let $m = 4$. Then $v = m^2 + m + 1 = 21$, $k = m + 1 = 5$, $\lambda = 1$. We apply the construction of Theorem 18 to produce a difference set with these parameters.

The field $GF[4^3] = GF[2^6]$ can be constructed from $GF[2]$ and the irreducible equation

$$x^6 + x^5 + x^3 + x^2 + 1 = 0$$

which has a root β, a primitive element of $GF[2^6]$. Equivalently, $GF[2^6]$ can be constructed from $GF[2^2]$, with elements $0, 1, \alpha, \alpha + 1$, and the irreducible equation

$$x^3+\alpha x^2+\alpha x+\alpha=0 \tag{3.2}$$

which also has β as a root. The elements of $GF[2^2]=F$ are thus 0, 1, $\beta^{21}=\alpha$, $\beta^{42}=\alpha^2=\alpha+1$, since $21=(2^6-1)/(2^2-1)$.

The set $\beta+F$ becomes

$$\{\beta, \beta+1, \beta+\beta^{21}, \beta+\beta^{42}\}$$

and we need to express these elements as powers of β. From (3.2), we have

$$\beta^3=\alpha\beta^2+\alpha\beta+\alpha,$$
$$\beta^4=\alpha\beta^3+\alpha\beta^2+\alpha\beta=\alpha(\alpha\beta^2+\alpha\beta+\alpha)+\alpha\beta^2+\alpha\beta=\beta^2+\beta+(\alpha+1),$$

and so on, as shown in Table 3.9. Now

$$\beta+1=\beta^8,$$
$$\beta+\beta^{21}=\beta+\alpha=\alpha(\,(\alpha+1)\beta+1)=\alpha\beta^6=\beta^{27},$$
$$\beta+\beta^{42}=\beta+\alpha^2=\beta+(\alpha+1)=\beta^{18},$$

so for this case D is the set of exponents of

$$\{1, \beta, \beta^8, \beta^{27}, \beta^{18}\}.$$

Table 3.9. Some elements of $GF[2^6]$

β	$=$		β	
β^2	$=$	β^2		
β^3	$=$	$\alpha\beta^2$	$+\alpha\beta$	$+\alpha$
β^4	$=$	β^2	$+\beta$	$+(\alpha+1)$
β^5	$=$	$(\alpha+1)\beta^2$	$+\beta$	$+\alpha$
β^6	$=$		$(\alpha+1)\beta$	$+1$
β^7	$=$	$(\alpha+1)\beta^2$	$+\beta$	
β^8	$=$		β	$+1$
β^9	$=$	β^2	$+\beta$	
β^{10}	$=$	$(\alpha+1)\beta^2$	$+\alpha\beta$	$+\alpha$
β^{11}	$=$	$(\alpha+1)\beta^2$	$+(\alpha+1)\beta$	$+1$
β^{12}	$=$	$\alpha\beta^2$		$+1$
β^{13}	$=$	$(\alpha+1)\beta^2$	$+\alpha\beta$	$+(\alpha+1)$
β^{14}	$=$	$(\alpha+1)\beta^2$	$+\alpha\beta$	$+1$
β^{15}	$=$	$(\alpha+1)\beta^2$		$+1$
β^{16}	$=$	β^2		$+1$
β^{17}	$=$	$\alpha\beta^2$	$+(\alpha+1)\beta$	$+\alpha$
β^{18}	$=$		β	$+(\alpha+1)$
β^{19}	$=$	β^2	$+(\alpha+1)\beta$	
β^{20}	$=$	β^2	$+\alpha\beta$	$+\alpha$
β^{21}	$=$			α

Table 3.10. (i, j) entry is $i-j \pmod{21}$

$-$	0	1	6	8	18
0	0	20	15	13	3
1	1	0	16	14	4
6	6	5	0	19	9
8	8	7	2	0	11
18	18	17	12	10	0

With the exponents reduced modulo 21, the difference set becomes

$$D = \{0, 1, 6, 8, 18\} \pmod{21};$$

the fact that D is indeed a difference set is checked easily, as in Table 3.10. Developing D modulo 21 leads to a $B[5, 1; 21]$ design. □

3.6. Recursive constructions for difference sets

If a $B[k, \lambda; v]$ design based on $X = Z_v$ can be developed from a family of starter blocks, it is said to be *cyclic* and when we need to emphasize its cyclic nature, we denote it by $CB[k, \lambda; v]$. In some circumstances, we can take starter blocks for two cyclic designs, and use them to construct starter blocks for a larger design; such constructions are called *recursive*. For convenience we assume that each starter block contains 0. Each starter block modulo v will be transformed into m starter blocks modulo mv, in such a way that the differences between corresponding elements of the m new starter blocks are all identical modulo v but distinct modulo mv. We look first at a small example.

Example 20. For a $CB[3, 1; 7]$ design we have the full starter block $\{0, 1, 3\}$, and for the trivial $CB[3, 1; 3]$ design, the short starter block $\{0, 1, 2\}$. From these designs based on Z_7 and Z_3 we construct a $CB[3, 1; 21]$ design.

From the block $\{0, 1, 3\}$, we construct three starter blocks $\{0, 1, 3\}$, $\{0, 8, 17\}$ and $\{0, 15, 10\}$ modulo 21; from the block $\{0, 1, 2\}$ we construct $\{0, 7, 14\}$ modulo 21. All differences not divisible by 7 are covered exactly once in the first, second, and third blocks, and all differences divisible by 7 in the fourth. This is checked in Table 3.11. Note that each multiple of 7 occurs three times as a difference in the fourth block; thus this is a short starter block giving rise only to $21/3 = 7$ blocks of the design. Notice also that in, say, the (1, 2) positions of the first, second and third difference tables, we have entries 20, 13, 6, all congruent to each other modulo 7. □

Table 3.11. Difference tables for $CB[3, 1; 21]$ design

−	0	1	3
0	0	20	18
1	1	0	19
3	3	2	0

−	0	8	17
0	0	13	4
8	8	0	12
17	17	9	0

−	0	15	10
0	0	6	11
15	15	0	5
10	10	16	0

−	0	7	14
0	0	14	7
7	7	0	14
14	14	7	0

In general, we have the following result.

Theorem 21. *Let m be relatively prime to $(k-1)!$. Suppose that there exist a $CB[k, \lambda; v]$ design containing only full starter blocks, and a $CB[k, \lambda; m]$ design. Then there exists a $CB[k, \lambda; mv]$ design.*

Proof. Let $\{S_1, S_2, \ldots, S_\alpha\}$ and $\{T_1, T_2, \ldots, T_\beta\}$, respectively, be sets of starter blocks for $CB[k, \lambda; v]$ and $CB[k, \lambda; m]$ designs, where m is relatively prime to $(k-1)!$ and each S_i is full, $i = 1, \ldots, \alpha$. We construct starter blocks for a $CB[k, \lambda; mv]$ design as follows.

For each $S_j = \{s_0 = 0, s_1, \ldots, s_{k-1}\}$, take m starter blocks

$$S_{ij} = \{s_{0,i} = 0, s_{1,i}, s_{2,i}, \ldots, s_{k-1,i}\},$$

where $s_{a,i} = s_a + aiv$, for $i = 0, 1, \ldots, m-1$, with arithmetic modulo mv.

For each $T_j = \{0, t_1, \ldots, t_{k-1}\}$, take one starter block

$$vT_j = \{0, vt_1, \ldots, vt_{k-1}\}.$$

We show that these $m\alpha + \beta$ starter blocks lead to a $CB[k, \lambda; mv]$. First, consider the difference

$$\Delta_{ab} = s_a - s_b$$

of elements in S_j. It corresponds to the difference

$$\Delta_{i,ab} = s_{a,i} - s_{b,i} = \Delta_{ab} + (a-b)iv$$

of the corresponding elements in S_{ij}. Thus as i increases by 1, $\Delta_{i,ab}$ increases by $(a-b)v$. Since m is relatively prime to $(k-1)!$, it must also be relatively prime to $(a-b)$. Thus as i runs from 0 through $m-1$, the differences $s_{a,i} - s_{b,i}$ run through all the elements of Z_{mv} congruent to Δ_{ab} modulo v. Hence all differences not divisible by v are covered λ times by these starter blocks. The remaining differences are covered by the starter blocks vT_j. □

In the case where k divides the number of varieties, we have a corresponding result. We look first at the parameters of a cyclic design with kv varieties. Suppose we write them as (kv, b, r, k, λ). Then using the identities of Section 1.2, we find that

$$b = vr, \quad r = \lambda(kv-1)/(k-1)$$

and hence

$$b = \lambda v(kv-1)/(k-1).$$

Since the map taking $i \to i+1 \pmod{kv}$ is an automorphism of the design, the blocks must either come from full starter blocks, with kv blocks per cycle, or from short starter blocks

$$\{0, v, 2v, \ldots, (k-1)v\}$$

with v blocks per cycle. If $\lambda = 1$, then kv does not divide b, so a short starter block must be present; since $\lambda = 1$, only one such short cycle can exist. Hence our set of starter blocks contains one short block, and a collection of full blocks.

Theorem 22. *Let m be relatively prime to $(k-1)!$ and suppose that there exist $CB[k, 1; kv]$ and $CB[k, 1; km]$ designs. Then there exists a $CB[k, 1; kmv]$ design.*

Proof. Let the full starter blocks of a $CB[k, 1; kv]$ design be S_1, $S_2, \ldots, S_\alpha$, and let the starter blocks of a $CB[k, 1; km]$ design be T_1, $T_2, \ldots, T_\beta$. We construct starter blocks of a $CB[k, 1; kmv]$ design as follows.

For each $S_j = \{0, s_1, \ldots, s_{k-1}\}$, take m starter blocks

$$S_{ij} = \{s_{0,i} = 0, s_{1,i}, s_{2,i}, \ldots, s_{k-1,i}\}$$

where $s_{a,i} = s_a + aikv$, for $i = 0, 1, \ldots, m-1$, with arithmetic modulo kmv.

For each $T_j = \{0, t_1, \ldots, t_{k-1}\}$ take one starter block $vT_j = \{0, vt_1, \ldots, vt_{k-1}\}$; one of these will be short.

The proof that these blocks cover each difference modulo kmv exactly once is similar to the proof of Theorem 21. The blocks S_{ij} cover the differences not divisible by v, the full blocks vT_j the differences divisible by v but not by mv, and the short block the differences divisible by mv. □

We consider an example of the construction of Theorem 22, with $k = 3$, $m = v = 5$.

Example 23. The blocks $\{0, 1, 4\}$, $\{0, 2, 8\}$, $\{0, 5, 10\}$ are starter blocks for a $CB[3, 1; 15]$ design; the last block is short. We take this design as both the $CB[k, 1; kv]$ and the $CB[k, 1; km]$ designs, so that $S_1 = T_1 = \{0, 1, 4\}$, $S_2 = T_2 = \{0, 2, 8\}$, $T_3 = \{0, 5, 10\}$. We obtain starter blocks for a $CB[3, 1; 75]$, namely

$\{0, 1, 4\}, \{0, 16, 34\}, \{0, 31, 64\}, \{0, 46, 19\}, \{0, 61, 49\}$, from S_1,
$\{0, 2, 8\}, \{0, 17, 38\}, \{0, 32, 68\}, \{0, 47, 23\}, \{0, 62, 53\}$, from S_2,
$\{0, 5, 20\}, \{0, 10, 40\}, \{0, 25, 50\}$, from T_1, T_2, T_3 respectively.

These blocks may be checked as for Example 20. □

If $\lambda > 1$, the construction of Theorem 22 cannot be used, since no short starter block need be present, as the following example shows.

Example 24. A $CB[3, 3; 9]$ design can be developed from the starter blocks

$$\{0, 1, 3\}, \{0, 1, 4\}, \{0, 3, 4\}, \{0, 2, 4\}.$$

If we attempt to apply the construction of Theorem 22, taking $k = m = v = 3$, and using these blocks as both S_1, S_2, S_3, S_4 and T_1, T_2, T_3, T_4, we obtain the starter blocks

$\{0, 1, 3\}, \{0, 10, 21\}, \{0, 19, 12\}$, from S_1,
$\{0, 1, 4\}, \{0, 10, 22\}, \{0, 19, 13\}$, from S_2,
$\{0, 3, 4\}, \{0, 12, 22\}, \{0, 21, 13\}$, from S_3,
$\{0, 2, 4\}, \{0, 11, 22\}, \{0, 20, 13\}$, from S_4,

and

$\{0, 3, 9\}, \{0, 3, 12\}, \{0, 9, 12\}, \{0, 6, 12\}$ from T_1, T_2, T_3, T_4 respectively.

But now the differences relatively prime to 9 or divisible by 9 are covered three times each, but the differences ± 3, ± 6, ± 12 (modulo 27) are covered six times each.

On the other hand, if we remove the starter blocks obtained from T_i, $i = 1, 2, 3, 4$, and adjoin the starter block $\{0, 9, 18\}$, *considered as a full block*, then we have 13 starter blocks as required, giving a $CB[3, 3; 27]$ design. □

3.7. References and comments

The earliest results on difference methods are those of Netto (1893); see Exercise 3.9. Further difference set constructions for block size $k = 3$ were given by Peltesohn (1938), Hwang and Lin (1974), and Bhattacharya (1943). Theorem 18 is due to Singer (1938). Bose (1939) studied difference set methods systematically; Theorems 4 and 5 are due to him. Sprott (1954) developed this idea further. Bruck (1955) dealt with difference sets in finite groups. Residue difference sets are discussed at length by Storer (1967), Hall (1956, 1967), and Raghavarao (1971), and cyclic difference sets in particular by Baumert (1971). See also, for example, Chowla (1949), Lehmer (1953), Stanton and Sprott (1958), and Whiteman (1962). R. M. Wilson (1972*b*) constructed many new difference sets. The recursive constructions of Section 5 are due to M. J. Colbourn and C. J. Colbourn (1984); for earlier constructions related to these see Butson (1963), Ganley and Spence (1975), Glynn (1978), Jeffcott and Spears (1977), and references there cited. The difference set in Exercise 3.15 is due to M. J. Colbourn and Mathon (1980). Exercise 3.17 is from D. J. Street (1982). For further constructions, see also Bhat-Nayak and Kane (1986). For relative difference sets, see Leonard (1986).

For an introduction to finite fields see, for instance, Stewart (1973) or Gilbert (1976); for an encyclopaedic treatment of the subject, see Lidl and Niederreiter (1983).

Exercises

3.1. Show that the given sets of blocks are starter blocks for designs with parameters as stated. Here 'xy mod a, b' is used as an abbreviation for (x, y), where x is taken modulo a and y modulo b.

(*a*) $B[4, 1; 25]$ — 00, 01, 10, 44; 00, 20, 02, 33; mod 5, 5.

(*b*) $B[4, 1; 37]$ — 00, 01, 12, 15; 01, 03, 08, 10; 0∞, 07, 15, 21 } mod 3, 11.
∞∞, 00, 10, 20; mod −, 11.
0∞, 1∞, 2∞, ∞∞.

(*c*) $B[4, 1; 40]$ — 0, 1, 26, 32; 0, 7, 19, 36; 0, 3, 16, 38; mod 40.
0, 10, 20, 30; mod 40 (short).

(*d*) $B[4, 2; 19]$ $\quad$ 0, 1, 3, 12; 0, 1, 5, 13; 0, 4, 6, 9; mod 19.
(*e*) $B[4, 2; 22]$ $\quad$ $0_0, 3_0, 9_0, 10_0; 0_0, 0_1, 2_1, 7_1; 0_0, 0_1, 9_1, 10_1;$
$0_0, 2_0, 5_1, 8_1; 0_0, 3_0, 4_1, 7_1; 0_0, 4_0, 3_1, 9_1;$
$0_0, 5_0, 2_1, 6_1$; mod 11.
(*f*) $B[4, 3; 12]$ $\quad$ 0, 1, 3, 7; 2, 4, 9, 10; ∞, 5, 6, 8; mod 11.

3.2. Let $v = p^n = 2m+1$. Let Q be the set of quadratic residues in $GF[v]$, and let $R = \theta Q$, where θ is a primitive element of the field. Show that Q and R are supplementary difference sets which may be developed to give a $B[m, m-1; 2m+1]$ design. By considering $-R$, show that if $m = 2t-1$, then Q and R are both difference sets, and the design is thus a union of two disjoint $B[2t-1, t-1; 4t-1]$ designs.
3.3. In Theorem 5, show that each element of X appears with ∞ in λ blocks.
3.4. Show that in the proof of Theorem 8, $\lambda_1 = t-1$ and $\lambda_2 = t$.
3.5. Let D be a (v, k, λ) difference set in an additive Abelian group G. Show that $G\backslash D$, the complement of D in G, is also a difference set, and find its parameters.
3.6. Let D be a difference set in a finite field, and let x be any non-zero element of the field. Show that xD is also a difference set, with the same parameters as D.
3.7. (a) Construct a $CB[5, 4; 11]$ design with repeated blocks.
(b) Construct two $CB[5, 2; 11]$ designs which have no block in common.
(c) Construct a $CB[5, 4; 11]$ design which has no repeated blocks, and which does not contain a $CB[5, 2; 11]$ design.
3.8. Apply Theorem 13 to construct a $B[3, 1; 13]$ design and a $B[3, 1; 31]$ design.
3.9. Let $v = 6t+1 = p^n$ for some prime p, and let θ be a primitive element of $GF[p^n]$. Show that the starter blocks

$$\{0, \theta^i, \theta^{t+i}\}, i = 0, 1, \ldots, t-1,$$

can be developed to give a $B[3, 1; v]$ design.
3.10. Prove Theorem 15, and apply it to construct a $B[4, 2; v]$ design for $v = 31, 43, 79$.
(Note: 3 is a primitive element for each of the fields $GF[31]$, $GF[43]$, $GF[79]$.)
3.11. Prove Theorem 16, and apply it to construct a $B[4, 3; 29]$ design.
3.12. Apply Theorem 18 with $m = 3$ to construct a $CB[4, 1; 13]$ design.
3.13. From Theorem 21, and the $CB[4, 1; 13]$ design of Exercise 3.12, construct a $CB[4, 1; 169]$ design.
3.14. From Theorem 22, using the $CB[3, 1; 15]$ design of Example 23, and the $CB[3, 1; 21]$ design of Example 20, construct a $CB[3, 1; 105]$ design.
3.15. Let $p = 12t+1$ be a prime, and suppose that $GF[p]$ has a primitive element θ such that $\theta \equiv 3 \pmod 4$. Show that the following $4t+1$ starter

blocks can be developed (modulo $v = 4p$) to give a $B[4, 1; v]$ design:

$$\{0, \theta^{4i}, \theta^{4i+3}, \theta^{4i+6}\}, i = 0, 1, \ldots, 3t-1;$$
$$\{0, \theta^{4i+1}, \theta^{4t+4i+1}, \theta^{8t+4i+1}\}, i = 0, 1, \ldots, t-1;$$
$$\{0, p, 2p, 3p\}.$$

Hence construct a $B[4, 1; 52]$ design.
(Hint: Show that each difference modulo p occurs four times, and that these four differences are congruent to 0, 1, 2, 3 (mod 4) respectively.)

3.16. Let k be odd, let $v = 2mk + 1$ be a prime or a prime power and let θ be a primitive element of $GF[v]$. Show that the sets

$$B_i = \{\theta^i, \theta^{i+2m}, \ldots, \theta^{i+2(k-1)m}\}, i = 0, 1, \ldots, m-1,$$

are m-supplementary difference sets. Hence construct a $B[5, 2; 11]$ design and a $B[3, 1; 13]$ design.

3.17. Let $q = 3^2$. Let Q be the set of non-zero quadratic residues in $GF[q]$ and let R be the set of the remaining non-zero residues in $GF[q]$. Verify that the blocks obtained by developing Q and R in $GF[q]$, together with the blocks

$$\{\infty, 0, 1, \theta^4\}, \{\infty, \theta^{4i+1}, \theta^{4i+1}+1, \theta^{4i+1}+\theta^4\}, \quad i = 0, 1,$$
$$\{\infty, 0, \theta, \theta^5\}, \{\infty, \theta^{4j}, \theta+\theta^{4j}, \theta^5+\theta^{4j}\}, \quad j = 0, 1,$$
$$\{\infty, 0, \theta+\theta^{4i}, \theta^4(\theta+\theta^{4i})\}, \quad i = 0, 1,$$

and

$$\{\infty, \theta^{4j}, \theta+\theta^{4i}+\theta^{4j}, \theta^4(\theta+\theta^{4i})+\theta^{4j}\}, \quad i = 0, 1; j = 0, 1$$

form an inversive plane with 10 points; that is, a $B[4, 4; 10]$ design in which each unordered triple of points appears in exactly one block.

4 Isomorphism and irreducibility

4.1. Motivation

Let X be a finite set, and let π be a one–one function on X or *permutation* on X. If $|X| = v$, then its elements may be ordered in $v!$ different ways, each of which corresponds to a permutation π. The set of all such permutations forms the *symmetric group* S_X on X, where the group operation is the usual composition of functions. We mentioned permutations in Section 1.1, but now consider them in more detail.

Example 1. Let $X = \{1, 2, 3, 4, 5, 6, 7\}$, and let π be the permutation which interchanges the elements 2 and 4, and leaves all other elements alone. The usual notation for a function would specify π by $1\pi = 1, 2\pi = 4, 3\pi = 3$, and so on. Instead we shall write permutations in *cycle notation*, so that $\pi = (24)$. This means that all the elements not appearing in the cycle are fixed (mapped to themselves), and that π takes 2 to 4, and 4 to 2. Similarly, $\sigma = (24)(56)$ fixes 1, 3, and 7, interchanges 2 with 4, and interchanges 5 with 6. As a last example $\tau = (1234567)$ takes 1 to 2, 2 to 3, and so on, finally taking 7 to 1.

Composing permutations, we find that applying first σ and then τ to X gives

$$\sigma\tau = (24)(56)(1234567) = (1257)(34),$$

whereas applying τ first, followed by σ, gives

$$\tau\sigma = (1234567)(24)(56) = (1467)(23).$$

□

Suppose that a design $D = (X, \mathscr{B})$ is based on the set X, and that a permutation π acts on X. How does π affect the design? If $B = \{b_1, b_2, \ldots, b_k\}$ is a block of the design, then we write $B\pi$ to mean the k-set

$$B\pi = \{b_1\pi, b_2\pi, \ldots, b_k\pi\},$$

which may or may not be a block of the design. Remember that we are not concerned with the order of elements within the block, so that if, for instance, $\pi = (b_1 b_2)$, then

$$B\pi = \{b_2, b_1, \ldots, b_k\} = B,$$

and we say that π *fixes* B. Neither are we concerned with the order of blocks within the set $\mathscr{B}$, so if, for every $B \in \mathscr{B}$, we have $B\pi \in \mathscr{B}$ also, then we say that π *fixes* the design.

Example 1 (continued). Let $\mathscr{B}$ be the set of blocks

$$B_1 = 124,\ B_2 = 235,\ B_3 = 346,\ B_4 = 457,\ B_5 = 561,\ B_6 = 672,\ B_7 = 713.$$

(This is the design (D4) of Chapter 1, except that we write 7 instead of 0.)

Now π fixes B_1, which contains both 2 and 4, and B_5 and B_7, which contain neither 2 nor 4. But it takes B_2, B_3, B_4, B_6 to 435, 326, 257, 674 respectively, and these new 3-sets are not blocks of $\mathscr{B}$. In other words, π maps the original design, $D = (X, \mathscr{B})$, to a new design, $E = (X, \mathscr{C})$, and $\mathscr{B} \cap \mathscr{C}$ contains three common blocks.

On the other hand, σ fixes B_1, B_5, and B_7, and permutes the remaining blocks among themselves, interchanging B_2 with B_3, and B_4 with B_6. Thus σ fixes D, and could be written as a permutation $(B_2B_3)(B_4B_6)$ of $\mathscr{B}$.

Again, τ fixes D, but it does not fix any individual block. It sends B_1 to B_2, B_2 to B_3, and so on, finally sending B_7 to B_1. As a permutation of $\mathscr{B}$, τ could be written $(B_1B_2B_3B_4B_5B_6B_7)$. □

If π is a permutation of the set X, which sends the design $D = (X, \mathscr{B})$ to the design $E = (X, \mathscr{C})$, then we say that D is *isomorphic* to E, and that π is an *isomorphism* between them. If π fixes the design D, so that $\mathscr{B} = \mathscr{C}$, we call π an *automorphism* of D. Strictly speaking, an isomorphism of designs consists of two permutations, one acting on the underlying set X, and the other acting on the set of k-sets of X (or, in the case of an automorphism, on the blocks in $\mathscr{B}$). But often we mention only the permutation on X itself, since it induces the permutation on the k-sets.

Suppose that π is an automorphism of D, so that for all $i \in \{1, 2, \ldots, b\}$ there exists some $j \in \{1, 2, \ldots, b\}$ such that $B_i\pi = B_j$. Then $B_i = B_j\pi^{-1}$, or, in other words, π^{-1} must also be an automorphism of D. Similarly if π and σ are automorphisms of D, then

$$B_i\pi\sigma = B_j\sigma = B_h,$$

for appropriate blocks of $\mathscr{B}$; so $\pi\sigma$ is also an automorphism of D. Since ι, the identity map, is always an automorphism, the set of all automorphisms of a design forms a group under the (associative) operation of composition. We denote this group by Aut D. It is a subgroup of S_X, and its properties tell us a great deal about the symmetry and structure of D, and are important in classifying designs.

Example 1 (continued). The permutations $\mu = (35)(67)$ and $\nu = (36)(75)$ are both automorphisms of D, and together they generate a group of order 4, namely,

$$K = \{(35)(67),\ (36)(75),\ (37)(56),\ (1)\},$$

where $(37)(56) = \mu\nu = \nu\mu$, and (1) is the usual notation for ι. (A cycle of length one is just a fixed element and usually, except for (1), we omit fixed elements.)

Now $\rho = (276)(435)$ is also an automorphism of D, and together with K generates a group H of order 24. Finally τ, together with H, generates a group G of order 168 (see Exercise 4.1), and $G \leq \operatorname{Aut} D$. □

In fact, in this case $G = \operatorname{Aut} D$, but in order to check this we need more information about permutation groups.

4.2. Orbits and stabilizers

If the group G acts on the set X, and if $x \in X$, then the *stabilizer* of x, G_x, is defined as the set of elements of G which fix x, so that

$$G_x = \{\pi \in G \mid x\pi = x\}.$$

If $\pi \in G_x$, then $x = x\pi^{-1}$, so $\pi^{-1} \in G_x$ also. If $\pi_1, \pi_2 \in G_x$, then $(x)\pi_1\pi_2 = x\pi_2 = x$; so $\pi_1\pi_2 \in G_x$. Since $\iota \in G_x$ for any x, this means that G_x is a subgroup of G.

The *orbit* of x under G, denoted by xG, is the set of all elements of X to which x may be mapped by some permutation belonging to G, so that

$$xG = \{y \in X \mid x\pi = y \text{ for some } \pi \in G\}.$$

We can define an equivalence relation on X by saying that the elements x and y of X are *equivalent* if and only if $x\pi = y$ for some $\pi \in G$; that is, $y \in xG$. The equivalence classes are just the orbits of x.

Example 2. Let $X = \{1, 2, 3\}$, and let G be S_3, the symmetric group of degree 3. Then

$$G = \{(1), (123), (132), (12), (13), (23)\}$$

and the stabilizers of the elements are

$$G_1 = \{(1), (23)\}, G_2 = \{(1), (13)\}, G_3 = \{(1), (12)\}$$

The orbit of each element is the whole of X. □

Example 3. Let $X = \{1, 2, 3, 4, 5\}$, and let G be the group generated by (123)(45), that is,

$$G = \{(1), (123)(45), (132), (45), (123), (132)(45)\}.$$

The stabilizers are

$$G_1 = G_2 = G_3 = \{(1), (45)\}, \text{ and } G_4 = G_5 = \{(1), (123), (132)\}.$$

The orbits are $1G = 2G = 3G = \{1, 2, 3\}$ and $4G = 5G = \{4, 5\}$. □

Notice that $|G| = 6$ in each of these examples and that in each case an element with an orbit of three elements has two permutations in its stabilizer, and an element with an orbit of two elements has three permutations in its stabilizer. This is not just coincidence.

Theorem 4. *If a permutation group G acts on a set X, then for every $x \in X$*

$$|G| = |G_x| . |xG|.$$

Proof. Let $G_x, G_x\pi_2, \ldots, G_x\pi_n$ be the cosets of G_x in G. We define a one–one correspondence, f, between the cosets of G_x and the elements of xG by

$$f\colon G_x\pi_i \mapsto x\pi_i.$$

We check that f is (i) well-defined, (ii) one–one, and (iii) onto.

(i) If π_i and π_j belong to the same coset of G_x in G, then $\pi_j = \pi\pi_i$, for some $\pi \in G_x$. Hence

$$x\pi_j = (x\pi)\pi_i = x\pi_i,$$

so the value of $x\pi_i$ is independent of the choice of coset representative.

(ii) If $x\pi_i = x\pi_j$, then $x\pi_j\pi_i^{-1} = x$, so $\pi_j\pi_i^{-1} = \pi$, for some $\pi \in G_x$. Hence $\pi_j = \pi\pi_i$, so $G_x\pi_i = G_x\pi_j$ and f is one-one.

(iii) If $y \in xG$, then by definition there exists π_i such that $x\pi_i = y$; so f is onto xG.

Hence G_x has n cosets in G if and only if xG contains n elements. Thus

$$|G|/|G_x| = n = |xG|$$

and the Theorem follows. □

Suppose that for any pair of elements, x and y, of X there exists a permutation π in G such that $x\pi = y$. Then $xG = X$, and G is said to be *transitive* on X.

Corollary 4.1. *If G is transitive on X, then for any $x \in X$, we have* $|G| = |X| . |G_x|$. □

Example 1 (continued). Since τ is a 7-cycle, Aut D is transitive on X; so by Corollary 4.1, $|\text{Aut } D| = 7 . |G_1|$.

Since $\rho \in G_1$, we see that G_1 has at most two orbits on $X \backslash \{1\} = \{2, 3, 4, 5, 6, 7\}$; these are $\{2, 6, 7\}$ and $\{3, 4, 5\}$. But $v \in G_1$ also, and thus 3 and 6 belong to the same orbit of Aut D, so $|G_1| = 6 . |G_{12}|$, where G_{12} is the subgroup of G_1 which fixes 2 as well.

Since $B_1 = 124$, any permutation fixing 1 and 2 must also fix 4. Thus G_{12} can only move 3, 5, 6, and 7, and a direct check shows that $G_{12} = K$, of order 4. Now

$$|\text{Aut } D| = 7 . 6 . 4 = 168,$$

and the group G that we found earlier is Aut D.

We may also regard Aut D and S_X as permutation groups acting on the set of 3-subsets of X, and thus on the set, $\mathscr{D}$, of all the $B[3, 1; 7]$ designs based on X. If π is a permutation on X, then $D\pi = D$ if and only if $\pi \in \text{Aut } D$, so that Aut D is the stabilizer of D. If $\pi \notin \mathrm{Aut}\, D$, then $D\pi = E$, where $E = (X, \mathscr{C})$ is isomorphic to D but $\mathscr{B} \neq \mathscr{C}$. Theorem 4 shows that each coset of Aut D in S_X corresponds to precisely one $B[3, 1; 7]$ design isomorphic to D and thus, by Corollary 4.1, that $|S_X| = |\mathscr{D}|.|\mathrm{Aut}\, D|$. That is,

$$|\mathscr{D}| = 7!/168 = 30$$

and there are 30 $B[3, 1; 7]$ designs isomorphic to D.

How many $B[3, 1; 7]$ designs are there altogether? We count directly, since each such design corresponds to a partition of the $\binom{7}{2} = 21$ edges of K_7 into seven edge-disjoint triangles, three of which are incident with each vertex.

One triangle must include the edge 67, and its remaining vertex may be chosen in five ways. Suppose that we have chosen 567. The elements 5, 6, and 7 must occur in two more blocks each, and the remaining elements of each of these blocks must correspond to a one-factor of the K_4 labelled 1, 2, 3, 4. Thus the blocks containing 5 could be

125, 345; or 135, 245; or 145, 235;

and similarly for 6 and 7. Since K_4 has three one-factors and they can be assigned to 5, 6, 7 in six ways, we have altogether $5 \times 6 = 30$ possible $B[3, 1; 7]$ designs on X.

Hence all $B[3, 1; 7]$ designs are isomorphic to D. □

4.3. Automorphisms of symmetric BIBDs

As we saw in Section 4.2, an automorphism of a BIBD can be regarded as a permutation acting on the set of varieties and on the set of blocks. If the design is symmetric, the permutation acts in two different ways on sets of size v. Another convenient way to represent a permutation in this context is as a *permutation matrix*; that is, a square matrix with precisely one 1 per row and column, and all other entries zero.

Example 5. The permutation (123) applied to the rows of I_3 leads to the matrix P, and applied to the columns of I_3 leads to Q; see Table 4.1. If M is any 3×3 matrix, then premultiplying M by P permutes the rows of M according to (123), and similarly postmultiplying M by Q permutes the columns. □

Table 4.1

$$\begin{bmatrix} 1 & 0 & 0 \\ 0 & 1 & 0 \\ 0 & 0 & 1 \end{bmatrix} \quad \begin{bmatrix} 0 & 0 & 1 \\ 1 & 0 & 0 \\ 0 & 1 & 0 \end{bmatrix} \quad \begin{bmatrix} 0 & 1 & 0 \\ 0 & 0 & 1 \\ 1 & 0 & 0 \end{bmatrix}$$

$$I_3 \qquad\qquad P \qquad\qquad Q$$

More formally, if σ is a permutation on a v-set, then $P = [p_{ij}]$ and $Q = [q_{ij}]$ are $v \times v$ matrices with

$$p_{ij} = \begin{cases} 1 & \text{if } j = i\sigma, \\ 0 & \text{otherwise} \end{cases} \quad \text{and} \quad q_{ij} = \begin{cases} 1 & \text{if } i = j\sigma, \\ 0 & \text{otherwise.} \end{cases}$$

Notice that $P^{-1} = P^T$ and $Q^{-1} = Q^T$. Also trace (P) = trace (Q) and each of them equals the number of points fixed by σ.

Theorem 6. *An automorphism of an SBIBD fixes equal numbers of points and blocks.*

Proof. Let A be the incidence matrix of the design, and let σ be an automorphism of the design. Then σ corresponds to a unique pair of matrices P and Q such that $PA = AQ$; here P permutes the points and Q the blocks of the design respectively. Since A is non-singular, we have $A^{-1}PA = Q$. Since

$$\text{trace }(Q) = \text{trace }(A^{-1}PA) = \text{trace }(A^{-1}AP) = \text{trace }(P),$$

the Theorem follows. □

Corollary 6.1. *An automorphism of an SBIBD has the same cycle structure, whether as a permutation of points or as a permutation of blocks. In other words, there is a one–one correspondence between cycles of points and cycles of blocks, and corresponding cycles have the same length.*

Proof. Let the automorphism σ correspond to permutations σ_X, acting on X, and $\sigma_{\mathscr{B}}$, acting on $\mathscr{B}$. For $d \geq 1$, let $f_X(d)$ and $f_{\mathscr{B}}(d)$ be the number of cycles of length d in the cycle representations of σ_X and $\sigma_{\mathscr{B}}$ respectively. By the Theorem, $f_X(1) = f_{\mathscr{B}}(1)$.

Now for any positive integer n, the number of fixed points of σ_X^n and $\sigma_{\mathscr{B}}^n$ are

$$\sum_{d|n} d \,.\, f_X(d) \quad \text{and} \quad \sum_{d|n} d \,.\, f_{\mathscr{B}}(d)$$

respectively, and by the Theorem, these two expressions are equal for any n. Hence by induction, $f_X(d) = f_{\mathscr{B}}(d)$ for all d. □

Example 1 (continued). We have already seen that

$$\sigma_X = (24)(56) \quad \text{and} \quad \sigma_{\mathscr{B}} = (B_2B_3)(B_4B_6),$$

and that

$$\tau_X = (1234567), \quad \text{and} \quad \tau_{\mathscr{B}} = (B_1B_2B_3B_4B_5B_6B_7).$$

Similarly,

$$\mu_X = (35)(67),\ \mu_{\mathscr{B}} = (B_3B_4)(B_5B_7),\ \nu_X = (36)(75),\ \nu_{\mathscr{B}} = (B_2B_6)(B_5B_7),$$

$$\rho_X = (276)(435),\ \rho_{\mathscr{B}} = (B_1B_7B_5)(B_2B_4B_3).$$

This example also illustrates what Theorem 6 does *not* say. G_{12}, the subgroup which fixes 1, 2, and consequently 4, consists of

$$\{(35)(67), (36)(75), (37)(56), (1)\}.$$

Considered as a group acting on $\mathscr{B}$, it consists of

$$\{(B_3B_4)(B_5B_7), (B_2B_6)(B_5B_7), (B_2B_6)(B_3B_4), (B_1)\}$$

and fixes just one block, B_1.

Similarly, G_1 fixes the point 1 when it acts on X, but fixes no block when it acts on $\mathscr{B}$. Hence although each individual automorphism fixes as many points as blocks, an automorphism *group* may not. □

Theorem 6 has another useful corollary, but first we need a preliminary result.

Lemma 7. *Let the permutation group G act on the set X, and for each* $\pi \in G$, *let* F_π *denote the set of points of X left fixed by* π, *so that*

$$F_\pi = \{x \in X \mid x\pi = x\}.$$

If there are exactly n orbits of X under G, then

$$n.|G| = \sum_{\pi \in G} |F_\pi|.$$

Proof. We take a permutation-point incidence matrix, $M = [m_{ij}]$, with $|G|$ rows and $|X|$ columns, where

$$m_{ij} = \begin{cases} 1 & \text{if } x_j\pi_i = x_j, \\ 0 & \text{otherwise,} \end{cases}$$

and sum the elements of M in two ways. Here

$$\sum_{j=1}^{|X|} m_{ij} = |F_{\pi_i}|$$

and

$$\sum_{i=1}^{|G|} m_{ij} = |G_x|;$$

so the sum of all the elements in M is

$$\sum_{\pi \in G} |F_\pi| = \sum_{x \in X} |G_x|.$$

All the elements of the same orbit have stabilizers of the same order, by Theorem 5. Thus if we choose $y_1, y_2, \ldots, y_n$ such that y_i belongs to the ith orbit, then

$$\sum_{x \in X} |G_x| = \sum_{i=1}^{n} \sum_{y \in y_i G} |G_y| = \sum_{i=1}^{n} |y_i G|.|G_y|.$$

By Theorem 4, $\sum_{i=1}^{n} |y_i G|.|G_y| = \sum_{i=1}^{n} |G| = n|G|$, and the Lemma follows. □

Corollary 6.2. *An automorphism group G of a symmetric design has as many orbits on points as on blocks.*

Proof. By Lemma 7, the number n_X of orbits of G acting on X is given by

$$n_X = \frac{1}{|G|} \cdot \sum_{\pi \in G} |F_\pi|, \text{ where } |F_\pi| = f_X(1),$$

and similarly if G acts on $\mathscr{B}$ the number of orbits is

$$n_{\mathscr{B}} = \frac{1}{|G|} \cdot \sum_{\pi \in G} |F_\pi|, \text{ where } |F_\pi| = f_{\mathscr{B}}(1).$$

By the Theorem, $f_X(1) = f_{\mathscr{B}}(1) = |F_\pi|$, so $n_X = n_{\mathscr{B}}$. □

Example 1 (concluded). G_{12}, acting on X, has orbits $\{1\}, \{2\}, \{4\}, \{3, 5, 6, 7\}$. G_{12}, acting on $\mathscr{B}$, has orbits $\{B_1\}, \{B_2, B_6\}, \{B_3, B_4\}, \{B_5, B_7\}$. In each case, it has four orbits. □

4.4. The multiplier theorem

Example 8. In Z_{13}, the difference set $\{0, 1, 3, 9\}$ gives rise to a $B[4, 1; 13]$ design, D. Certainly, the map ϕ where $x\phi = x + 1 \pmod{13}$ is an automorphism of D, of order 13. Another automorphism is ψ, where $x\psi = 3x \pmod{13}$. This acts on Z_{13} as the permutation

$$(1, 3, 9)\ (2, 6, 5)\ (4, 12, 10)\ (7, 8, 11),$$

and similarly on the set of blocks. Thus if B_i denotes $B_0 + i \pmod{13}$, then $3B_i = B_{3i}$. □

An integer t is a *multiplier* of the difference set $\{a_1, a_2, \ldots, a_k\} \pmod{v}$ if and only if the map taking x to $xt \pmod{v}$ is an automorphism of the $SB[k, \lambda; v]$ design developed from the difference set. This amounts to saying that

$$\begin{aligned} t\{a_1, a_2, \ldots, a_k\} &= \{ta_1, ta_2, \ldots, ta_k\} = \{a_1 + s, a_2 + s, \ldots, a_k + s\} \\ &= \{a_1, a_2, \ldots, a_k\} + s, \end{aligned} \tag{4.1}$$

for some s, (where the arithmetic is carried out modulo v). Notice in particular that 1 must occur as a difference, and since

$$1 \equiv a_i - a_j \equiv t(a_h - a_l) \pmod v$$

for some h, i, j, l, we must have $(t, v) = 1$. Also, the multipliers of the difference set form a multiplicative group modulo v.

All known difference sets (mod v) have a non-trivial multiplier, and knowledge of a multiplier is often very useful in constructing a difference set, or proving its non-existence.

To study multipliers of cyclic difference sets modulo v, we associate with each set $B \subseteq Z_v$ a polynomial in x and x^{-1}, with integer coefficients. Thus if $B = \{a_1, a_2, \ldots, a_k\}$, its associated polynomial, usually called the Hall polynomial, is

$$\theta(x) = x^{a_1} + x^{a_2} + \ldots + x^{a_k},$$

and

$$\theta(x).\theta(x^{-1}) = \sum_{i=1}^{k} \sum_{j=1}^{k} x^{a_i - a_j}.$$

Notice that $x^i \equiv x^j (\bmod x^v - 1)$ if and only if $i \equiv j \pmod v$.

Now B is a difference set if and only if the values of $a_i - a_j$ include 0 (k times) and each non-zero integer (modulo v) λ times. Thus

$$\theta(x).\theta(x^{-1}) \equiv k + \lambda(x + \ldots + x^{v-1}) \pmod{x^v - 1}$$

for a difference set. If we let $n = k - \lambda$, and $T(x) = 1 + x + \ldots + x^{v-1}$, this equation becomes

$$\theta(x).\theta(x^{-1}) \equiv n + \lambda T(x) \pmod{x^v - 1}. \tag{4.2}$$

If t is a multiplier of B, then eqn (4.1) can be rewritten as

$$\theta(x^t) \equiv x^s \theta(x) \pmod{x^v - 1}. \tag{4.3}$$

We shall also make use of the observations that

$$x^v - 1 = (x - 1)T(x), \tag{4.4}$$

$$x^i T(x) \equiv T(x) \pmod{x^v - 1}, \tag{4.5}$$

$$T(x^{-1}) \equiv T(x) \pmod{x^v - 1}, \tag{4.6}$$

$$x^{-v} - 1 = -x^{-v}(x^v - 1), \tag{4.7}$$

and for any prime p and any polynomials $F(x)$, $G(x)$,

$$(F(x) + G(x))^p \equiv F(x)^p + G(x)^p \equiv F(x^p) + G(x^p) \pmod p. \tag{4.8}$$

We are now ready to prove

Theorem 9. *Let $B = \{a_1, a_2, \ldots, a_k\} \subseteq Z_v$ be a difference set, where $k(k-1) = \lambda(v-1)$. Let $n = k - \lambda$, and let p be a prime such that $p \mid n$, $p \nmid v$, and $p > \lambda$. Then p is a multiplier of B.*

Proof. Since $p|n$, we have by eqns (4.2) and (4.4) that

$$\begin{aligned}\theta(x).\theta(x^{-1}) &= n+\lambda T(x)+U_1(x).(x^v-1)\\ &= n+T(x)\ (\lambda+(x-1)U_1(x))\\ &= p(n/p)+T(x).V_1(x).\end{aligned} \tag{4.9}$$

Multiplying eqn (4.9) by $\theta(x)^{p-1}$ shows that

$$\begin{aligned}\theta(x)^p.\theta(x^{-1}) &= p\theta(x)^{p-1}(n/p)+T(x)\,\theta(x)^{p-1}V_1(x)\\ &= pU_2(x)+T(x).V_2(x).\end{aligned} \tag{4.10}$$

By eqn (4.8) we see that

$$\theta(x)^p \equiv \theta(x^p)\ (\text{mod } p)$$

and hence from eqn (4.10) that

$$\theta(x^p).\theta(x^{-1}) = pU_3(x)+T(x).V_3(x) \tag{4.11}$$

for polynomials $U_3(x)$, $V_3(x)$. By eqn (4.5) we have

$$T(x).V_3(x) \equiv T(x).V_3(1)\ (\text{mod } x^v-1)$$

and so eqn (4.11) becomes

$$\theta(x^p).\theta(x^{-1}) \equiv pU_3(x)+V_3(1).T(x)\ (\text{mod } x^v-1). \tag{4.12}$$

Substitute $x = 1$ in eqn (4.12). Since $\theta(1) = k$, and $T(1) = v$, we have

$$k^2 = pU_3(1)+vV_3(1). \tag{4.13}$$

But $k(k-1) = \lambda(v-1)$, so

$$k^2 = k-\lambda+\lambda v = n+\lambda v. \tag{4.14}$$

By eqns (4.13) and (4.14),

$$n+\lambda v = pU_3(1)+vV_3(1).$$

Since $p|n$, this means that $V_3(1).v \equiv \lambda v(\text{mod } p)$. Since $p\nmid v$, we have $V_3(1) \equiv \lambda(\text{mod } p)$, and hence $V_3(1) = \lambda+pm$ for some integer m. Then from eqn (4.12),

$$\begin{aligned}\theta(x^p).\theta(x^{-1}) &\equiv pU_3(x)+(\lambda+pm).T(x) \quad (\text{mod } x^v-1)\\ &\equiv p.S(x)+\lambda T(x) \quad (\text{mod } x^v-1),\end{aligned} \tag{4.15}$$

where $S(x) = U_3(x)+mT(x)$.

Now let $x = 1$ in eqn (4.15). As before, $k^2 = pS(1)+\lambda v$, and by eqn (4.14),

$$pS(1) = n. \tag{4.16}$$

Replace x by x^{-1} in eqn (4.15); by (4.6)

$$\theta(x^{-p}).\theta(x) \equiv p.S(x^{-1})+\lambda T(x)\ (\text{mod } x^v-1). \tag{4.17}$$

Since B is a difference set and $p\nmid v$, the set $pB = \{pa_1, \ldots, pa_k\}$ is also a difference set, and thus

$$\theta(x^p).\theta(x^{-p}) \equiv n + \lambda T(x) \pmod{x^v - 1}. \tag{4.18}$$

We now combine four congruences: (4.2), (4.18), (4.15), and (4.17). Multiplying (4.2) and (4.18) gives

$$\theta(x).\theta(x^{-1}).\theta(x^p).\theta(x^{-p}) \equiv (n + \lambda T(x))^2 \pmod{x^v - 1}.$$

Multiplying (4.15) and (4.17) gives

$$\theta(x).\theta(x^{-1}).\theta(x^p).\theta(x^{-p})$$
$$\equiv (p.S(x) + \lambda.T(x))\,(p.S(x^{-1}) + \lambda.T(x)) \pmod{x^v - 1}.$$

Thus

$$(p.S(x) + \lambda.T(x))\,(p.S(x^{-1}) + \lambda.T(x)) \equiv (n + \lambda T(x))^2 \pmod{x^v - 1}. \tag{4.19}$$

But, by (4.5), $p.S(x).T(x) \equiv p.S(1).T(x) \pmod{x^v - 1}$ and hence, by eqn (4.16), $p.S(x).T(x) \equiv nT(x) \pmod{x^v - 1}$. This reduces eqn (4.19) to

$$p^2 S(x) S(x^{-1}) \equiv n^2 \pmod{x^v - 1}. \tag{4.20}$$

Now consider eqn (4.15). Every coefficient in $\theta(x^p).\theta(x^{-1})$ is a non-negative integer, regardless of reduction modulo $x^v - 1$. Thus

$$\theta(x^p).\theta(x^{-1}) \equiv c_0 + c_1 x + \ldots + c_{v-1} x^{v-1} \pmod{x^v - 1},$$

where $c_i \geq 0$ for $i = 0, 1, \ldots, v-1$, and from (4.15) $c_i \equiv \lambda \pmod{p}$ for $i = 0, 1, \ldots, v-1$.

So far, we have not used the hypothesis that $p > \lambda$, but now we must. The assumption that $p > \lambda$ forces $c_i \geq \lambda$ for each i, so

$$\theta(x^p).\theta(x^{-1}) - \lambda T(x) \equiv pS(x) \pmod{x^v - 1}$$

has non-negative coefficients; that is,

$$S(x) \equiv d_0 + d_1 x + \ldots + d_{v-1} x^{v-1} \pmod{x^v - 1}$$

where $d_i \geq 0$ for $i = 0, 1, \ldots, v-1$. Now suppose that $S(x)$ has at least two positive coefficients, say $d_h > 0$ and $d_l > 0$. Then in eqn (4.20), the left side has a term $d_h d_l x^{h-l}$ with $d_h d_l > 0$, whereas x^{h-l} has coefficient zero on the right side. Hence for some s, $0 \leq s \leq v-1$, we have

$$S(x) \equiv d_s x^s \pmod{x^v - 1}$$

and

$$pS(x) \equiv pd_s x^s \equiv nx^s \pmod{x^v - 1} \tag{4.21}$$

by eqn (4.16). By (4.21) and (4.15) we have

$$\theta(x^p).\theta(x^{-1}) \equiv nx^s + \lambda T(x) \pmod{x^v - 1}$$

and hence

$$\theta(x^p).\theta(x).\theta(x^{-1}) \equiv nx^s\theta(x) + \lambda\theta(x).T(x) \pmod{x^v - 1}. \qquad (4.22)$$

Now $\theta(x).T(x) \equiv \theta(1).T(x) \equiv kT(x) \pmod{x^v - 1}$ and this, together with eqn (4.2), simplifies eqn (4.22) to

$$\theta(x^p)\,(n + \lambda T(x)) \equiv nx^s\theta(x) + \lambda kT(x) \pmod{x^v - 1}.$$

But $\lambda\theta(x^p).T(x) \equiv \lambda kT(x) \pmod{x^v - 1}$, giving

$$n\theta(x^p) \equiv nx^s\theta(x) \pmod{x^v - 1}.$$

Finally, dividing by n, we have

$$\theta(x^p) \equiv x^s.\theta(x)$$

which shows, by comparison with eqn (4.3), that p is a multiplier. □

If t is a multiplier of the difference set B, then the mapping $x \mapsto tx$ is an automorphism of the design D developed from B, and this mapping fixes the element 0. Since D is an SBIBD we may apply Theorem 6 to obtain

Corollary 6.3. *If t is a multiplier of the difference set B, then it induces an automorphism of the design D developed from B which fixes at least one block of D.* □

Sometimes we can say more.

Theorem 10. *Let $B = \{a_1, a_2, \ldots, a_k\} \subseteq Z_v$ be a difference set, where $k(k-1) = \lambda(v-1)$. If $(v, k) = 1$, then there is a block of the design developed from B which is fixed by every multiplier.*

Proof. The blocks of the design are $B = B_0, B_1, \ldots, B_{v-1}$, where $B_i = B + i \pmod{v}$. For the block $B_i = \{a_1 + i, a_2 + i, \ldots, a_k + i\}$, the sum of the elements is $(a_1 + a_2 + \ldots + a_k) + ki$. Since $(v, k) = 1$, there is precisely one value i such that $\sum_{j=1}^{k} a_j + ki \equiv 0 \pmod{v}$. For this value of i, the block B_i is fixed by every multiplier. □

Example 11. Let $v = 37$, $k = 9$, $\lambda = 2$, so that $n = 7$. By Theorem 9, 7 is a multiplier; by Corollary 6.3, it fixes a difference set. This suggests that we should consider the powers of 7 (mod 37), which are

1, 7, 12, 10, 33, 9, 26, 34, 16,

and do indeed form the required difference set. □

Example 12. Let $v = 91$, $k = 45$, $\lambda = 22$, so that $n = 23$. By Theorem 9, 23 is a multiplier; by Corollary 6.3, it fixes a difference set. We consider the distribution of the elements of this difference set $B = \{a_1, a_2, \ldots, a_{45}\}$ among the residue classes modulo 7; let α_j be the number of the a_i's congruent to j (modulo 7), $j = 0, 1, \ldots, 6$. Since $23 \equiv 2 \pmod 7$, we have

$$\alpha_1 = \alpha_2 = \alpha_4 = x,$$

$$\alpha_3 = \alpha_5 = \alpha_6 = y.$$

Since $k = 45$, we also have

$$\alpha_0 + 3x + 3y = 45. \tag{4.23}$$

There are 13 elements congruent to 1 (mod 7) among 0, 1, ..., 90, and each of these 13 must occur 22 times among differences of B. Hence

$$\sum_{j=0}^{6} \alpha_j \alpha_{j+1} = 13 \times 22 = 286,$$

so that

$$\alpha_0 (x + y) + x^2 + 3xy + y^2 = 286. \tag{4.24}$$

Now, working modulo 91, the period of 23 is 6, and thus α_0 is divisible by 6. Since only 13 elements congruent to 0 (mod 7) exist (mod 91), we have $\alpha_0 = 0$, 6, or 12. But eqns (4.23) and (4.24) have no simultaneous solution in integers in any of these cases, and hence no cyclic difference set with these parameters can exist. □

4.5. Disjoint designs

In the rest of this chapter, we show some applications of isomorphisms in constructing sets of designs with special properties.

Example 13. Consider the two $B[3, 1; 9]$ designs, D_1 and D_2, based on $X = \{1, 2, 3, 4, 5, 6, 7, 8, 9\}$, with blocks as listed in Table 4.2. D_1 and D_2 have one common block, namely 123. Suppose that we want to produce two designs with these parameters but with no blocks in common. We leave D_1 unchanged, and try to find a permutation, σ, such that $D_2\sigma$ has no block in common with D_1.

Consider the blocks containing 1. The block 147 appears in D_1, and the pair 47 belongs to the block 457 in D_2. Thus, if we applied $\sigma = (15)$ to D_2, we would map 457 to 147 and still have a block in common with D_1. Similarly, since 168 appears in D_1, and 689 in D_2, choosing $\sigma = (19)$ would not remove the common block. Again, 156, 179 appear in D_2, and 456, 789 in D_1, so $\sigma = (14)$ or (18) applied to D_2 would still leave a block in common. Certainly (12) or (13) leaves 123 unchanged. But either (16) or (17) applied to D_2 produces a design with no block in common with D_1, as shown in Table 4.2. □

Table 4.2

123	123	236	237
147	148	468	478
159	156	156	456
168	179	679	179
258	259	259	259
267	278	278	128
249	246	124	246
369	367	137	136
348	349	349	349
357	358	358	358
456	457	457	145
789	689	189	689
D_1	D_2	$D_2(16)$	$D_2(17)$

Two designs $D_i = (X, \mathscr{B}_i)$, $i = 1, 2$, based on the same set X are said to be *disjoint* if they have no common block; that is, if $\mathscr{B}_1 \cap \mathscr{B}_2 = \varnothing$. Suppose that D_1 and D_2 have the same parameters. (We make no assumption as to whether or not they are isomorphic.) Does there exist a permutation α on X such that D_1 and $D_2\alpha$ have no common blocks, or in other words, such that $\mathscr{B}_1 \cap \mathscr{B}_2\alpha = \varnothing$? If $\lambda = 1$ and $k \geq 3$, then such a permutation must exist; we describe an algorithm for finding it.

We may suppose without loss of generality that $\{1, 2, 3, \ldots, k\} \in \mathscr{B}_1 \cap \mathscr{B}_2$, and that the designs D_1 and D_2 contain the blocks shown in Table 4.3. Consider the $r-1$ blocks of D_1, other than $\{1, 2, 3, \ldots, k\}$, which contain 1. If

$$\{1, a_{i2}, a_{i3}, \ldots, a_{ik}\} \in \mathscr{B}_1,$$

then it may happen that

$$\{y_i, a_{i2}, a_{i3}, \ldots, a_{ik}\} \in \mathscr{B}_2.$$

We assume that this is true of $n-1$ such blocks, where $1 \leq n \leq r$, and that these blocks are listed in D_1 immediately following $\{1, 2, 3, \ldots, k\}$. A similar statement is true for $m-1$ blocks of $\mathscr{B}_2$, with $1 \leq m \leq r$.

We now define the *spread* of 1 with respect to the block $\{1, 2, 3, \ldots, k\}$ to be

$$\mathscr{S}(1) = \{1, 2, 3, \ldots, k, x_2, \ldots, x_m, y_2, \ldots, y_n\}$$

and note that

$$|\mathscr{S}(1)| \leq k + 2(r-1).$$

Since $\lambda = 1$, we have

$$r - 1 = (v-k)/(k-1)$$

so that

$$v - |\mathscr{S}(1)| \geq (v-k)(k-3)/(k-1) \geq 0.$$

Hence for $k \geq 4$, there exists $x \in X \backslash \mathscr{S}(1)$, but for $k = 3$ it is possible that $X = \mathscr{S}(1)$.

Table 4.3

	1	2	3	...	k	1	2	3	...	k	
↑	1	a_{22}	a_{23}	...	a_{2k}	1	b_{22}	b_{23}	...	b_{2k}	↑
	1	a_{32}	a_{33}	...	a_{3k}	1	b_{32}	b_{33}	...	b_{3k}	
$r-1$	⋮				⋮	⋮				⋮	$r-1$
↓	1	a_{r2}	a_{r3}		a_{rk}	1	b_{r2}	b_{r3}		b_{rk}	↓
↑	x_2	b_{22}	b_{23}	...	b_{2k}	y_2	a_{22}	a_{23}	...	a_{2k}	↑
$m-1$					⋮					⋮	$n-1$
↓	x_m	b_{m2}	b_{m3}	...	b_{mk}	y_n	a_{n2}	a_{n3}	...	a_{nk}	↓
		⋮						⋮			
		D_1						D_2			

Lemma 14. *Suppose that there is an element x, such that $x \in X \backslash \mathscr{S}(1)$. Then the permutation $(1x)$ applied to the blocks of D_2 gives*

$$|\mathscr{B}_1 \cap \mathscr{B}_2(1x)| < |\mathscr{B}_1 \cap \mathscr{B}_2|.$$

Proof. The permutation $(1x)$ takes $\{1, 2, 3, \ldots, k\}$ to $\{x, 2, 3, \ldots, k\}$ without introducing any new common blocks. □

In the case $k = 3$, no such x may be available, and we shall call a triple

$$\{1, 2, 3\} \in \mathscr{B}_1 \cap \mathscr{B}_2$$

good if and only if it contains a point whose spread is not the whole of X; otherwise we say it is *bad*.

Lemma 15. *The triple $\{1, 2, 3\} \in \mathscr{B}_1 \cap \mathscr{B}_2$ is good provided that there exist triples $\{x, y, a\} \in \mathscr{B}_1$ and $\{x, y, b\} \in B_2$, such that $\{a, b\} \subseteq \{1, 2, 3\}$.*

Proof. This follows since it implies that $m < r$ or $n < r$ (or both). Note that we may have $a = b$. □

Lemma 16. *If the triple $\{1, 2, 3\} \in \mathscr{B}_1 \cap \mathscr{B}_2$, and if it is a bad triple, then*

$$|\mathscr{B}_1 \cap \mathscr{B}_2(1x)| \leq |\mathscr{B}_1 \cap \mathscr{B}_2|$$

for $x \notin \{1, 2, 3\}$.

Proof. Since $\mathscr{S}(1) = \mathscr{S}(2) = \mathscr{S}(3) = X$, at most one new common triple is introduced. Since $x \notin \{1, 2, 3\}$, the common triple $\{1, 2, 3\}$ is removed. □

Lemma 17. *Suppose that the triple* $\{1, 2, 3\} \in \mathscr{B}_1 \cap \mathscr{B}_2$. *If every triple in* $\mathscr{B}_1 \cap \mathscr{B}_2$ *is bad, then there exists* $x \in X$ *such that* $\mathscr{B}_1 \cap \mathscr{B}_2(1x)$ *contains a good triple.*

Proof. We list triples in the two designs in Table 4.4, showing the proof. □

Table 4.4. The proof of Lemma 17

	$\mathscr{B}_1$	$\mathscr{B}_2$	
A:	123	123	$\{1, 2, 3\}$ is a common block.
B:	123	123 145	Without loss of generality, $\{1, 4, 5\} \in \mathscr{B}_2$.
C:	123 456	123 145	Since $\mathscr{S}(1) = X$, without loss of generality $\{4, 5, 6\} \in B_1$.
D:	123 456	123 145 167	Without loss of generality, $\{1, 6, 7\} \in \mathscr{B}_2$.
E:	123 456 678	123 145 167	Since $\mathscr{S}(1) = X$, without loss of generality $\{6, 7, 8\} \in \mathscr{B}_1$.
			Now apply the permutation (18) to $\mathscr{B}_2$.
	$\mathscr{B}_1$	$\mathscr{B}_2(18)$	
F:	123 456 678	238 458 678	We have $\{6, 7, 8\} \in \mathscr{B}_1 \cap \mathscr{B}_2(18)$. Since $\{4, 5, 6\} \in \mathscr{B}_1$ and $\{4, 5, 8\} \in \mathscr{B}_2(18)$, by Lemma 15 $\{6, 7, 8\}$ is a good triple.

Thus, if we are given block designs $(X, \mathscr{B}_1)$ and $(X, \mathscr{B}_2\}$, the algorithm summarized in Fig. 4.1 applies to X a permutation α such that $\mathscr{B}_1 \cap \mathscr{B}_2\alpha = \varnothing$. (Note that in Fig. 4.1, the procedures of Lemma 14 and Lemma 17 are denoted by A and B respectively.)

4.6. Irreducible designs: tests and constructions

A design $D = (X, \mathscr{B})$ is *irreducible* if and only if it cannot be written as the union of two smaller designs; that is, if there do not exist designs $D_1 = (X, \mathscr{B}_1)$ and $D_2 = (X, \mathscr{B}_2)$ such that $\mathscr{B} = \mathscr{B}_1 \cup \mathscr{B}_2$. We have already considered reducibility briefly in Section 2.3.

```
if k ≥ 4  then
     while 𝓑₁ ∩ 𝓑₂ ≠ ∅ do
          A
  else {k = 3}
     while 𝓑₁ ∩ 𝓑₂ ≠ ∅ do
        begin
          if {no good triple} then B;
          {good triple exists}
          A
        end
```

Fig. 4.1 An algorithm for making two designs disjoint, when $\lambda = 1$.

We prove first that only a finite number of irreducible designs can be based on a given finite set. We make use of the following result.

Lemma 18. *Let T be a finite set, let $N_0 = \{0, 1, 2, \ldots\}$, and let $P = \{f \mid f: T \to N_0\}$. Let the partial order $\leq$ be defined on P by*

$$f \leq g \quad \text{if and only if} \quad f(t) \leq g(t) \quad \text{for all } t \in T.$$

Then any infinite subset of P contains two elements which are comparable under $\leq$.

Proof. Let $K = \{v_1, v_2, \ldots\} \subseteq P$, where K is infinite, and consider the first co-ordinate of each v_i. Since an infinite sequence of non-negative integers has a non-decreasing subsequence, we can choose a sequence of elements of K whose first co-ordinates are non-decreasing. From them, we choose a sequence of elements of K whose second co-ordinates are also non-decreasing, and so on.

Since only $|T|$ co-ordinates are involved, we eventually have $v \leq w$ for some $v, w \in K$. □

Theorem 19. *Let X be a finite set. Then the number of irreducible PBDs based on X is finite.*

Proof. We apply Lemma 18 to the set, 2^X, of all subsets of X. Any PBD may be regarded as a function $f: 2^X \to N_0$, where for any subset $A \subseteq X, f(A)$ is the number of times that A appears as a block of the PBD.

If f, g are PBDs, $f \neq g$, and $f \leq g$, then $g - f$ is also a PBD with

$$\text{index } (g - f) = \text{index } (g) - \text{index } (f).$$

Hence no two irreducible PBDs are comparable, so the set of irreducible PBDs is finite by Lemma 18. □

Corollary 19.1. *The number of irreducible BIBDs based on a given finite set is finite.* □

Suppose that we are given a BIBD which has the parameters of a multiple design, and that we wish to test it for irreducibility. Various graph-theoretic methods are available.

The *associated (multi)graph* of a $B[k, \lambda; v]$ design has b vertices, one for each block; an edge joins vertices i and j for each pair of elements common to blocks i and j. Figure 4.2 shows the associated graph of a reducible $B[3, 2; 7]$ design; Fig. 4.3 shows that of an irreducible $B[3, 2; 9]$ design.

A reducible $B[k, 2; v]$ design must be the union of two $B[k, 1; v]$ designs, giving a bipartite graph. The converse is also easy to prove, leading to

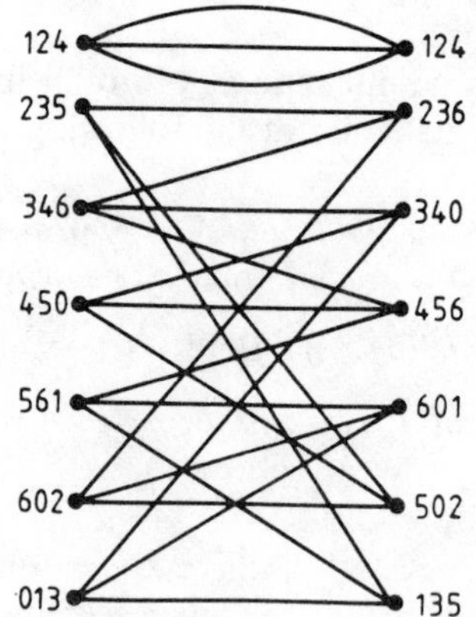

Fig. 4.2

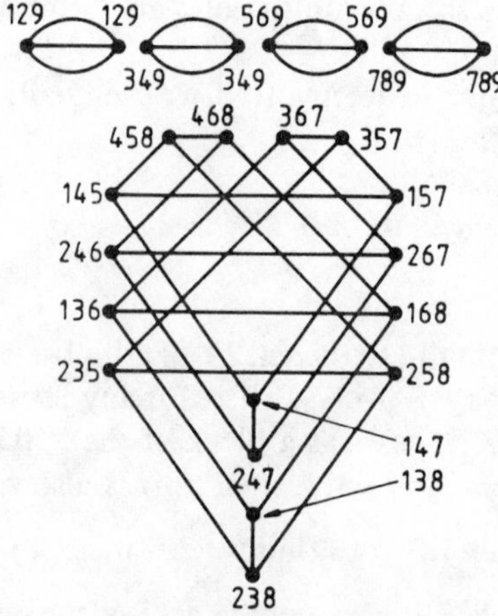

Fig. 4.3

Theorem 20. *A $B[k, 2; v]$ is irreducible if and only if its associated graph contains an odd cycle.* □

The *adjacency (multi)graph* $A(x)$ of an element x in a $B[k, \lambda; v]$ design has $v-1$ vertices, one for each element other than x, and an edge joining vertices i and j for each occurrence of elements i and j together in an *x-block*; that is, a block containing x. If $\lambda = 2$, and $A(x)$ has an odd cycle for some x, then the corresponding design is irreducible. Figure 4.4 shows the adjacency graphs of elements in the $B[3, 2; 9]$ design specified in Fig. 4.3; $A(1)$ and $A(2)$ contain odd cycles. If we place the block 138 in one subdesign and the block 136 in the other, then the block 168 cannot be placed in either, showing again that the design is irreducible.

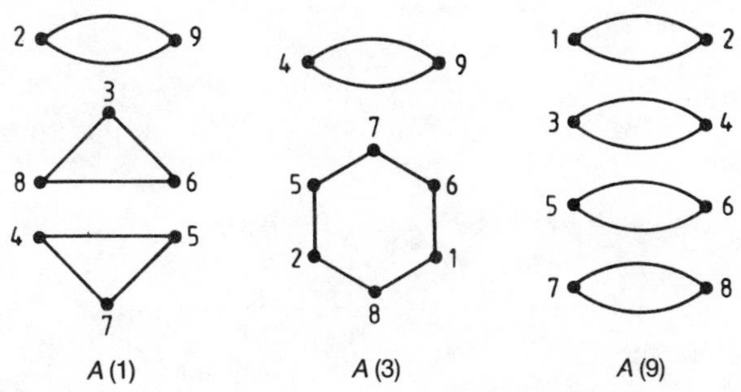

Fig. 4.4 $A(2)$ is isomorphic to $A(1)$; $A(x)$ is isomorphic to $A(3)$ for $x = 4, 5, 6, 7, 8$.

But consider the $B[3, 2; 9]$ design specified by Table 4.5. The adjacency graphs of its elements are shown in Fig. 4.5, and all of them contain only even cycles. Nevertheless, this design is irreducible. To understand why, we introduce the idea of a *chain*. For $\lambda \geq 2$, and $k \geq 3$, a chain of x-blocks, denoted by $x-(x_1, x_2, \ldots, x_{n+1})$, is a sequence of x-blocks

$$xx_1x_2 \ldots, xx_2x_3 \ldots, xx_3x_4 \ldots, \ldots, xx_nx_{n+1} \ldots,$$

where we list just three elements of each block. Suppose that $\lambda = 2$, and that a $B[k, 2; v]$ design contains a chain $x-(x_1, x_2, x_3, x_4, \ldots, x_{n-1}, x_n, x_{n+1})$ of x-blocks, and a chain $y-(x_1, x_2, y_3, y_4, \ldots, y_{m-1}, x_n, x_{n+1})$ of y-blocks. If the design is reducible, it must be a union of two $B[k, 1; v]$ designs, say A and B. If we allocate the block $xx_1x_2 \ldots$ to A, then we must allocate $yx_1x_2 \ldots$ to B, $yx_2y_3 \ldots$ to A, $xx_2x_3 \ldots$ to B, and so on, as shown in Table 4.6. To ensure that one of $xx_nx_{n+1} \ldots$ and $yx_nx_{n+1} \ldots$ is allocated to A and the other to B, we must have $m \equiv n \pmod 2$. But the $B[3, 2; 9]$ design of Table 4.5 has chains of opposite parity, as shown in Table 4.7, and is thus irreducible.

Table 4.5

129	139	145	167	178	156	128	134
368	458	236	249	237	248	347	256
257	467	789	358	469	359	689	579

Table 4.6

(A)	$xx_1x_2\ldots$	(B)	$yx_1x_2\ldots$
(B)	$xx_2x_3\ldots$	(A)	$yx_2y_3\ldots$
(A)	$xx_3x_4\ldots$	(B)	$yy_3y_4\ldots$
$\vdots$	$\vdots$	$\vdots$	$\vdots$
	$xx_{n-1}x_n\ldots$		$yy_{m-1}x_n$
	$xx_nx_{n+1}\ldots$		yx_nx_{n+1}

Table 4.7

(A)	698	(B)	798
(B)	683	(A)	781
(A)	632	(B)	716
		(A)	764
		(B)	743
		(A)	732

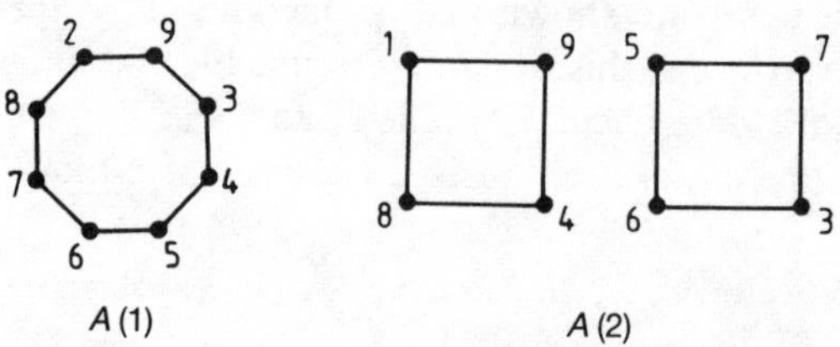

Fig. 4.5 $A(x)$ is isomorphic to $A(1)$ for $x = 3, 4, 5, 6, 7, 8$.

A *λ-arc* is a generalization of a chain. Suppose that an element x occurs in blocks $B_1, B_2, \ldots, B_r$ of a $B[k, \lambda; v]$ design and that, for some q, the blocks $B_1, B_2, \ldots, B_q$ contain the elements $y_1, y_2, \ldots, y_q$ precisely λ times each, and each block B_i, $i = 1, \ldots, q$, contains λ of the elements y_i, $i = 1, \ldots, q$. Then these q blocks form a λ-arc for the element x. For example, if $\lambda = 3$, $q = 4$, then the four blocks

$$xy_1y_2y_3\ldots,$$
$$xy_1y_2y_4\ldots,$$
$$xy_1y_3y_4\ldots,$$
$$xy_2y_3y_4\ldots,$$

form a 3-arc for x. For q blocks to form a λ-arc, we must necessarily have $q \le r$ and $\lambda+1 \le k$.

Theorem 21. *Suppose that a $B[k, \lambda; v]$ design D has q blocks which form a λ-arc for some element x, and let $d = gcd\,(q, \lambda)$. Then any $B[k, i; v]$ subdesign has index $i = m(\lambda/d)$ for some integer m.*

Proof. Since $d = gcd\,(q, \lambda)$, we can choose integers s and t such that

$$qs + \lambda t = d. \tag{4.25}$$

Suppose that D is the union of two subdesigns D_1 and D_2, with indices i and $\lambda - i$ respectively, where w blocks of the λ-arc are contained in D_1 and $q-w$ in D_2. As shown in Fig. 4.6, the w blocks of the λ-arc contained in D_1 account for the occurrences of q elements i times each in the x-blocks. Thus $w\lambda = qi$ and $w\lambda s = qis = (d-\lambda t)i$ from (4.25). Hence $\lambda(ws+it) = id$ and $i = (ws+it)(\lambda/d)$, where the integer $m = ws+it$. □

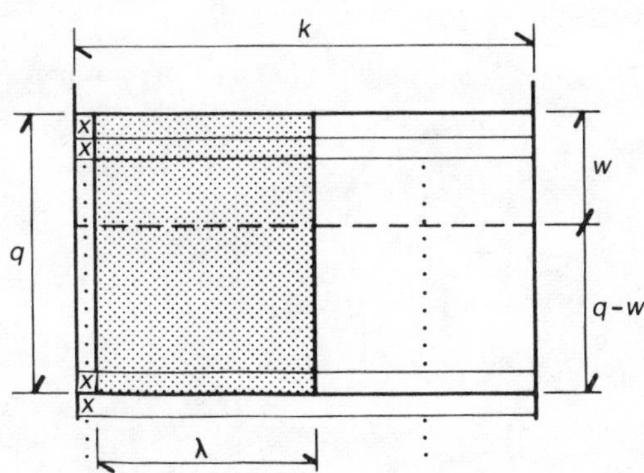

Fig. 4.6 The shaded area contains q elements, λ times each.

Corollary 21.1. *If $d = 1$, then $i = m\lambda$ and since $i \le \lambda$ we have $i = 0$ or $i = \lambda$. Hence D is irreducible.* □

Now, given a $B[k, 1; v]$ design, we can construct a $B[k, 2; v]$ design which contains a 2-arc consisting of three blocks and is therefore irreducible.

Theorem 22. *Suppose that there exists a $B[k, 1; v]$ design, $k > 2$. Then there exists an irreducible $B[k, 2; v]$ design, unless $v = 7$ and $k = 3$.*

Proof. Let $D_1 = (X, \mathscr{B}_1)$ be any block design with $k \geq 3$, and let $\{1, 2, 3, v_4, \ldots, v_k\}$ be a block of D_1. Then without loss of generality, D_1 has blocks of the form shown in the left column of Table 4.8. Since $\lambda = 1$, the elements

$$1, 2, 3, 4, 5, 6, v_4, \ldots, v_k, s_4, \ldots, s_k, t_4, \ldots, t_k$$

are all distinct. Now element 5 occurs once each with 1, 2, 3, and again in at least one of the remaining $b - 3r + 2$ blocks, since $r \geq 4$. Suppose that one of these other blocks containing 5 is $\{5, 7, y, u_4, \ldots, u_k\}$, where clearly

$$\{7, y\} \cap \{1, 2, 3, 4, s_4, \ldots, s_k\} = \varnothing.$$

It may happen that $y = 6$.

Now we define the permutation β by

$$\beta = (34)\,(57)\,(6y)\,(s_4 t_4 u_4) \ldots (s_k t_k u_k),$$

except that if $y = 6$, the transposition $(6y)$ is trivial, and if $t_i = u_i$, then $(s_i t_i u_i)$ is replaced by $(s_i t_i)$. The blocks of $D_2 = D_1\beta$ are shown in the right column of Table 4.8.

If we take all the blocks in D_1 and in D_2, delete the four blocks which are starred in Table 4.8, and replace them with

$$134t_4 \ldots t_k,\ 175s_4 \ldots s_k,\ 645s_4 \ldots s_k,\ 637t_4 \ldots t_k,$$

Table 4.8 If $k = 3$, then s_i, t_i, u_i, v_i are not present

	$123v_4 \ldots v_k$		$124v_4 \ldots v_k$	
$r-1$	$145s_4 \ldots s_k$	$(*)$	$137t_4 \ldots t_k$	$(*)$
	$1 \ldots$		$1 \ldots$	
	$\vdots$		$\vdots$	
	1		1	
$r-1$	$2 \ldots$		$2 \ldots$	
	$\vdots$		$\vdots$	
	2		2	
$r-1$	$346t_4 \ldots t_k$	$(*)$	$43yu_4 \ldots u_k$	
	$3 \ldots$		$4 \ldots$	
	$\vdots$		$\vdots$	
	3		4	
$b-3r+2$	$57yu_4 \ldots u_k$		$756s_4 \ldots s_k$	$(*)$
	$\vdots$		$\vdots$	
	D_1		$D_2 = D_1\beta$	

then we have a design with $\lambda = 2$, which is irreducible since it contains the three blocks

$$123v_4 \ldots v_k, 124v_4 \ldots v_k, 134t_4 \ldots t_k,$$

which form a 2-arc. □

In general, such a design will have repeated blocks, but in some circumstances we may apply the algorithm of Fig. 4.1 to remove the blocks of $\mathscr{B}_2$ which are common to $\mathscr{B}_1$. For consider the set of elements

$$Y = \{1, 2, 3, 4, 5, 6, 7, y, s_4, \ldots, s_k, t_4, \ldots, t_k, u_4, \ldots, u_k, v_4, \ldots, v_k\}.$$

Since $\lambda = 1$ for each of the designs D_1 and D_2, no block chosen entirely from elements of Y can occur in both designs if $k \geq 4$. In other words, if

$$\{x_1, x_2, \ldots, x_k\} \in \mathscr{B}_1 \cap \mathscr{B}_2,$$

then at least one $x_i \notin Y$. Now $\mathscr{S}(x_i)$, the spread of x_i, has cardinality

$$|\mathscr{S}(x_i)| \leq k + 2(r-1) = k + 2(v-k)/(k-1)$$

and

$$|Y| \leq 8 + 4(k-3) = 4(k-1),$$

so

$$|\mathscr{S}(x_i) \cup Y| \leq k + 2(r-1) + 4(k-1) = (2v + 5k^2 - 11k + 4)/(k-1).$$

Hence

$$v - |S(x_i) \cup Y| \geq [v(k-3) - (5k^2 - 11k + 4)]/(k-1),$$

so that for v large enough, we can choose the permutation to remove the common block $\{x_1, x_2, \ldots, x_k\}$ from $\mathscr{B}_2$ without disturbing any of the starred blocks in Table 4.8. For example, if $k = 4$, then we need $v > 40$ to ensure that $v - |\mathscr{S}(x_i) \cup Y| > 0$. Thus we have

Corollary 22.1. *From any given balanced incomplete block design $D = (X, \mathscr{B})$, with $k \geq 4$, $\lambda = 1$, and $v > (5k^2 - 11k + 4)/(k-3)$, an irreducible quasi-multiple design with $\lambda = 2$ and no repeated blocks may be constructed.* □

4.7. Irreducible designs from difference set methods

As in Section 3.1, a family $\{B_1, B_2, \ldots, B_n\}$ of k-subsets of Z_v are said to be starter blocks (or supplementary difference sets) if developing them by addition modulo v in the usual way leads to a $B[k, \lambda; v]$ design. The family is a *minimal family* if no proper subset of $\{B_1, B_2, \ldots, B_n\}$ leads to a balanced incomplete block design when developed modulo v. In other words, the family of starter blocks is minimal precisely when it leads to a balanced incomplete block design which cannot be partitioned into a union of proper subdesigns, each consisting of complete cyclic sets of v blocks.

A design developed from starter blocks must have the cycle $(0, 1, 2, \ldots, v-1)$ as an automorphism, and the family is minimal if the design contains no subdesigns with the same automorphism. However, a design developed from a minimal family of starter blocks may be reducible as the following example shows.

Example 23. Let $B_1 = \{0, 1, 2\}$, $B_2 = \{0, 3, 6\}$, $B_3 = \{0, 4, 8\}$, $B_4 = \{0, 1, 3\}$, $B_5 = \{0, 3, 5\}$, $B_6 = \{0, 4, 5\}$ be starter blocks modulo 10. These generate a $B[3, 4; 10]$ design and are a minimal family. Nevertheless, the design can be decomposed into two $B[3, 2; 10]$ designs. The blocks of one of these designs are given in Table 4.9. □

Table 4.9. A $B[3, 2; 10]$ design contained in the $B[3, 4; 10]$ design of Example 23

012	036	159	124	035	045
234	258	371	346	257	267
456	470	593	568	479	489
678	692	715	780	691	601
890	814	937	902	813	823

Example 24. The sets $B_1 = \{0, 1, 3, 4\}$, $B_2 = \{0, 5, 7, 12\}$, $B_3 = \{0, 4, 6, 9\}$ form a family of starter blocks modulo 13 corresponding to a $B[4, 3; 13]$. Table 4.10 shows the number of times $\pm i$, $i = 1, \ldots, 6$ occurs as a difference of two elements from B_j, $j = 1, 2, 3$. Certainly the family is minimal, and in this case, the design is irreducible. For suppose it were reducible. Then it would contain a $B[4, 1; 13]$ with a_1, a_2 and a_3 blocks belonging to the cycles of B_1, B_2 and B_3 respectively.

Now the a_1, a_2 and a_3 blocks that together form a $B[4, 1; 13]$ design contain every unordered pair of elements just once and therefore constitute a family of starter blocks modulo 13. When developed cyclically these yield a $B[4, 13; 13]$ design which contains a_1 copies of the cycle from B_1, a_2 copies of the cycle from B_2 and a_3 copies of the cycle from B_3. By counting the number of times elements differing by $\pm i$, $i = 1, \ldots, 6$ occur together we obtain (from Table 4.10) the simultaneous equations listed in Table 4.11. The only solution these equations have is

$$a_1 = a_2 = a_3 = \frac{13}{3}$$

which for integral a_1, a_2, a_3 is impossible. Therefore the $B[4, 3; 13]$ design is irreducible. □

Table 4.10. Number of times that i occurs as a difference, modulo 13, of two elements from B_j

i	B_1	B_2	B_3
± 1	2	1	0
± 2	1	1	1
± 3	2	0	1
± 4	1	0	2
± 5	0	2	1
± 6	0	2	1

Table 4.11. Equations for number of blocks from each cycle in a subdesign

$$\begin{aligned} 2a_1 + a_2 \phantom{{}+a_3} &= 13, \\ a_1 + a_2 + a_3 &= 13, \\ 2a_1 \phantom{{}+a_2} + a_3 &= 13, \\ a_1 \phantom{{}+a_2} + 2a_3 &= 13, \\ 2a_2 + a_3 &= 13, \\ 2a_2 + a_3 &= 13. \end{aligned}$$

We may generalize to some extent as shown by

Theorem 25. *Let $B_1, B_2, \ldots, B_m$ be a family of k-subsets of Z_v which, as starter blocks modulo v, generate a $B[k, \lambda; v]$ design. If this design contains a $B[k, \mu; v]$ subdesign, then λ divides both $m\mu v$ and $m\mu k$.*

Proof. In any $B[k, \lambda; v]$ design the number of blocks is $\lambda v(v-1)/k(k-1)$. Let the subdesign have a_i blocks in the cycle of B_i. Then a count of blocks gives

$$a_1 + a_2 + \ldots + a_m = \frac{v(v-1)}{k(k-1)}\mu = \frac{v(v-1)}{k(k-1)}\lambda \cdot \frac{\mu}{\lambda} = mv\frac{\mu}{\lambda},$$

since each B_i provides v blocks in the $B[k, \lambda; v]$ design.

In any $B[k,\lambda;v]$ design each symbol occurs in exactly $\lambda(v-1)/(k-1)$ blocks. But in the cyclically generated design each B_i generates a cycle of blocks containing each symbol k times. Therefore $\lambda(v-1)/(k-1) = mk$. For the subdesign the number of incidences per symbol is $\mu(v-1)/(k-1)$; so $m\mu k/\lambda$ must be an integer. □

Note that it is not necessary to have a minimal family of starter blocks for Theorem 25 to apply.

This test quickly decides the case of Example 24 where for a subdesign to exist we must have $4.13.\mu/3$ an integer. The smallest value for μ is 3 corresponding to the original $B[4, 3; 13]$ design.

For the case of just two starter blocks, we can say much more.

Theorem 26. *Let B_1 and B_2 be minimal starter blocks, modulo v, for a $B[k, \lambda; v]$ design. Then any subdesign must be symmetric and must contain half the blocks of each of the cycles of B_1 and B_2.*

Proof. Suppose that $\pm i$ occurs α_i times as a difference of elements in B_1 and β_i times as a difference of elements from B_2. Then

$$\alpha_i + \beta_i = \lambda \quad \text{for} \quad i = 1, 2, \ldots, [v/2]. \tag{4.26}$$

Since neither of B_1 and B_2 taken separately is a difference set there is at least one value of i for which $\alpha_i \neq \beta_i$.

If a_1 blocks of the first cycle and a_2 blocks of the second cycle are used to make a $B[k, \mu; v]$ design then (by Theorem 25)

$$a_1 + a_2 = 2\mu v/\lambda. \tag{4.27}$$

This collection of blocks treated as starter blocks generates a $B[k, v\mu; v]$ design. If we count the number of occurrences of pairs of elements differing by $\pm i$ (modulo v), we find that

$$a_1\,\alpha_i + a_2\,\beta_i = \mu v. \tag{4.28}$$

If λ is eliminated between (4.26) and (4.27), then

$$(\alpha_i + \beta_i)(a_1 + a_2) = 2\mu v. \tag{4.29}$$

If v is eliminated from (4.28) and (4.29), we find that

$$(a_1 - a_2)(\alpha_i - \beta_i) = 0$$

for all i. But $\alpha_i \neq \beta_i$ for at least one i; so

$$a_1 = a_2 = \mu v/\lambda.$$

Furthermore, for any balanced incomplete block design, Fisher's inequality holds and demands that the number of blocks be not less than the number of elements. The blocks not taken from each cycle also form a design to which Fisher's inequality also applies. Therefore $a_1 = a_2 = v/2$. Thus the subdesign has equal numbers of elements and blocks and is therefore symmetric with $\mu = \frac{1}{2}\lambda$. □

Corollary 26.1. *If a $B[k, \lambda; v]$ design is generated by a minimal pair of starter blocks, B_1 and B_2, and λ is odd, then the design is not reducible.* □

It is also possible for a design generated from starter blocks to contain a subdesign with different numbers of blocks from different cycles.

Example 27. $B_1 = \{0, 1, 3\}$, $B_2 = \{0, 1, 4\}$, $B_3 = \{0, 3, 4\}$, $B_4 = \{0, 2, 4\}$ form a minimal family of starter blocks modulo 9, and generate a $B[3, 3; 9]$ design. This design has a $B[3, 1; 9]$ subdesign, consisting of three blocks from the first cycle (013, 346, 670), six from the third (145, 256, 478, 580, 712, 823) and three from the fourth (024, 357, 681). Since 014 belongs to no resolution class, the remaining blocks form an irreducible $B[3, 2; 9]$ design. □

4.8. References and comments

For a detailed account of automorphism groups of designs, see Biggs and White (1979). For symmetric BIBDs in particular, see Lander (1983). Theorem

6 is due to Parker (1957). Lemma 7 is generally known as Burnside's lemma; see Burnside (1911). Theorem 9, the Multiplier Theorem, appeared in Hall and Ryser (1951). Multipliers are also discussed in Lander (1983), Hall (1967), Mann (1965), and Baumert (1971).

The algorithm in Section 4.5 for making designs disjoint is that of Teirlinck (1977). Theorem 19 is due to D. G. Hoffman (1984). Kramer (1974) gave constructions for irreducible designs. Theorems 21 and 22 are from Billington (1982). Corollary 22.1 is an application by Lindner and A. P. Street (1986) of Teirlinck's algorithm. The techniques of Section 4.7 are those of Bhat-Nayak and Kane (1984) and Breach and A. P. Street (1985). For irreducible designs see also A. P. Street (1985), references there cited, and Bhat-Nayak and Kane (1986).

Exercises

4.1. Show that the permutations μ, ν, ρ, and τ, of Example 1, generate a group of order 168.

4.2. Consider the design D_1 of Table 4.2.

(i) Show that $\sigma = (24583796)$, $\tau = (123)(465)$, $\rho = (14)(25)(36)$ are all automorphisms of D_1.

(ii) Show that $G = \text{Aut}\, D_1$ is transitive on the set $X = (1, 2, 3, 4, 5, 6, 7, 8, 9)$, and that G_1 is transitive on the set $X \backslash \{1\}$.

(iii) Show that the pointwise stabilizer of $\{7, 8\}$ contains a (proper or improper) subgroup isomorphic to S_3.

(iv) Hence show that $432 \,|\, |G|$.

(v) Suppose that we delete from K_9 the four triples (or triangles) of the design D_1 which contain the vertex 9. This leaves a graph isomorphic to a copy of K_8 from which a one-factor has been deleted. Show that there are precisely 840 distinct designs, based on the set X, with the same parameters as D_1.

(vi) From (iv) and (v) show that $|G| = 432$ and that all designs with the same parameters as D_1 are isomorphic to D_1.

4.3. Consider the quadratic residues in $GF[103]$. Show that they form a difference set ($k = 51$, $\lambda = 25$), for which 2 is a multiplier.

4.4. Suppose that -1 is a multiplier of a difference set $D \subseteq Z_v$.

(i) Apply Corollary 6.3 to show that λ is even.

(ii) Next show that v is even.

(iii) Finally show that $\lambda \geq k-1$, and that D is a trivial difference set.

4.5. Show that an irreducible PBD based on the set $\{1, 2, 3, 4\}$ has $\lambda = 1$ or 2.

4.6. Construct an irreducible $B[4, 2; 13]$ design by starting from the difference set $\{0, 1, 3, 9\}$ in Z_{13} and applying the method of Corollary 22.1.

4.7. Consider the set $D = \{0, 1, 4, 14, 16\} \subseteq Z_{21}$. Show that D is a difference set, and that 4 is a multiplier of D. Find the permutation of elements and blocks induced by the multiplier 4.

4.8. (i) Show that the automorphisms of the field $GF[2^3]$ form a group of order 3 generated by the mapping $\alpha: x \to x^3$ for every $x \in GF[2^3]$.

(ii) Generalize the result of (i) by showing that the automorphisms of $GF[p^n]$ form a cyclic group of order n generated by the mapping $\alpha: x \to x^p$ for every $x \in GF[p^n]$.

5 Latin squares and triple systems

5.1. Basic properties of Latin squares

A *Latin square of order n* is an $n \times n$ array based on some set of n symbols, such that each row and each column contains each symbol exactly once. The nature of the symbols is immaterial; unless otherwise specified they are assumed to be $1, 2, \ldots, n$. Table 5.1 shows some small examples of Latin squares. For every positive integer n, there is a Latin square of order n. For instance if we define an $n \times n$ array to have as its (i, j) entry

$$l_{ij} = j - i + 1 \pmod{n}, \quad 1 \leq l_{ij} \leq n,$$

then we obtain a Latin square; A and C are examples of such squares of orders 3 and 5 respectively.

Table 5.1

$n = 2$

1	2
2	1

$n = 3$

A

1	2	3
3	1	2
2	3	1

B

1	3	2
3	2	1
2	1	3

$n = 4$

C

4	3	2	1
3	4	1	2
2	1	4	3
1	2	3	4

$n = 5$

D

1	2	3	4	5
5	1	2	3	4
4	5	1	2	3
3	4	5	1	2
2	3	4	5	1

These squares are often used in experimental design because the varieties are grouped into complete replicates in two different ways, by rows and by columns. The effect of this double grouping is to allow the elimination of errors caused by differences between rows and between columns, and the experimental material should be arranged so that these differences represent major sources of variation. For example, in agriculture such differences might be variations of soil fertility within a field, or variations in conditions along and across a greenhouse bench, though applications of such designs are certainly not restricted to agriculture.

One obvious way to construct a Latin square is to write down the group table of a group of the appropriate order. Suppose that we consider the two groups of order 4, Z_4 under addition (modulo 4) and $Z_2 \times Z_2$ under componentwise addition (modulo 2); their group tables are shown in Table 5.2.

By deleting the borders of the table and replacing symbols appropriately, we obtain the squares E and F respectively, which are Latin squares based on $\{1, 2, 3, 4\}$. The Latin property follows from the group axioms. Closure ensures that we obtain only n symbols from a group of order n. The existence of a group identity, e, and of an inverse for each element means that if (writing multiplicatively)

$$g_1 g_2 = g_1 g_3,$$

then

$$g_1^{-1}(g_1 g_2) = g_1^{-1}(g_1 g_3).$$

But the associative law ensures that

$$(g_1^{-1} g_1) g_2 = (g_1^{-1} g_1) g_3$$

so that

$$g_2 = e g_2 = e g_3 = g_3$$

and hence no element is repeated in any row of the table. Similarly, no element is repeated in any column, and the body of the table is a Latin square.

Table 5.2

$+_4$	0	1	2	3
0	0	1	2	3
1	1	2	3	0
2	2	3	0	1
3	3	0	1	2

Group table of Z_4

$+_2$	00	01	10	11
00	00	01	10	11
01	01	00	11	10
10	10	11	00	01
11	11	10	01	00

Group table of $Z_2 \times Z_2$

4	1	2	3
1	2	3	4
2	3	4	1
3	4	1	2

E (from Z_4)

4	1	2	3
1	4	3	2
2	3	4	1
3	2	1	4

F (from $Z_2 \times Z_2$).

But these axioms are stronger than they need to be, if all we are trying to do with them is construct a Latin square. If we simply require that a set Q be closed under an operation $*$, and that left and right cancellation laws should apply, so that

$$qq_1 = qq_2 \quad \text{implies that} \quad q_1 = q_2$$

and

$$q_1 q = q_2 q \quad \text{implies that} \quad q_1 = q_2,$$

then we have a *quasigroup* $(Q, *)$; its quasigroup table is a Latin square and any Latin square, equipped with a border row and column, defines a quasigroup.

Thus on the set $Q = \{1, 2, 3\}$, the squares A and B of Table 5.1 define the quasigroups Q_A and Q_B respectively, where in Q_A

$$1*1 = 2*2 = 3*3 = 1,$$
$$1*2 = 2*3 = 3*1 = 2,$$
$$1*3 = 2*1 = 3*2 = 3,$$

and in Q_B

$$1*1 = 2*3 = 3*2 = 1,$$
$$1*3 = 2*2 = 3*1 = 2,$$
$$1*2 = 2*1 = 3*3 = 3.$$

Permuting rows, columns or symbols of a Latin square preserves the Latin property, but we make the convention that the border row and column are in natural order.

The properties of the Latin square reflect those of the quasigroup: Q_A has a left identity, 1, since row 1 of A is in its natural order, but no right identity since no column is in its natural order; Q_B has neither a left nor a right identity. B (but not A) is symmetric, corresponding to the fact that Q_B (but not Q_A) is commutative. The main diagonal of B contains the elements 1, 2, 3 in their natural order, corresponding to the *idempotent* property of Q_B, that is, in Q_B $x*x = x$. In general, a Latin square L_n of order n in which

$$l_{ii} = i, \; i = 1, 2, \ldots, n,$$

is also said to be idempotent, since it defines an idempotent quasigroup.

A set of positions in a Latin square of order n is said to be a *transversal* if it contains one position in each row, one position in each column, and if the positions contain between them each member of the set $\{1, 2, \ldots, n\}$ precisely once. A *transversal square* is a Latin square which contains a transversal. Certainly, an idempotent square has a transversal, namely its main diagonal. But a transversal square need not be idempotent. For example A, in Table 5.1, has three transversals: positions $(1,1)$, $(2,3)$, $(3,2)$ contain symbols 1, 2, 3 respectively; so do positions $(2,2)$, $(3,1)$, $(1,3)$ and again positions $(3,3)$, $(1,2)$, $(2,1)$. B, D (in Table 5.1) and F (in Table 5.2) are also transversal squares; the square of order 2 (in Table 5.1) and E (in Table 5.2) are not. Given a transversal square of order n, we can permute its columns to bring the transversal positions onto the main diagonal, and then permute the elements of the square so that the diagonal elements are in their natural order. In other words, from a transversal square, we can obtain by permutation one which is idempotent.

From a transversal square L of order n, we may construct a Latin square M of order $n+1$ in the following way. Suppose that a transversal of L occupies positions

$$(1, j_1), (2, j_2), \ldots, (n, j_n).$$

Then we define the entries of M by:

$$m_{i,j_i} = n+1, \quad \text{for} \quad i = 1, \ldots, n;$$
$$m_{n+1,n+1} = n+1;$$
$$m_{i,n+1} = l_{i,j_i} = m_{n+1,j_i}, \quad \text{for } i = 1, \ldots, n;$$
$$m_{ij} = l_{ij} \quad \text{for all other pairs } (i, j).$$

Since the entries replaced by $n+1$ included all the elements $1, 2, \ldots, n$, M is a Latin square, called the *prolongation* of L. Notice that if L is idempotent, and some transversal in L, other than the diagonal, is replaced, then M is still idempotent. In other words, a Latin square of order n with two transversals gives a transversal square of order $n+1$.

The Latin square C in Table 5.1 is a prolongation of the square B, using the main diagonal of B as the transversal.

Latin squares are closely connected with the triple systems or $B[3, \lambda; v]$ designs which we discuss in more detail in the next section.

5.2. Existence criteria for triple systems

We know from Theorem 1.2.1 that if a block design exists with parameters v, b, r, k, then

$$vr = bk \tag{5.1}$$

and that if it is balanced, with index λ, then

$$\lambda(v-1) = r(k-1), \tag{5.2}$$

so that (5.1) and (5.2) are necessary conditions for the existence of a BIBD. But we also know from Example 2.4.11, that no $SB[7, 2; 22]$ design exists, although its parameters satisfy (5.1) and (5.2), so these conditions are not sufficient.

If $k = 2$ and we have a $B[2, \lambda; v]$ design, then by eqn (5.2)

$$r = \lambda(v-1)$$

and by eqn (5.1)

$$b = \frac{vr}{2} = \frac{\lambda v(v-1)}{2} = \lambda\binom{v}{2}.$$

There is only one way to construct a design with these parameters: simply write down λ copies each of the $\binom{v}{2}$ subsets of size 2 chosen from the set of varieties. Thus if $k = 2$, conditions (5.1) and (5.2) are also sufficient for the existence of the design.

If $k = 3$, and we have a $B[3, \lambda; v]$ design, then by eqn (5.2)

$$r = \lambda(v-1)/2$$

so that if λ is odd, then v must be odd, but if λ is even, this equation gives no information about v. Similarly by eqn (5.1)

$$b = \frac{vr}{3} = \frac{\lambda v(v-1)}{6},$$

and the restrictions on v are conveniently stated as follows, where (x, y) denotes the greatest common divisor of x and y.

(i) If $\lambda \equiv 1$ or 5 (mod 6), so that $(\lambda, 6) = 1$, then $v \equiv 1$ or 3 (mod 6).
(ii) If $\lambda \equiv 2$ or 4 (mod 6), so that $(\lambda, 6) = 2$, then $v \equiv 0$ or 1 (mod 3).
(iii) If $\lambda \equiv 3$ (mod 6), so that $(\lambda, 6) = 3$, then $v \equiv 1$ (mod 2).
(iv) If $\lambda \equiv 0$ (mod 6), so that $(\lambda, 6) = 6$, then v is unrestricted.

Also we need $v \geq 3$.

If $(\lambda, 6) = d$, say, and $\lambda = md$ for some integer m, then we can certainly form a $B[3, \lambda; v]$ design by simply taking an m-multiple of a $B[3, d; v]$ design, that is, by taking m copies of each block. Thus to show that, for $v \geq 3$, the necessary conditions (i) to (iv) are also sufficient, we need only to construct the following designs:

a $B[3, 1; v]$ for all $v \equiv 1$ or 3 (mod 6);
a $B[3, 2; v]$ for all $v \equiv 0$ or 4 (mod 6);
a $B[3, 3; v]$ for all $v \equiv 5$ (mod 6);
a $B[3, 6; v]$ for all $v \equiv 2$ (mod 6), $v \geq 8$.

We construct each of these designs by constructing first an appropriate Latin square and from it the triple system.

Construction 1: $v = 6n + 3$, $\lambda = 1$.
Let L be a symmetric idempotent Latin square of order $2n + 1$, such as

$$L = [l_{ij}],\ l_{ij} = (n+1)(i+j) \,(\text{mod } 2n+1),\ 1 \leq l_{ij} \leq 2n+1.$$

Notice here that $2(n+1) \equiv 1$ (mod $2n + 1$), so that $n + 1 = 2^{-1}$ in the ring Z_{2n+1}. In other words, we are effectively finding l_{ij} by averaging i and j in Z_{2n+1}. Thus L defines a symmetric, idempotent quasigroup $(Q, *)$ where $Q = \{1, 2, \ldots, 2n+1\}$.

We now take three copies of each element of $\{1, 2, \ldots, 2n+1\}$, and subscript them accordingly. Thus our set of $v = 6n + 3$ varieties consists of

$$\{1_1, 1_2, 1_3, 2_1, 2_2, 2_3, \ldots, (2n+1)_1, (2n+1)_2, (2n+1)_3\}.$$

The blocks of our $B[3, 1; 6n+3]$ design are given in Table 5.3; see also Fig. 5.1.

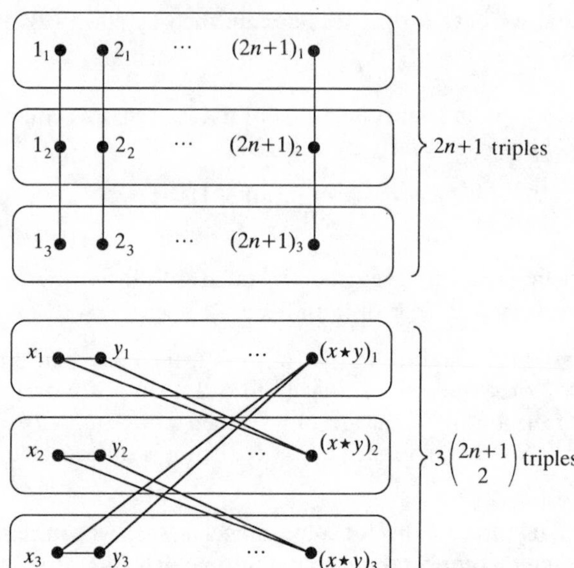

Fig. 5.1. The triples of Construction 1.

Table 5.3

$$x_1, x_2, x_3, \qquad 1 \le x \le 2n+1;$$

$$\left.\begin{array}{l} x_1, y_1, (x * y)_2 \\ x_2, y_2, (x * y)_3 \\ x_3, y_3, (x * y)_1 \end{array}\right\} \; 1 \le x < y \le 2n+1.$$

We have altogether

$$b = 2n+1+3\binom{2n+1}{2} = 6n^2+5n+1 = \frac{(6n+3)(6n+2)}{6} \text{ blocks,}$$

containing each element

$$r = 1+2n+n = (6n+2)/2 \text{ times,}$$

and we need only check that the design is balanced.

Two elements x_i, x_j occur together in the block $\{x_1, x_2, x_3\}$; two elements x_i, y_i occur together in the block $\{x_i, y_i, (x * y)_{i+1 \,(\mathrm{mod}\, 3)}\}$; any other pair of elements has the form u_h, v_j, where $u \neq v$, $h \neq j$. This means that either $h = j+1 \pmod 3$ or $j = h+1 \pmod 3$. Renaming the elements as x_i, $z_{i+1 (\mathrm{mod}\, 3)}$, we recall that the Latin property of the square L guarantees the existence of a unique y such that $x * y = z$, and that the elements therefore occur together in the block $\{x_i, y_i, (x*y)_{i+1 \pmod 3}\}$. Thus each pair of elements occur together in at least one block.

But each block contains 3 pairs, and thus the whole design contains $3(6n^2+5n+1)$ pairs, that is, $\binom{6n+3}{2}$ pairs. But this is the total number of available pairs, and since each occurs at least once, none can be repeated. Thus the design is balanced, with $\lambda = 1$. □

Construction 2: $v = 6n+1, \lambda = 1$.
A Latin square which is idempotent and symmetric must have odd order (see Exercise 5.1); for this construction we need a symmetric Latin square of order $2n$, and we require it to be *half-idempotent*, that is, its diagonal has entries $1, 2, \ldots, n, 1, 2, \ldots, n$. We take the group table for the addition of integers (modulo $2n$) and rename its elements to give $M = [m_{ij}]$, in the following way:

$$\text{if } i+j \text{ is even, then } 1 \le m_{ij} \le n,\ m_{ij} = \tfrac{1}{2}(i+j) \pmod n;$$
$$\text{if } i+j \text{ is odd, then } n+1 \le m_{ij} \le 2n,\ m_{ij} = \tfrac{1}{2}(i+j+1) \pmod n).$$

M defines a symmetric quasigroup $\{Q, *\}$ on $Q = \{1, 2, \ldots, 2n\}$.

We now take three copies each of the elements of Q, labelled by subscripts, together with an additional element, ∞, so that our set of $v = 6n+1$ varieties consists of

$$\{1_1, 1_2, 1_3, 2_1, 2_2, 2_3, \ldots, (2n)_1, (2n)_2, (2n)_3, \infty\}.$$

The blocks of our $B[3, 1; 6n+1]$ design are given in Table 5.4; see also Fig. 5.2.

Table 5.4

$$x_1, x_2, x_3, \qquad 1 \le x \le n;$$

$$\left.\begin{array}{l} \infty, x_1, (x-n)_2 \\ \infty, x_2, (x-n)_3 \\ \infty, x_3, (x-n)_1 \end{array}\right\}, \quad n+1 \le x \le 2n;$$

$$\left.\begin{array}{l} x_1, y_1, (x*y)_2 \\ x_2, y_2, (x*y)_3 \\ x_3, y_3, (x*y)_1 \end{array}\right\}, \quad 1 \le x < y \le 2n.$$

Checking the parameters of this design is left as an exercise; the proof follows that for Construction 1. □

Example 3. Let $n = 2$. The group table for the addition of integers (modulo 4) is given in Table 5.5. (Note: in Table 5.2, we have the same group table but with the elements represented as 0, 1, 2, 3 instead of 1, 2, 3, 4.) The Latin square, M, is obtained as above:

$$\text{if } i+j \equiv 2 \pmod 4, \text{ then } m_{ij} = 1; \quad \text{if } i+j \equiv 4 \pmod 4, \text{ then } m_{ij} = 2;$$
$$\text{if } i+j \equiv 1 \pmod 4, \text{ then } m_{ij} = 3; \quad \text{if } i+j \equiv 3 \pmod 4, \text{ then } m_{ij} = 4.$$

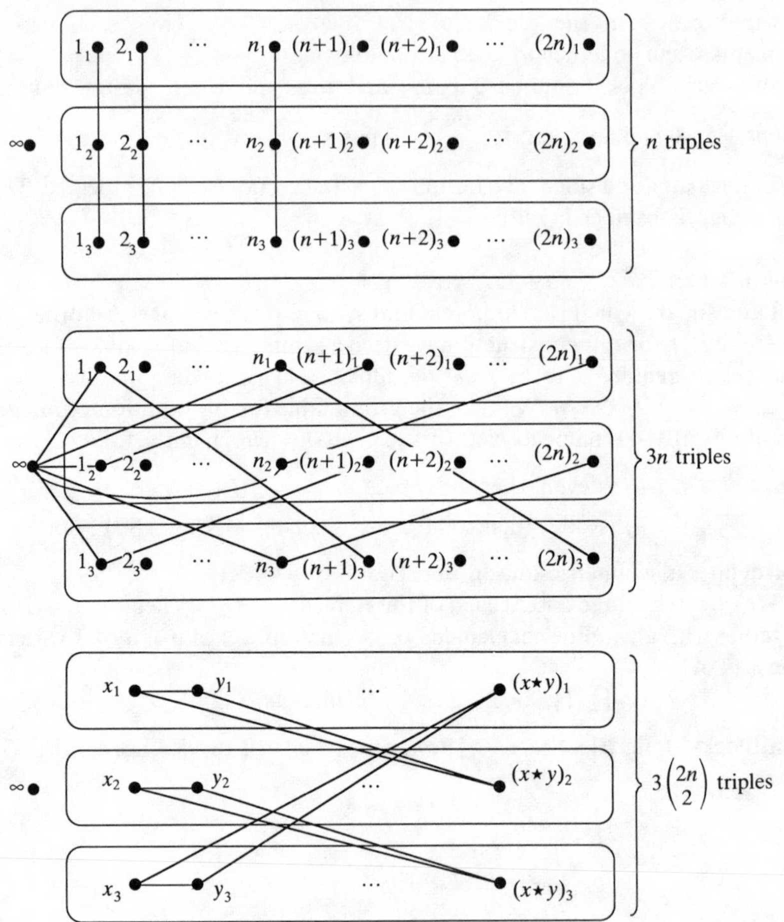

Fig. 5.2. The triples of Construction 2.

Then Construction 2 gives a $B[3, 1; 13]$ design with blocks as in Table 5.6. □

Table 5.5

$+_4$	1	2	3	4
1	2	3	4	1
2	3	4	1	2
3	4	1	2	3
4	1	2	3	4

$$M = \begin{bmatrix} 1 & 4 & 2 & 3 \\ 4 & 2 & 3 & 1 \\ 2 & 3 & 1 & 4 \\ 3 & 1 & 4 & 2 \end{bmatrix}$$

Table 5.6

$1_1, 1_2, 1_3,$	$1_1, 2_1, 4_2,$	$2_1, 3_1, 3_2,$
$2_1, 2_2, 2_3,$	$1_2, 2_2, 4_3,$	$2_2, 3_2, 3_3,$
	$1_3, 2_3, 4_1,$	$2_3, 3_3, 3_1,$
$\infty, 3_1, 1_2,$	$1_1, 3_1, 2_2,$	$2_1, 4_1, 1_2,$
$\infty, 3_2, 1_3,$	$1_2, 3_2, 2_3,$	$2_2, 4_2, 1_3,$
$\infty, 3_3, 1_1,$	$1_3, 3_3, 2_1,$	$2_3, 4_3, 1_1,$
$\infty, 4_1, 2_2,$	$1_1, 4_1, 3_2,$	$3_1, 4_1, 4_2,$
$\infty, 4_2, 2_3,$	$1_2, 4_2, 3_3,$	$3_2, 4_2, 4_3,$
$\infty, 4_3, 2_1,$	$1_3, 4_3, 3_1,$	$3_3, 4_3, 4_1,$

To deal with the case $\lambda = 2$, we need an idempotent Latin square of every order $n > 2$. For odd order $2n+1$, we have already constructed such a square, namely L in Construction 1, which also happens to be symmetric. For even order $2n$, we use this construction to make first an idempotent Latin square L, of order $2n-1$, for which

$$l_{ij} = n(i+j) \pmod{2n-1},\ 1 \le l_{ij} \le 2n-1;$$

then we construct the prolongation, M, of L, using the transversal in positions $(1, 2), (2, 3), \ldots, (2n-2, 2n-1), (2n-1, 1)$ of L. The new square, M, is also idempotent.

Example 4. The idempotent symmetric Latin square L, of order 5, and the idempotent square M, of order 6, constructed as above are given in Table 5.7. □

Table 5.7

L:

1	4	2	5	3
4	2	5	3	1
2	5	3	1	4
5	3	1	4	2
3	1	4	2	5

L with elements replaced:

1	6	2	5	3
4	2	6	3	1
2	5	3	6	4
5	3	1	4	6
6	1	4	2	5

L extended to M:

1	6	2	5	3	4
4	2	6	3	1	5
2	5	3	6	4	1
5	3	1	4	6	2
6	1	4	2	5	3
3	4	5	1	2	6

Construction 5: $v = 3n, \lambda = 2$.
For $n = 2$, we have no idempotent Latin square, but we have already constructed a $B[3, 2; 6]$ design (Example 2.3.8).

For $n > 2$, let L be an idempotent Latin square of order n, defining a quasigroup $(Q, *)$, and consider the symbols $\{x_1\}, \{x_2\}, \{x_3\}, 1 \le x \le n$, as before. A $B[3, 2; 3n]$ design on these $v = 3n$ varieties is given by the blocks in Table 5.8. Checking the design is again left as an exercise. □

Table 5.8

$$x_1, x_2, x_3, \qquad 1 \le x \le n, \quad \text{taken twice each;}$$

$$\left.\begin{matrix} x_1, y_1, (x*y)_2 \\ x_2, y_2, (x*y)_3 \\ x_3, y_3, (x*y)_1 \end{matrix}\right\}, \qquad 1 \le x \le n, \quad 1 \le y \le n, \quad x \ne y.$$

Example 6. Table 5.9 shows idempotent Latin squares L, M for the case $n = 4$. Table 5.10 shows the blocks obtained from M, by Construction 5, forming a $B[3, 2; 12]$ design. □

Table 5.9

1	3	2	
3	2	1	
2	1	3	
	L		
1	4	2	3
3	2	4	1
4	1	3	2
2	3	1	4
	M		

Table 5.10

$1_1, 1_2, 1_3$	$1_1, 2_1, 4_2$	$1_2, 2_2, 4_3$	$1_3, 2_3, 4_1$
$2_1, 2_2, 2_3$	$1_1, 3_1, 2_2$	$1_2, 3_2, 2_3$	$1_3, 3_3, 2_1$
$3_1, 3_2, 3_3$	$1_1, 4_1, 3_2$	$1_2, 4_2, 3_3$	$1_3, 4_3, 3_1$
$4_1, 4_2, 4_3$	$2_1, 1_1, 3_2$	$2_2, 1_2, 3_3$	$2_3, 1_3, 3_1$
	$2_1, 3_1, 4_2$	$2_2, 3_2, 4_3$	$2_3, 3_3, 4_1$
	$2_1, 4_1, 1_2$	$2_2, 4_2, 1_3$	$2_3, 4_3, 1_1$
$1_1, 1_2, 1_3$	$3_1, 1_1, 4_2$	$3_2, 1_2, 4_3$	$3_3, 1_3, 4_1$
$2_1, 2_2, 2_3$	$3_1, 2_1, 1_2$	$3_2, 2_2, 1_3$	$3_3, 2_3, 1_1$
$3_1, 3_2, 3_3$	$3_1, 4_1, 2_2$	$3_2, 4_2, 2_3$	$3_3, 4_3, 2_1$
$4_1, 4_2, 4_3$	$4_1, 1_1, 2_2$	$4_2, 1_2, 2_3$	$4_3, 1_3, 2_1$
	$4_1, 2_1, 3_2$	$4_2, 2_2, 3_3$	$4_3, 2_3, 3_1$
	$4_1, 3_1, 1_2$	$4_2, 3_2, 1_3$	$4_3, 3_3, 1_1$

Construction 7: $v = 3n + 1, \lambda = 2$.
For $n = 2$, we have as before no idempotent Latin square, but we have already constructed a $B[3, 2; 7]$ design (Example 2.3.9).

For $n > 2$, let L be an idempotent Latin square of order n and consider the symbols $\{x_1\}, \{x_2\}, \{x_3\}, 1 \le x \le n$, together with one additional symbol, ∞. A $B[3, 2; 3n + 1]$ design on these $v = 3n + 1$ varieties is given by the triples in Table 5.11. The design has

$$b = 4n + 3n(n-1) = 2(3n+1)3n/6 \text{ blocks,}$$

and each element occurs in

$$r = 3 + 2(n-1) + (n-1) = 3n \text{ blocks.}$$

Each pair of elements occurs in at least two blocks: x_i, x_j in the blocks $\{x_1, x_2, x_3\}$ and $\{x_i, x_j, \infty\}$; x_i, y_i in the blocks $\{x_i, y_i, (x*y)_{i+1\,(\mathrm{mod}\,3)}\}$ and $\{y_i, x_i, (y*x)_{i+1\,(\mathrm{mod}\,3)}\}$; any other pair of elements, renamed x_i, $z_{i+1\,(\mathrm{mod}\,3)}$, in the blocks $\{x_i, y_i, (x*y)_{i+1\,(\mathrm{mod}\,3)}\}$ where $x*y=z$, and $\{w_i, x_i, (w*x)_{i+1\,(\mathrm{mod}\,3)}\}$ where $w*x=z$. The total number of pairs in the design is $3b = 3n(3n+1)$; the total number of distinct pairs chosen from $3n+1$ elements is $(3n+1)3n/2$. Since each pair occurs at least twice, no pair can occur more than twice. Hence the design is balanced with $\lambda = 2$. □

Table 5.11

$$\left.\begin{matrix} x_1 & x_2 & x_3 \\ \infty & x_1 & x_2 \\ \infty & x_2 & x_3 \\ \infty & x_3 & x_1 \end{matrix}\right\} 1 \le x \le n, \qquad \left.\begin{matrix} x_1 & y_1 & (x*y)_2 \\ x_2 & y_2 & (x*y)_3 \\ x_3 & y_3 & (x*y)_1 \end{matrix}\right\} 1 \le x \le n,\ 1 \le y \le n,\ x \ne y.$$

Construction 8: $v = 6n+5$, $\lambda = 3$.
Here we take as blocks all the triples in arithmetic progression (modulo v); that is,

$$\{x, x+y, x+2y\},\ 1 \le x \le 6n+5,\ 1 \le y \le 3n+2,$$

where addition is taken modulo v. There are $b = (6n+5)(3n+2)$ such blocks. Each element occurs exactly $r = 3(3n+2)$ times. The number of pairs chosen from the v elements is $(6n+5)(3n+2)$, and the number of pairs covered by all the blocks is $3(6n+5)(3n+2)$. Now any two elements i, j, that differ by y occur together in the three blocks

$$\{i, i+y=j, i+2y\},\ \{i-y,\ i,\ i+y=j\},\ (j,\ j+(3n+2)y,\ j+(6n+4)y=j-y=i\},$$

so no pair can occur less than three times, hence $\lambda = 3$. □

Construction 9: $v = 3n+2$, $\lambda = 6$.
We start from an idempotent Latin square, L, of order n and base a $B[3, 6; 3n+2]$ design on the symbols

$$\{1_1, 1_2, 1_3, 2_1, 2_2, 2_3, \ldots, n_1, n_2, n_3, \infty_1, \infty_2\}.$$

Its triples are given in Table 5.12. Checking is again left as an exercise. □

From Constructions 1, 2, 5, 7, 8, and 9, and the m-multiple construction, we have the following result.

Theorem 10. *A $B[3, \lambda; v]$ design exists if and only if*

$$\lambda(v-1) \equiv 0 \pmod 2 \textit{ and } \lambda v(v-1) \equiv 0 \pmod 6. \qquad \square$$

Table 5.12

$$\left.\begin{array}{lll} x_1 & y_1 & (x*y)_2 \\ x_2 & y_2 & (x*y)_3 \\ x_3 & y_3 & (x*y)_1 \end{array}\right\} 1 \leq x \leq n,\ 1 \leq y \leq n,\ x \neq y, \text{ taken three times each;}$$

$$\left.\begin{array}{lllllll} \infty_1 & x_1 & x_2 & \quad & \infty_2 & x_1 & x_2 \\ \infty_1 & x_2 & x_3 & \quad & \infty_2 & x_2 & x_3 \\ \infty_1 & x_3 & x_1 & \quad & \infty_2 & x_3 & x_1 \end{array}\right\} \quad 2 \leq x \leq n, \text{ taken three times each;}$$

$$\begin{array}{lllllllllll} \infty_1 & \infty_2 & 1_1 & \quad & \infty_1 & 1_1 & 1_2 & \quad & \infty_2 & 1_1 & 1_2 \\ \infty_1 & \infty_2 & 1_2 & \quad & \infty_1 & 1_2 & 1_3 & \quad & \infty_2 & 1_2 & 1_3 \\ \infty_1 & \infty_2 & 1_3 & \quad & \infty_1 & 1_3 & 1_1 & \quad & \infty_2 & 1_3 & 1_1 \end{array} \qquad 1_1\ 1_2\ 1_3 \left.\vphantom{\begin{array}{l} a \\ b \end{array}}\right\} \begin{array}{l} \text{taken} \\ \text{twice each} \end{array}$$

5.3. Simple triple systems

Several of the constructions in the previous section lead to designs with repeated blocks. A design with no repeated blocks is said to be *simple* and such designs are of particular interest in some contexts; see Chapter 8. In this section we consider some constructions for simple $B[3, 2; v]$ designs.

Construction 11: $v = 6n$, $\lambda = 2$.
For $n = 1$, $v = 6$, we already have a simple triple system with $\lambda = 2$; see Example 2.3.8.

For $n = 2$, $v = 12$, we use first the simple $B[3, 2; 4]$ design, namely

123, 124, 134, 234,

together with corresponding designs based on the sets {5, 6, 7, 8} and {9, A, B, C} respectively. Remaining blocks must have one element in each of these three sets. The design is given in Table 5.13.

For $n \geq 3$ we start from an idempotent Latin square of order n. For each x, $1 \leq x \leq n$, we take six symbols x_i, $0 \leq i \leq 5$, and construct a simple $B[3, 2; 6]$ design, based on the set $\{x_0, x_1, x_2, x_3, x_4, x_5\}$; this gives altogether $10n$ blocks. The remaining $12n(n-1)$ blocks are now defined in terms of the Latin square. The blocks are listed in Table 5.14. □

Table 5.13. A simple $B[3, 2; 12]$ design

123	567	9AB	159	25A	35B	45C	15B	25C	359	45A
124	568	9AC	16A	26B	36C	469	16C	269	36A	46B
134	578	9BC	17B	27C	379	47A	179	27A	37B	47C
234	678	ABC	18C	289	38A	48B	18A	28B	38C	489

Construction 12: $v = 6n + 3$, $\lambda = 2$.
We take an idempotent symmetric Latin square of order $2n + 1$, and for each x,

$1 \le x \le 2n+1$, we take three symbols x_i, $i = 1, 2, 3$. Blocks are given in Table 5.15; checking the properties of the design is left as an exercise. □

Table 5.14. A simple $B[3,2;6n]$ design, $n \ge 3$

$$\left.\begin{array}{lll} x_1 & x_2 & x_3 \\ x_1 & x_3 & x_4 \\ x_1 & x_4 & x_5 \\ x_1 & x_5 & x_0 \\ x_1 & x_2 & x_0 \\ x_2 & x_3 & x_5 \\ x_2 & x_4 & x_5 \\ x_2 & x_4 & x_0 \\ x_3 & x_4 & x_0 \\ x_3 & x_5 & x_0 \end{array}\right\} 1 \le x = n. \qquad \left.\begin{array}{lll} x_i & y_i & (x*y)_{i+2} \\ x_i & y_i & (x*y)_{i+3} \\ x_i & y_{i+1} & (x*y)_{i+2} \\ x_i & y_{i+1} & (x*y)_{i+3} \\ x_{i+1} & y_i & (x*y)_{i+2} \\ x_{i+1} & y_i & (x*y)_{i+3} \\ x_{i+1} & y_{i+1} & (x*y)_{i+2} \\ x_{i+1} & y_{i+1} & (x*y)_{i+3} \end{array}\right\} \begin{array}{l} 1 \le x < y \le n, \\ i = 0, 2, 4. \end{array}$$

Table 5.15. A simple $B[3,2;6n+3]$ design, $n \ge 1$; addition and subtraction are taken modulo $2n+1$

$$\left.\begin{array}{lll} x_1 & x_2 & x_3 \\ x_1 & x_2 & (x+1)_3 \end{array}\right\} 1 \le x \le 2n+1.$$

$$\left.\begin{array}{lll} x_1 & y_1 & (x*y)_2 \\ x_1 & y_1 & ((x*y)+1)_3 \\ x_2 & y_2 & (x*y)_3 \\ x_2 & y_2 & (x*y)_1 \\ x_3 & y_3 & (x*y)_1 \\ x_3 & y_3 & ((x*y)-1)_2 \end{array}\right\} \quad 1 \le x < y \le 2n+1.$$

Example 13. We take $n = 1$, $v = 9$, $\lambda = 2$, and start from the square B of Table 5.1. The blocks of the simple $B[3,2;9]$ design are given in Table 5.16. □

Table 5.16. A simple $B[3, 2; 9]$ design

$1_1\ 1_2\ 1_3$	$1_1\ 2_1\ 3_2$	$1_1\ 3_1\ 2_2$	$2_1\ 3_1\ 1_2$
$1_1\ 1_2\ 2_3$	$1_1\ 2_1\ 1_3$	$1_1\ 3_1\ 3_3$	$2_1\ 3_1\ 2_3$
$2_1\ 2_2\ 2_3$	$1_2\ 2_2\ 3_3$	$1_2\ 3_2\ 2_3$	$2_2\ 3_2\ 1_3$
$2_1\ 2_2\ 3_3$	$1_2\ 2_2\ 3_1$	$1_2\ 3_2\ 2_1$	$2_2\ 3_2\ 1_1$
$3_1\ 3_2\ 3_3$	$1_3\ 2_3\ 3_1$	$1_3\ 3_3\ 2_1$	$2_3\ 3_3\ 1_1$
$3_1\ 3_2\ 1_3$	$1_3\ 2_3\ 2_2$	$1_3\ 3_3\ 1_2$	$2_3\ 3_3\ 3_2$

The only remaining case is $v \equiv 1 \pmod 3$. We give a construction which depends on the application of a *skew idempotent* Latin square; that is, an idempotent Latin square $L = [l_{ij}]$ such that $l_{ij} \ne l_{ji}$ unless $i = j$. No such square can exist of order 2 or 3; for all larger orders we proceed to construct such squares.

If n is odd, we define a Latin square $L = [l_{ij}]$ of order n by

$$l_{ij} = 2i - j \pmod n,\ 1 \le l_{ij} \le n. \tag{5.3}$$

Certainly L is idempotent. It is also skew unless for some $i, j, i \ne j$, we have

$$2i - j = 2j - i \pmod n,$$

which can only happen if $n = 3m$, m odd. For $m = 3$, $n = 9$ we have the skew idempotent square of Table 5.17. For $m > 3$, we factor

$$n = 3m = 3^a b,$$

where $a \ge 1$, $(b, 6) = 1$. By (5.3), we already have a skew idempotent square S, say, of order b, and by Construction 1, we have symmetric idempotent squares of orders 3^a and b respectively, say A and B. The required skew idempotent square L of order $3^a b$ is a $3^a \times 3^a$ array of $b \times b$ subsquares:

the ith diagonal subsquare is $S + (i-1)b$;

if $i \ne j$, then let $k = i - j \pmod{3^a}$, $0 \le k \le 3^a - 1$, and the (i, j) subsquare is $((B + k) \pmod b) + (a_{ij} - 1)b$.

Example 14 shows the construction for the smallest case.

Table 5.17. A skew idempotent Latin square of order 9

1	8	7	6	9	3	5	4	2
4	2	9	8	3	7	1	5	6
5	6	3	2	7	9	4	1	8
3	5	6	4	8	2	9	7	1
2	9	4	7	5	1	8	6	3
7	1	8	5	2	6	3	9	4
9	4	2	1	6	8	7	3	5
6	3	5	9	1	4	2	8	7
8	7	1	3	4	5	6	2	9

Example 14. Let $n = 15$, so that $a = 1$, $b = 5$. Then S, A, B are idempotent Latin squares of orders 5, 3, 5 respectively, S is skew, A and B are symmetric; see Table 5.18. L is the skew idempotent Latin square of order 15 constructed from S, A and B by the method given above. □

We now have skew idempotent Latin squares of all odd orders. If $n = 4$, we have the skew idempotent square L shown in Table 5.19. If n is even, $n \ge 6$, $n \ne 10$, we take a skew idempotent Latin square, L, of order $n-1$, constructed as above, and we construct its prolongation, M, of order n in the following way. If $n - 1 \not\equiv 0 \pmod 3$, then L is given by eqn (5.3), and we work with the transversal in positions

$$(1, 2), (2, 3), \ldots, (n-2, n-1), (n-1, 1).$$

Table 5.18. Construction of a skew idempotent Latin square of order 15

1	5	4	3	2
3	2	1	5	4
5	4	3	2	1
2	1	5	4	3
4	3	2	1	5

S

1	3	2
3	2	1
2	1	3

A

1	4	2	5	3
4	2	5	3	1
2	5	3	1	4
5	3	1	4	2
3	1	4	2	5

B

1	5	4	3	2	D	B	E	C	F	7	A	8	6	9
3	2	1	5	4	B	E	C	F	D	A	8	6	9	7
5	4	3	2	1	E	C	F	D	B	8	6	9	7	A
2	1	5	4	3	C	F	D	B	E	6	9	7	A	8
4	3	2	1	5	F	D	B	E	C	9	7	A	8	6
C	F	D	B	E	6	A	9	8	7	3	1	4	2	5
F	D	B	E	C	8	7	6	A	9	1	4	2	5	3
D	B	E	C	F	A	9	8	7	6	4	2	5	3	1
B	E	C	F	D	7	6	A	9	8	2	5	3	1	4
E	C	F	D	B	9	8	7	6	A	5	3	1	4	2
8	6	9	7	A	2	5	3	1	4	B	F	E	D	C
6	9	7	A	8	5	3	1	4	2	D	C	B	F	E
9	7	A	8	6	3	1	4	2	5	F	E	D	C	B
7	A	8	6	9	1	4	2	5	3	C	B	F	E	D
A	8	6	9	7	4	2	5	3	1	E	D	C	B	F

L

If $n-1=3^a b$, where $a \geq 1$, $(b, 6) = 0$, then L is a $3^a \times 3^a$ array of $b \times b$ submatrices, and we work with the transversal consisting of the main diagonals of the submatrices in positions $(1, 2), (2, 3), \ldots, (b-1, b), (b, 1)$.

Table 5.19

1	4	2	3
3	2	4	1
4	1	3	2
2	3	1	4

From these skew idempotent Latin squares, we can construct our remaining class of simple triple systems.

For $n = 2$ and 3, we have no skew idempotent squares available. However, we already have a simple $B[3, 2; 7]$ design (Example 2.3.9) and a simple $B[3, 2; 10]$ design (Example 4.7.23).

For $n \geq 4$, we take a skew idempotent square of order n, and use Construction 7. The fact that the underlying Latin square is skew ensures that no block is repeated.

5.4. References and comments

The book by Dénes and Keedwell (1974) gives a detailed treatment of Latin squares. The sufficiency of the condition $v \equiv 1$ or 3 (mod 6) for Steiner triple systems (that is, with $\lambda = 1$) was proved by several people working independently last century; see Moore (1893) and Reiss (1858). Bose (1939) settled the case $\lambda = 2$ and Hanani (1961) the remaining cases; these constructions were recursive and in some cases quite long. Again, the existence of simple $B[3, 2; v]$ designs was proved by van Buggenhaut (1974) but his proof depended on those of Doyen (1972) and Hanani (1963). Constructions given in this chapter are direct: Construction 1 is due to Bose (1939) and Construction 2 to Skolem (1958), and both have been considerably simplified (and pictured) by Lindner (1980). Remaining constructions are those of Stinson and W. D. Wallis (1983*a*, *b*) with some modifications. The square of Table 5.17 is due to Mendelsohn (1971). Chapter 3 contained various difference set constructions for triple systems; recursive methods can also be used as, for example, in Stanton and Goulden (1981).

Exercises

5.1. A Latin square which is idempotent and symmetric must have odd order. Why?

5.2. Complete the proof that the design of Construction 2 is, in fact, a $B[3, 1; v]$, $v = 6n + 1$.

5.3. Consider the idempotent Latin square L given in Table 5.20. If we attempted to extend it to a Latin square of order 10, by replacing $l_{i,\,i+1}$ by the symbol 0 and adding an extra row and column (as in Example 4), would the

Table 5.20

1	3	2	7	9	8	4	6	5
3	2	1	9	8	7	6	5	4
2	1	3	8	7	9	5	4	6
7	6	5	4	3	2	1	9	8
6	8	4	3	5	1	9	2	7
5	4	9	2	1	6	8	7	3
4	9	8	1	6	5	7	3	2
9	5	7	6	2	4	3	8	1
8	7	6	5	4	3	2	1	9

L

resulting array be a Latin square? On what property of the original square does this construction depend?

5.4. When does the design of Construction 7 have repeated blocks? What property of the square L would ensure that the design was simple?

5.5. Show that the design of Construction 9 is in fact a $B[3, 6; v]$, $v = 3n + 2$.

5.6. Consider the Latin square in Table 5.21; it is neither idempotent nor symmetric. If we apply Construction 1 to this square, do we obtain a $B[3, 1; 15]$ design? Why?

Table 5.21

1	5	4	3	2
4	3	2	1	5
2	1	5	4	3
5	4	3	2	1
3	2	1	5	4

6 Mutually orthogonal Latin squares

6.1. Orthogonality

Consider the Latin square L in Table 6.1. Not only is it idempotent, so that its main diagonal is a transversal, but it can be partitioned into four disjoint transversals with positions as follows:

$$T_1 = \{(1,3), (2,4), (3,1), (4,2)\}, \quad T_2 = \{(1,2), (2,1), (3,4), (4,3)\},$$
$$T_3 = \{(1,4), (2,3), (3,2), (4,1)\}, \quad T_4 = \{(1,1), (2,2), (3,3), (4,4)\}.$$

If we now construct a 4×4 square by writing i in each position belonging to T_i, we obtain the Latin square S in Table 6.1.

Table 6.1

1	3	4	2	4	2	1	3	1	2	3		1	3	2
4	2	1	3	2	4	3	1	3	1	2		3	2	1
2	4	3	1	1	3	4	2	2	3	1		2	1	3
3	1	2	4	3	1	2	4							
	L				S				A				B	
			$n = 4$								$n = 3$			

Similarly, we could partition L into the transversals

$$\{(1,1), (2,4), (3,2), (4,3)\}, \quad \{(1,3), (2,2), (3,4), (4,1)\},$$
$$\{(1,4), (2,1), (3,3), (4,2)\}, \quad \{(1,2), (2,3), (3,1), (4,4)\},$$

and, writing 1, 2, 3, 4 respectively in these positions, obtain the Latin square L^T. The decomposition of a Latin square of order n into n disjoint transversals leads us to the idea of *orthogonality*. (A similar decomposition of B enables us to construct A, and vice versa.)

If we have two $n \times n$ arrays, H and K, we may superimpose them on each other or in other words construct an $n \times n$ array of ordered pairs where (h, k) occupies position (i, j) of the array of pairs if and only if h occupies position (i, j) of H and k position (i, j) of K. Two $n \times n$ arrays, each containing n copies of each symbol $1, 2, \ldots, n$, are said to be *orthogonal* if superimposing one on the other gives an array with every possible ordered pair occurring exactly once.

Thus, if R is the $n \times n$ array in which every entry of row i equals i, and if $C = R^T$, then R and C are mutually orthogonal, and a Latin square of order n may be defined as an array orthogonal both to R and to C. Table 6.2 shows R, C and some superimposed squares for $n = 3$; A and B are the squares from Table 6.1, and we see that all nine ordered pairs 11, 12, 13, 21, 22, 23, 31, 32, 33 appear when A and B are superimposed. Thus A and B are not only orthogonal to R and to C but also to each other.

Table 6.2

1	1	1		1	2	3		(1, 1)	(1, 2)	(1, 3)
2	2	2		1	2	3		(2, 1)	(2, 2)	(2, 3)
3	3	3		1	2	3		(3, 1)	(3, 2)	(3, 3)
	R				C				superimposed	

11	12	13		11	22	33		11	32	23
23	21	22		13	21	32		33	21	12
32	33	31		12	23	31		22	13	31
	R with A superimposed				C with A superimposed				B with A superimposed	

Just as two sources of variation in an experiment were controlled by the use of one Latin square, so three sources can be controlled by the use of a pair of orthogonal Latin squares. Here the varieties are assigned to the elements of one of the squares, and the three sources of variation to the rows, the columns, and the elements of the second square respectively. The elements of the second square could, however, be a second set of varieties; see Chapters 9 and 10.

Lemma 1. *If H is orthogonal to K, then K is orthogonal to H.* □

The Latin square S in Table 6.1 is a symmetric array, orthogonal to the square L. Notice also that L is orthogonal to its transpose; such a square is said to be *self-orthogonal*. Since S is symmetric, this means that L, L^T and S form a set of three *mutually orthogonal Latin squares* (MOLS).

The property of being a Latin square is obviously not arithmetical in any way, so that permuting the symbols 1, 2, . . . , n makes no difference to the Latin property. Thus, if we take the square A of Table 6.1 and interchange the symbols 2 and 3, we still get a Latin square as shown in Table 6.3. Further, the permuted square is still orthogonal to B, just as A was.

A Latin square is said to be *in standard form* (or *standardized*) if its first row contains the symbols 1, 2, . . . , n in their natural order. Any Latin square may be put into standard form by permuting its elements. Thus, to put L into standard form we must replace 2 by 4, 3 by 2 and 4 by 3, as shown in Table 6.4. Similarly, L^T and S may be standardized. Notice that the standard forms of L, L^T and S are still mutually orthogonal; permuting symbols does not change this property.

Table 6.3

1	3	2		11	33	22
2	1	3		23	12	31
3	2	1		32	21	13
$A(23)$				$A(23)$ superimposed on B		

We shall let $N(n)$ denote the largest possible number of MOLS of order n. We make the convention that $N(1) = \infty$. We have already seen that a Latin square exists for every order; so $N(n) \geq 1$. Further, for $n > 1$, we can easily obtain an upper bound for $N(n)$.

Table 6.4

1	3	4	2		1	2	3	4
4	2	1	3		3	4	1	2
2	4	3	1		4	3	2	1
3	1	2	4		2	1	4	3
L					$L(243)$			
1	4	2	3		1	2	3	4
3	2	4	1		4	3	2	1
4	1	3	2		2	1	4	3
2	3	1	4		3	4	1	2
L^T					$L^T(234)$			
4	2	1	3		1	2	3	4
2	4	3	1		2	1	4	3
1	3	4	2		3	4	1	2
3	1	2	4		4	3	2	1
S					$S(134)$			

Theorem 2. *For $n > 1$, $N(n) \leq n - 1$.*

Proof. Suppose that $L_1, L_2, \ldots, L_n$ are n MOLS of order n, and further suppose that they are all standardized, so that the ordered pair (j, j), $j = 1, 2, \ldots, n$, occurs in the $(1, j)$ position when any two of these squares are superimposed. Now let a_i be the (2, 1) element of the square L_i. Since the (1, 1) element is 1, we know that $a_i \neq 1$. Since the pairs (j, j) have already occurred in the first row of two superimposed squares, we know that $a_1, a_2, \ldots, a_n$ are all distinct. Hence from the set $\{2, 3, \ldots, n\}$ we must choose n distinct elements, clearly an impossibility. Hence $N(n) \leq n - 1$. □

Thus $N(2) = 1$, and since Table 6.1 shows a pair of orthogonal Latin squares of order 3, $N(3) = 2$. Euler conjectured that $N(n) = 1$ for all $n \equiv 2 \pmod 4$. Subsequently he was shown to be correct for $n = 6$ (see Section 6.3) but it is now known that $N(10) \geq 2$, and that $N(n) \geq 3$ for all $n \notin \{2, 3, 6, 10\}$, as shown in Chapter 7. (It is not yet known whether $N(10)$ is in fact equal to 2.)

If $N(n) = n-1$, then a set of $n-1$ MOLS exists and is said to be a *complete set*. A complete set of MOLS of order n is equivalent to a $B[n, 1; n^2+n]$ design and can always be extended to a $B[n+1, 1; n^2+n+1]$ design, as discussed in Chapter 8.

If n is a prime power, then we can easily construct a complete set of MOLS of order n, by using properties of finite fields.

6.2. Some constructions of MOLS

Using the field of order $n = p^k$, we now construct a complete set of MOLS of order n.

Theorem 3. *Let $n = p^k$ for some prime p. In the field $GF[p^k]$, let the elements be labelled*

$$f_0 = 0, f_1 = 1, f_2, \ldots, f_{n-1},$$

where $f_2, \ldots, f_{n-1}$ occur in some arbitrary but fixed order. Let the $n \times n$ arrays $A_1, A_2, \ldots, A_{n-1}$ be defined by

$$A_m = [a_{ij}^{(m)}];\ i, j = 0, 1, \ldots, n-1;\ m = 1, \ldots, n-1;$$

where

$$a_{ij}^{(m)} = f_m f_i + f_j.$$

Then $A_1, A_2, \ldots, A_{n-1}$ are a set of $n-1$ MOLS of order n.

Proof. We check first that each A_m is a Latin square. Suppose that two elements in row i of A_m are equal, that is,

$$a_{ij}^{(m)} = a_{ik}^{(m)}, \quad j \neq k.$$

Then

$$f_m f_i + f_j = f_m f_i + f_k$$

so that

$$f_j = f_k,$$

a contradiction. Similarly, if two elements in column j of A_m are equal, then

$$f_m f_i + f_j = f_m f_k + f_j, \quad i \neq k,$$

so that

$$f_i = f_k \quad \text{as} \quad f_m \neq 0,$$

again a contradiction. Thus $A_1, A_2, \ldots, A_{n-1}$ are Latin squares.

To finish the proof we must show that A_l, A_m are orthogonal whenever $1 \leq l < m \leq n-1$. Suppose not. Then for some i, j, u, v such that $(i, j) \neq (u, v)$, we must have

$$(a_{ij}^{(l)}, a_{ij}^{(m)}) = (a_{uv}^{(l)}, a_{uv}^{(m)}).$$

This means that both

$$f_l f_i + f_j = f_l f_u + f_v$$

and

$$f_m f_i + f_j = f_m f_u + f_v.$$

But since $f_l \neq f_m$, these force $f_i = f_u$ and hence $i = u$. Thus $f_j = f_v$ so $j = v$ and we have our contradiction. Hence the squares form a complete set of MOLS. □

Example 4. We use the method of Theorem 3 to construct three MOLS of order 4. Suppose that the field $GF[2^2]$ is represented as usual; see Table 3.5. Let the ordering be

$$f_0 = 0, f_1 = 1, f_2 = \alpha, f_3 = \alpha + 1,$$

so that

$$a_{ij}^{(1)} = 1 . f_i + f_j = f_i + f_j,$$
$$a_{ij}^{(2)} = \alpha f_i + f_j,$$
$$a_{ij}^{(3)} = (\alpha + 1) f_i + f_j.$$

Since the rows are to be numbered 0 to 3, and $f_0 = 0$, we have

$$a_{0j} = f_j, \quad \text{for each square.}$$

Thus all the squares are standardized. For square A_1, row 1 is just

$$a_{1j}^{(1)} = f_1 + f_j = 1 + f_j.$$

Continuing we find the squares listed in Table 6.5. □

Table 6.5

0	1	α	$\alpha+1$	0	1	α	$\alpha+1$	0	1	α	$\alpha+1$
1	0	$\alpha+1$	α	α	$\alpha+1$	0	1	$\alpha+1$	α	1	0
α	$\alpha+1$	0	1	$\alpha+1$	α	1	0	1	0	$\alpha+1$	α
$\alpha+1$	α	1	0	1	0	$\alpha+1$	α	α	$\alpha+1$	0	1
	A_1				A_2				A_3		

If n is a prime power, then certainly a complete set of MOLS of order n exists. But this does *not* mean that any given Latin square of order n must necessarily be extendible to a complete set of MOLS, nor even that it must have any Latin square orthogonal to it. For example, there is no Latin square orthogonal to L_1 of Table 6.6. To see this, we suppose there is; without loss of generality we can assume that it is standardized and hence of the form L_2 or L_3.

Table 6.6

$$\begin{array}{cccc} 1&2&3&4\\ 2&4&1&3\\ 3&1&4&2\\ 4&3&2&1 \end{array} \qquad \begin{array}{cccc} 1&2&3&4\\ 3&a&b&c\\ d&e&f&g\\ h&i&j&k \end{array} \qquad \begin{array}{cccc} 1&2&3&4\\ 4&p&q&r\\ s&t&u&v\\ w&x&y&z \end{array}$$

L_1 $\qquad$ L_2 $\qquad$ L_3

In L_2, if $a = 4$, then the ordered pair (4, 4) occurs in both (1, 4) and (2, 2) positions of L_1 with L_2. Hence $a = 1$, forcing $b = 4$, $c = 2$. Now (3, 2) has occurred in the (2, 4) position, so $d = 4$. This forces $e = 3$, $i = 4$ and the pair (3, 4) has occurred in the (3, 1) and (4, 2) positions. A similar argument shows that L_3 cannot be completed.

We now have a technique for producing MOLS of prime-power order. For composite orders, we have the following theorem.

Theorem 5. *$N(mn) \geq \min(N(m), N(n))$ for $m, n > 1$.*

Proof. Suppose that M_1, M_2 are orthogonal Latin squares of order m, and N_1, N_2 are orthogonal Latin squares of order n. Let N_i^r be the array obtained from N_i by adding $(r-1)n$ to each entry, for $r = 1, 2, \ldots, m$. Now construct the $mn \times mn$ array L_1 by replacing the entry r in M_1 by the block N_1^r, and similarly construct L_2 by replacing r in M_2 by the block N_2^r. We refer to this as the *direct product* but denote it by $L_i = M_i \underline{\times} N_i$, as a reminder that m different versions of N_i appear.

For each element a in L_1 there are unique numbers $r_a \in \{1, 2, \ldots, m\}$ and $s_a \in \{1, 2, \ldots, n\}$ such that $a = (r_a - 1)n + s_a$. Then a occurs in position (i, j) of L_1 if and only if

$$i = (r_i - 1)n + s_i, \quad j = (r_j - 1)n + s_j,$$

s_a occurred in position (s_i, s_j) of N_1 and r_a in position (r_i, r_j) of M_1. With this, and the similar expression for elements of L_2, it is routine to check that L_1 and L_2 are orthogonal Latin squares. This argument extends to a set of any number of squares of each order. □

Corollary 5.1. *Let $n = p_1^{e_1} p_2^{e_2} \ldots p_h^{e_h}$ be the factorization of n into prime powers, where $e_1, e_2, \ldots, e_h \geq 1$, and $p_1, p_2, \ldots, p_h$ are distinct primes. Then*

$$N(n) \geq \min(p_1^{e_1} - 1, p_2^{e_2} - 1, \ldots, p_h^{e_h} - 1). \qquad \square$$

Unfortunately, if $n = 2(2t + 1)$, then Corollary 5.1 gives no useful information about $N(n)$.

Example 6. Suppose we take as examples some of the Latin squares in Table 6.1: $M_1 = A$, $M_2 = B$, $N_1 = L$, $N_2 = S$, say. Then we construct a pair of orthogonal Latin squares of order 12, as shown in Table 6.7 where we write out the array L_1 in full. □

Table 6.7

$$N_1 = N_1^1 = \begin{matrix} 1 & 3 & 4 & 2 \\ 4 & 2 & 1 & 3 \\ 2 & 4 & 3 & 1 \\ 3 & 1 & 2 & 4 \end{matrix}; \quad N_1^2 = \begin{matrix} 5 & 7 & 8 & 6 \\ 8 & 6 & 5 & 7 \\ 6 & 8 & 7 & 5 \\ 7 & 5 & 6 & 8 \end{matrix}; \quad N_1^3 = \begin{matrix} 9 & 11 & 12 & 10 \\ 12 & 10 & 9 & 11 \\ 10 & 12 & 11 & 9 \\ 11 & 9 & 10 & 12 \end{matrix}$$

$$N_2 = N_2^1 = \begin{matrix} 4 & 2 & 1 & 3 \\ 2 & 4 & 3 & 1 \\ 1 & 3 & 4 & 2 \\ 3 & 1 & 2 & 4 \end{matrix}; \quad N_2^2 = \begin{matrix} 8 & 6 & 5 & 7 \\ 6 & 8 & 7 & 5 \\ 5 & 7 & 8 & 6 \\ 7 & 5 & 6 & 8 \end{matrix}; \quad N_2^3 = \begin{matrix} 12 & 10 & 9 & 11 \\ 10 & 12 & 11 & 9 \\ 9 & 11 & 12 & 10 \\ 11 & 9 & 10 & 12 \end{matrix}$$

$$L_1 = \begin{matrix} N_1^1 & N_1^2 & N_1^3 \\ N_1^3 & N_1^1 & N_1^2 \\ N_1^2 & N_1^3 & N_1^1 \end{matrix} \qquad L_2 = \begin{matrix} N_2^1 & N_2^3 & N_2^2 \\ N_2^3 & N_2^2 & N_2^1 \\ N_2^2 & N_2^1 & N_2^3 \end{matrix}$$

$$L_1 = \begin{matrix}
1 & 3 & 4 & 2 & 5 & 7 & 8 & 6 & 9 & 11 & 12 & 10 \\
4 & 2 & 1 & 3 & 8 & 6 & 5 & 7 & 12 & 10 & 9 & 11 \\
2 & 4 & 3 & 1 & 6 & 8 & 7 & 5 & 10 & 12 & 11 & 9 \\
3 & 1 & 2 & 4 & 7 & 5 & 6 & 8 & 11 & 9 & 10 & 12 \\
9 & 11 & 12 & 10 & 1 & 3 & 4 & 2 & 5 & 7 & 8 & 6 \\
12 & 10 & 9 & 11 & 4 & 2 & 1 & 3 & 8 & 6 & 5 & 7 \\
10 & 12 & 11 & 9 & 2 & 4 & 3 & 1 & 6 & 8 & 7 & 5 \\
11 & 9 & 10 & 12 & 3 & 1 & 2 & 4 & 7 & 5 & 6 & 8 \\
5 & 7 & 8 & 6 & 9 & 11 & 12 & 10 & 1 & 3 & 4 & 2 \\
8 & 6 & 5 & 7 & 12 & 10 & 9 & 11 & 4 & 2 & 1 & 3 \\
6 & 8 & 7 & 5 & 10 & 12 & 11 & 9 & 2 & 4 & 3 & 1 \\
7 & 5 & 6 & 8 & 11 & 9 & 10 & 12 & 3 & 1 & 2 & 4
\end{matrix}$$

We note here that $N(12) \geq 5$, so that the direct product construction need not give the best bound for $N(n)$. A set of five MOLS of order 12 is given by the following, where we use the set $\{0, 1, \ldots, 11\}$ as symbols. Let

$$X = \begin{bmatrix} 0 & 1 & 2 & 3 & 4 & 5 \\ 1 & 2 & 3 & 4 & 5 & 0 \\ 2 & 3 & 4 & 5 & 0 & 1 \\ 3 & 4 & 5 & 0 & 1 & 2 \\ 4 & 5 & 0 & 1 & 2 & 3 \\ 5 & 0 & 1 & 2 & 3 & 4 \end{bmatrix}, \qquad Y = \begin{bmatrix} 1 & 2 \\ 2 & 1 \end{bmatrix},$$

and let the first square of the family be $L_1 = Y \underline{\times} X$. Thus the first row of L_1 is

$$0, 1, 2, 3, 4, 5, 6, 7, 8, 9, 10, 11,$$

and the remaining squares are obtained from L_1 by column permutation, and have first rows as follows:

L_2: 0, 6, 8, 2, 7, 1, 9, 11, 4, 10, 5, 3;
L_3: 0, 3, 6, 1, 9, 11, 2, 8, 5, 4, 7, 10;
L_4: 0, 8, 1, 11, 5, 9, 3, 10, 2, 7, 6, 4;
L_5: 0, 4, 11, 10, 2, 7, 8, 6, 9, 1, 3, 5.

As we have already stated, complete sets of MOLS are related to certain kinds of BIBDs, namely the affine and projective planes. But this is not the only relation between Latin squares and other designs. Their relation with transversal designs is discussed in the next section. Other constructions involving Latin squares will be discussed in Chapters 8, 9, 11, and 14.

6.3. Transversal designs

Much work dealing with Latin squares is most conveniently stated in terms of transversal designs, which we now define. A *transversal design* with k groups of size n and index λ, denoted by $T[k, \lambda; n]$, is a triple $(X, \mathscr{G}, \mathscr{A})$ where:

(i) X is a set of kn varieties;
(ii) $\mathscr{G} = \{G_1, G_2, \ldots, G_k\}$ is a family of k n-sets (or *groups*) which form a partition of X;
(iii) $\mathscr{A}$ is a family of k-sets (or *blocks*) of varieties such that each k-set in $\mathscr{A}$ intersects each group G_i in precisely one variety, and any pair of varieties which belong to different groups occur together in precisely λ blocks in $\mathscr{A}$.

There is an unfortunate clash of terminology here: the groups of a transversal design have no operation defined on them and are simply subsets partitioning the underlying set.

Example 7. Table 6.8 shows the groups and blocks of a $T[4, 1; 3]$ design. □

Table 6.8

Groups	Blocks
	B_1: x_{11} x_{21} x_{31} x_{41}
G_1: x_{11} x_{12} x_{13}	B_2: x_{11} x_{22} x_{32} x_{42}
G_2: x_{21} x_{22} x_{23}	B_3: x_{11} x_{23} x_{33} x_{43}
G_3: x_{31} x_{32} x_{33}	B_4: x_{12} x_{21} x_{32} x_{43}
G_4: x_{41} x_{42} x_{43}	B_5: x_{12} x_{22} x_{33} x_{41}
	B_6: x_{12} x_{23} x_{31} x_{42}
	B_7: x_{13} x_{21} x_{33} x_{42}
	B_8: x_{13} x_{22} x_{31} x_{43}
	B_9: x_{13} x_{23} x_{32} x_{41}

Several important properties of transversal designs follow immediately from counting arguments.

Lemma 8. *A $T[k, \lambda; n]$ design has λn^2 blocks.*

Proof. The $\binom{n}{2}$ unordered pairs of varieties common to one group must not occur in any block of the design. Thus, of the $\binom{kn}{2}$ possible pairs, precisely $k\binom{n}{2}$ are excluded. Since the design has index λ, there must be

$$\lambda\left(\binom{kn}{2}-k\binom{n}{2}\right)=\lambda n^2\binom{k}{2}$$

occurrences of pairs in $\mathscr{A}$. Since each block covers $\binom{k}{2}$ distinct unordered pairs, there must be λn^2 blocks. □

Lemma 9. *If $k \leqslant n$ and $2 \leqslant n$, then any block of a $T[k, l; n]$ design is disjoint from at least one other block.*

Proof. By Lemma 8, there are n^2 blocks of size k in the design. Since each of the kn varieties must occur once with each of $(k-1)n$ other varieties, each variety occurs in n blocks. Hence each of the k varieties belonging to the given block A, say, belongs to $n-1$ other blocks, and thus A intersects $k(n-1)$ other blocks. This means that A is disjoint from $n^2-k(n-1)-1$ blocks, and since $k \leq n$,

$$n^2-k(n-1)-1 \geq n^2-n(n-1)-1=n-1.$$

By hypothesis, $n \geq 2$, so $n-1 \geq 1$ and the Lemma follows. □

Example 10. Table 6.9 shows the blocks and groups of a $T[3, 1; 4]$ design. Notice that here $n^2-k(n-1)-1=16-9-1=6$, and for example, B_1 is disjoint from B_6, B_7, B_{10}, B_{12}, B_{15}, B_{16}, altogether six blocks. □

Table 6.9

Group				
G_1:	x_{11}	x_{12}	x_{13}	x_{14}
G_2:	x_{21}	x_{22}	x_{23}	x_{24}
G_3:	x_{31}	x_{32}	x_{33}	x_{34}

Block				Block			
B_1:	x_{11}	x_{21}	x_{31}	B_9 :	x_{13}	x_{21}	x_{33}
B_2:	x_{11}	x_{22}	x_{32}	B_{10}:	x_{13}	x_{22}	x_{34}
B_3:	x_{11}	x_{23}	x_{33}	B_{11}:	x_{13}	x_{23}	x_{31}
B_4:	x_{11}	x_{24}	x_{34}	B_{12}:	x_{13}	x_{24}	x_{32}
B_5:	x_{12}	x_{21}	x_{32}	B_{13}:	x_{14}	x_{21}	x_{34}
B_6:	x_{12}	x_{22}	x_{33}	B_{14}:	x_{14}	x_{22}	x_{31}
B_7:	x_{12}	x_{23}	x_{34}	B_{15}:	x_{14}	x_{23}	x_{32}
B_8:	x_{12}	x_{24}	x_{31}	B_{16}:	x_{14}	x_{24}	x_{33}

Our next result shows the relation between transversal designs and Latin squares.

Theorem 11. *The existence of a $T[k, 1; n]$ design is equivalent to the existence of $k-2$ MOLS of order n.*

Proof. Let a $T[k, 1; n]$ design have groups $G_1, G_2, \ldots, G_k$, with varieties so labelled that

$$G_h = \{x_{h1}, x_{h2}, \ldots, x_{hn}\}, \quad h = 1, 2, \ldots, k.$$

For $1 \le h \le k-2$, we define the $n \times n$ array A^h in the following way. For $1 \le i, j \le n$, there exists a unique block B in the design which contains both $x_{k-1,i}$ and $x_{k,j}$, since $\lambda = 1$.

Since B is a block of a transversal design, it also contains precisely one element of G_h, say

$$B \cap G_h = \{x_{hm}\}.$$

Then we define

$$a^h_{ij} = m.$$

We verify first that A^h is a Latin square. Suppose that $a^h_{ij} = a^h_{il} = m$, $j \ne l$; that is, that two entries in row i are equal. This means that there are blocks B_1 and B_2, say, such that

$$\{x_{h,m}, x_{k-1,i}, x_{k,j}\} \subseteq B_1$$

and

$$\{x_{h,m}, x_{k-1,i}, x_{k,l}\} \subseteq B_2.$$

Since $x_{kj} \ne x_{kl}$, and only one element of each group can occur in each block, we have $B_1 \ne B_2$ and $\lambda \ge 2$, which is a contradiction. Similarly, no two entries in the same column are equal.

Now we check that A^h and A^l are mutually orthogonal. For suppose that

$$a^h_{ij} = a^h_{uv} = p \quad \text{and} \quad a^l_{ij} = a^l_{uv} = q.$$

Then there are blocks B_1, B_2, B_3, B_4 such that

$$\{x_{hp}, x_{k-1,i}, x_{kj}\} \subseteq B_1, \quad \{x_{hp}, x_{k-1,u}, x_{kv}\} \subseteq B_2,$$

$$\{x_{lq}, x_{k-1,i}, x_{kj}\} \subseteq B_3, \quad \{x_{lq}, x_{k-1,u}, x_{kv}\} \subseteq B_4.$$

Since $\lambda = 1$, we must have $B_1 = B_3$ and $B_2 = B_4$. Thus

$$\{x_{hp}, x_{lq}, x_{k-1,i}, x_{kj}\} \subseteq B_1$$

and

$$\{x_{hp}, x_{lq}, x_{k-1,u}, x_{kv}\} \subseteq B_2.$$

But now x_{hp}, x_{lq} occur together in two blocks, which is a contradiction. Hence from a $T[k, 1; n]$ we can construct $k-2$ MOLS of order n. Since the construction can be reversed, the two structures are equivalent. □

We are effectively using the groups G_{k-1} and G_k to define a co-ordinate system. Then each of $G_1, \ldots, G_{k-2}$ defines the elements of an array.

Example 12. From the $T[4, 1; 3]$ design of Example 7, we construct two orthogonal Latin squares of order 3, namely A^1 and A^2. For example, to find the elements of position (1, 2), consider the elements x_{31}, x_{42} of the design. They occur together in block B_6, with x_{12} and x_{23}. Thus the (1, 2) element of A^1 is 2 and of A^2 is 3. Table 6.10 shows the construction. □

A *pairwise balanced design* (PBD), as defined in Section 1.2, is a design in which each unordered pair of elements occurs in precisely λ blocks, but block size is not required to be constant. Suppose we have a transversal design $T[k, \lambda; n]$.

If $\lambda = 1$, then the groups $G_1, G_2, \ldots, G_k$ may be taken as blocks to give a pairwise balanced design $(X, \mathcal{G} \cup \mathcal{A})$. If $\lambda > 1$, we need λ copies of each group to ensure pairwise balance. In either case the pairwise balanced design has block sizes k and n and replication number $n + \lambda$.

Table 6.10

Position	Co-ordinate elements	Elements determining entries	Arrays
1, 1	$x_{31}, x_{41} \in B_1$	with x_{11}, x_{21}	
1, 2	$x_{31}, x_{42} \in B_6$	with x_{12}, x_{23}	
1, 3	$x_{31}, x_{43} \in B_8$	with x_{13}, x_{22}	
2, 1	$x_{32}, x_{41} \in B_9$	with x_{13}, x_{23}	$\begin{bmatrix} 1 & 2 & 3 \\ 3 & 1 & 2 \\ 2 & 3 & 1 \end{bmatrix}$ $\begin{bmatrix} 1 & 3 & 2 \\ 3 & 2 & 1 \\ 2 & 1 & 3 \end{bmatrix}$
2, 2	$x_{32}, x_{42} \in B_2$	with x_{11}, x_{22}	
2, 3	$x_{32}, x_{43} \in B_4$	with x_{12}, x_{21}	
3, 1	$x_{33}, x_{41} \in B_5$	with x_{12}, x_{22}	A^1 A^2
3, 2	$x_{33}, x_{42} \in B_7$	with x_{13}, x_{21}	
3, 3	$x_{33}, x_{43} \in B_3$	with x_{11}, x_{23}	

6.4. Non-existence of orthogonal Latin squares of order 6

In this section we prove that no pair of orthogonal Latin squares of order 6 can exist or, in other words, that $N(6) = 1$. By Theorem 11, it is sufficient to prove that no $T[4, 1; 6]$ design exists. We suppose there does, and arrive at a contradiction.

A $T[4, 1; 6]$ design is a triple $(X, \mathcal{G}, \mathcal{A})$ where $|X| = 24$, $\mathcal{G}$ is a partition of X into

$$X = G_1 \cup G_2 \cup G_3 \cup G_4, \qquad |G_i| = 6,$$

and $\mathcal{A}$ is a family of 36 blocks, $|B_i| = 4$.

We consider the PBD $P = (X, \mathcal{G} \cup \mathcal{A})$; the design P has blocks

B_1, B_2, B_3, B_4, each of size 6, and $B_5, \ldots, B_{40}$, each of size 4. Each element of X occurs in six blocks of size 4 and one block of size 6, so the replication number is $r = 7$ for each $x_i \in X$. We assume without loss of generality that the groups are as shown in Table 6.11.

Table 6.11. The groups of the design P

$B_1 = G_1$:	1	2	3	4	5	6
$B_2 = G_2$:	7	8	9	10	11	12
$B_3 = G_3$:	13	14	15	16	17	18
$B_4 = G_4$:	19	20	21	22	23	24

Let M be the incidence matrix of P, so that M is 24×40 and

$$m_{ij} = \begin{cases} 1 & \text{if } x_i \in B_j, \\ 0 & \text{if } x_i \notin B_j. \end{cases}$$

The row sum for any row of M is $r = 7$. We consider the ith row as a vector, $\boldsymbol{r}_i$, over $GF\,[2]$, with 40 components, so that the row space, U, of M is a subspace of a 40-dimensional vector space, V, over $GF\,[2]$. We are going to prove that

$$21 \leq \operatorname{rank} M \leq 20,$$

and hence that no $T\,[4, 1; 6]$ design exists.

We assume the usual inner product on V so that

$$\langle (v_1, v_2, \ldots, v_{40}), (w_1, w_2, \ldots, w_{40}) \rangle = \sum_{i=1}^{40} v_i w_i,$$

where addition and multiplication are those of $GF\,[2]$. Further, we define the orthogonal complement of a subspace W to be $W^{\perp}$, where

$$W^{\perp} = \{ v \in V \mid \langle w, v \rangle = 0 \quad \text{for all } w \in W \}$$

and the usual proof shows that

$$\dim W + \dim W^{\perp} = \dim V.$$

But note that $W \cap W^{\perp}$ need not be just the zero subspace; for example, any vector containing an even number of non-zero entries is orthogonal to itself.

Since the row space, U, of M is spanned by $\boldsymbol{r}_1, \boldsymbol{r}_2, \ldots, \boldsymbol{r}_{24}$, we may define its orthogonal complement by

$$U^{\perp} = \{ \boldsymbol{u} \in V \mid \langle \boldsymbol{u}, \boldsymbol{r}_i \rangle = 0,\ i = 1, \ldots, 24 \}.$$

Now in the design P, $r = 7$, so

$$\langle \boldsymbol{r}_i, \boldsymbol{r}_i \rangle = 1 \pmod 2, \quad i = 1, \ldots, 24,$$

and $\lambda = 1$, so

$$\langle \boldsymbol{r}_i, \boldsymbol{r}_j \rangle = 1,\ i \neq j,\ i, j = 1, \ldots, 24.$$

Hence for all $h, i, j \in \{1, \ldots, 24\}$, we have

$$\langle r_i, r_j + r_h \rangle = 0$$

and thus

$$\{r_1 + r_2, r_1 + r_3, \ldots, r_1 + r_{24}\} \subseteq U^{\perp}.$$

Now $\langle r_1, r_1 + r_2, r_1 + r_3, \ldots, r_1 + r_{24} \rangle = U$ and thus

$$\dim U^{\perp} \geq \dim \langle r_1 + r_2, \ldots, r_1 + r_{24} \rangle \geq (\dim U) - 1.$$

But

$$\dim U + \dim U^{\perp} = 40,$$

forcing

$$\dim U \leq 20.$$

In other words,

$$\operatorname{rank} M \leq 20. \tag{6.1}$$

Since the spanning set of U has 24 vectors, but its dimension is at most 20, there must be at least four linearly independent linear equations relating $r_1, \ldots, r_{24}$. Such an equation can be written as

$$\sum_{i \in I} r_i = \mathbf{0}, \tag{6.2}$$

for some index set $I \subseteq \{1, 2, \ldots, 24\}$. This means that if we define a set $Y \subseteq X$ by

$$Y = \{x_i \in X \mid i \in I\},$$

then $|B_j \cap Y|$ is even, for $j = 1, \ldots, 40$. We call Y an *even subset* of X. Examples of even subsets include $B_1 \cup B_2$, $B_1 \cup B_3$, $B_1 \cup B_4$, since for $i \leq 4$ and $j \geq 5$, we have

$$|B_i \cap B_j| = 1.$$

This gives us three independent equations of the form of (6.2); for instance, the equation corresponding to $Y = B_1 \cup B_2$ is

$$\sum_{i=1}^{12} r_i = \mathbf{0}.$$

As M has rank at most 20 there is at least one more independent equation.

We pursue the properties of even subsets further. If Y is an even subset of X, then so is its complement, $X \setminus Y$. Hence we may assume that $|Y| \leq 12$. Also, if Y_1 and Y_2 are even subsets of X, then so is their symmetric difference, $Y_1 + Y_2$, defined by

$$Y_1 + Y_2 = (Y_1 \cup Y_2) \setminus (Y_1 \cap Y_2).$$

Now we define an *even sub-PBD* of P to be the design $(Y, \{Y \cap B_i : i = 1, \ldots, 40\})$, for Y an even subset of X. Note that $|Y \cap B_i| \in \{0, 2, 4, 6\}$ and that for $i \geq 5$, $|Y \cap B_i| \in \{0, 2, 4\}$. We let b_j denote the number of blocks of size j in the even sub-PBD, let $|Y| = m$, and observe that for each element of Y, the replication number is still $r = 7$ in the even sub-PBD. Counting the total number of blocks, we have

$$b_0 + b_2 + b_4 + b_6 = 40. \tag{6.3}$$

Since $r = 7$ for each point in Y, we have

$$2b_2 + 4b_4 + 6b_6 = 7m, \tag{6.4}$$

and since $\lambda = 1$ for pairs of points in Y,

$$b_2 + 6b_4 + 15b_6 = m(m-1)/2. \tag{6.5}$$

From (6.5) and (6.6), we find that

$$b_4 + 3b_6 = m(m-8)/8, \tag{6.6}$$

implying that $m \geq 8$, $m \equiv 0 \pmod 4$. But by assumption, $m \leq 12$. Hence

$$m = |Y| = 8 \text{ or } 12. \tag{6.7}$$

Suppose first that $m = 12$. The 12 points of Y must be distributed across the four blocks B_1, B_2, B_3, B_4 since these were the groups of the $T[4, 1; 6]$ design. Table 6.12 shows the possible intersection sizes of Y with these four blocks. Case (i) has already been dealt with, in finding the equations earlier. Since Y and $B_1 \cup B_2$ are both even subsets, $Y + (B_1 \cup B_2)$ is also an even subset; Table 6.13 shows the distribution of $Y + (B_1 \cup B_2)$ across the blocks B_1, B_2, B_3, B_4 in cases (ii) to (v). Case (ii) would lead us to an even subset of size 4, and is thus impossible; the remaining cases lead to even subsets of size 8.

Table 6.12

	B_1	B_2	B_3	B_4
(i)	6	6	0	0
(ii)	6	4	2	0
(iii)	6	2	2	2
(iv)	4	4	4	0
(v)	4	4	2	2

Table 6.13

	B_1	B_2	B_3	B_4
(ii)	0	2	2	0
(iii)	0	4	2	2
(iv)	2	2	4	0
(v)	2	2	2	2

From now on we assume that $m = 8$. From eqn (6.6), this means that

$$b_4 + 3b_6 = 0,$$

so

$$b_4 = b_6 = 0.$$

Hence from eqns (6.4) or (6.5), $b_2 = 28$, and from (6.3), $b_0 = 12$. Thus Y has no three points in any one block, and hence $|Y \cap B_i| = 2$, $i = 1, 2, 3, 4$. We may as well consider that $Y = \{5, 6, 11, 12, 17, 18, 23, 24\}$; see Table 6.11.

Now let $Q = (X \setminus Y, \mathscr{B}')$ be the PBD formed from P by deleting the points of Y from the set X and from every block in $\mathscr{B}$. Thus $B' \in \mathscr{B}'$ if and only if $B' = B \setminus Y$ for some $B \in \mathscr{B} = (\mathscr{G} \cup \mathscr{A})$. The PBD Q has four blocks of size 4 (from B_1 to B_4), together with another 12 blocks of size 4 and 24 blocks of size 2 (from B_5 to B_{40}). We study the structure of Q by defining a graph G, with 16 vertices $X \setminus Y$ (the points of Q) and 24 edges, namely the size 2 blocks of Q.

Lemma 13. *G is 3-regular; that is, there are exactly three edges incident with each vertex of G.*

Proof. Choose $i \in X \setminus Y$. Suppose that i occurs in α blocks of size 2 and β blocks of size 4 in Q. Then counting the replications of i, we see that

$$\alpha + \beta = 7$$

and counting the number of pairs containing i,

$$\alpha + 3\beta = 15,$$

so $\alpha = 3$, $\beta = 4$. This means each vertex i has three edges incident with it. □

Lemma 14. *Any point, i, of G is adjacent to precisely one point from each of the three groups of $\mathscr{G}$ not containing i.*

Proof. Without loss of generality, we may assume that $i = 1$, and that the blocks of P containing 1 have the form shown in Table 6.14. Each dash indicates a point chosen from $\{7, 8, 9, 10, 13, 14, 15, 16, 19, 20, 21, 22\}$, one from each of B_2, B_3, B_4 in each block. Thus the points marked with an arrow must be chosen, one each from B_2, B_3, B_4. When the points of Y are deleted these give the blocks of size 2 corresponding to the edges of G. □

Table 6.14

1 2 3 4 5 6		1 — 11 17
1 — — —	blocks of P containing	1 — 12 23
1 — — —	no elements of Y	1 — 18 24
1 — — —		↑

Lemma 15. *G has no triangles.*

Proof. If G contains a triangle, it must consist of three points from different groups of $\mathscr{G}$, say 1, 7, 13. Then we may assume that the design P has blocks 1, 7, 17, 23 and 1, 13, 11, 24. What is the block containing 7 and 13? It must be one of: 7, 13, 5, 23; 7, 13, 5, 24; 7, 13, 6, 23; 7, 13, 6, 24. But any of these leads to a repeated pair. □

Lemma 16. *If the point $i \in X \setminus Y$ has neighbours x, y, z in G, then the triple $\{x, y, z\}$ does not occur in any block of P.*

Proof. For suppose it did. Without loss of generality, we take

$$1, 7;\ 1, 13;\ 1, 19$$

as edges of G, and 2, 7, 13, 19 as a block of P. We may now assume that Q has blocks

2, 7, 13, 19;	1, 8, 15, 22;
2, 8, 14, 20;	1, 9, 16, 20;
2, 9, 15, 21;	1, 10, 14, 21;

besides six more blocks, each containing one point from $\{3, 4\}$ and one from $\{7, 13, 19\}$, thus

$$\left.\begin{array}{ll} 3,\ 7,\ -\ - & 4,\ 7,\ -\ - \\ 3,\ 13,\ -\ - & 4,\ 13,\ -\ - \\ 3,\ 19,\ -\ - & 4,\ 19,\ -\ -. \end{array}\right\} \quad (*)$$

Now, the elements which have not yet appeared in a block with 2 are 10, 16, 22. Hence they must be neighbours of 2 in G. Thus the pairs 10, 16; 10, 22; 16, 22; must occur in Q in blocks of $(*)$. But this forces a repeated pair. □

Corollary 16.1. *If the point $i \in X \setminus Y$ has neighbours x, y, z in G, then the pairs xy, yz, zx occur in three different blocks of P.* □

We are now in a position to finish the proof. Suppose that adjacencies in G for points of the first group in G are as shown in Fig. 6.1. By Corollary 16.1, the point 1 occurs with exactly one pair from each triple of neighbours in any given block. If, say, there is a block 1, 8, 14, 21, then there must also be blocks 1, 9, 15, 22 and 1, 10, 16, 20. Where now does the pair 7, 13 occur?

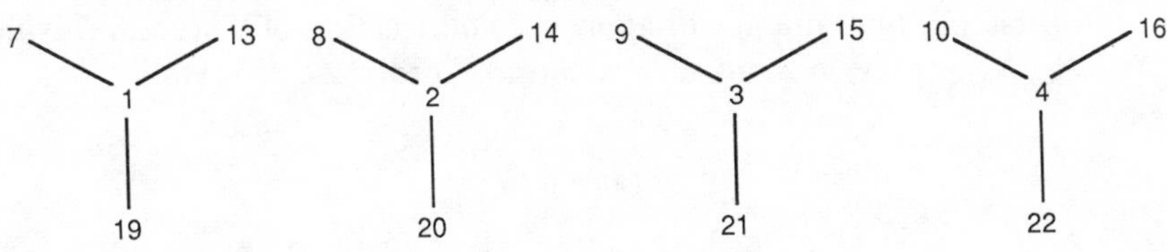

Fig. 6.1

If we have a block 2, 7, 13, i, then $i \in B_4$ and must, by Lemma 16, be either 21 or 22. In either case, this forces blocks 2, 9, 15, j and 2, 10, 16, k, causing repeated pairs. An exactly similar argument shows that the pair 7, 13 cannot occur with 3 or 4 either, proving that the pair 7, 13 never occurs. Hence no fourth linearly independent even subset of X exists. Since only three linearly independent linear equations relate the 24 rows of the matrix M, we see that

$$\operatorname{rank} M \geq 21. \tag{6.8}$$

But now eqns (6.1) and (6.8) are contradictory, which proves

Theorem 17. *No $T[4,1;6]$ design exists.* □

Corollary 17.1. *No pair of orthogonal Latin squares of order* 6 *exists, that is,* $N(6) = 1$. □

6.5. Almost orthogonal Latin squares of order 6

In proving Corollary 17.1, we had to consider the structure of a $T[4,1;6]$ design quite carefully. This suggests that perhaps we can come close to constructing orthogonal Latin squares of order 6.

Theorem 18. *Let L_1 and L_2 be Latin squares of order* 6. *Then superimposing L_2 on L_1 yields at most* 34 *of the* 36 *distinct ordered pairs of elements from* $\{1, 2, 3, 4, 5, 6\}$.

Proof. By Corollary 17.1, at least one of the 36 distinct ordered pairs does not appear when L_2 is superimposed on L_1. Suppose that only one such pair is missing, say (1, 1). Then without loss of generality, we may assume that the pair (1, 2) appears twice. But now 2, which appears six times in L_2, is paired twice with 1 in L_1, and hence with only four of the remaining five elements. Thus $(b, 2)$ is also missing, for some $b \in \{2, 3, 4, 5, 6\}$, proving the Theorem. □

There are in fact two 6×6 Latin squares which, when superimposed, yield all but two of the 36 distinct ordered pairs; see Table 6.15.

Suppose that we choose two orthogonal Latin squares of order 4, say L and S, of Table 6.1, and attempt to apply the construction of Theorem 5 with $M_1, M_2, N_1 = L, N_2 = S$, to give squares of order 24. Everything works,

Table 6.15

5	6	3	4	1	2
2	1	6	5	3	4
6	5	1	2	4	3
4	3	5	6	2	1
1	4	2	3	5	6
3	2	4	1	6	5

M_1

1	2	5	6	3	4
6	5	1	2	4	3
4	3	6	5	1	2
5	6	4	3	2	1
2	4	3	1	5	6
3	1	2	4	6	5

M_2

except that the identical 2×2 subsquares in the bottom right corners of M_1 and M_2 give rise to identical 8×8 subsquares in $M_1 \underline{\times} N_1$ and $M_2 \underline{\times} N_2$. But if we delete these 8×8 subsquares, and substitute for them two orthogonal 8×8 Latin squares on the appropriate elements, then we obtain two orthogonal Latin squares of order 24. One of them is shown in Table 6.16; see also Exercise 6.8. We are regarding M_1 and M_2 as *frames*, that is, as incomplete arrays which may be used in a recursive construction.

Table 6.16. L_1, from M_1 and N_1; the 8×8 square (shown in italics) was constructed independently of this method

17	19	20	18	21	23	24	22	9	11	12	10	13	15	16	14	1	3	4	2	5	7	8	6
20	18	17	19	24	22	21	23	12	10	9	11	16	14	13	15	4	2	1	3	8	6	5	7
18	20	19	17	22	24	23	21	10	12	11	9	14	16	15	13	2	4	3	1	6	8	7	5
19	17	18	20	23	21	22	24	11	9	10	12	15	13	14	16	3	1	2	4	7	5	6	8
5	7	8	6	1	3	4	2	21	23	24	22	17	19	20	18	9	11	12	10	13	15	16	14
8	6	5	7	4	2	1	3	24	22	21	23	20	18	17	19	12	10	9	11	16	14	13	15
6	8	7	5	2	4	3	1	22	24	23	21	18	20	19	17	10	12	11	9	14	16	15	13
7	5	6	8	3	1	2	4	23	21	22	24	19	17	18	20	11	9	10	12	15	13	14	16
21	23	24	22	17	19	20	18	1	3	4	2	5	7	8	6	13	15	16	14	9	11	12	10
24	22	21	23	20	18	17	19	4	2	1	3	8	6	5	7	16	14	13	15	12	10	9	11
22	24	23	21	18	20	19	17	2	4	3	1	6	8	7	5	14	16	15	13	10	12	11	9
23	21	22	24	19	17	18	20	3	1	2	4	7	5	6	8	15	13	14	16	11	9	10	12
13	15	16	14	9	11	12	10	17	19	20	18	21	23	24	22	5	7	8	6	1	3	4	2
16	14	13	15	12	10	9	11	20	18	17	19	24	22	21	23	8	6	5	7	4	2	1	3
14	16	15	13	10	12	11	9	18	20	19	17	22	24	23	21	6	8	7	5	2	4	3	1
15	13	14	16	11	9	10	12	19	17	18	20	23	21	22	24	7	5	6	8	3	1	2	4
1	3	4	2	13	15	16	14	5	7	8	6	9	11	12	10	*17*	*24*	*18*	*23*	*22*	*19*	*21*	*20*
4	2	1	3	16	14	13	15	8	6	5	7	12	10	9	11	*22*	*18*	*24*	*19*	*17*	*23*	*20*	*21*
2	4	3	1	14	16	15	13	6	8	7	5	10	12	11	9	*21*	*23*	*19*	*24*	*20*	*18*	*17*	*22*
3	1	2	4	15	13	14	16	7	5	6	8	11	9	10	12	*18*	*22*	*17*	*20*	*24*	*21*	*19*	*23*
9	11	12	10	5	7	8	6	13	15	16	14	1	3	4	2	*20*	*19*	*23*	*18*	*21*	*24*	*22*	*17*
12	10	9	11	8	6	5	7	16	14	13	15	4	2	1	3	*23*	*21*	*20*	*17*	*19*	*22*	*24*	*18*
10	12	11	9	6	8	7	5	14	16	15	13	2	4	3	1	*24*	*17*	*22*	*21*	*18*	*20*	*23*	*19*
11	9	10	12	7	5	6	8	15	13	14	16	3	1	2	4	*19*	*20*	*21*	*22*	*23*	*17*	*18*	*24*

6.6. References and comments

Hedayat and Shrikhande (1971) survey the relation between Latin squares, sets of MOLS and other designs used in experimental work. More generally, Cochran and G. M. Cox (1957), D. R. Cox (1958) and Pearce (1983), amongst others, discuss the choice of appropriate designs for use in experiments.

The book of Dénes and Keedwell (1974) gives an enormous amount of information on Latin squares and a detailed bibliography. For information on transversal designs, see Hanani (1975). The proof of non-existence of orthogonal Latin squares of order 6 is due to Stinson (1984); Betten (1983) gives another proof. The construction of five MOLS of order 12 is due to Johnson, Dulmage, and Mendelsohn (1961), and the construction of two almost orthogonal Latin squares of order 6 to Horton (1974). For an almost self-orthogonal Latin square of order 6, see Heinrich (1977). The Latin square of side 8 shown in italics in Table 6.16 was constructed by Franklin (1984*a*).

Two distinct sets of MOLS of the same order need not be isomorphic. Equivalence of such sets is discussed by Dénes and Keedwell (1974).

For a recent example of the use of frames, see C. J. Colbourn, Manson, and W. D. Wallis (1984).

Sometimes the same plots are used for two non-interacting sets of varieties. Each set of varieties is arranged as a BIBD and each pair of varieties, with one from the first set and the other from the second set, appears exactly once in some plot. Such designs are discussed by Preece (1966, 1976), Seberry (1979), and D. J. Street (1981). An example is given in Exercise 6.9.

Exercises

6.1. Prove Lemma 1. Is orthogonality a transitive property, that is, if H is orthogonal to K, and K to L, must H be orthogonal to L?

6.2. Construct the second square, L_2, of Example 6.

6.3. Show that for any value of n, there exists a $T[3,1;n]$ design.

6.4. If $k = n+1$, can a block of a $T[k, 1; n]$ design be disjoint from any other block? What if $k = n+2$?

6.5. Does there exist a $T[6, 1; 5]$ design?

6.6. Let V be an n-dimensional vector space over $GF[2]$. Show that V has an $(n-1)$-dimensional subspace, W, such that $\langle w, w \rangle = 0$ for every $w \in W$. Is it true that every $(n-1)$-dimensional subspace of V has this property?

6.7. Let V be a $2n$-dimensional vector space over $GF[2]$. Show that V has an n-dimensional subspace, W, such that $W = W^{\perp}$. Is it true that every n-dimensional subspace of V has this property?

6.8. Given that the 8×8 italicized subsquare in Table 6.16 is self-orthogonal, construct a Latin square L_2 orthogonal to L_1 of order 24 from M_2 and N_2.

6.9. Consider the starter blocks $\{(\infty, 0), (1, 3), (2, 6), (4, 5)\}$ and $\{(0, 0), (1, 4), (2, 1), (4, 2)\}$ to be developed modulo 7 by adding $(i, i), 0 \leq i \leq 6$, in turn. Verify that if the second element in each pair is ignored, the blocks form a $B[4, 3; 8]$ design, that if the first element in each pair is ignored, the blocks form a $B[4,2; 7]$ design and that, in the 14 blocks as given, each variety in the first design appears exactly once with each variety in the second design.

7 Further results on Latin squares

7.1. Recursive constructions of transversal designs

To construct sets of MOLS for certain orders, we use recursive methods. We state them in terms of transversal designs, making use of the correspondence between a $T[k, 1; n]$ design and a set of $k-2$ MOLS of order n. We define the symbol $T(k, \lambda)$ by

$$T(k, \lambda) = \{n \mid a \ T[k, \lambda; n] \text{ design exists}\}.$$

The simplest recursive construction we use shows that if $s \leq t$ and if $T[k+1, 1; t]$, $T[k, 1; s]$, $T[k, 1; m]$ and $T[k, 1; m+1]$ designs exist, then so does a $T[k, 1; mt+s]$ design.

Theorem 1. *Let s, m, t be given integers such that*

$$0 \leq s \leq t, \quad 1 \leq m, \quad t \in T(k+1, 1), \quad \text{and} \quad s, m, m+1 \in T(k, 1).$$

Then

$$mt + s \in T(k, 1).$$

Proof. Let $(X, \mathscr{G}, \mathscr{A})$ denote a $T[k+1, 1; t]$ design, where

$$\mathscr{G} = \{G_1, G_2, \ldots, G_k, H\},$$

and the varieties are so labelled that

$$G_i = \{x_1^i, x_2^i, \ldots, x_t^i\}, \ i = 1, \ldots, k,$$
$$H = \{y_1, y_2, \ldots, y_t\}.$$

Since $s \leq t$, we may choose an s-subset $S \subseteq H$, and label H so that

$$S = \{y_1, y_2, \ldots, y_s\}.$$

We construct a $T[k, 1; mt+s]$ design. It has, as its $k(mt+s)$ varieties,

(i) m copies each of the kt varieties

$$x_j^i, \ i = 1, \ldots, k; \ j = 1, \ldots, t,$$

labelled as x_{jl}^i, $l = 1, \ldots, m$,

and

(ii) k copies each of the s varieties

$$y_p, \ p = 1, \ldots, s,$$

labelled as y_{pq}, $q = 1, \ldots, k$.

Its groups G_i^* are given by

$$G_i^* = \{x_{jl}^i \mid j = 1, \ldots, t; l = 1, \ldots, m\} \cup \{y_{pi} \mid p = 1, \ldots, s\},$$

for $i = 1, \ldots, k$.

To find the blocks of the new design, consider a block $A \in \mathscr{A}$ of the original $T[k+1, 1; t]$ design. Since A is a block of a transversal design, it must have the form

$$A = \{x_{j_1}^1, x_{j_2}^2, \ldots, x_{j_k}^k, y_p\},$$

where $j_1, j_2, \ldots, j_k, p \in \{1, 2, \ldots, t\}$; if $p \leq s$, then $A \cap S = \{y_p\}$; if $p > s$, then $A \cap S = \varnothing$.

Consider first a block A which is disjoint from S. We construct a $T[k, 1; m]$ design on the set of varieties

$$\{x_{j_i, l}^i \mid i = 1, \ldots, k; l = 1, \ldots, m\},$$

whose ith group is the intersection of this set with G_i^*, that is,

$$G_i' = \{x_{j_i, l}^i \mid l = 1, \ldots, m\}, \text{ for each } i = 1, \ldots, k.$$

The set of blocks of this $T[k, 1; m]$ design will be denoted by $\mathscr{B}(A)$.

Now consider a block A which intersects S. We construct a $T[k, 1; m+1]$ design on the set of varieties

$$\{x_{j_i, l}^i \mid i = 1, \ldots, k; l = 1, \ldots, m\} \cup \{y_{pq} \mid q = 1, \ldots, k\},$$

whose ith group is, similarly,

$$G_i' = \{x_{j_i, l}^i \mid l = 1, \ldots, m\} \cup \{y_{pi}\}, \text{ for each } i = 1, \ldots, k.$$

We choose the labelling of this design so that it contains a block

$$A^* = \{y_{p1}, y_{p2}, \ldots, y_{pk}\},$$

and we denote by $\mathscr{B}(A)$ the set of all blocks of the $T[k, 1; m+1]$ design other than A^*.

Finally, we construct a $T[k, 1; s]$ design on the set

$$\{y_{pq} \mid p = 1, \ldots, s; q = 1, \ldots, k\}$$

with groups

$$H_q = \{y_{1q}, y_{2q}, \ldots, y_{sq}\}, \text{ for each } q = 1, \ldots, k.$$

(Note that $H_i \subseteq G_i^*$, for each i.) Its set of blocks is denoted by $\mathscr{C}$.

We now combine these designs to form a $T[k, 1; mt+s]$ design with the varieties and groups listed earlier and the set of blocks

$$(\bigcup_{A \in \mathscr{A}} \mathscr{B}(A)) \cup \mathscr{C}.$$

It follows from the construction that no two elements of the same group can occur together in the same block. To complete the proof we must show that two elements of different groups must occur together in a unique common block. Four cases arise.

(a) The elements are $x^a_{j_a l_a}$ and $x^b_{j_b l_b}$, where the block A of the original design containing $x^a_{j_a}$ and $x^b_{j_b}$ was disjoint from S. These new elements occur together in precisely one block of $\mathscr{B}(A)$.

(b) The elements are $x^a_{j_a l_a}$ and $x^b_{j_b l_b}$, where the block A of the original design containing $x^a_{j_a}$ and $x^b_{j_b}$ intersected S in y_p. These two elements occur together in precisely one block of $\mathscr{B}(A)$, since the discarded block A^* contains only elements of the form y_{pq}.

(c) The elements are $x^a_{j_a l_a}$ and y_{pq}, again occurring together in precisely one block of $\mathscr{B}(A)$, where $x^a_{j_a}$ and y_p occur together in the block A of the original design.

(d) The elements are y_{fa} and y_{gb}, where $f \neq g$ but a and b may or may not be distinct. The only blocks of the new design which contain more than one element of this form are those of $\mathscr{C}$; precisely one such block contains y_{fa} and y_{gb}. □

Example 2. We choose $s = 2$, $k = t = 3$, $m = 4$, and from $T[4, 1; 3]$, $T[3, 1; 4]$ and $T[3, 1; 5]$ designs we construct a $T[3, 1; 14]$ design.

The groups and blocks of the original $T[4, 1; 3]$ design are shown in Table 7.1; the starred blocks are those disjoint from $S = \{y_1, y_2\}$.

To get the $k(mt + s) = 42$ varieties of the $T[3, 1; 14]$ design, we need 4 copies of the 9 varieties of $G_1 \cup G_2 \cup G_3$, and 3 copies of the 2 varieties in S. These are placed in the groups G_1^*, G_2^*, G_3^* shown in Table 7.2.

Table 7.1. $T[4, 1; 3]$ design

G_1: x^1_1 x^1_2 x^1_3
G_2: x^2_1 x^2_2 x^2_3
G_3: x^3_1 x^3_2 x^3_3
H: $\underbrace{y_1 \quad y_2}_{S}$ y_3

x^1_1	x^2_1	x^3_1	y_1	A_1
x^1_1	x^2_2	x^3_3	y_2	A_2
x^1_1	x^2_3	x^3_2	y_3	A_3^*
x^1_2	x^2_1	x^3_2	y_3	A_4^*
x^1_2	x^2_2	x^3_1	y_1	A_5
x^1_2	x^2_3	x^3_3	y_2	A_6
x^1_3	x^2_1	x^3_3	y_2	A_7
x^1_3	x^2_2	x^3_2	y_3	A_8^*
x^1_3	x^2_3	x^3_1	y_1	A_9

Table 7.2. Groups of the $T[3, 1; 14]$ design

G_1^*:	x^1_{11}	x^1_{12}	x^1_{13}	x^1_{14}	x^1_{21}	x^1_{22}	x^1_{23}	x^1_{24}	x^1_{31}	x^1_{32}	x^1_{33}	x^1_{34}	y_{11}	y_{21}
G_2^*:	x^2_{11}	x^2_{12}	x^2_{13}	x^2_{14}	x^2_{21}	x^2_{22}	x^2_{23}	x^2_{24}	x^2_{31}	x^2_{32}	x^2_{33}	x^2_{34}	y_{12}	y_{22}
G_3^*:	x^3_{11}	x^3_{12}	x^3_{13}	x^3_{14}	x^3_{21}	x^3_{22}	x^3_{23}	x^3_{24}	x^3_{31}	x^3_{32}	x^3_{33}	x^3_{34}	y_{13}	y_{23}

Table 7.3. $T[3, 1; 5]$ design from A_1

The groups	The blocks				
G_1': $x_{11}^1\ x_{12}^1\ x_{13}^1\ x_{14}^1\ y_{11}$	→ $y_{11}\ y_{12}\ y_{13}$	$x_{14}^1\ y_{12}\ x_{14}^3$	$x_{13}^1\ y_{12}\ x_{13}^3$	$x_{12}^1\ y_{12}\ x_{12}^3$	$x_{11}^1\ y_{12}\ x_{11}^3$
G_2': $x_{11}^2\ x_{12}^2\ x_{13}^2\ x_{14}^2\ y_{12}$	$y_{11}\ x_{14}^2\ x_{14}^3$	$x_{14}^1\ x_{14}^2\ x_{13}^3$	$x_{13}^1\ x_{14}^2\ x_{12}^3$	$x_{12}^1\ x_{14}^2\ x_{11}^3$	$x_{11}^1\ x_{14}^2\ y_{13}$
G_3': $x_{11}^3\ x_{12}^3\ x_{13}^3\ x_{14}^3\ y_{13}$	$y_{11}\ x_{13}^2\ x_{13}^3$	$x_{14}^1\ x_{13}^2\ x_{12}^3$	$x_{13}^1\ x_{13}^2\ x_{11}^3$	$x_{12}^1\ x_{13}^2\ y_{13}$	$x_{11}^1\ x_{13}^2\ x_{14}^3$
	$y_{11}\ x_{12}^2\ x_{12}^3$	$x_{14}^1\ x_{12}^2\ x_{11}^3$	$x_{13}^1\ x_{12}^2\ y_{13}$	$x_{12}^1\ x_{12}^2\ x_{14}^3$	$x_{11}^1\ x_{12}^2\ x_{13}^3$
	$y_{11}\ x_{11}^2\ x_{11}^3$	$x_{14}^1\ x_{11}^2\ y_{13}$	$x_{13}^1\ x_{11}^2\ x_{14}^3$	$x_{12}^1\ x_{11}^2\ x_{13}^3$	$x_{11}^1\ x_{11}^2\ x_{12}^3$

The groups

The blocks; the arrowed block is deleted to give $\mathscr{B}(A_1)$.

Table 7.4. $T[3, 1; 4]$ design from A_3

The groups	The blocks of $\mathscr{B}(A_3)$			
G_1': $x_{11}^1\ x_{12}^1\ x_{13}^1\ x_{14}^1$	$x_{11}^1\ x_{31}^2\ x_{21}^3$	$x_{12}^1\ x_{31}^2\ x_{22}^3$	$x_{13}^1\ x_{31}^2\ x_{23}^3$	$x_{14}^1\ x_{31}^2\ x_{24}^3$
G_2': $x_{31}^2\ x_{32}^2\ x_{33}^2\ x_{34}^2$	$x_{11}^1\ x_{32}^2\ x_{22}^3$	$x_{12}^1\ x_{32}^2\ x_{23}^3$	$x_{13}^1\ x_{32}^2\ x_{24}^3$	$x_{14}^1\ x_{32}^2\ x_{21}^3$
G_3': $x_{21}^3\ x_{22}^3\ x_{23}^3\ x_{24}^3$	$x_{11}^1\ x_{33}^2\ x_{23}^3$	$x_{12}^1\ x_{33}^2\ x_{24}^3$	$x_{13}^1\ x_{33}^2\ x_{21}^3$	$x_{14}^1\ x_{33}^2\ x_{22}^3$
	$x_{11}^1\ x_{34}^2\ x_{24}^3$	$x_{12}^1\ x_{34}^2\ x_{21}^3$	$x_{13}^1\ x_{34}^2\ x_{22}^3$	$x_{14}^1\ x_{34}^2\ x_{23}^3$

The groups

The blocks of $\mathscr{B}(A_3)$

Table 7.5. $T[3, 1; 2]$ design from S

The groups	The blocks of $\mathscr{C}$	
H_1': $y_{11}\ y_{21}$	$y_{11}\ y_{12}\ y_{13}$	$y_{21}\ y_{12}\ y_{23}$
H_2': $y_{12}\ y_{22}$	$y_{11}\ y_{22}\ y_{23}$	$y_{21}\ y_{22}\ y_{13}$
H_3'' $y_{13}\ y_{23}$		

The groups

The blocks of $\mathscr{C}$

Corresponding to each of the three blocks disjoint from S, namely A_3, A_4, A_8, we construct a $T[3, 1; 4]$ design; all these blocks are used, giving $3 \times 16 = 48$ blocks. Corresponding to each of the six blocks which intersect S, namely $A_1, A_2, A_5, A_6, A_7, A_9$, we construct a $T[3, 1; 5]$ design, which has 25 blocks by Lemma 6.3.9; we discard one block from each of these designs, keeping $6 \times 24 = 144$ of these blocks. Corresponding to the set S, we construct a $T[3, 1; 2]$ design, giving the last four blocks. Thus we have altogether

$$144 + 48 + 4 = 196 = (14)^2$$

blocks, as required for a $T[3, 1; 14]$ design. Tables 7.3, 7.4, and 7.5 show some of the blocks. □

The next construction is a generalization of the last.

Theorem 3. *Let s_1, s_2, m, t be given integers such that*

$0 \le s_2 \le s_1 \le t$, $1 \le m$, $t \in T(k+2, 1)$, *and* $s_1, s_2, m, m+1, m+2 \in T(k, 1)$.

Then

$$mt + s_1 + s_2 \in T(k, 1).$$

Proof. Let $(X, \mathscr{G}, \mathscr{A})$ denote a $T[k+2, 1; t]$ design where

$$\mathscr{G} = \{G_1, G_2, \ldots, G_k, H_1, H_2\}$$

and the varieties are so labelled that

$$G_i = \{x_1^i, x_2^i, \ldots, x_t^i\}, i = 1, 2, \ldots, k,$$
$$H_h = \{y_1^h, y_2^h, \ldots, y_t^h\}, h = 1, 2,$$

where $S_h \subseteq H_h$, and

$$S_1 = \{y_1^1, \ldots, y_{s_1}^1\}, S_2 = \{y_1^2, \ldots, y_{s_2}^2\}.$$

Again, we take m copies each of the kt varieties x_j^i, $i = 1, \ldots, k$; $j = 1, \ldots, t$, labelled as x_{jl}^i, $l = 1, \ldots, m$, and k copies each of the $(s_1 + s_2)$ varieties y_p^h, $h = 1, 2$; $p = 1, 2, \ldots, s_h$, labelled as y_{pq}^h, $q = 1, \ldots, k$. These are the varieties of our $T[k, 1; mt + s_1 + s_2]$ design. Its groups are $G_1^*, G_2^*, \ldots, G_k^*$, where

$$G_i^* = \{x_{jl}^i \mid j = 1, \ldots, t;\ l = 1, \ldots, m\} \cup \{y_{pi}^1 \mid p = 1, \ldots, s_1\} \cup \{y_{pi}^2 \mid p = 1, \ldots, s_2\},$$

for $i = 1, 2, \ldots, k$.

To find the blocks of the new design, consider a block A of the original $T[k+2, 1; t]$ design. A has the form

$$A = \{x_{j_1}^1, x_{j_2}^2, \ldots, x_{j_k}^k, y_{p_1}^1, y_{p_2}^2\},$$

where $j_1, j_2, \ldots, j_k, p_1, p_2 \in \{1, 2, \ldots, t\}$. This time three cases arise.

(a) $A \cap (S_1 \cup S_2) = \varnothing$. We construct a $T[k, 1; m]$ design on the set

$$\{x_{j_i,l}^i \mid i = 1, \ldots, k;\ l = 1, \ldots, m\},$$

with groups

$$G_i' = \{x_{j_i,l}^i \mid l = 1, \ldots, m\}, \quad \text{for } i = 1, \ldots, k.$$

The set of blocks of this design is $\mathscr{B}(A)$.

(b) $A \cap (S_1 \cup S_2) = \{y_{p_h}^h\}$ for $h = 1$ or 2. We construct a $T[k, 1; m+1]$ design on the set

$$\{x^i_{j_i,l} | i = 1, \ldots, k; l = 1, \ldots, m\} \cup \{y^h_{p_h q} | q = 1, \ldots, k\}$$

with groups

$$G'_i = \{x^i_{j_i l} | l = 1, \ldots, m\} \cup \{y^h_{p_h i}\}, \quad \text{for } i = 1, \ldots, k.$$

We choose the labelling of the design so that it has the block

$$A^* = \{y^h_{p_h 1}, y^h_{p_h 2}, \ldots, y^h_{p_h k}\},$$

and let $\mathscr{B}(A)$ denote the set of all blocks of this design except for A^*.

(c) $A \cap (S_1 \cup S_2) = \{y^1_{p1}, y^2_{p2}\}$. This time we construct a $T[k, 1; m+2]$ design on the set

$$\{x^i_{j_i l} | i = 1, \ldots, k; l = 1, \ldots, m\} \cup \{y^h_{p_h q} | h = 1, 2; q = 1, \ldots, k\},$$

with groups

$$G'_i = \{x^i_{j_i l} | l = 1, \ldots, m\} \cup \{y^1_{p_1 i}, y^2_{p_2 i}\}, \ i = 1, \ldots, k.$$

Since $k \leq m+2$ (see Exercise 7.3), we can apply Lemma 6.3.10, showing that any given block of this design has another block disjoint from it. We choose the labelling so that the design has blocks

$$A^*_1 = \{y^1_{p_1 1}, y^1_{p_1 2}, \ldots, y^1_{p_1 k}\}$$

and

$$A^*_2 = \{y^2_{p_2 1}, y^2_{p_2 2}, \ldots, y^2_{p_2 k}\},$$

and let $\mathscr{B}(A)$ denote the set of all the blocks of this design except A^*_1 and A^*_2.

Finally we construct a $T[k, 1; s_1]$ design on the set

$$\{y^1_{pq} | p = 1, \ldots, s_1; q = 1, \ldots, k\}$$

with groups

$$H^1_q = \{y^1_{1q}, \ldots, y^1_{s_1 q}\} \quad \text{for } q = 1, \ldots, k,$$

where we denote its set of blocks by $\mathscr{C}_1$, and similarly a $T[k, 1; s_2]$ design on the set

$$\{y^2_{pq} | p = 1, \ldots, s_2; q = 1, \ldots, k\}$$

with groups

$$H^2_q = \{y^2_{1q}, \ldots, y^2_{s_2 q}\}, \quad \text{for } q = 1, \ldots, k,$$

with set of blocks $\mathscr{C}_2$.

Now we take the $T[k, 1; mt + s_1 + s_2]$ design with varieties and groups as listed, and blocks

$$\left(\bigcup_{A \in \mathscr{A}} \mathscr{B}(A)\right) \cup \mathscr{C}_1 \cup \mathscr{C}_2.$$

The proof that every pair of varieties occurs in a unique block is left as Exercise 7.2. □

7.2. Sets of three MOLS: small orders

In the remainder of this chapter we shall prove that there exist sets of three MOLS of order n, for $n \neq 2, 3, 6, 10$, that is, that $N(n) \geq 3$ for all other values of n. So far we have seen that (Corollary 6.2.5.1)

$$N(2) = N(6) = 1$$

and that

$$N(3) = 2,\ N(4) = 3,\ N(9) = 8. \tag{7.1}$$

Further, by Corollary 6.2.5.1, we know that if

$$n = p_1^{e_1} p_2^{e_2} \ldots p_h^{e_h},$$

then

$$N(n) \geq \min(p_1^{e_1} - 1,\ p_2^{e_2} - 1,\ \ldots,\ p_h^{e_h} - 1).$$

So if $gcd(n, 6) = 1$, then

$$N(n) \geq 5 - 1 = 4. \tag{7.2}$$

Similarly, if n is odd and $9|n$, or if $3 \nmid n$ and $4|n$, or if $36|n$, then again

$$N(n) \geq 3. \tag{7.3}$$

But if $n \equiv 2 \pmod 4$ and/or $n \equiv 3$ or $6 \pmod 9$, then Corollary 6.2.5.1 is not enough to tell us whether or not $N(n) \geq 3$.

For $n = 10$ it is known only that $N(n) \geq 2$; Table 7.6 shows a self-orthogonal Latin square of side 10 constructed in the following way. We take the initial row (row 0) of a 9×9 array L to be

0 2 4 7 1 8 5 3 6

Table 7.6

0	2	4	7	1	8	*5*	3	6
7	1	3	5	8	2	0	*6*	4
5	8	2	4	6	0	3	1	*7*
8	6	0	3	5	7	1	4	2
3	*0*	7	1	4	6	8	2	5
6	4	*1*	8	2	5	7	0	3
4	7	5	*2*	0	3	6	8	1
2	5	8	6	*3*	1	4	7	0
1	3	6	0	7	*4*	2	5	8

(a)

0	2	4	7	1	8	x	3	6	5
7	1	3	5	8	2	0	x	4	6
5	8	2	4	6	0	3	1	x	7
x	6	0	3	5	7	1	4	2	8
3	x	7	1	4	6	8	2	5	0
6	4	x	8	2	5	7	0	3	1
4	7	5	x	0	3	6	8	1	2
2	5	8	6	x	1	4	7	0	3
1	3	6	0	7	x	2	5	8	4
8	0	1	2	3	4	5	6	7	x

(b)

and then define rows 1 to 8 by

$$l_{ij} = l_{0,j-i} + i \pmod 9$$

giving the array of Table 7.6(a), a *cyclic* Latin square of order 9. (The terminology here is confusing: a *circulant* array has $l_{ij} = l_{0,j-i}$.)

To construct our 10×10 self-orthogonal Latin square K, we adjoin a new symbol, x, to the set of integers modulo 9 and define the entries of K as follows:

$$k_{ij} = \begin{cases} l_{ij}, & \text{if } i \neq 9, j \neq 9 \text{ and } j \not\equiv i+6 \pmod 9; \\ l_{i,i+6} & \text{if } j = 9 \text{ and } i \neq 9; \\ l_{j-3,j} & \text{if } i = 9 \text{ and } j \neq 9; \\ x & \text{if } j \equiv i+6 \pmod 9 \text{ or if } i = j = 9. \end{cases}$$

K is shown in Table 7.6(b). Less formally, we think of the construction in the following way. We remove from L the broken diagonal shown in italics in Table 7.6(a), starting from (0, 6) in L, and copy its elements into the corresponding positions in the last row and column of K. Thus, the element in position $(i, i+6)$ of L is copied into positions $(i, 9)$ and $(9, i+6)$ of L. Position (9, 9) of K, and each position of the vacant broken diagonal of L, is now filled with a new symbol, x. Finally, the modified array L is copied into rows and columns 0 through 8 of K.

This square, K, is generally denoted by the sequence

$$0 \quad 2 \quad 4 \quad 7 \quad 1 \quad 8 \quad (x5) \quad 3 \quad 6,$$

and is called a *bordered* cyclic Latin square of order 10. A similar construction starting from the sequence

$$0 \quad 9 \quad 5 \quad 8 \quad 10 \quad 2 \quad 4 \quad 6 \quad 12 \quad 3 \quad 11 \quad 7 \quad (x1)$$

gives a bordered cyclic Latin square of order 14 which is also self-orthogonal. However, a set of three MOLS of order 14 is now known; see Table 7.7.

We have already seen, in Section 6.2, that $N(12) \geq 5$. There are still several other special cases that we need, in order to start our recursive constructions.

First, $N(15) \geq 4$. Our squares with rows and columns 0 to 14 are:

(i) a Latin square C with constant diagonals, and initial row

$$0, 1, 2, \ldots, 14,$$

thus

$$c_{i,j} = j - i \pmod{15};$$

(ii) a symmetric cyclic Latin square S with initial row

$$0, 7, 4, 1, 12, 14, 11, 10, 3, 5, 9, 8, 13, 2, 6$$

and

$$s_{ij} = s_{0,j-1} + i \pmod{15};$$

Table 7.7. Three MOLS of order 14

1	2	3	4	5	6	7	8	9	A	B	C	D	E
2	1	6	A	4	D	8	C	E	5	7	9	B	3
3	4	1	7	B	5	E	9	D	2	6	8	A	C
4	D	5	1	8	C	6	2	A	E	3	7	9	B
5	C	E	6	1	9	D	7	3	B	2	4	8	A
6	B	D	2	7	1	A	E	8	4	C	3	5	9
7	A	C	E	3	8	1	B	2	9	5	D	4	6
8	7	B	D	2	4	9	1	C	3	A	6	E	5
9	6	8	C	E	3	5	A	1	D	4	B	7	2
A	3	7	9	D	2	4	6	B	1	E	5	C	8
B	9	4	8	A	E	3	5	7	C	1	2	6	D
C	E	A	5	9	B	2	4	6	8	D	1	3	7
D	8	2	B	6	A	C	3	5	7	9	E	1	4
E	5	9	3	C	7	B	D	4	6	8	A	2	1

1	2	3	4	5	6	7	8	9	A	B	C	D	E
4	A	7	1	3	8	C	2	6	B	E	5	9	D
5	E	B	8	1	4	9	D	3	7	C	2	6	A
6	B	2	C	9	1	5	A	E	4	8	D	3	7
7	8	C	3	D	A	1	6	B	2	5	9	E	4
8	5	9	D	4	E	B	1	7	C	3	6	A	2
9	3	6	A	E	5	2	C	1	8	D	4	7	B
A	C	4	7	B	2	6	3	D	1	9	E	5	8
B	9	D	5	8	C	3	7	4	E	1	A	2	6
C	7	A	E	6	9	D	4	8	5	7	1	B	3
D	4	8	B	2	7	A	E	5	9	6	3	1	C
E	D	5	9	C	3	8	B	2	6	A	7	4	1
2	1	E	6	A	D	4	9	C	3	7	B	8	5
3	6	1	2	7	B	E	5	A	D	4	8	C	9

1	2	3	4	5	6	7	8	9	A	B	C	D	E
E	3	B	D	8	A	6	5	2	C	9	4	7	1
2	1	4	C	E	9	B	7	6	3	D	A	5	8
3	9	1	5	D	2	A	C	8	7	4	E	B	6
4	7	A	1	6	E	3	B	D	9	8	5	2	C
5	D	8	B	1	7	2	4	C	E	A	9	6	3
6	4	E	9	C	1	8	3	5	D	2	B	A	7
7	8	5	2	A	D	1	9	4	6	E	3	C	B
8	C	9	6	3	B	E	1	A	5	7	2	4	D
9	E	D	A	7	4	C	2	1	B	6	8	3	5
A	6	2	E	B	8	5	D	3	1	C	7	9	4
B	5	7	3	2	C	9	6	E	4	1	D	8	A
C	B	6	8	4	3	D	A	7	2	5	1	E	9
D	A	C	7	9	5	4	E	B	8	3	6	1	2

(iii) a self-orthogonal cyclic Latin square A, with initial row

$$0, 2, 11, 6, 10, 13, 5, 14, 4, 7, 12, 1, 9, 8, 3$$

and

$$a_{ij} = a_{0,\, j-i} + i \pmod{15};$$

(iv) A^T, the transpose of A.

Next, $N(46) \geq 4$. We shall prove this by a difference set construction which produces a $T[6, 1; 46]$ design. We start from Z_{37}, the field of order 37, and let C_0 denote its set of quartic residues, so that

$$C_0 = \{7, 12, 10, 33, 9, 26, 34, 16, 1\}.$$

We now choose a set X, labelled by the elements of C_0, so that

$$X = \{x_c \mid c \in C_0\},$$

and we base our transversal design on the set

$$S = Z_{37} \cup X.$$

As the group G_i of the design, for each $i = 1, \ldots, 6$, we take the set S_i; that is, the elements of S with subscript i. Thus, in each case,

$$G_i = \{0_i, 1_i, 2_i, \ldots, 36_i, x_{7,\, i}, x_{12,\, i}, \ldots, x_{1,\, i}\}.$$

We define addition in the following way:
if $g \in Z_{37}$, then

$$h_i + g = (h+g)_i \quad \text{for } h \in Z_{37}$$

where $h+g$ is evaluated modulo 37, and

$$x_{c,\, i} + g = x_{c,\, i} \quad \text{for } x_c \in X$$

so the elements of X behave like infinity elements. Further if $s_1, \ldots, s_l \in S$, then we define

$$(s_1, \ldots, s_l) + g = (s_1 + g, \ldots, s_l + g).$$

The $(46)^2 = 2116$ blocks are chosen as follows:

(i) the 81 blocks of a $T[6, 1; 9]$ design with group i the set X_i, that is

$$\{x_{7,\, i}, x_{12,\, i}, \ldots, x_{1,\, i}\} \subseteq G_i;$$

(ii) the 37 blocks of the form

$$\{0_1, 0_2, 0_3, 0_4, 0_5, 0_6\} + g,$$

for $g = 0, 1, \ldots, 36$;

(iii) the $6 \times 9 \times 37$ blocks of the form

$$\{x_{c,\, 1}, c_2, 13c_3, 21c_4, 14c_5, 34c_6\} + g,$$
$$\{34c_1, x_{c,\, 2}, c_3, 13c_4, 21c_5, 14c_6\} + g,$$
$$\{14c_1, 34c_2, x_{c,\, 3}, c_4, 13c_5, 21c_6\} + g,$$

$$\{21c_1, 14c_2, 34c_3, x_{c,4}, c_5, 13c_6\} + g,$$
$$\{13c_1, 21c_2, 14c_3, 34c_4, x_{c,5}, c_6\} + g,$$
$$\{c_1, 13c_2, 21c_3, 14c_4, 34c_5, x_{c,6}\} + g,$$

for $g \in Z_{37}$, $c \in C_0$. (Note that if $c = 7$, for instance, then $13c_3$ denotes $91_3 \equiv 17_3 \pmod{37}$.)

Now by Theorem 6.3.12, the existence of this $T[6, 1; 46]$ design is equivalent to the existence of a set of four MOLS of order 46.

The remaining orders of MOLS needed to complete the proof are $n = 18$, 22, 26, 30, 34, 38, 42. For each order we construct a self-orthogonal Latin square, L, and a symmetric Latin square, S, orthogonal to L (and thus to L^T). Each square consists of four subarrays, thus

$$L = \begin{bmatrix} L_1 & L_2 \\ L_3 & L_4 \end{bmatrix} \quad \text{and} \quad S = \begin{bmatrix} S_1 & S_2 \\ S_3 & S_4 \end{bmatrix},$$

where for $n = 18, 22, 26, 30$, we have L_1 and S_1 each 4×4, and for $n = 34$, 38, 42, we have L_1 and S_1 each 8×8. The remaining subarrays fill in the $n \times n$ squares. Tables 7.8 and 7.9 show L_1, S_1 for the two cases.

Table 7.8

W	Y	Z	X
Z	X	W	Y
X	Z	Y	W
Y	W	X	Z

L_1

Z	X	W	Y
X	Z	Y	W
W	Y	Z	X
Y	W	X	Z

S_1

for $n = 18$, 22, 26, 30

Table 7.9

W	Z	y	x	Y	w	z	X
y	X	W	Y	x	z	w	Z
x	z	Y	W	y	X	Z	w
z	x	w	Z	X	y	W	Y
Z	W	X	z	w	Y	x	y
X	y	Z	w	z	x	Y	W
w	Y	z	X	Z	W	y	x
Y	w	x	y	W	Z	X	z

L_1

z	W	X	Y	Z	w	x	y
W	z	w	x	y	Z	X	Y
X	w	z	W	Y	x	y	Z
Y	x	W	z	X	y	Z	w
Z	y	Y	X	z	W	w	x
w	Z	x	y	W	z	Y	X
x	X	y	Z	w	Y	z	W
y	Y	Z	w	x	X	W	z

S_1

for $n = 34$, 38, 42

For $n = 18, 22, 26, 30$, the rows and columns of L and S are labelled by W, X, Y, Z, 0, 1, . . . , $n-5$, where the rows and columns of L_4 and S_4 are those labelled 0, 1, . . . , $n-5$. These matrices (L_4, S_4) are cyclic for each case, so that

$$a_{ij} = a_{0,\, j-i} + i \pmod{n-4},$$

with the following conventions:

if $g, h \in Z_{n-4}$, then the addition is simply modulo $n-4$;
$W+g = W, \qquad Z+g = Z;$
addition involving X or Y is defined differently in the two matrices, so that in L, we have

$$X+g = X \quad \text{and} \quad Y+g = Y,$$

whereas in S we have

$$X+1 = Y \quad \text{and } Y+1 = X, \text{ or in other words, } X+g = \begin{cases} Y, & \text{if } g \text{ is odd,} \\ X, & \text{if } g \text{ is even.} \end{cases}$$

The remaining matrices are L_2 and S_2 (each $4 \times (n-4)$) and L_3 and S_3 (each $(n-4) \times 4$); they contain only elements of Z_{n-4} and are defined by:

$$a_{ij} = a_{i0} + j \pmod{n-4}, \quad \text{for } L_2, S_2;$$
$$a_{ij} = a_{0j} + i \pmod{n-4}, \quad \text{for } L_3, S_3.$$

Thus, in each case we need only specify the initial columns of L_2 and S_2, and the initial rows of L_3, S_3, L_4, S_4. These can be chosen as in Table 7.10. Note that the first column of S_2 is the transpose of the first row of S_3. The resulting squares are shown in Table 7.11 for $n = 18$.

Table 7.12 shows the corresponding initial rows and columns for $n = 34, 38, 42$, working modulo $n-8$ since the subsquares are of order 8. For the remaining eight elements of the set $\mathscr{E} = \{W, X, Y, Z, w, x, y, z\}$, we have in the square L

$$E+g = E \text{ for all } E \in \mathscr{E}, \text{ for all } g \in Z_{n-8},$$

but in the square S we have

$$W+g = W, z+g = z, \quad \text{for all } g \in Z_{n-8},$$
$$X+1 = w, w+1 = X,$$
$$Y+1 = x, x+1 = Y,$$
$$Z+1 = y, y+1 = Z.$$

7.3. Sets of three MOLS: the recursive construction

We now have all the ingredients from which to construct sets of three MOLS of order n, $n \notin \{2, 3, 6, 10\}$. First from Theorem 1, we have

Corollary 1.1. *If* $0 \leq s \leq t$, *then*

$$N(mt+s) \geq \min\{N(m), N(m+1), N(t)-1, N(s)\}.$$

Table 7.10. Initial rows and columns for constructing sets of three MOLS, $n = 18, 22, 26, 30$

Initial columns for L_2, S_2:

L_2	S_2	L_2	S_2	L_2	S_2	L_2	S_2
13	0	7	11	10	18	13	22
3	10	5	0	20	5	8	7
12	8	4	6	7	20	15	3
2	1	16	1	3	4	18	8
$n = 18$		$n = 22$		$n = 26$		$n = 30$	

Initial rows for L_3, S_3, L_4, S_4:

$n = 18$:

L_3:	8	5	4	6
S_3:	0	10	8	1

L_4:	0	7	13	12	11	10	2	1	9	Z	Y	X	W	3
S_4:	Z	X	13	6	9	7	4	W	12	2	5	3	11	Y

$n = 22$:

L_3:	2	3	10	12
S_3:	11	0	6	1

L_4:	0	9	17	16	15	14	8	6	11	1	4	7	13	Z	Y	X	W	5
S_4:	Z	X	17	10	2	13	9	12	4	W	14	5	3	8	16	7	15	Y

$n = 26$:

L_3:	3	6	4	2																	
S_3:	18	5	20	4																	
L_4:	0	9	21	20	19	18	15	12	10	13	16	1	11	14	8	7	5	Z	Y	X	W	17
S_4:	Z	X	21	9	16	13	7	17	0	11	3	W	15	2	14	10	1	8	12	6	19	Y

$n = 30$:

L_3:	2	3	6	16																					
S_3:	22	7	3	8																					
L_4:	0	7	25	24	23	22	4	11	15	18	21	12	14	1	8	20	19	13	17	5	10	Z	Y	X	W	9
S_4:	Z	X	25	9	2	20	17	19	18	13	5	1	0	W	14	16	21	4	10	12	11	15	24	6	23	Y

Table 7.11. MOLS of order 18 with $a = 10$, $b = 11$, $c = 12$, $d = 13$

	W	X	Y	Z	0	1	2	3	4	5	6	7	8	9	a	b	c	d
W	W	Y	Z	X	d	0	1	2	3	4	5	6	7	8	9	a	b	c
X	Z	X	W	Y	3	4	5	6	7	8	9	a	b	c	d	0	1	2
Y	X	Z	Y	W	c	d	0	1	2	3	4	5	6	7	8	9	a	b
Z	Y	W	X	Z	2	3	4	5	6	7	8	9	a	b	c	d	0	1
0	8	5	4	6	0	7	d	c	b	a	2	1	9	Z	Y	X	W	3
1	9	6	5	7	4	1	8	0	d	c	b	3	2	a	Z	Y	X	W
2	a	7	6	8	W	5	2	9	1	0	d	c	4	3	b	Z	Y	X
3	b	8	7	9	X	W	6	3	a	2	1	0	d	5	4	c	Z	Y
4	c	9	8	a	Y	X	W	7	4	b	3	2	1	0	6	5	d	Z
5	d	a	9	b	Z	Y	X	W	8	5	c	4	3	2	1	7	6	0
6	0	b	a	c	1	Z	Y	X	W	9	6	d	5	4	3	2	8	7
7	1	c	b	d	8	2	Z	Y	X	W	a	7	0	6	5	4	3	9
8	2	d	c	0	a	9	3	Z	Y	X	W	b	8	1	7	6	5	4
9	3	0	d	1	5	b	a	4	Z	Y	X	W	c	9	2	8	7	6
a	4	1	0	2	7	6	c	b	5	Z	Y	X	W	d	a	3	9	8
b	5	2	1	3	9	8	7	d	c	6	Z	Y	X	W	0	b	4	a
c	6	3	2	4	b	a	9	8	0	d	7	Z	Y	X	W	1	c	5
d	7	4	3	5	6	c	b	a	9	1	0	8	Z	Y	X	W	2	d

L, self-orthogonal.

	W	X	Y	Z	0	1	2	3	4	5	6	7	8	9	a	b	c	d
W	Z	X	W	Y	0	1	2	3	4	5	6	7	8	9	a	b	c	d
X	X	Z	Y	W	a	b	c	d	0	1	2	3	4	5	6	7	8	9
Y	W	Y	Z	X	8	9	a	b	c	d	0	1	2	3	4	5	6	7
Z	Y	W	X	Z	1	2	3	4	5	6	7	8	9	a	b	c	d	0
0	0	a	8	1	Z	X	d	6	9	7	4	W	c	2	5	3	b	Y
1	1	b	9	2	X	Z	Y	0	7	a	8	5	W	d	3	6	4	c
2	2	c	a	3	d	Y	Z	X	1	8	b	9	6	W	0	4	7	5
3	3	d	b	4	6	0	X	Z	Y	2	9	c	a	7	W	1	5	8
4	4	0	c	5	9	7	1	Y	Z	X	3	a	d	b	8	W	2	6
5	5	1	d	6	7	a	8	2	X	Z	Y	4	b	0	c	9	W	3
6	6	2	0	7	4	8	b	9	3	Y	Z	X	5	c	1	d	a	W
7	7	3	1	8	W	5	9	c	a	4	X	Z	Y	6	d	2	0	b
8	8	4	2	9	c	W	6	a	d	b	5	Y	Z	X	7	0	3	1
9	9	5	3	a	2	d	W	7	b	0	c	6	X	Z	Y	8	1	4
a	a	6	4	b	5	3	0	W	8	c	1	d	7	Y	Z	X	9	2
b	b	7	5	c	3	6	4	1	W	9	d	2	0	8	X	Z	Y	a
c	c	8	6	d	b	4	7	5	2	W	a	0	3	1	9	Y	Z	X
d	d	9	7	0	Y	c	5	8	6	3	W	b	1	4	2	a	X	Z

S, symmetric.

Table 7.12. Initial rows and columns for constructing sets of three MOLS, $n = 34, 38, 42$

Initial columns for L_2, S_2:

L_2	S_2	L_2	S_2	L_2	S_2
3	0	3	4	12	18
15	19	15	13	31	0
18	13	8	19	22	16
23	12	11	3	3	10
7	10	29	12	6	22
25	7	22	15	14	15
4	2	12	16	15	3
10	1	5	25	9	26
$n = 34$		$n = 38$		$n = 42$	

Initial rows for L_3, S_3, L_4, S_4:

$n = 34$

L_3:	13	7	4	2	3	6	10	8
S_3:	0	19	13	12	10	7	2	1

$n = 38$

L_3:	2	3	4	6	7	8	10	11
S_3:	4	13	19	3	12	15	16	25

$n = 42.$

L_3:	5	8	7	2	10	11	13	3
S_3:	18	0	16	10	22	15	3	26

$n = 34.$

L_4:	0	25	23	22	21	18	17	15	20	5	16	12	14	1	19	24	z	11	w	9	y	x	Z	Y	X	W
S_4:	z	25	22	18	9	8	23	X	14	Y	21	Z	16	W	4	y	11	x	6	w	17	3	5	15	20	24

$n = 38.$

L_4:	0	29	27	26	25	22	24	16	21	5	20	13	18	17	28	1	23	14	19	9	z	15	y	12	w	x	Z	Y	X	W
S_4:	z	29	26	21	11	10	20	y	8	X	2	Y	9	6	1	W	17	23	27	x	22	w	0	Z	14	5	7	18	24	28

$n = 42.$

L_4:	0	33	30	32	27	31	25	28	24	20	6	12	17	26	21	23	18	1	22	29	19	4	15	9	14	16	z	w	y	Y	x	X	Z	W
S_4:	z	33	25	31	8	29	11	13	20	y	17	30	14	x	1	X	9	W	27	w	21	Y	2	19	7	Z	12	6	5	24	4	28	23	32

Proof. Let $k-2$ denote the minimum on the right-hand side. Then:

$$\begin{aligned} N(m) &\geq k-2, && \text{so } m \in T(k, 1); \\ N(m+1) &\geq k-2, && \text{so } m+1 \in T(k, 1); \\ N(t)-1 &\geq k-2, && \text{so } t \in T(k+1, 1); \\ N(s) &\geq k-2, && \text{so } s \in T(k, 1). \end{aligned}$$

But now by Theorem 1, $mt+s \in T(k, 1)$, so

$$N(mt+s) \geq k-2,$$

as required. □

Similarly, from Theorem 3 there follows

Corollary 3.1. *If* $0 \leq s_2 \leq s_1 \leq t$, *and* $k \leq m+2$, *then*

$$N(mt+s_1+s_2) \geq \min\{N(m), N(m+1), N(m+2), N(t)-2, N(s_1), N(s_2)\}.$$

Proof. Again let $k-2$ denote the minimum on the right-hand side. As in Corollary 1.1 we know that:

$$m \in T(k, 1);\ m+1 \in T(k, 1);\ m+2 \in T(k, 1);$$
$$t \in T(k+2, 1);\ s_1 \in T(k, 1);\ s_2 \in T(k, 1).$$

Since $k \leq m+2$ we have, by Theorem 3, that

$$mt+s_1+s_2 \in T(k, 1),$$

so $N(mt+s_1+s_2) \geq k-2$ as required. □

Suppose that we choose $m=4$ and $t=6r\pm 1$, so that $(t, 6)=1$. By Corollary 1.1 and eqns (7.1) and (7.2), we have, for $0 \leq s \leq t$,

$$N(4t+s) \geq \min\{3, N(s)\}.$$

Since $m=4$ and we are dealing with values of t (modulo 6), we must work modulo 24. By eqns (7.2) and (7.3), we need not consider those values of n congruent to 1, 4, 5, 7, 8, 11, 13, 16, 17, 19, 20, or 23 (modulo 24); we already know that $N(n) \geq 3$ in these cases. The remaining twelve cases are set out in Table 7.13, where $n=4t+s$. The value of t is chosen in each case to ensure that $N(s) \geq 3$. (Remember that $N(1)=\infty$.) Since we must have $s \leq t$, there is in each case a minimum value of r to which Corollary 1.1 applies, and the cases which are missing, and must thus be dealt with separately, are listed in the final column.

We already know that $N(2)=N(6)=1$, and that $N(3)=2$. Further, we know that $N(10) \geq 2$, but we have no further information for this value. By Theorem 6.2.3, we have $N(9)=8$ and $N(27)=26$. In Section 6.2, we saw that $N(12) \geq 5$, and in Section 7.2, that $N(15) \geq 4$, $N(46) \geq 4$ and that $N(n) \geq 3$ for

Table 7.13. Parameters for recursive constructions for Theorem 4

n	t	s	minimum r	missing values of n
$24r$	$6r-1$	4	1	—
$24r+2$	$6r-5$	22	5	2, 26, 50, 74, 98
$24r+3$	$6r-1$	7	2	3, 27
$24r+6$	$6r-5$	26	6	6, 30, 54, 78, 102, 126
$24r+9$	$6r+1$	5	1	9
$24r+10$	$6r-5$	30	6	10, 34, 58, 82, 106, 130
$24r+12$	$6r+1$	8	2	12, 36
$24r+14$	$6r-1$	18	4	14, 38, 62, 86
$24r+15$	$6r+1$	11	2	15, 39
$24r+18$	$6r-1$	22	4	18, 42, 66, 90
$24r+21$	$6r+5$	1	0	—
$24r+22$	$6r+1$	18	4	22, 46, 70, 94

Table 7.14. Parameters for recursive constructions from Corollary 1.1

m	t	s	$n = mt + s$
4	8	7	39
7	7	1	50
7	7	5	54
7	9	7	70
7	11	1	78
7	11	5	82
7	11	9	86
4	19	18	94
7	13	7	98
7	13	11	102
7	15	1	106
7	17	11	130

Table 7.15. Parameters for recursive constructions from Corollary 3.1

m	t	s_1	s_2	$n = mt + s_1 + s_2$
7	8	1	1	58
7	8	5	1	62
7	8	5	5	66
7	9	7	4	74

$n = 14, 18, 22, 26, 30, 34, 38, 42$. By Corollary 6.2.5.1, $N(36) \geq 3$. By Theorem 6.2.5, since $N(18) \geq 3$, we have $N(90) \geq 3$ and $N(126) \geq 3$. Of the remaining sixteen cases, twelve can be dealt with using Corollary 1.1 and four using Corollary 3.1; appropriate choices for m, t, s, s_1, s_2 are shown in Tables 7.14 and 7.15.

This completes the proof of

Theorem 4. *For* $n \neq 2, 3, 6, 10$, $N(n) \geq 3$. □

7.4. An asymptotic existence result

We prove one further consequence of Corollary 1.1, namely

Theorem 5. *As $n \to \infty$, $N(n) \to \infty$.*

Proof. Let x be an arbitrarily large positive number, and let m be defined by

$$m+1 = \prod_{p \leq x} p^x, p \text{ prime}. \tag{7.4}$$

Then by Corollary 6.2.5.1, we know that

$$N(m+1) \geq 2^x - 1 \geq x \tag{7.5}$$

and, since each prime factor of m is larger than x, that

$$N(m) \geq x. \tag{7.6}$$

Next we define

$$a = m^m \prod_{\substack{q \nmid n \\ q \leq x}} q^m, \; q \text{ prime}, \tag{7.7}$$

and note that although a is defined in terms of n, it is bounded above by a number dependent only on x.

Now consider the interval $\left[\dfrac{n}{(m+1)a}, \dfrac{n-1}{ma}\right]$. If n is large enough, then we can choose b such that

$$\frac{n}{(m+1)a} < b < \frac{n-1}{ma} \tag{7.8}$$

and

$$b \equiv 1 \pmod{m!}, \tag{7.9}$$

so that each prime factor of b is larger than m. Let $t = ab$. Then, by Theorem 6.2.5 and eqns (7.7) and (7.9),

$$N(t) \geq \min\{N(a), N(b)\} \geq \min\{2^m - 1, m\} \geq m. \tag{7.10}$$

Let $s = n - mt$. By eqns (7.4), (7.7), and (7.9), each prime divisor of s is larger than x, so

$$N(s) \geq x, \tag{7.11}$$

and by (7.8), we have $mt + 1 < n < (m+1)t$, so that

$$1 < s < t. \tag{7.12}$$

But now by eqns (7.5), (7.6), (7.10), and (7.11), and Corollary 1.1, we have

$$N(n) \geq x$$

for arbitrary x and sufficiently large n. □

7.5. References and comments

The general theorem (of which our Theorems 1 and 3 are special cases) is due to R. M. Wilson (1974*a*) and the proof to Paul Schellenberg; see W. D. Wallis (1984). It generalizes results of Bose, Shrikhande, and Parker (1960). The construction of three MOLS of order 14 is due to Todorov (1985). The other constructions given for sets of MOLS are due to Franklin (1984*a*) for orders 10, 14, and 15, to Johnson, Dulmage, and Mendelsohn (1961) for order 12, to R. M. Wilson (1974*b*) for order 46, and to Wang (1978) for the remaining orders. Wang's construction is described by W. D. Wallis (1984) in terms of spouse-avoiding mixed doubles tournaments. Schellenberg, van Rees, and Vanstone (1978) gave a different construction for four MOLS of order 15. Further constructions of orthogonal and almost orthogonal Latin squares are given by Franklin (1984*b*, *c*). Theorem 5 was proved by Chowla, Erdös, and Straus (1960) who also gave lower bounds for $N(n)$. Since then R. M. Wilson (1974*a*) and Beth (1983) have improved these bounds. An application of orthogonal Latin squares in compiler testing is given by Mandl (1985).

Exercises

7.1. Could the construction of Theorem 1 be applied to give a $T[4, 1; 14]$ design?

7.2. Complete the proof of Theorem 3 by showing that every two distinct varieties occur together in a unique block.

7.3. In the proof of Theorem 3, we have used the fact that $k \leq m+2$. How do we know that this is so?

7.4. How would you construct the $T[6, 1; 9]$ design needed in the construction of a $T[6, 1; 46]$ design?

7.5. In our recursive proof that $N(m) \geq 3$ for $n \geq 11$, we have applied Corollary 1.1 with $m = 4$. If we tried to apply Corollary 3.1 instead, what would be the smallest value of m that we could choose?

8 Resolvable designs and finite geometries

8.1. Definitions and examples

A balanced incomplete block design is said to be *resolvable* if the set of b blocks can be partitioned into t classes such that each variety appears in exactly one block of each class. Thus $b = t\beta$ and $v = k\beta$, where β is the number of blocks in each class of the partition. The classes will be called the *resolution classes* of the design.

A resolvable design is said to be *affine resolvable* if any two blocks from different resolution classes intersect in q_2 varieties.

A design is said to be *α-resolvable* if the set of blocks can be partitioned into classes such that each variety appears α times in each class. An α-resolvable design is called *affine α-resolvable* if any two blocks in the same class intersect in q_1 varieties and any two blocks in different classes intersect in q_2 varieties.

Example 1. The $B[3, 1; 9]$ design of Table 8.1 is affine resolvable with $q_2 = 1$. □

Example 2. The $B[3, 1; 15]$ design of Table 8.2 is resolvable but not affine resolvable. □

Example 3. The $B[3, 2; 9]$ design of Table 8.3 is 2-resolvable but not resolvable. □

Vertical lines separate the classes in each table; within each class, a row defines a block.

Table 8.1. A $B[3, 1; 9]$ design

1	2	3	1	4	7	1	5	9	1	6	8
4	5	6	2	5	8	2	6	7	2	4	9
7	8	9	3	6	9	3	4	8	3	5	7

Table 8.2. A $B[3, 1; 15]$ design

1 2 3	1 4 5	1 6 7	1 8 9	1 10 11	1 12 13	1 14 15
4 8 12	2 8 10	2 9 11	2 12 15	2 13 14	2 4 6	2 5 7
5 10 14	3 13 15	3 12 14	3 5 6	3 4 7	3 9 10	3 8 11
6 11 13	6 9 14	4 10 15	4 11 14	5 9 12	5 11 15	4 9 13
7 9 15	7 11 12	5 8 13	7 10 13	6 8 15	7 8 14	6 10 12

Table 8.3. A $B[3, 2; 9]$ design

1 2 9	1 2 9	1 3 8	1 4 7	1 5 7	1 6 8	1 4 5	1 3 6
4 5 8	4 6 8	2 4 7	2 3 8	2 5 8	2 6 7	2 4 6	2 3 5
3 6 7	3 5 7	5 6 9	5 6 9	3 4 9	3 4 9	7 8 9	7 8 9

□

Resolvability was the basis of the schoolgirls problem proposed by the Reverend T. P. Kirkman in 1850:

A schoolmistress has fifteen girl pupils and she wishes to take them on a daily walk. The girls are to walk in five rows of three girls each. It is required that no two girls should walk in the same row more than once per week. Can this be done? This amounts to asking if there exists a resolvable $B[3, 1; 15]$ design. The answer is yes, and one possible solution is given in Example 2. Indeed, the more general question

'For which values of v does a resolvable $B[3, 1; v]$ design exist?'

has also been answered; we discuss this in Chapter 13. Notice that an automorphism of a $B[3, 1; 15]$ design may or may not preserve resolution classes.

The use of resolvable and α-resolvable designs in a statistical context was discussed by Fisher and Yates. Yates showed that, if certain assumptions were valid, then analysing a resolvable design as a randomized complete block experiment gives an unbiased estimate of the error for treatment comparisons. Although an incomplete block design which is not resolvable may be less efficient than randomized complete blocks, a resolvable design must always be as efficient as randomized complete blocks.

Another advantage of resolvable designs concerns management. It is often not possible to sow or harvest the complete design in a single session. If the design is resolvable then one or more complete replicates can be dealt with at each session and possible differences caused by different times of sowing or harvesting are then balanced for all varieties.

8.2. Parameters of resolvable BIBDs

Consider an α-resolvable BIBD. Then the set of blocks is partitioned into t classes, each of β blocks, such that each variety appears α times in each class. Thus

$$v\alpha = \beta k, \quad \beta t = b \quad \text{and} \quad r = t\alpha.$$

Recall that $A \times B$ denotes the Kronecker product of the matrices A and B (see Lemma 2.6.22). In this section, l_i denotes the ith column of $\begin{bmatrix} 1 \; . \; . \; . \; 1 \\ -I_{t-1} \end{bmatrix} \times j_\beta$.

Theorem 4. *In an α-resolvable BIBD, $b \geq v+t-1$.*

Proof. Let A be the incidence matrix of an α-resolvable BIBD with the blocks ordered so that the first β blocks are those of the first resolution class and so on. Then

$$A\,l_i = \mathbf{0}$$

since each entry of $A\,l_i$ is the difference between the number of times a particular variety appears in the first resolution class and the number of times it appears in the $(i+1)$st resolution class; that is, $\alpha-\alpha=0$. Thus $\rho(A)$, the rank of A, satisfies

$$\rho(A) \leq b-(t-1) = b-t+1.$$

From Theorem 2.2.3 we know that $\rho(AA^T)=v$. But $\rho(A)=\rho(AA^T)$ so

$$v \leq b-t+1. \qquad \square$$

Lemma 5. *In an affine α-resolvable BIBD,*

$$q_1 = k(\alpha-1)/(\beta-1)$$

and

$$q_2 = k\alpha/\beta = k^2/v.$$

Proof. There are k varieties in any block of the design and each of these k varieties appears $\alpha-1$ times in the other blocks of the resolution class. Each of the other $\beta-1$ blocks in the resolution class intersects the original block in q_1 elements. Thus $k(\alpha-1)=(\beta-1)q_1$.

Consider a block B and a resolution class not containing B. Each of the β blocks in this class intersects B in q_2 elements. But each of the k elements of B occurs in α blocks of the class. Thus $q_2\beta = k\alpha$ and $q_2 = k\alpha/\beta$. Since $\beta = v\alpha/k$ we have $q_2 = k^2/v$. $\square$

The next result will help us to characterize affine α-resolvable block designs. First, however, we give an example.

Example 6. Consider the matrix

$$M = I_t \times \{(x-y)I_\beta + (y-z)J_\beta\} + zJ_t \times J_\beta.$$

The expansion of M, with $t=2$ and $\beta=3$, appears in Table 8.4. We can see that

$$Mj_6 = (x+2y+3z)j_6,$$
$$M(1,1,1,-1,-1,-1)^T = (x+2y-3z)\,(1,1,1,-1,-1,-1)^T.$$

Let e_i denote a column vector of dimension βt with a 1 in the ith position and zeroes elsewhere. Then

$$M(e_1-e_2)=(x-y)(e_1-e_2),\quad M(e_1-e_3)=(x-y)(e_1-e_3),$$
$$M(e_4-e_5)=(x-y)(e_4-e_5),\quad M(e_4-e_6)=(x-y)(e_4-e_6).$$

Thus $(x+2y+3z)$ is an eigenvalue with multiplicity one, $(x+2y-3z)$ is an eigenvalue with multiplicity one and $(x-y)$ is an eigenvalue with multiplicity four. □

Table 8.4. The matrix M when $t=2$ and $\beta=3$

$$I_2\times\{(x-y)I_3+(y-z)J_3\}+zJ_2\times J_3$$

$$=\begin{bmatrix}1&0\\0&1\end{bmatrix}\times\left\{(x-y)\begin{bmatrix}1&0&0\\0&1&0\\0&0&1\end{bmatrix}+(y-z)\begin{bmatrix}1&1&1\\1&1&1\\1&1&1\end{bmatrix}\right\}+z\begin{bmatrix}1&1\\1&1\end{bmatrix}\times\begin{bmatrix}1&1&1\\1&1&1\\1&1&1\end{bmatrix}$$

$$=\begin{bmatrix}1&0\\0&1\end{bmatrix}\times\begin{bmatrix}x-z&y-z&y-z\\y-z&x-z&y-z\\y-z&y-z&x-z\end{bmatrix}+zJ_6$$

$$=\left[\begin{array}{ccc|ccc}x&y&y&z&z&z\\y&x&y&z&z&z\\y&y&x&z&z&z\\\hline z&z&z&x&y&y\\z&z&z&y&x&y\\z&z&z&y&y&x\end{array}\right]$$

Any matrix of the form given in Example 6 can be partitioned into a $t\times t$ array of submatrices, each of size $\beta\times\beta$; each off-diagonal submatrix has every entry equal to z; each diagonal submatrix is of the same form as the matrix AA^T of Theorem 1.2.2, with x on the diagonal and y off the diagonal.

Lemma 7. *Let $N=N(m,n)=(m-n)I_\beta+nJ_\beta$, and let $P=pJ_\beta$. If Q is a partitioned matrix of order βt with submatrices N on the main diagonal, and submatrices P elsewhere, then*

$$\det Q=(m-n)^{t(\beta-1)}\{m+(\beta-1)n+\beta(t-1)p\}\{m+(\beta-1)n-\beta p\}^{t-1}. \tag{8.1}$$

Proof. As in the proof of Theorem 1.2.2,

$$\det N(m,n)=(m-n)^{\beta-1}\{m+(\beta-1)n\}, \tag{8.2}$$

and

$$\begin{aligned}\det(N+qP)&=\det N(m+qp,n+qp)\\&=(m-n)^{\beta-1}\{m+(\beta-1)n+\beta qp\}.\end{aligned} \tag{8.3}$$

Now

$$\det Q = \det \begin{bmatrix} N & P & P & \dots & P \\ P & N & P & \dots & P \\ P & P & N & \dots & P \\ \vdots & \vdots & \vdots & \ddots & \vdots \\ P & P & P & \dots & N \end{bmatrix}$$

$$= \det \begin{bmatrix} N & P-N & P-N & \dots & P-N \\ P & N-P & 0 & \dots & 0 \\ P & 0 & N-P & \dots & 0 \\ \vdots & \vdots & \vdots & \ddots & \vdots \\ P & 0 & 0 & \dots & N-P \end{bmatrix}$$

$$= \det \begin{bmatrix} N+(t-1)P & 0 & 0 & \dots & 0 \\ P & N-P & 0 & \dots & 0 \\ P & 0 & N-P & \dots & 0 \\ \vdots & \vdots & \vdots & \ddots & \vdots \\ P & 0 & 0 & \dots & N-P \end{bmatrix}$$

$$= \det[N+(t-1)P].\det(N-P)^{t-1}$$

$$= (m-n)^{\beta-1}\{m+(\beta-1)n+\beta(t-1)p\}.\{(m-n)^{\beta-1}[m+(\beta-1)n-\beta p]\}^{t-1},$$

by eqns (8.2) and (8.3), giving eqn (8.1). □

Lemma 8. *Let M be a symmetric matrix of order βt. The necessary and sufficient conditions that M be of the form*

$$M = I_t \times \{(x-y)I_\beta + (y-z)J_\beta\} + zJ_t \times J_\beta$$

are given by

(i) $\rho_0 = x+(\beta-1)y+\beta(t-1)z$ *is an eigenvalue of M of multiplicity* 1 *with corresponding eigenvector* $j_{\beta t}$;

(ii) $\rho_1 = x+(\beta-1)y-\beta z$ *is an eigenvalue of M of multiplicity $t-1$ with a complete set of eigenvectors* $l_1, l_2, \dots, l_{t-1}$;

(iii) $\rho_2 = x-y$ *is an eigenvalue of M of multiplicity* $t(\beta-1)$.

Proof. Since M has order βt, (i), (ii), and (iii) account for all its eigenvalues.

(a) Suppose that

$$M = I_t \times \{(x-y)I_\beta + (y-z)J_\beta\} + zJ_t \times J_\beta.$$

Then

$$Mj_{\beta t} = (x+(\beta-1)y+\beta(t-1)z)j_{\beta t}$$

and

$$Ml_i = (x+(\beta-1)y-\beta z)l_i, \quad i = 1, \dots, t-1.$$

Now we apply Lemma 7 with $n = x-\eta$, $p = z$. Thus, by eqn (8.1),

$$\det(M-\eta I) = (x+(\beta-1)y+\beta(t-1)z-\eta)(x-y-\eta)^{\beta-1}$$
$$.\{(x+(\beta-1)y-\beta z-\eta)(x-y-\eta)^{\beta-1}\}^{t-1}.$$

Hence M satisfies (i), (ii) and (iii).

(b) Now suppose that M is a symmetric matrix satisfying (i), (ii) and (iii). Since M is a symmetric matrix, it must have an orthonormal set of eigenvectors. Suppose that these are

$$\boldsymbol{k}_1, \boldsymbol{v}_1, \boldsymbol{v}_2, \ldots, \boldsymbol{v}_{t-1}, \boldsymbol{w}_1, \ldots, \boldsymbol{w}_{(\beta-1)t},$$

corresponding to the eigenvalues ρ_0, ρ_1 and ρ_2 respectively. Then

$$M = \rho_0 \boldsymbol{k}_1\boldsymbol{k}_1^T + \rho_1 \sum_{i=1}^{t-1} \boldsymbol{v}_i\boldsymbol{v}_i^T + \rho_2 \sum_{j=1}^{t(\beta-1)} \boldsymbol{w}_j\boldsymbol{w}_j^T$$
$$= \rho_0 \boldsymbol{k}_1\boldsymbol{k}_1^T + \rho_1 \sum_{i=1}^{t-1} \boldsymbol{v}_i\boldsymbol{v}_i^T + \rho_2 (I - \boldsymbol{k}_1\boldsymbol{k}_1^T - \sum_{i=1}^{t-1} \boldsymbol{v}_i\boldsymbol{v}_i^T)$$

(since the matrix $U = [\boldsymbol{k}_1\ \boldsymbol{v}_1\ \boldsymbol{v}_2 \ldots \boldsymbol{v}_{t-1}\ \boldsymbol{w}_1 \ldots \boldsymbol{w}_{(\beta-1)t}]$ satisfies $UU^T = I$). Now we know that $\boldsymbol{k}_1 = j_{\beta t}/\sqrt{\beta t}$ and that $\boldsymbol{v}_1, \ldots, \boldsymbol{v}_{t-1}$ are obtained from $\boldsymbol{l}_1$, $\boldsymbol{l}_2, \ldots, \boldsymbol{l}_{t-1}$ by orthonormalization. Hence M is uniquely determined by knowing ρ_0, ρ_1, ρ_2 and $\boldsymbol{k}_1$, $\boldsymbol{v}_1, \ldots, \boldsymbol{v}_{t-1}$ so M is of the required form. □

Example 6 (continued). Orthonormalizing eigenvectors we find that in this case the matrix U of Lemma 8 is

$$U = 1/\sqrt{6}\begin{bmatrix} 1 & 1 & \sqrt{3} & 1 & 0 & 0 \\ 1 & 1 & -\sqrt{3} & 1 & 0 & 0 \\ 1 & 1 & 0 & -2 & 0 & 0 \\ 1 & -1 & 0 & 0 & \sqrt{3} & 1 \\ 1 & -1 & 0 & 0 & -\sqrt{3} & 1 \\ 1 & -1 & 0 & 0 & 0 & -2 \end{bmatrix}.$$

□

Theorem 9. *An α-resolvable BIBD is affine α-resolvable if and only if*

$$b = v + t - 1.$$

Proof. We shall work with both AA^T (as we did before) and with $A^TA = M$, say. Then M is a $b \times b$ matrix, with the entry m_{ij} equal to the number of elements common to blocks i and j of the design.

(i) Suppose that A is the incidence matrix of an affine α-resolvable BIBD with the blocks ordered into resolution classes as in the proof of Theorem 4.

Thus we have

$$m_{ij} = \begin{cases} k, & i = j, \\ q_1, & i \neq j, \quad \text{blocks } i \text{ and } j \text{ in the same resolution class,} \\ q_2, & i \neq j, \quad \text{blocks } i \text{ and } j \text{ in different resolution classes.} \end{cases}$$

More formally,

$$M = I_t \times ((k - q_1) I_\beta + (q_1 - q_2) J_\beta) + q_2 J_t \times J_\beta.$$

Hence from Lemma 8 we know that:

$\rho_0 = k + (\beta - 1) q_1 + \beta(t - 1) q_2$ is an eigenvalue with multiplicity 1;

$\rho_1 = k + (\beta - 1) q_1 - \beta q_2 = 0$ (by Lemma 5) is an eigenvalue with multiplicity $t - 1$;

$\rho_2 = k - q_1$ is an eigenvalue with multiplicity $t(\beta - 1)$.

Thus the rank of $M = A^T A$ is $b - t + 1$. As $\rho(A) = \rho(AA^T) = \rho(A^T A)$ we have that $v = b - t + 1$.

(ii) Suppose that A is the incidence matrix of an α-resolvable design with $v = b - t + 1$. Then the $t - 1$ vectors l_i, $i = 1, \ldots, t - 1$, of Theorem 4 and Lemma 8 are a set of $t - 1$ linearly independent eigenvectors associated with the eigenvalue 0. Let η be an eigenvalue of AA^T with corresponding eigenvector $\mathbf{x}$. Then $AA^T\mathbf{x} = \eta\mathbf{x}$ and $A^T A(A^T\mathbf{x}) = \eta A^T\mathbf{x}$ so the non-zero eigenvalues of AA^T and $A^T A$ are the same. We know from Theorem 2.2.3 that the eigenvalues of AA^T are rk and $r - \lambda$ with multiplicities 1 and $v - 1 = b - t$ respectively. Thus, to use the result of Lemma 8, we need only show that these three eigenvalues give three consistent equations to determine x, y, and z. We have

$$\begin{aligned} \rho_0 &= rk = x + (\beta - 1) y + \beta(t - 1) z, \\ \rho_1 &= 0 = x + (\beta - 1) y - \beta z, \\ \rho_2 &= r - \lambda = x - y, \end{aligned}$$

a system of rank three with unique solutions $x = k$, $y = q_1$, $z = q_2$. Then $A^T A$ satisfies the conditions of Lemma 8 and the design is affine α-resolvable. □

Corollary 9.1. *In an affine α-resolvable BIBD, $q_1 = k + \lambda - r$.*

Proof. As the unique eigenvalue of $A^T A$ with multiplicity $b - t = t(\beta - 1)$ is both $r - \lambda$ and $k - q_1$, the result follows. □

Example 10. Consider the $B[3, 1; 9]$ design of Example 1. Then the matrices A and $A^T A$ are those given in Table 8.5. In the notation of Lemma 8, $x = 3$, $y = 0$ and $z = 1$. □

Table 8.5

$$
A = \left[\begin{array}{ccc|ccc|ccc|ccc}
1 & & & 1 & & & 1 & & & 1 & & \\
1 & & & & 1 & & & 1 & & & 1 & \\
1 & & & & & 1 & & & 1 & & & 1 \\
\hline
 & 1 & & 1 & & & & & 1 & & 1 & \\
 & 1 & & & 1 & & 1 & & & & & 1 \\
 & 1 & & & & 1 & & 1 & & 1 & & \\
\hline
 & & 1 & 1 & & & & 1 & & & & 1 \\
 & & 1 & & 1 & & & & 1 & 1 & & \\
 & & 1 & & & 1 & 1 & & & & 1 &
\end{array}\right]
\qquad
A^T A = \left[\begin{array}{ccc|ccc|ccc|ccc}
3 & 0 & 0 & 1 & 1 & 1 & 1 & 1 & 1 & 1 & 1 & 1 \\
0 & 3 & 0 & 1 & 1 & 1 & 1 & 1 & 1 & 1 & 1 & 1 \\
0 & 0 & 3 & 1 & 1 & 1 & 1 & 1 & 1 & 1 & 1 & 1 \\
\hline
1 & 1 & 1 & 3 & 0 & 0 & 1 & 1 & 1 & 1 & 1 & 1 \\
1 & 1 & 1 & 0 & 3 & 0 & 1 & 1 & 1 & 1 & 1 & 1 \\
1 & 1 & 1 & 0 & 0 & 3 & 1 & 1 & 1 & 1 & 1 & 1 \\
\hline
1 & 1 & 1 & 1 & 1 & 1 & 3 & 0 & 0 & 1 & 1 & 1 \\
1 & 1 & 1 & 1 & 1 & 1 & 0 & 3 & 0 & 1 & 1 & 1 \\
1 & 1 & 1 & 1 & 1 & 1 & 0 & 0 & 3 & 1 & 1 & 1 \\
\hline
1 & 1 & 1 & 1 & 1 & 1 & 1 & 1 & 1 & 3 & 0 & 0 \\
1 & 1 & 1 & 1 & 1 & 1 & 1 & 1 & 1 & 0 & 3 & 0 \\
1 & 1 & 1 & 1 & 1 & 1 & 1 & 1 & 1 & 0 & 0 & 3
\end{array}\right]
$$

Lemma 11. *In an affine resolvable BIBD we have*

$$v = \beta^2(\gamma(\beta-1)+1) = \beta k, \quad b = \beta(\gamma\beta^2+\beta+1) = \beta r, \quad \lambda = \beta\gamma+1,$$

for some integer γ.

Proof. As $\alpha = 1$, $t = r$ and so by Theorem 9, $b = v+r-1$. Now, we know that $\lambda(v-1) = r(k-1)$ and that $b = t\beta = r\beta$, so

$$b = r+v-1 = r+\frac{r(k-1)}{\lambda} = r\left(1+\frac{k-1}{\lambda}\right) = r\beta.$$

Hence

$$\beta = 1+\frac{k-1}{\lambda}$$

or, rearranging, $\lambda = (k-1)/(\beta-1)$ and λ is an integer. From Lemma 5 we know that $q_2 = k^2/v = k/\beta$ and q_2 is an integer. Hence

$$k = \lambda(\beta-1)+1 \equiv 0 \pmod{\beta}$$

or, equivalently, $\lambda \equiv 1 \pmod{\beta}$. Let $\lambda = \beta\gamma+1$. Then

$$k = (\beta\gamma+1)(\beta-1)+1 = \beta\gamma(\beta-1)+\beta-1+1 = \beta(\gamma(\beta-1)+1),$$

$v = \beta k$, $b = \beta r$, and $b-r = v-1$, by Theorem 9. So

$$\begin{aligned}
r &= (v-1)/(\beta-1) \\
&= (\beta^2(\gamma(\beta-1)+1)-1)/(\beta-1) \\
&= (\beta^2\gamma(\beta-1)+\beta^2-1)/(\beta-1) \\
&= \beta^2\gamma+\beta+1, \text{ as required.}
\end{aligned}$$

□

Theorem 12. *Any $B[k, 1; k^2]$ design is affine resolvable and can be extended to an $SB[k+1, 1; k^2+k+1]$ design.*

Proof. A $B[k, 1; k^2]$ design has
$$r = (k^2-1)/(k-1) = k+1 \text{ and } b = k^2(k+1)/k = k(k+1).$$

Consider any block B_0 of the design. As $\lambda = 1$ it intersects every other block of the design in either 0 or 1 elements. Each of the k elements in B_0 appears in $r-1 = k$ other blocks and these blocks are all distinct. Hence there are

$$k^2+k-(k^2+1) = k-1$$

blocks of the design disjoint from B_0. Now consider an element in one of these blocks. It appears once with each of the k elements of B_0 and so appears once in a block containing no element of B_0. Thus, each of the k^2-k elements not in B_0 appears in a block disjoint from B_0. Hence the $(k-1)$ blocks disjoint from B_0 are also pairwise disjoint and so form a resolution class. As B_0 was any block of the design, each block of the design belongs to a unique resolution class, showing the design is resolvable. By construction any two blocks in different resolution classes intersect in one element so the design is in fact affine resolvable. There are $k+1$ resolution classes.

To extend the design adjoin the symbol ∞_i to each block of the ith resolution class, and adjoin the block $\{\infty_1\ \infty_2 \ldots \infty_{k+1}\}$ to the design. This gives a new design, with parameters, V, B, R, K, Λ, where

$$V = k^2+(k+1),\ B = (k^2+k)+1,\ K = k+1.$$

Each of the original elements appears in $k+1$ blocks; each of the new elements appears in the k blocks of a resolution class and in the new block. Hence

$$R = k+1.$$

Similarly $\Lambda = 1$, since each new element appears once with each of the original elements in the blocks of one resolution class, each pair of new elements occurs once (in the new block) and each pair of the original elements appears once (as it did before).

Hence the resulting design is an $SB[k+1, 1; k^2+k+1]$ design. □

Example 13. Let $k = 2$. Then a $B[2, 1; 4]$ design is affine resolvable with $q_2 = 1$, and resolution classes as shown in Table 8.6. Adjoining ∞_1, ∞_2, ∞_3 gives an $SB[3, 1; 7]$ design. □

Table 8.6

$12\,\infty_1$	$13\,\infty_2$	$14\,\infty_3$	$\infty_1\,\infty_2\,\infty_3$
$34\,\infty_1$	$24\,\infty_2$	$23\,\infty_3$	

The designs of Theorem 12 are, in fact, finite planes: a $B[n, 1; n^2]$ design is called an *affine plane of order n*, and its extension, an $SB[n+1, 1; n^2+n+1]$ design, is a *projective plane of order n*.

8.3. Constructions for resolvable designs

The first construction is based on the correspondence between a complete set of mutually orthogonal Latin squares (MOLS) of order n and an affine plane of order n. A complete set of MOLS was constructed in Theorem 6.2.3.

Theorem 14. *An affine plane of order n exists if and only if there is a complete set of MOLS of order n.*

Proof. Let A be an $n \times n$ array with entries $1, 2, \ldots, n^2$ in some order. Let $L_3, L_4, \ldots, L_{n+1}$ be a complete set of MOLS of order n. Let R be the $n \times n$ array with each entry of row i equal to i, and let $C = R^T$. To construct the design, first superimpose $L_1 (= R), L_2 (= C), L_3, \ldots, L_n, L_{n+1}$, in turn on A to form in each case n^2 ordered pairs $([l]_i, a)$, where $[l]_i$ is an element of L_i. Then, as the jth block of the ith resolution class, take the set of values a for which $[l]_i$ takes the value j.

Clearly $v = n^2$, $k = n$. Since $1 \le i \le n+1$ and $1 \le j \le n$, we have $b = n(n+1)$ and $r = n+1$. Also there are $t = r$ resolution classes of $\beta = k$ blocks each.

Consider two varieties x and y, $1 \le x, y \le n^2$. If x and y occur together in two blocks of the design then the two blocks come from R or C and a Latin square, or from two Latin squares. In the first case the Latin square has the same element twice in some row or column, a contradiction. In the second case the two Latin squares, when superimposed, have the same ordered pair appearing twice. Thus they are not orthogonal, again a contradiction. Hence each pair of points appears at most once in the design. Counting pairs shows each pair of points appears at least once in the design so $\lambda = 1$.

This construction can be reversed, to give a complete set of MOLS from an affine plane: see Exercise 8.3. □

Example 15. Let $n = 4$. The arrays for the construction of Theorem 14 appear in Table 8.7 where L_3, L_4, L_5 come from Table 6.5 with elements relabelled. The resulting design is listed in Table 8.8 as D_2. □

The next two constructions are recursive.

Theorem 16. *Suppose that there exists a $(v_1, b_1, r_1, k_1, \lambda_1)$ BIBD, D_1, and a $(k_2 v_1, r_2 v_1, r_2, k_2, \lambda_2)$ resolvable BIBD, D_2. Then there exists an α-resolvable BIBD, D, with parameters*

$$v = k_2 v_1, b = b_1 r_2, r = r_1 r_2, k = k_1 k_2, \alpha = r_1, \beta = b_1, t = r_2,$$
$$\lambda = r_1 \lambda_2 + \lambda_1 (r_2 - \lambda_2).$$

Table 8.7

$$A = \begin{matrix} 1 & 2 & 3 & 4 \\ 5 & 6 & 7 & 8 \\ 9 & A & B & C \\ D & E & F & G \end{matrix} \quad L_1 = R = \begin{matrix} 1 & 1 & 1 & 1 \\ 2 & 2 & 2 & 2 \\ 3 & 3 & 3 & 3 \\ 4 & 4 & 4 & 4 \end{matrix} \quad L_2 = C = \begin{matrix} 1 & 2 & 3 & 4 \\ 1 & 2 & 3 & 4 \\ 1 & 2 & 3 & 4 \\ 1 & 2 & 3 & 4 \end{matrix} \quad L_3 = \begin{matrix} 1 & 2 & 3 & 4 \\ 2 & 1 & 4 & 3 \\ 3 & 4 & 1 & 2 \\ 4 & 3 & 2 & 1 \end{matrix} \quad L_4 = \begin{matrix} 1 & 2 & 3 & 4 \\ 3 & 4 & 1 & 2 \\ 4 & 3 & 2 & 1 \\ 2 & 1 & 4 & 3 \end{matrix} \quad L_5 = \begin{matrix} 1 & 2 & 3 & 4 \\ 4 & 3 & 2 & 1 \\ 2 & 1 & 4 & 3 \\ 3 & 4 & 1 & 2 \end{matrix}$$

Table 8.8

$$\begin{matrix} 12 & 13 & 14 \\ 23 & 24 & 34 \end{matrix}$$

D_1

$$\begin{matrix} T_{11} = 1\ 2\ 3\ 4 & T_{21} = 1\ 5\ 9\ D & T_{31} = 1\ 6\ B\ G & T_{41} = 1\ 7\ C\ E & T_{51} = 1\ 8\ A\ F \\ T_{12} = 5\ 6\ 7\ 8 & T_{22} = 2\ 6\ A\ E & T_{32} = 2\ 5\ C\ F & T_{42} = 2\ 8\ B\ D & T_{52} = 2\ 7\ 9\ G \\ T_{13} = 9\ A\ B\ C & T_{23} = 3\ 7\ B\ F & T_{33} = 3\ 8\ 9\ E & T_{43} = 3\ 5\ A\ G & T_{53} = 3\ 6\ C\ D \\ T_{14} = D\ E\ F\ G & T_{24} = 4\ 8\ C\ G & T_{34} = 4\ 7\ A\ D & T_{44} = 4\ 6\ 9\ F & T_{54} = 4\ 5\ B\ E \end{matrix}$$

D_2

$$\begin{matrix} T_{11}T_{12} & T_{11}T_{13} & T_{11}T_{14} & T_{21}T_{22} & T_{21}T_{23} & T_{21}T_{24} & T_{31}T_{32} & T_{31}T_{33} & T_{31}T_{34} \\ T_{12}T_{13} & T_{12}T_{14} & T_{13}T_{14} & T_{22}T_{23} & T_{22}T_{24} & T_{23}T_{24} & T_{32}T_{33} & T_{32}T_{34} & T_{33}T_{34} \end{matrix}$$

$$\begin{matrix} T_{41}T_{42} & T_{41}T_{43} & T_{41}T_{44} & T_{51}T_{52} & T_{51}T_{53} & T_{51}T_{54} \\ T_{42}T_{43} & T_{42}T_{44} & T_{43}T_{44} & T_{52}T_{53} & T_{52}T_{54} & T_{53}T_{54} \end{matrix}$$

D in terms of T_{ij}

$$\begin{matrix} 1\,2\,3\,4\,5\,6\,7\,8 & 1\,2\,3\,4\,9\,A\,B\,C & 1\,2\,3\,4\,D\,E\,F\,G \\ 5\,6\,7\,8\,9\,A\,B\,C & 5\,6\,7\,8\,D\,E\,F\,G & 9\,A\,B\,C\,D\,E\,F\,G \end{matrix}$$

The first resolution class of
D in terms of the varieties

Proof. Let T_{ij} denote the jth block of the ith resolution class of D_2, for $i = 1, \ldots, r_2; j = 1, \ldots, v_1$. Construct r_2 copies of D_1, using as varieties the v_1 blocks, $T_{i1}, T_{i2}, \ldots, T_{iv_1}, i = 1, 2, \ldots, r_2$.

The design D is based on the varieties of D_2. There are $r_2 b_1$ blocks constructed on the set $\{T_{ij}\}$. Since D_2 is resolvable, $T_{ij} \cap T_{im} = \emptyset, j \neq m$. Also $|T_{ij}| = k_2$. Thus, each block constructed on $\{T_{ij}\}$ contains $k_1 k_2$ of the varieties of D_2. The $r_2 b_1$ such blocks are the blocks of D. Each variety appears in r_2 of the T_{ij} and each T_{ij} appears r_1 times in D_1; thus, each variety appears $r_2 r_1$ times in D. Again each of the r_2 copies of D_1 has each variety r_1 times so $t = r_2, \alpha = r_1, \beta = b_1$. Finally, consider any two varieties x and y. There are λ_2 resolution classes in which x and y appear together in a block and $r_2 - \lambda_2$ in which x and y appear in different blocks. In the first case x and y appear together r_1 times in the copy of D_1 corresponding to each resolution class, and in the second case x and y appear together λ_1 times in the copy of D_1 corresponding to each resolution class. Thus $\lambda = \lambda_2 r_1 + (r_2 - \lambda_2)\lambda_1$ in D. □

Example 17. Let D_1 be a (4, 6, 3, 2, 1) BIBD and let D_2 be a (16, 20, 5, 4, 1) resolvable BIBD. Then the method of Theorem 16, which is illustrated in Table 8.8, gives a 3-resolvable design with

$$v = 16,\ b = 30,\ r = 15,\ k = 8,\ \alpha = 3,\ \beta = 6,\ t = 5,\ \lambda = 7. \qquad \square$$

Corollary 16.1. *Suppose that designs D_1 and D_2 exist, where D_1 is an SBIBD with parameters $v_1 = b_1 = \beta$, $r_1 = k_1 = \alpha$, λ_1 and D_2 is an affine resolvable BIBD with parameters*

$$v_2 = k_2 v_1 = \beta k_2 = \beta^2((\beta - 1)\gamma + 1), \quad b_2 = r_2 v_1 = \beta r_2 = \beta(\beta^2\gamma + \beta + 1),$$
$$\lambda_2 = \beta\gamma + 1.$$

Then there exists an affine α-resolvable BIBD, D, with parameters

$$v = \beta k_2,\ b = \beta r_2,\ r = r_1 r_2,\ k = k_1 k_2,\ \alpha,\ \beta,\ t = r_2,\ \lambda = r_1\lambda_2 + \lambda_1(r_2 - \lambda_2).$$

Proof. This follows directly from Theorem 16, since any two blocks of D_1 intersect in precisely λ_1 varieties. □

It is worth noting that designs constructed from a set of t supplementary difference sets are at least t-resolvable.

8.4. Affine and projective planes over finite fields

There are other ways of constructing affine and projective planes, besides the method of Theorem 14. We now define an affine plane in terms of a vector space over a finite field.

Let $V_2(q)$ denote the two dimensional vector space over $GF[q]$ and let W be a subspace of $V_2(q)$. A *coset* of W is a set of the form $W+v$, where $v \in V_2(q)$. Let $AG(2, q)$ be the set of all cosets of subspaces of $V_2(q)$. Then $AG(2,q)$ is called an *affine plane over* $GF[q]$. The *points* of $AG(2, q)$ are the cosets of the subspace of dimension 0 of $V_2(q)$, and the *lines* of $AG(2,q)$ are the cosets of the subspaces of dimension 1 of $V_2(q)$. A point and a line are said to be *incident* if the point is a subset of the line. Two lines are said to be *parallel* if no point is incident with both.

A line is the set of all points (x, y) such that $\alpha x + \beta y = \gamma$ for $\alpha, \beta, \gamma \in GF[q]$, with α, β, γ fixed, and α, β not both zero. If $\alpha \neq 0$, then

$$x = \alpha^{-1}\gamma - \alpha^{-1}\beta y.$$

As y runs through the elements of $GF[q]$, x is uniquely determined. If $\alpha = 0$ then $\beta \neq 0$ so $y = \beta^{-1}\gamma$ and x can be any element of $GF[q]$. In either case there are q points on a line.

If $\gamma = 0$ then the line passes through the origin. Otherwise γ^{-1} exists and the equation of the line can be written

$$\gamma^{-1}\alpha x + \gamma^{-1}\beta y = 1.$$

Since $(\gamma^{-1}\alpha, \gamma^{-1}\beta) \neq (0, 0)$ there are $q^2 - 1$ lines not through the origin.

The number of lines through the origin is $q+1$, since either $\alpha \neq 0$ and the line can be written as $x + \alpha^{-1}\beta y = 0$, or $\alpha = 0$ and the line can be written as $y = 0$. These lines correspond to the $q+1$ subspaces of dimension 1 of $V_2(q)$.

The number of lines through a point (x, y) is the sum of the number of solution pairs (α, β) of $\alpha x + \beta y = 1$, and the number of solution pairs of $\alpha x + \beta y = 0$ with $\alpha = 1$ and with $\alpha = 0$. Thus in all cases there are $q+1$ lines through a point.

If (x_1, y_1) and (x_2, y_2) are two distinct points, and we want a line passing through both, then we have for α, β, the equations

$$\alpha x_1 + \beta y_1 = \gamma,$$
$$\alpha x_2 + \beta y_2 = \gamma,$$

with $\gamma = 1$ or 0. These are equivalent to

$$\alpha(x_1 y_2 - x_2 y_1) = \gamma(y_2 - y_1),$$
$$\beta(x_1 y_2 - x_2 y_1) = \gamma(x_1 - x_2).$$

Since $(x_1 - x_2, y_1 - y_2) \neq (0, 0)$, and $(\alpha, \beta) \neq (0, 0)$, we have

$$x_1 y_2 - x_2 y_1 = 0$$

if and only if $\gamma = 0$. In all cases a unique line is determined, which has the equation

$$(y_2 - y_1)x + (x_1 - x_2)y = y_2 x_1 - y_1 x_2. \tag{8.4}$$

There are $q+1$ lines through each of the q^2 points, giving $q^2(q+1)$

point–line incidences in all. Since there are q points per line, there are $q(q+1)$ lines in all. This leads us to

Theorem 18. *With points as varieties and lines as blocks, an $AG(2, q)$ is a $B[q, 1; q^2]$ design.* □

Example 19. Let $GF[4] = \{0, 1, \alpha, \alpha+1\}$, where $\alpha^2 = \alpha + 1$. The points and lines of $AG(2, 4)$ are given in Table 8.9. With points as varieties and lines as blocks we see that $AG(2, 4)$ is a $B[4, 1; 16]$ design. □

Table 8.9. Points and lines of $AG(2, 4)$; β denotes $\alpha+1$; xy denotes (x, y) in lists of points

Points: $00, 01, 0\alpha, 0\beta, 10, 11, 1\alpha, 1\beta, \alpha 0, \alpha 1, \alpha\alpha, \alpha\beta, \beta 0, \beta 1, \beta\alpha, \beta\beta$.

Lines:

$\{(x,y) \mid x = c\}$	$\{(x,y) \mid y = c\}$	$\{(x,y) \mid x + y = c\}$
$00, 01, 0\alpha, 0\beta$	$00, 10, \alpha 0, \beta 0$	$00, 11, \alpha\alpha, \beta\beta$
$10, 11, 1\alpha, 1\beta$	$01, 11, \alpha 1, \beta 1$	$01, 10, \alpha\beta, \beta\alpha$
$\alpha 0, \alpha 1, \alpha\alpha, \alpha\beta$	$0\alpha, 1\alpha, \alpha\alpha, \beta\alpha$	$0\alpha, 1\beta, \alpha 0, \beta 1$
$\beta 0, \beta 1, \beta\alpha, \beta\beta$	$0\beta, 1\beta, \alpha\beta, \beta\beta$	$0\beta, 1\alpha, \alpha 1, \beta 0$

$\{(x,y) \mid x + \alpha y = c\}$	$\{(x,y) \mid x + (\alpha+1)y = c\}$
$00, 1\beta, \alpha 1, \beta\alpha$	$00, 1\alpha, \alpha\beta, \beta 1$
$01, 1\alpha, \alpha 0, \beta\beta$	$01, 1\beta, \alpha\alpha, \beta 0$
$0\alpha, 11, \alpha\beta, \beta 0$	$0\alpha, 10, \alpha 1, \beta\beta$
$0\beta, 10, \alpha\alpha, \beta 1$	$0\beta, 11, \alpha 0, \beta\alpha$

All known finite affine planes have prime-power order, and since we know that $GF[q]$ exists if and only if q is a prime power, we know that $AG(2, q)$ exists if and only if q is a prime power. But there are also finite affine planes of prime-power order q, which are quite different from $AG(2, q)$ in structure, and it is possible that affine planes of composite orders may exist.

We have a rather similar construction for projective planes over finite fields. Let $V_3(q)$ denote the three-dimensional vector space over $GF[q]$ and let $PG(2, q)$ be the set of all subspaces of $V_3(q)$. Then $PG(2, q)$ is called the *projective plane over* $GF[q]$. For any subspace M of $V_3(q)$ we define the *projective dimension* of M, pdim M, by pdim $M = \dim M - 1$. The *points* of $PG(2, q)$ are the elements of pdim 0 and the *lines* are the elements of pdim 1. A line and a point are said to be *incident* if the subspace corresponding to the line contains the subspace corresponding to the point.

Notice that if (x_1, x_2, x_3) represents a point, then $(\eta x_1, \eta x_2, \eta x_3)$, $\eta \in GF[q] \backslash \{0\}$, represents the same point.

Example 20. The points and lines of $PG(2, 3)$ are given in Table 8.10. We have omitted $(0, 0, 0)$ from every subspace. We also give a relabelling of the points so that the correspondence between $PG(2, 3)$ and the design $(D2)$ of Chapter 1 is obvious. □

Theorem 21. *In* $PG(2, q)$:

(i) *there are* q^2+q+1 *points and* q^2+q+1 *lines;*

(ii) *any two points determine a unique line;*

(iii) $q+1$ *points are incident with each line and* $q+1$ *lines are incident with each point.*

Proof. (i) There are (q^3-1) non-zero elements in $V_3(q)$ and each of these has $(q-1)$ non-zero multiples. Hence there are $(q^3-1)/(q-1) = q^2+q+1$ subspaces of $V_3(q)$, of dimension 1, which are the points of $PG(2, q)$.

A subspace of $V_3(q)$ of dimension 2 has two independent elements in its basis. There are $(q^3-1)(q^3-1-(q-1))$ pairs of independent elements in $V_3(q)$ and the number of pairs of independent elements in a two-dimensional subspace is $(q^2-1)(q^2-1-(q-1))$. Hence there are

$$\frac{(q^3-1)(q^3-q)}{(q^2-1)(q^2-q)} = q^2+q+1$$

subspaces of $V_3(q)$ of dimension 2, which are the lines of $PG(2, q)$.

(ii) Any two points are linearly independent and so determine a unique subspace of dimension 2, or line of $PG(2, q)$.

(iii) Given that a particular point must be included in a basis there are $(q^3-q)/(q^2-q) = q+1$ two-dimensional subspaces containing that

Table 8.10. Points and lines of $PG(2, 3)$; lines are given as sets of points; $x_1\, x_2\, x_3$ denotes (x_1, x_2, x_3)

Points	Lines
{001, 002} = 001 = ∞	{001, 010, 011, 012} = ∞, 0, 1, 2
{010, 020} = 010 = 0	{001, 100, 101, 102} = ∞, 00, 01, 02
{100, 200} = 100 = 00	{001, 110, 111, 112} = ∞, 10, 11, 12
{011, 022} = 011 = 1	{001, 120, 121, 122} = ∞, 20, 21, 22
{101, 202} = 101 = 01	{010, 100, 110, 120} = 0, 00, 10, 20
{110, 220} = 110 = 10	{010, 101, 111, 121} = 0, 01, 11, 21
{111, 222} = 111 = 11	{010, 102, 112, 122} = 0, 02, 12, 22
{012, 021} = 012 = 2	{011, 100, 111, 122} = 1, 00, 11, 22
{102, 201} = 102 = 02	{011, 101, 112, 120} = 1, 01, 12, 20
{120, 210} = 120 = 20	{011, 102, 110, 121} = 1, 02, 10, 21
{112, 221} = 112 = 12	{012, 100, 112, 121} = 2, 00, 12, 21
{121, 212} = 121 = 21	{012, 101, 110, 122} = 2, 01, 10, 22
{122, 211} = 122 = 22	{012, 102, 111, 120} = 2, 02, 11, 20

point, or lines of $PG(2, q)$ through the point. The number of elements in a 2-dimensional subspace is q^2 and so the number of points on a line of $PG(2, q)$ is $(q^2-1)/(q-1) = q+1$. □

Thus, $PG(2, q)$ is an $SB[q+1, 1; q^2+q+1]$ design, or projective plane of order q. In Theorem 3.5.18, we constructed some designs with these parameters. Later in this chapter (Theorem 31) we shall see a generalization of this result which shows that the design is in fact $PG(2, q)$. But just as with affine planes, there are projective planes of order q of structure different from $PG(2, q)$.

$PG(2, q)$ contains four points, no three of which are collinear; see Exercise 8.10.

Example 22. In Table 8.11 we list the lines of $PG(2, 3)$ which contain two of the points 100, 010, 001, and 111. □

Table 8.11

Pair	Line
100, 010	010, 100, 110, 120
100, 001	001, 100, 101, 102
100, 111	011, 100, 111, 122
010, 001	001, 010, 011, 012
010, 111	010, 101, 111, 121
001, 111	001, 110, 111, 112

The residual design constructed from a $PG(2, q)$ by deleting any line (block) and the points (varieties) that it contains is an $AG(2, q)$.

We can represent the extension from an affine plane to a projective plane by extending the pairs to triples in the following way. Let the points of $AG(2, q)$ be

$$(1, x_1, x_2), \quad x_1, x_2 \in GF[q],$$

and let the points to be adjoined be denoted by

$$(0, y_1, y_2), \quad y_1, y_2 \in GF[q]; \quad y_1, y_2 \text{ not both zero},$$

where $(0, y_1, y_2)$ and $(0, z_1, z_2)$ represent the same point if $y_1 = \eta z_1, y_2 = \eta z_2$, for some $\eta \in GF[q]$, $\eta \neq 0$. Then $(q^2-1)/(q-1) = q+1$ is the number of adjoined points. If we regard $(1, x_1, x_2)$ as the representative of $\{(\eta, \eta x_1, \eta x_2) | \eta \in GF[q]\}$ then the correspondence between this extension of $AG(2, q)$ and our original definition of $PG(2, q)$ is clear.

Example 23. The design of Table 8.10 illustrates this type of extension. □

8.5. Affine geometries

We extend the concept of an affine plane to an affine geometry by increasing the dimension of the underlying vector space.

Let $V_n(q)$ denote the n-dimensional vector space over $GF[q]$ and let $AG(n, q)$ be the set of all cosets of subspaces of $V_n(q)$. Then $AG(n, q)$ is called the *affine geometry of dimension n over* $GF[q]$. The *points* of $AG(n, q)$ are the cosets of the subspace of dimension 0, the *lines* are the cosets of the subspaces of dimension 1, the *planes* are the cosets of the subspaces of dimension 2 and the *hyperplanes* are the cosets of the subspaces of dimension $n-1$. In general, the elements of $AG(n, q)$ of dimension m are called the *m-flats* of $AG(n, q)$. Two flats are said to be *incident* if they have at least one common point and to be *parallel* if they have no common points.

Recall that the vectors in an m-dimensional coset of an n-dimensional vector space are all the solutions of a set of $(n-m)$ consistent, independent equations. This will prove to be the most useful representation in Chapter 9.

Example 24. The points and flats of $AG(3, 2)$ are given in Table 8.12. □

Lemma 25. *The number of points in an m-flat of* $AG[n, q]$ *is* q^m. □

There are q^{n-1} points of $AG(n, q)$ incident with a hyperplane of $AG(n, q)$. Consider the hyperplane

$$\{(x_1, x_2, \ldots, x_n) | \eta_1 x_1 + \ldots + \eta_n x_n = \theta\}.$$

Table 8.12. Points and flats of $AG(3, 2)$

Points: 000, 001, 010, 011, 100, 101, 110, 111.

Lines:

000, 001	000, 010	000, 011	000, 100
010, 011	001, 011	001, 010	001, 101
100, 101	100, 110	100, 111	010, 110
110, 111	101, 111	101, 110	011, 111

000, 101	000, 110	000, 111
001, 100	001, 111	001, 110
010, 111	010, 100	010, 101
110, 011	011, 101	100, 011

Planes (or Hyperplanes):

000, 001, 010, 011	000, 001, 100, 101	000, 010, 100, 110
100, 101, 110, 111	010, 011, 110, 111	001, 011, 101, 111

000, 001, 110, 111	000, 010, 101, 111
100, 101, 010, 011	100, 110, 001, 011

000, 100, 011, 111	000, 011, 101, 110
010, 110, 001, 101	100, 111, 001, 010

Then by fixing $\eta_1, \eta_2, \ldots, \eta_n$ and letting θ take on each of the q values in $GF[q]$ we get a set of q hyperplanes which partition the points of $AG(n, q)$. Such a set of q hyperplanes is called a *pencil* and is denoted by $\mathscr{P}(\eta_1, \eta_2, \ldots, \eta_n)$. As

$$\mathscr{P}(\eta_1, \eta_2, \ldots, \eta_n) = \mathscr{P}(v\eta_1, v\eta_2, \ldots, v\eta_n)$$

for all $v \in GF[q]$, $v \neq 0$, we will assume that the first non-zero η_i is 1.

In Table 8.12 the planes (or hyperplanes) of $AG(3, 2)$ are grouped into the seven pencils. These are, in order: $\mathscr{P}(100)$, $\mathscr{P}(010)$, $\mathscr{P}(001)$, $\mathscr{P}(110)$, $\mathscr{P}(101)$, $\mathscr{P}(011)$, and $\mathscr{P}(111)$.

The next result concerns the number of flats of $AG(n, q)$. For ease of stating the result let

$$\phi(n, m, q) = \begin{cases} \dfrac{(q^n - 1)(q^n - q) \ldots (q^n - q^{m-1})}{(q^m - 1)(q^m - q) \ldots (q^m - q^{m-1})}, & 1 \leq m \leq n, \\ 1, & m = 0. \end{cases}$$

Theorem 26.

(i) *The number of distinct m-flats in $AG(n, q)$ is $q^{n-m}\phi(n, m, q)$.*

(ii) *The number of distinct m-flats passing through a fixed point of $AG(n, q)$ is $\phi(n, m, q)$.*

(iii) *The number of distinct m-flats passing through two fixed points of $AG(n, q)$ is $\phi(n-1, m-1, q)$.*

Proof. (i) We begin by counting the number of m-dimensional subspaces of $V_n(q)$. The first element of the basis can be chosen in $(q^n - 1)$ ways, the second in $(q^n - q)$ ways, the third in $(q^n - q^2)$ ways, and so on. Hence there are $(q^n - 1)(q^n - q) \ldots (q^n - q^{m-1})$ ordered m-tuples of independent points in $V_n(q)$. The number of distinct bases for a particular subspace of dimension m is $(q^m - 1)(q^m - q) \ldots (q^m - q^{m-1})$. Hence there are $\phi(n, m, q)$ m-dimensional subspaces of $V_n(q)$. As each subspace has q^{n-m} cosets, the number of distinct m-flats in $AG(n, q)$ is $q^{n-m}\phi(n, m, q)$.

(ii) Given a particular point, P say, of $AG(n, q)$, each subspace of $V_n(q)$ has a unique coset containing P. Hence the number of distinct m-flats containing P is $\phi(n, m, q)$.

(iii) The number of subspaces of $V_n(q)$ of dimension m containing a particular point, P, is

$$(q^n - q)(q^n - q^2) \ldots (q^n - q^{m-1})/(q^m - q) \ldots (q^m - q^{m-1}) = \phi(n-1, m-1, q)$$

and each of these subspaces has a unique coset which contains a particular point Q. Hence there are $\phi(n-1, m-1, q)$ distinct m-flats of $AG(n, q)$ which contain both P and Q. □

Corollary 26.1. *With points as varieties and m-flats as blocks, $AG(n, q)$ is a block design with*

$$v = q^n, b = q^{n-m}\phi(n, m, q), r = \phi(n, m, q), k = q^m, \lambda = \phi(n-1, m-1, q). \quad \square$$

Example 27. In $AG(3, 2)$ the points and lines form a $B[2, 1; 8]$ design and the points and hyperplanes form a $B[4, 3; 8]$ design. □

8.6. Projective geometries

Let $V_{n+1}(q)$ denote the $(n+1)$-dimensional vector space over $GF[q]$ and let $PG(n, q)$ be the set of all subspaces of $V_{n+1}(q)$. Then $PG(n, q)$ is called the *projective geometry of dimension n over* $GF[q]$. The *points* of $PG(n, q)$ are the subspaces of pdim 0, the *lines* are the subspaces of pdim 1, the *planes* are the subspaces of pdim 2 and the *hyperplanes* are the subspaces of pdim $(n-1)$. Two spaces of $PG(n, q)$ are said to be *incident* if they have some common points.

Example 28. The spaces of $PG(3, 2)$ are given in Table 8.13. (We have again omitted (0000) from every subspace.) □

Theorem 29.

(i) *The number of distinct spaces of* pdim *m in* $PG(n, q)$ *is* $\phi(n+1, m+1, q)$.

(ii) *The number of distinct spaces of* pdim *m in* $PG(n, q)$ *containing a particular point is* $\phi(n, m, q)$.

(iii) *The number of distinct spaces of* pdim *m in* $PG(n, q)$ *containing two particular points is* $\phi(n-1, m-1, q)$.

Proof. See Exercise 8.13. □

Corollary 29.1. *With points as varieties and spaces of* pdim *m as blocks,* $PG(n, q)$ *is a block design with*

$$v = (q^{n+1}-1)/(q-1), \quad b = \phi(n+1, m+1, q), \quad r = \phi(n, m, q),$$
$$k = (q^{m+1}-1)/(q-1), \quad \lambda = \phi(n-1, m-1, q). \quad \square$$

Example 30. In $PG(3, 2)$ the points and lines form a $B[3, 1; 15]$ design and the points and hyperplanes form a $B[7, 3; 15]$ design. □

Indeed, we can say more about the structure of the designs formed from the hyperplanes, generalizing Theorem 3.5.18.

Table 8.13. Points and spaces of $PG(3, 2)$

Points: 0001, 0010, 0011, 0100, 0101, 0110, 0111,
1000, 1001, 1010, 1011, 1100, 1101, 1110, 1111.

Lines:

0001, 0010, 0011	0010, 0100, 0110	0011, 0101, 0110
0001, 0100, 0101	0010, 0101, 0111	0011, 1000, 1011
0001, 0110, 0111	0010, 1000, 1010	0011, 1001, 1010
0001, 1000, 1001	0010, 1001, 1011	0011, 1100, 1111
0001, 1010, 1011	0010, 1100, 1110	0011, 1101, 1110
0001, 1100, 1101	0010, 1101, 1111	0100, 1000, 1100
0001, 1110, 1111	0011, 0100, 0111	0100, 1001, 1101

0100, 1010, 1110	0110, 1001, 1111
0100, 1011, 1111	0110, 1010, 1100
0101, 1000, 1101	0110, 1011, 1101
0101, 1001, 1100	0111, 1000, 1111
0101, 1010, 1111	0111, 1001, 1110
0101, 1011, 1110	0111, 1010, 1101
0110, 1000, 1110	0111, 1011, 1100

Planes (or hyperplanes):

0001, 0010, 0011, 0100, 0101, 0110, 0111
0001, 0010, 0011, 1000, 1001, 1010, 1011
0001, 0010, 0011, 1100, 1101, 1110, 1111
0001, 0100, 0101, 1000, 1001, 1100, 1101
0001, 0100, 0101, 1010, 1011, 1110, 1111
0001, 0110, 0111, 1000, 1001, 1110, 1111
0001, 0110, 0111, 1010, 1011, 1100, 1101
0010, 0100, 0110, 1000, 1010, 1100, 1110
0010, 0100, 0110, 1001, 1011, 1101, 1111
0010, 0101, 0111, 1000, 1010, 1101, 1111
0010, 0101, 0111, 1001, 1011, 1100, 1110
0011, 0100, 0111, 1000, 1011, 1100, 1111
0011, 0100, 0111, 1001, 1010, 1101, 1110
0011, 0101, 0110, 1000, 1011, 1101, 1110
0011, 0101, 0110, 1001, 1010, 1100, 1111

Theorem 31. *With the points of $PG(n, q)$ as varieties and the hyperplanes as blocks we have an*

$$SB[(q^n-1)/(q-1), (q^{n-1}-1)/(q-1); (q^{n+1}-1)/(q-1)]$$

design. The design is cyclic and the points in any hyperplane determine a difference set.

Proof. The first part follows from Corollary 29.1. The hard part is to show that the design is cyclic.

We have already seen that every element of $GF[q] = GF[p^a]$ can be written as an a-tuple with elements from $GF[p]$. In the same way we can set up a correspondence between the points of $PG(n, q)$ and the elements of $GF[q^{n+1}]$. Let x be a primitive element of $GF[q^{n+1}]$. Then

$$x^j = (y_0, y_1, \ldots, y_n) = \mathbf{y}, \quad y_i \in GF[q].$$

If $\mathbf{y}$ and $\mathbf{z}$ represent the same point then $\mathbf{z} = \eta \mathbf{y}$ for some $\eta \in GF(q)$, $\eta \neq 0$. Let $v = (q^{n+1}-1)/(q-1)$. Then the elements of $GF[q^{n+1}]$ which are also elements of $GF[q]$ are $1, x^v, \ldots, x^{(q-2)v}$ and 0. Hence if $\mathbf{y}$ corresponds to x^j and $\mathbf{z}$ corresponds to x^l then $\mathbf{y}$ and $\mathbf{z}$ represent the same point if and only if

$$x^l = x^{sv}x^j = x^{sv+j},$$

for some s, $0 \leq s \leq (q-2)$; that is, if and only if $j \equiv l \pmod{v}$.

Let σ be a mapping of $GF[q^{n+1}]$ given by $\sigma(0) = 0$ and $\sigma(x^i) = x^{i+1}$. This mapping is both one-one and onto; also it maps subspaces of dimension m onto subspaces of dimension m since if $\mathbf{u}_1, \ldots, \mathbf{u}_m$ are linearly independent vectors then

$$\sigma\left(\sum_{i=1}^{m} a_i \mathbf{u}_i\right) = \sum_{i=1}^{m} a_i \sigma(\mathbf{u}_i) = 0$$

if and only if $\sum_i a_i \mathbf{u}_i = 0$. As $1, x, \ldots, x^{v-1}$ correspond to the v different points σ permutes the points in a single cycle.

If σ^t maps the points of a hyperplane onto themselves then σ^t must permute points in a cycle of length w and so w must divide k. Thus $\sigma^w = 1$ and $\sigma^v = 1$, so w must also divide v. But $(v, k) = 1$ since $v - qk = 1$. Then $w = 1$ and the only power of σ which fixes a hyperplane is $\sigma^v = 1$. Thus σ must permute the hyperplanes in a cycle of length v and the result follows. □

Example 32. We can obtain a (15, 7, 3) difference set from the first hyperplane in Table 8.13 as follows. Let x be a primitive element of $GF[16]$ where $x^4 = 1 + x$. Write the elements of the hyperplane as polynomials in x and then as powers of x. The indices of x form a difference set. This is set out in Table 8.14. □

In Exercise 8.15 we give two (31, 15, 7) difference sets. One has been derived by the method of Theorem 31 and one has not. Thus, the parameters alone do not determine the structure of a design.

Again, we can obtain an $AG(n, q)$ from a $PG(n, q)$ by deleting a hyperplane. For ease of description assume we delete the hyperplane $x_0 = 0$ and let $(1, x_1, \ldots, x_n)$ represent the point

$$\{\eta(1, x_1, \ldots, x_n) \mid \eta \in GF[q], \eta \neq 0\}$$

Table 8.14

Hyperplane	Polynomial in x	Power of x	Difference set
0001	1	x^0	0
0010	x	x	1
0011	$x+1$	x^4	4
0100	x^2	x^2	2
0101	x^2+1	x^8	8
0110	x^2+x	x^5	5
0111	x^2+x+1	x^{10}	10

in $PG(n, q)$. Then the points of the affine geometry are $(x_1, x_2, \ldots, x_n)$ and the flats of $AG(n, q)$ can be determined from the spaces of $PG(n, q)$.

Example 33. We delete the hyperplane $x_0 = 0$ from the $PG(3, 2)$ given in Table 8.13. The resulting $AG(3, 2)$ is given in Table 8.12. □

8.7. Other resolvable designs

In this section we look briefly at the construction of designs which are resolvable but which may lack some other desirable properties such as pairwise balance or constant block size.

Historically, the first resolvable designs considered in this category were the lattice designs.

A *two-dimensional lattice design* is an arrangement of k^2 varieties into blocks of size k such that the set of blocks can be partitioned into $t \leq k+1$ resolution classes of k blocks each and such that any two varieties appear together in at most one block. If $t = k+1$ then we have an affine plane. Otherwise we have some resolution classes of an affine plane, although it may not be possible to embed them in an affine plane. Hence the method of construction of Theorem 14 can be used: a lattice design with t resolution classes corresponds to a set of $t-2$ MOLS of side k. A *simple lattice* is one with $t = 2$, a *triple lattice* has $t = 3$ and a *quadruple lattice* has $t = 4$. A *balanced lattice* is one with $t = k+1$. A *three-dimensional lattice design* is an arrangement of k^3 varieties into blocks of size k such that the set of blocks can be partitioned into $t \leq k^2+k+1$ resolution classes of k^2 blocks each and such that any two varieties appear together in at most one block.

We will consider another method of construction of lattice designs in Section 9.5.

A *rectangular lattice design* is an arrangement of $k(k-l)$ varieties into blocks of size $k-l$ such that the set of blocks can be partitioned into $t \leq (v-1)/(k-1)$ resolution classes of k blocks each and such that any two varieties appear together in at most one block. Again we proceed as in Theorem 14. Here we let A be a $k \times k$ array with entries $1, 2, \ldots, k(k-l)$ in some order and such that each row and column of A has l empty cells. Then the rows and columns of A give the blocks of two resolution classes. We can construct a third resolution class by superimposing on A a Latin square of order k in which l entries have been deleted from each row, each column and the set of occurrences of each symbol. However, we need not be quite so fussy, as the next example shows.

Example 34. We construct a rectangular lattice design in which $k = 3$, $l = 1$, $v = 3(3-1) = 6$, and $t = 5$. The array A, the five resolution classes and the 'partial Latin squares' corresponding to the final three classes all appear in Table 8.15. Note that P_1, P_2 and P_3 are mutually orthogonal, but none of them can be completed to a Latin square. Note that '*' denotes an empty cell. □

Table 8.15

1	2	*
3	*	4
*	5	6

A

12	13	14	15	16
34	25	26	24	23
56	46	35	36	45

Resolution classes

$$P_1 = \begin{matrix} 1 & 2 & * \\ 3 & * & 1 \\ * & 3 & 2 \end{matrix} \qquad P_2 = \begin{matrix} 1 & 2 & * \\ 3 & * & 2 \\ * & 1 & 3 \end{matrix} \qquad P_3 = \begin{matrix} 1 & 2 & * \\ 2 & * & 3 \\ * & 3 & 1 \end{matrix}$$

The following result classifies the resolvable designs with two replicates. It is a natural extension of earlier work (see Exercise 8.17). A *non-binary design* is one in which a variety may appear more than once in a block, so the corresponding incidence matrix does not just contain 0s and 1s.

Theorem 35. *Every resolvable design with two replications is equivalent to a symmetric block design, which is not necessarily resolvable and which may not be binary.*

Proof. Consider a $(k\beta, 2\beta, 2, k)$ resolvable design D. (As λ may not be constant we omit it from the set of parameters.) Let $B_1, B_2, \ldots, B_\beta$ and $C_1, C_2, \ldots, C_\beta$ be the blocks of the two resolution classes. Construct a design D^* with varieties $B_1, B_2, \ldots, B_\beta$ and blocks determined as follows: B_j occurs in the ith block m times if and only if $|B_j \cap C_i| = m$. Thus D^* has β varieties, β blocks, each block is of size k (since $|C_i| = k$), and each B_j appears k times (since $|B_j| = k$ in D) but some of these occurrences may be in the same block.

Now suppose that we start with a (β, β, k, k) design D^*. Construct a graph from D^* as follows: take one vertex for each variety and one for each block of D^* and draw an edge from a vertex representing a variety to one representing a block if the variety is an element of the block. Thus each vertex has k edges incident with it. Now label the edges $1, 2, \ldots, k\beta$ and construct 2β blocks by listing the elements on the edges adjacent to each vertex. The design has two resolution classes and each block is of size k so the result follows. □

Example 36. Consider the (5, 5, 3, 3) design, D^*: 123, 145, 134, 235, 245. The graph corresponding to this design is given in Fig. 8.1 and the equivalent (15, 10, 2, 3) design in Table 8.16. □

Table 8.16

1 2 3	1 4 7
4 5 6	2 10 13
7 8 9	3 8 11
10 11 12	5 9 14
13 14 15	6 12 15

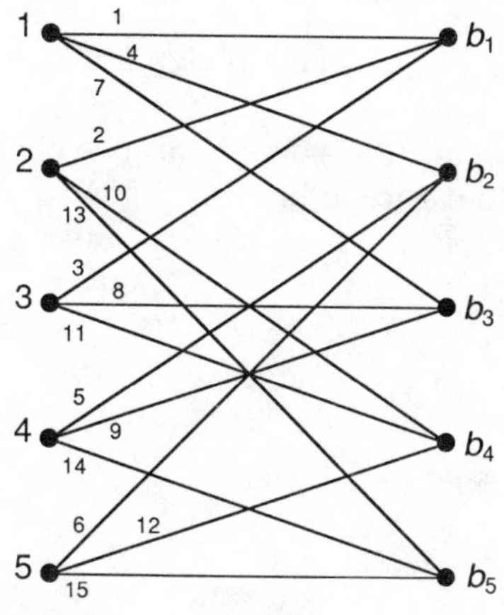

Fig. 8.1

Our final example of a resolvable design was developed for the field trials required for statutory variety testing in the United Kingdom. In these designs the number of varieties and the number of replications are fixed. The number of varieties is so large the blocks must be incomplete and for management reasons the designs must be resolvable.

An *$\alpha(g_1, g_2, \ldots, g_n)$-design* is a design on $v = sk$ varieties with r replicates of each variety and with blocks of size k arranged in r resolution classes of s blocks each. Any pair of varieties appear together in either g_1 or g_2 or ... or g_n blocks. The blocks in each resolution class are obtained from one initial block as shown in the following example.

Example 37. An $\alpha(0, 1, 2)$-design with $v = 20$, $s = 4$, $k = 5$, $r = 3$ is given in Table 8.17. (Rows are blocks.) We can see that, for example, varieties 0 and 1 never appear together, 0 and 4 appear together in one block and 4 and 9 appear together in two blocks.

Let σ_i denote the permutation $(4(i-1), 4(i-1)+1, 4(i-1)+2, 4(i-1)+3)$, $1 \leq i \leq k$. Then, given the first block in a replicate (or resolution class), the element in position i of block j of that class is found by applying σ_i^{j-1} to the ith element of the first block. □

Table 8.17

Replicate 1	Replicate 2	Replicate 3
0 4 8 12 16	0 5 10 15 19	0 6 11 13 18
1 5 9 13 17	1 6 11 12 16	1 7 8 14 19
2 6 10 14 18	2 7 8 13 17	2 4 9 15 16
3 7 11 15 19	3 4 9 14 18	3 5 10 12 17

Theorem 38. *If an $\alpha(0, 1)$-design exists then $k \leq s$, provided that $r \geq 2$.*

Proof. As the design is an $\alpha(0, 1)$-design, any two blocks intersect in at most one element. Consider a particular block B. It intersects $k(r-1)$ blocks in one element each and these blocks are all from the other $r-1$ resolution classes. Thus $k(r-1) \leq s(r-1)$ or $k \leq s$. □

Theorem 39. *The following four series of $\alpha(0, 1)$-designs are known:*

(i) $r = 2$, $k \leq s$;
(ii) $r = 3$, *s odd*, $k \leq s$;
(iii) $r = 3$, *s even*, $k \leq s-1$;
(iv) $r = 4$, $s \equiv 1$ *or* 5 (mod 6), $k \leq s$.

Proof. Starter blocks for designs in which k is equal to its maximum possible value are given in Table 8.18. To obtain the complete design proceed as follows: let $\sigma_i = (s(i-1), s(i-1)+1, \ldots, s(i-1)+s-1)$, $1 \leq i \leq k$. Then the element in position i in the jth block in a resolution class is obtained by applying σ_i^{j-1} to the element in position i of the appropriate starter block. One way to obtain designs for smaller values of k is to use only the first k positions in each starter block. □

Example 40. We construct an $\alpha(0, 1)$-design with $r = 3$, $s = 6$ and $k = 4$. The design is given in Table 8.19. □

No $\alpha(0, 1)$-designs are known with $r = 3$, s even and $k = s$. When $r = 4$, no general solutions are known for s even, or a multiple of three, although individual solutions are known within the required range. The gaps left by the

Table 8.18. Starter blocks for Theorem 39 with $s_1 = s/2$, $s_2 = (s+1)/2$

0	s	$2s$	$\dots$	$(s-1)s$
0	$s+1$	$2s+2$	$\dots$	$(s-1)s+s-1$

(i)

0	s	$2s$	$\dots$	$(s-1)s$
0	$s+1$	$2s+2$	$\dots$	$(s-1)s+(s-1)$
0	$s+s-1$	$2s+s-2$	$\dots$	$(s-1)s+1$

(ii)

0	s	$2s$	$3s$	$4s$	$\dots$	$(s-3)s$	$(s-2)s$
0	$s+1$	$2s+2$	$3s+3$	$4s+4$	$\dots$	$(s-3)s+(s-3)$	$(s-2)s+(s-2)$
0	$s+s_1$	$2s+1$	$3s+s_1+1$	$4s+2$	$\dots$	$(s-3)s+(s_1-1)$	$(s-2)s+s_1+s_1-1$

(iii)

0	s	$2s$	$3s$	$4s$	$\dots$	$(s-2)s$	$(s-1)s$
0	$s+1$	$2s+2$	$3s+3$	$4s+4$	$\dots$	$(s-2)s+s-2$	$(s-1)s+s-1$
0	$s+s-1$	$2s+s-2$	$3s+s-3$	$4s+s-4$	$\dots$	$(s-2)s+2$	$(s-1)s+1$
0	$s+s_2$	$2s+1$	$3s+s_2+1$	$4s+2$	$\dots$	$(s-2)s+s_2+s_2-2$	$(s-1)s+s_2-1$

(iv)

Table 8.19

0	6	12	18		0	7	14	21		0	9	13	22
1	7	13	19		1	8	15	22		1	10	14	23
2	8	14	20		2	9	16	23		2	11	15	18
3	9	15	21		3	10	17	18		3	6	16	19
4	10	16	22		4	11	12	19		4	7	17	20
5	11	17	23		5	6	13	20		5	8	12	21

$\alpha(0, 1)$-designs have been filled by $\alpha(0, 1, 2)$-designs. As no general constructions have been given we will not discuss these further.

There is another construction available when all the factors of v are outside the range of acceptable block sizes; this is discussed in Exercise 8.20.

8.8. References and comments

A number of early references to the concept of resolvability can be found in Preece (1982). Yates (1940*a*) discusses the statistical advantages of resolvable designs. The terms *resolvable* and *affine resolvable* were introduced by Bose

(1942) and the concepts extended to *α-resolvability* by Bose and Shrikhande (1960).

Lemma 8, Theorems 4, 9, and 16, and Corollary 16.1 are due to Shrikhande and Raghavarao (1963). For related work see also Kageyama (1976) and Kageyama and Tsuji (1977). The result about a symmetric matrix and its eigenvectors (used in proving Lemma 8) can be found, for example, in Theorem 9.9.2 of Hohn (1973). Lemma 11 is given in Raghavarao (1971). The most recent list of designs is due to Mathon and Rosa (1985), and includes information on resolvability.

As mentioned in Chapter 1, finite planes over fields of order q were first studied by von Staudt (1856) and more extensively by Veblen and Bussey (1906) and Veblen and Wedderburn (1907). Carmichael (1937) gives an elegant treatment of finite geometries; see also Dembowski (1968). A detailed account of projective planes is given by Hughes and Piper (1973) and of projective geometries over finite fields by Hirschfeld (1979). Theorem 31 is due to Singer (1938); the method of proof given here is due to Hall (1967). See also Batten (1986) and Beutelspacher (1986).

Note that in three and higher dimensions, the only finite projective geometries are those constructed from the finite fields. It is only the projective planes which may have different structures; for accessible examples see Lorimer (1980), and Room and Kirkpatrick (1971).

Lattice designs were introduced by Yates (1940*b*) and extended by Harshbarger (1947, 1950). Theorem 35 is due to Patterson and E. R. Williams (1975) and α-designs were introduced by them in 1976. Resolvable cyclic designs have been considered by David (1967), John (1966) and John, Wolock, and David (1972). Some of these designs are α-designs in which the efficiency may not be as high as one with the same parameters constructed by Patterson and E. R. Williams (see Exercise 8.9). Exercise 8.1 is due to Bose (1942). Exercise 8.17 is from Bose and Nair (1962).

Shrikhande (1976) and Baker (1983) give a survey of results on affine resolvable BIBDs and the relationship between resolvable BIBDs and self-orthogonal Latin squares. The problem initially raised by Kirkman (1850*b*) was solved by Ray-Chaudhuri and Wilson (1971); see Section 13.4. Resolvable designs with variance-balance are given by Mukerjee and Kageyama (1985).

Exercises

8.1. Use the method of proof of Theorem 2.1.2(ii) to show that in a resolvable BIBD $b \geq v+t-1$.

(Hint: Let B_{ij} denote the jth block of the ith resolution class and let $|B_{11} \cap B_{ij}| = l_{ij}$. Consider the value of $\sum_i \sum_j (l_{ij}-\bar{l})^2$, where $\bar{l} = \sum_i \sum_j l_{ij}/(b-1)$.)

8.2. Construct a $B[4, 1; 16]$ design. To each block of a resolution class adjoin a new point and then construct a final block containing the new points. What are the parameters of this design?

8.3. Complete the proof of Theorem 14.

8.4. Let D be an $SB[2m-1, m-1; 4m-1]$ design – that is, an Hadamard design – based on the set X. Let $X^* = X \cup \{x\}$, where $x \notin X$, and let D^* be a design based on X^* and constructed from D in the following way. If B is a block of D, then $B^* = B \cup \{x\}$ and $\bar{B} = X \backslash B$ are both blocks of D^*. This accounts for all the blocks of D^*.

Show that D^* is an affine resolvable BIBD.

8.5. (a) Let $v_1 = 4m+3$ be a prime or a prime power. Show that an affine $(2m+1)$-resolvable BIBD exists with parameters

$$v = (4m+3)^2 = v_1^2, \quad b = v_1(v_1+1), \quad r = (v_1^2-1)/2, \quad k = v_1(v_1-1)/2,$$
$$\lambda = (v_1+1)(v_1-2)/4.$$

Construct the design with $m = 1$.

(b) Let $v_1 = 4m+1$ be a prime or a prime power. Show that a $4m$-resolvable BIBD exists with parameters

$$v_2 = v_1^2, \; b = 2v_1(v_1+1), \; r = v_1^2-1, \; k = v_1(v_1-1)/2, \; \lambda = (v_1+1)(v_1-2)/2.$$

Construct the design with $m = 1$.

8.6. Show that a subspace of $V_2(q)$ of dimension 1 may be written as either

$$\{(x, y) \mid x+\eta y = 0\} \quad \text{or} \quad \{(x, y) \mid y = 0\}.$$

8.7. Construct $AG(2, 3)$ and $AG(2, 5)$.

8.8. From the affine planes of Exercise 8.7, construct two MOLS of order 3 and four MOLS of order 5.

8.9. Construct $PG(2, 4)$ and $PG(2, 2)$.

8.10. Show that $PG(2, q)$ contains four points, no three of which are collinear.

8.11. Extend the planes constructed in Exercise 8.7 to projective planes.

8.12. Construct $AG(3, 3)$. Use this geometry to illustrate Corollary 26.1.

8.13. Prove Theorem 29.

8.14. Extend $AG(3, 3)$ to get $PG(3, 3)$. Use this to illustrate Corollary 26.1 and Theorem 31.

8.15. Consider the two (31, 15, 7) difference sets

$$(1, 2, 3, 4, 6, 8, 12, 15, 16, 17, 23, 24, 27, 29, 30)$$

and

$$(1, 2, 4, 5, 7, 8, 9, 10, 14, 16, 18, 19, 20, 25, 28).$$

By considering the intersection of triples of blocks, or otherwise, show that the first design comes from a projective geometry and that the second does not.

8.16. Construct a triple lattice on 9 varieties. Can a triple lattice on 9 varieties always be extended to a balanced lattice? Construct a triple lattice on 16 varieties which cannot be extended to a quadruple lattice.

8.17. Suppose that an $SB[k, \lambda; v]$ design exists. Then there exists a resolvable design with two replicates and with mv varieties and blocks of size m where

$m = kp + (v - k)q$ for two integers p and q. Prove this result by considering the incidence matrix of the $SB[k, \lambda; v]$ and replacing each 1 by a set of p elements and each 0 by a set of q elements. Construct the design when $k = 2$, $\lambda = 1$, $v = 3$, $p = 2$ and $q = 1$.

8.18. Use the method of Theorem 35 to construct a (21, 14, 2, 3) resolvable design.

8.19. Complete the proof of Theorem 39.

8.20. Suppose $v = s_1k_1 + s_2k_2$ where $s_1, s_2, k_1, k_2 \in N$ and $|k_1 - k_2| = 1$. Each replicate (or resolution class) consists of s_1 blocks of size k_1 and s_2 of size k_2. To construct such a design construct an α-design on $v + s_2$ varieties with $s = s_1 + s_2$ blocks of k_1 plots in each replicate. Delete a set of s_2 varieties, no two of which occur together in a block. Hence find a design on 17 varieties with $s_1 = 1$, $s_2 = 3$.

8.21. Let $v = 24$, $r = k = 4$ and consider the design obtained from the starter block (0, 1, 10, 3) cyclically. Show that this design in an $\alpha(0, 1)$ design.

9 Symmetrical factorial designs

9.1. Motivation

In the designs we have considered so far, a number of treatments (or varieties) have been compared and these treatments have not necessarily been related to each other. In this chapter and the next we look at experiments in which there is an underlying structure to the set of treatments. We begin with a small example.

Suppose we are interested in the yield of one particular variety of wheat under different growing conditions. We will consider four conditions determined by the amounts of two different nutrients – nitrogen and phosphorus. We call the nutrients the *factors* under consideration. Each nutrient is applied at one of two *levels*, say high and low, represented by 1 and 0 respectively. We are interested in the effect of nitrogen at the two different levels of phosphorus, the effect of phosphorus at the different levels of nitrogen and the way the two nutrients interact with respect to the yield. Let us denote the two factors by the letters $\mathcal{N}$ and $\mathcal{P}$. The four treatment combinations are given in Table 9.1. Each letter sequence is used to denote both the treatment combination and the sum of the yields of the plots receiving that combination. Sometimes we regard the letters as commuting variables.

The effect of nitrogen at the low level of phosphorus is represented by n-(1) and at the high level of phosphorus by np-p. We define the *main effect* of nitrogen to be the average of these two effects. Thus, if we denote the main effect of nitrogen by $\mathcal{N}$, then

$$\begin{aligned}\mathcal{N} &= \tfrac{1}{2}((np - p) + (n - (1)))\\ &= \tfrac{1}{2}(np - p + n - (1)).\end{aligned}$$

Similarly,

$$\mathcal{P} = \tfrac{1}{2}(np - n + p - (1)).$$

Table 9.1. Treatment combinations

Level of $\mathcal{N}$	Level of $\mathcal{P}$	Binary sequence	Letter sequence
Low	Low	00	(1)
Low	High	01	p
High	Low	10	n
High	High	11	np

If the two factors were acting independently we would expect the two estimates of the effect of nitrogen to be the same with

$$np - p = n - (1).$$

Hence, the difference between these two estimates is a measure of how much the two factors are interacting. Thus, the *interaction* of the factors $\mathcal{N}$ and $\mathcal{P}$, denoted by $\mathcal{NP}$, is estimated by

$$\begin{aligned}\mathcal{NP} &= \tfrac{1}{2}((np - p) - (n - (1))) \\ &= \tfrac{1}{2}(np - p - n + (1)) \\ &= \tfrac{1}{2}((np - n) - (p - (1))).\end{aligned}$$

We see this is also the difference between the two estimates of the effect of phosphorus.

We generalize these ideas in the remainder of this chapter.

9.2. Notation

A *factorial experiment* is an experiment in which several *factors* (such as varieties, fertilizers, or antibiotics) are applied to each experimental unit and each factor is applied at two, or more, *levels*. The levels may be *quantitative* (as with amounts of fertilizers) or *qualitative* (where the levels refer to different varieties of wheat, say), but in either case are represented by the elements of a finite set, usually by 0, 1, 2, . . . , $s_i - 1$, where the ith factor occurs at s_i levels.

A factorial experiment in which n factors are tested, and in which the ith factor has s_i levels, is called an $s_1 \times s_2 \times \ldots \times s_n$ factorial experiment. If $s_1 = s_2 = \ldots = s_n = s$ then we have an s^n *symmetrical factorial experiment*. We will denote the factors by capital letters ($\mathcal{A}$, $\mathcal{B}$, $\mathcal{C}$, . . .). The particular combination of factor levels applied to a plot (experimental unit) will be represented either by an n-tuple, where the ith co-ordinate in the n-tuple denotes the level of the ith factor, or by a sequence of lower-case letters where the letter i has as a superscript the level of the ith factor. Often, we do not write i^0 – we omit the letter and assume the corresponding factor is present at level 0.

Example 1. Consider a 2^3 factorial experiment involving the three factors $\mathcal{A}$, $\mathcal{B}$, and $\mathcal{C}$. There are eight possible treatment combinations; these are listed in Table 9.2. □

In the remainder of this chapter we assume that the number of levels, s, is a prime or a prime power. When we represent the treatment combinations as an n-tuple we will use the elements of $GF[s]$ to represent the levels of the factors. Hence there is a correspondence between the treatment combinations and the points of $AG(n, s)$. We will use this in the sequel. When we represent the treatment combinations as a sequence of lower-case letters the superscripts

Table 9.2. Treatments in a 2^3 factorial design

Sequence	Binary triple	Factors at each level: Low	High
(1)	000	$\mathscr{A}, \mathscr{B}, \mathscr{C}$	—
a	100	$\mathscr{B}, \mathscr{C}$	$\mathscr{A}$
b	010	$\mathscr{A}, \mathscr{C}$	$\mathscr{B}$
ab	110	$\mathscr{C}$	$\mathscr{A}, \mathscr{B}$
c	001	$\mathscr{A}, \mathscr{B}$	$\mathscr{C}$
ac	101	$\mathscr{B}$	$\mathscr{A}, \mathscr{C}$
bc	011	$\mathscr{A}$	$\mathscr{B}, \mathscr{C}$
abc	111	—	$\mathscr{A}, \mathscr{B}, \mathscr{C}$

will be elements of $\{0, 1, 2, \ldots, s-1\}$. Thus there is a correspondence with the elements of the direct product of n copies of the cyclic group on s elements.

We begin by considering the 2^n factorial design in some detail.

9.3. The 2^n factorial design

In this section we suppose that a 2^n factorial design is carried out either as a completely randomized design or as a randomized complete block design; incomplete blocks are considered subsequently. Then we can estimate the effect of each factor, independently of the others (the main effect), and the effect of the interaction of two (or more) factors (the interaction effect).

As before, we will call the two levels of each factor the high level and the low level. Then we can calculate the effect of a factor, for a particular combination Π of levels of the other factors, as the difference of the yields with $\mathscr{A}$ at a high level and with $\mathscr{A}$ at a low level, both in the presence of the same combination Π of levels of the other factors. The main effect of that factor is the average effect over all the possible combinations Π.

The interaction effect of the two factors $\mathscr{A}$ and $\mathscr{B}$ is half the difference between the main effect of $\mathscr{A}$ at the high level of $\mathscr{B}$ and the main effect of $\mathscr{A}$ at the low level of $\mathscr{B}$ (or equivalently, as we shall see, half the difference between the main effect of $\mathscr{B}$ at the high level of $\mathscr{A}$ and the main effect of $\mathscr{B}$ at the low level of $\mathscr{A}$). Higher order interactions are defined recursively.

Example 2. Consider a 2^3 factorial experiment involving the three factors $\mathscr{A}$, $\mathscr{B}$ and $\mathscr{C}$. In Table 9.3 we give the effect of $\mathscr{A}$ at each of the four possible combinations of levels of the other factors.

Table 9.3. Effect of $\mathscr{A}$

Level of $\mathscr{B}$	Level of $\mathscr{C}$	Effect of $\mathscr{A}$
Low	Low	$a-(1)$
Low	High	$ac-c$
High	Low	$ab-b$
High	High	$abc-bc$

Thus the main effect of $\mathscr{A}$, denoted by $\mathscr{A}$, is equal to

$$\tfrac{1}{4}((a-(1))+(ac-c)+(ab-b)+(abc-bc))$$
$$=\tfrac{1}{4}(abc+ab+ac+a-(bc+b+c+(1))).$$

Next, we calculate the estimate of the interaction effect of $\mathscr{A}$ and $\mathscr{B}$, denoted by $\mathscr{A}\mathscr{B}$, in two ways.

First, we note that the main effect of $\mathscr{A}$ at the high level of $\mathscr{B}$ is

$$\tfrac{1}{2}((abc+ab)-(bc+b))$$

and at the low level of $\mathscr{B}$ is

$$\tfrac{1}{2}((ac+a)-(c+(1))).$$

Half the difference between these is $\mathscr{A}\mathscr{B}$ and is

$$\tfrac{1}{2}(\tfrac{1}{2}((abc+ab)-(bc+b))-\tfrac{1}{2}((ac+a)-(c+(1))))$$
$$=\tfrac{1}{4}((abc+ab+c+(1))-(ac+a+bc+b)).$$

Secondly, we note that the main effect of $\mathscr{B}$ at the high level of $\mathscr{A}$ is

$$\tfrac{1}{2}((abc+ab)-(ac+a))$$

and at the low level of $\mathscr{A}$ is

$$\tfrac{1}{2}((bc+b)-(c+(1))).$$

Half the difference between these is

$$\tfrac{1}{2}(\tfrac{1}{2}((abc+ab)-(ac+a))-\tfrac{1}{2}((bc+b)-(c+(1))))$$
$$=\tfrac{1}{4}((abc+ab+c+(1))-(ac+a+bc+b))$$

which is $\mathscr{A}\mathscr{B}$, just as before.

There are three ways we can calculate $\mathscr{A}\mathscr{B}\mathscr{C}$: as half the difference between $\mathscr{A}\mathscr{B}$ at the high level of $\mathscr{C}$ and $\mathscr{A}\mathscr{B}$ at the low level of $\mathscr{C}$, half the difference between $\mathscr{A}\mathscr{C}$ at high $\mathscr{B}$ and $\mathscr{A}\mathscr{C}$ at low $\mathscr{B}$ or half the difference between $\mathscr{B}\mathscr{C}$ at high $\mathscr{A}$ and $\mathscr{B}\mathscr{C}$ at low $\mathscr{A}$. We work out the first of these and leave the other two as exercises. The estimate of the effect $\mathscr{A}\mathscr{B}$ at the high level of $\mathscr{C}$ is

$$\tfrac{1}{2}((abc+c)-(ac+bc))$$

and at the low level of $\mathscr{C}$ is

$$\tfrac{1}{2}((ab+(1))-(a+b)).$$

Thus the estimate of the $\mathscr{ABC}$ effect is

$$\tfrac{1}{4}(((abc+c)-(ac+bc))-((ab+(1))-(a+b)))$$
$$=\tfrac{1}{4}((abc+c+a+b)-(ac+bc+ab+(1))).$$

We can write the coefficients of the three main effects and the four interaction effects as a 7×8 matrix. This is given in Table 9.4, omitting the factor of $\frac{1}{4}$.

Consider the $\mathscr{A}$ effect and write out the treatments in two sets, one consisting of those treatments with a coefficient of 1, the other those with a coefficient of -1. This gives

$$\mathscr{A}_1 = \{100, 110, 101, 111\} \quad \text{and} \quad \mathscr{A}_0 = \{000, 010, 001, 011\}.$$

We see that $\mathscr{A}_0$ and $\mathscr{A}_1$ are the hyperplanes in the pencil $\mathscr{P}(1,0,0)$ of $AG(3,2)$. Similarly, the partition for $\mathscr{B}$ corresponds to the hyperplanes of the pencil $\mathscr{P}(0,1,0)$ and of $\mathscr{AB}$ to the pencil $\mathscr{P}(1,1,0)$. □

We will often refer to a main effect as an interaction effect involving one factor, for symmetry when stating results. Also we let e_i denote a row vector with a 1 in the ith position and zeroes elsewhere.

Table 9.4. Coefficients of effects in a 2^3 design

(1)	a	b	ab	c	ac	bc	abc	Effect
-1	1	-1	1	-1	1	-1	1	$\mathscr{A}$
-1	-1	1	1	-1	-1	1	1	$\mathscr{B}$
1	-1	-1	1	1	-1	-1	1	$\mathscr{AB}$
-1	-1	-1	-1	1	1	1	1	$\mathscr{C}$
1	-1	1	-1	-1	1	-1	1	$\mathscr{AC}$
1	1	-1	-1	-1	-1	1	1	$\mathscr{BC}$
-1	1	1	-1	1	-1	-1	1	$\mathscr{ABC}$

Lemma 3. *The interaction effect of factors $i_1, i_2, \ldots, i_t$ in a 2^n factorial experiment is estimated by $1/2^{n-1}$ times the difference between the sum of the treatments with $x_{i_1} + \ldots + x_{i_t} \equiv t \pmod 2$ and the sum of the treatments with $x_{i_1} + \ldots + x_{i_t} \equiv t+1 \pmod 2$, that is, between the flats of the pencil*

$$\mathscr{P}(e_{i_1} + e_{i_2} + \ldots + e_{i_t}).$$

Proof. We will ignore the factor of $1/2^{n-1}$ in the proof. The main effect of factor i_1 is the difference between the sum of the treatments receiving factor i_1 at the high level (and so with $x_{i_1} = 1$) and the sum of the treatments receiving factor i_1 at the low level (and so with $x_{i_1} = 0$). Thus the result holds for $t = 1$.

The interaction between factors i_1 and i_2 is the difference between the main effect of factor i_1 at the high level of i_2 and the main effect of i_1 at the low level of i_2. If we let

$$W_{gh} = \{(x_1, x_2, \ldots, x_n) | x_{i_1} = g, x_{i_2} = h\}$$

and

$$\sum_{W_{gh}} (x_1, \ldots, x_n) = \sum_{(x_1, \ldots, x_n) \in W_{gh}} (x_1, \ldots, x_n)$$

then the interaction between factors i_1 and i_2 is

$$\left(\sum_{W_{11}} (x_1, \ldots, x_n) - \sum_{W_{01}} (x_1, \ldots, x_n)\right) - \left(\sum_{W_{10}} (x_1, \ldots, x_n) - \sum_{W_{00}} (x_1, \ldots, x_n)\right)$$

$$= \left(\sum_{W_{11}} (x_1, \ldots, x_n) + \sum_{W_{00}} (x_1, \ldots, x_n)\right) - \left(\sum_{W_{01}} (x_1, \ldots, x_n) + \sum_{W_{10}} (x_1, \ldots, x_n)\right).$$

This is the difference between the sum of treatments with $x_{i_1} + x_{i_2} \equiv 0 \pmod 2$ and the sum of treatments with $x_{i_1} + x_{i_2} \equiv 1 \pmod 2$. The result follows by induction. □

We can state Lemma 3 informally as follows: in estimating an interaction effect all treatment combinations which intersect that effect in an even number of letters have the same sign (and so do all combinations intersecting it in an odd number of letters).

In a 2^n factorial design there are $2^n - 1$ effects of interest (n involving one factor, $\binom{n}{2}$ involving two, and so on). From Lemma 3 we can see that each effect has coefficients which sum to zero.

Lemma 4. *The vectors of coefficients of any two distinct effects in a 2^n factorial design are orthogonal.*

Proof. Consider two distinct effects, one the interaction of factors i_1, $i_2, \ldots, i_t$ and the other the interaction of factors $j_1, j_2, \ldots, j_s$. Let

$$W_m = \{(x_1, \ldots, x_n) | x_{i_1} + \ldots + x_{i_t} \equiv t + m \,(\text{mod} 2)\}, \quad m = 0, 1,$$

$$V_m = \{(x_1, \ldots, x_n) | x_{j_1} + \ldots + x_{j_s} \equiv s + m (\text{mod} 2)\}, \quad m = 0,1.$$

As $\{i_1, i_2, \ldots, i_t\} \neq \{j_1, j_2, \ldots, j_s\}$, the two equations defining $W_m \cap V_p$ have rank 2 and hence

$$|W_m \cap V_p| = 2^{n-2}.$$

In the interaction involving $i_1, \ldots, i_t$ the coefficient of the elements in W_0 is 1 and in W_1 is -1, and, similarly, for elements in V_0 and V_1 in the interaction involving $j_1, \ldots, j_s$. Thus, the product of the coefficients of the entries of $V_0 \cap W_0$ and $V_1 \cap W_1$ is 1 and the product of the coefficients of the entries of $V_0 \cap W_1$ and $V_1 \cap W_0$ is -1. Hence, the sum of the products of the coefficients is $2^{n-2}(2-2) = 0$ as required. □

Recall that the m-dimensional vector space over Q, the rationals, is the set

$$\{(x_1, x_2, \ldots, x_m) | x_i \in \mathrm{Q}\} = \mathrm{Q}^m, \text{ say.}$$

Clearly $\boldsymbol{e}_1, \boldsymbol{e}_2, \ldots, \boldsymbol{e}_m$ form an orthogonal basis for Q^m. We write

$$\mathrm{Q}^m = \langle \boldsymbol{e}_1, \boldsymbol{e}_2, \ldots, \boldsymbol{e}_m \rangle.$$

Multiplying all the elements of this basis by a non-singular transformation produces another basis for Q^m. Any such transformation is represented by a non-singular $m \times m$ matrix. With the usual 'dot', or scalar, product as the inner product Q^m is an inner product space.

Example 2 (continued). Let H be the 7×8 matrix given in Table 9.4 and let

$$H_A = \begin{bmatrix} j_8^T \\ H \end{bmatrix}$$

so that H_A is 8×8. Now $H_A^T H_A = 8I$, showing that H_A is a Hadamard matrix and is non-singular. Thus $H_A \boldsymbol{e}_i$, $i = 1, 2, \ldots, 8$ form a basis for Q^8. □

In general, the set of the $2^n - 1$ effects and j_{2^n} is an orthogonal basis for the 2^n-dimensional inner product space over Q.

9.4. 2^n factorial designs in incomplete blocks

Suppose that we wish to conduct a 2^n factorial experiment and that it is not possible to get enough homogeneous plots to conduct the experiment as a randomized complete block design. The blocks must, therefore, be incomplete and we will assume that each block is of the same size, say 2^m, that the design is resolvable and that there are r resolution classes (or replicates).

We begin by considering the possible allocations of treatment combinations to blocks within one resolution class.

Example 5. Suppose we must use blocks of size 4 to conduct a 2^3 experiment. Two possible allocations are given in Table 9.5.

In layout 1 the difference between block totals is

$$(a + ab + ac + abc) - (bc + b + c + (1)),$$

which is, ignoring the factor of $\frac{1}{4}$, the main effect of $\mathcal{A}$. Hence we know that all the other effects are orthogonal to this contrast between block totals. (Recall,

Table 9.5. Two possible layouts for a 2^3 experiment in blocks of size 4

Layout 1	Block 1	(1)	b	c	bc
	Block 2	a	ab	ac	abc
Layout 2	Block 1	(1)	a	b	c
	Block 2	ab	ac	bc	abc

from Chapter 1, that a contrast of the block totals, $B_1, \ldots, B_b$, is a linear parametric function of the form $\sum_i \lambda_i B_i$, where $\sum_i \lambda_i = 0$.)

In layout 2 the difference between block totals is

$$(ab+ac+bc+abc)-(a+b+c+(1))=z,$$

which is, again ignoring the factor of $\frac{1}{4}$, one-half of the sums of the main effects of $\mathscr{A}$, $\mathscr{B}$ and $\mathscr{C}$ minus the interaction effect $\mathscr{A}\mathscr{B}\mathscr{C}$. That is,

$$z=\tfrac{1}{2}(\mathscr{A}+\mathscr{B}+\mathscr{C}-\mathscr{A}\mathscr{B}\mathscr{C}).$$

Hence $\mathscr{A}\mathscr{B}$, $\mathscr{A}\mathscr{C}$, and $\mathscr{B}\mathscr{C}$ are orthogonal to this contrast between block totals but $\mathscr{A}$, $\mathscr{B}$, $\mathscr{C}$ and $\mathscr{A}\mathscr{B}\mathscr{C}$ are not.

Thus, in the first case the one-dimensional space over Q that corresponds to block contrasts can be viewed as a subspace of Q^8 generated by $\mathscr{A}$, while in the second case the space corresponding to block contrasts is generated by $\frac{1}{2}(\mathscr{A}+\mathscr{B}+\mathscr{C}-\mathscr{A}\mathscr{B}\mathscr{C})$. □

The space Q^{2^n} can be written as

$$\mathrm{Q}^{2^n}=\langle \boldsymbol{j}, \boldsymbol{h}_1, \boldsymbol{h}_2, \ldots, \boldsymbol{h}_{2^n-1}\rangle$$

where $\boldsymbol{h}_1, \ldots, \boldsymbol{h}_{2^n-1}$ are the coefficient vectors for the 2^n-1 effects. We call $\langle \boldsymbol{h}_1, \boldsymbol{h}_2, \ldots, \boldsymbol{h}_{2^n-1}\rangle$ the *space generated by the treatment effects*. Note that it consists of all the vectors orthogonal to $\boldsymbol{j}$; that is, it is the hyperplane orthogonal to $\boldsymbol{j}$.

Consider the space generated by the vectors representing the contrasts between block totals (called the *block contrast subspace* for brevity). It is a subspace of $\langle \boldsymbol{h}_1, \ldots, \boldsymbol{h}_{2^n-1}\rangle$, since all elements in it are orthogonal to $\boldsymbol{j}$, and it has dimension $b-1$. If the block contrast subspace has a basis consisting of $b-1$ $\boldsymbol{h}_i$'s then we say that these $b-1$ effects have been *completely confounded* with blocks. Otherwise the subspace has a basis consisting of linear combinations of at least b $\boldsymbol{h}_i$'s and these effects are said to be *partly confounded* with blocks.

To give an easy analysis, within each replicate we choose to confound an effect either completely or not at all. This also means that a systematic method for the determination of the blocks in each replicate can be given, based on the effects to be confounded.

Note that traditionally an effect which is completely confounded in each replicate is said to be *totally confounded* and an effect which is completely confounded in some replicates but not in others is said to be *partially confounded*.

Example 5 (continued). Let $\boldsymbol{h}_i$ be the ith row of the matrix in Table 9.4. Then in layout 1 the block contrast subspace is generated by $\boldsymbol{h}_1$. In layout 2 the block contrast subspace is generated by $\frac{1}{4}(\boldsymbol{h}_1+\boldsymbol{h}_2+\boldsymbol{h}_4-\boldsymbol{h}_7)$ and so is a subspace of the space spanned by $\boldsymbol{h}_1$, $\boldsymbol{h}_2$, $\boldsymbol{h}_4$ and $\boldsymbol{h}_7$. Hence, in layout 1, $\mathscr{A}$ is completely confounded with blocks whereas in the second case $\mathscr{A}$, $\mathscr{B}$, $\mathscr{C}$ and $\mathscr{A}\mathscr{B}\mathscr{C}$ are each partly confounded with blocks. □

The next result concerns the structure of the b blocks in a replicate in which exactly $b-1$ effects are completely confounded with blocks.

Lemma 6. *Suppose that in a 2^n factorial experiment conducted in 2^{n-m} blocks of size 2^m, exactly $2^{n-m}-1$ effects are confounded with blocks. Then the blocks are parallel m-flats of $AG(n, 2)$ and there are $n-m$ independent, confounded effects, the others being determined from them by linear combination.*

Proof. All the treatment combinations in one block have the same sign in every contrast of block totals. Thus the treatment combinations in one block must lie in the same hyperplane for each of the confounded effects and so satisfy $2^{n-m}-1$ equations.

Now suppose that the interaction effects of factors $i_1, i_2, \ldots, i_t$ and $j_1, j_2, \ldots, j_s$ are confounded with blocks. If $\{i_1, i_2, \ldots, i_t\}$ and $\{j_1, j_2, \ldots, j_s\}$ intersect in u elements, then renumber, if necessary, so that $i_1=j_1$, $i_2=j_2, \ldots, i_u=j_u$. Let

$$W_m=\{(x_1, \ldots, x_n)|x_{i_1}+\ldots+x_{i_t}=m\}, \quad m=0, 1,$$

$$V_m=\{(x_1, \ldots, x_n)|x_{i_1}+\ldots+x_{i_u}+x_{j_{u+1}}+\ldots+x_{j_s}=m\}, \quad m=0, 1.$$

Then $W_0\cap V_0$, $W_0\cap V_1$, $W_1\cap V_0$ and $W_1\cap V_1$ are equal to unions of blocks. But

$$(W_0\cap V_0)\cup(W_1\cap V_1)=\{(x_1, \ldots, x_n)|x_{i_{u+1}}+\ldots+x_{i_t}+x_{j_{u+1}}+\ldots+x_{j_s}=0\}$$

and

$$(W_0\cap V_1)\cup(W_1\cap V_0)=\{(x_1, \ldots, x_n)|x_{i_{u+1}}+\ldots+x_{i_t}+x_{j_{u+1}}+\ldots+x_{j_s}=1\}$$

so the interaction effect of factors $i_{u+1}, \ldots, i_t, j_{u+1}, \ldots, j_s$ is also confounded with blocks.

Thus the set of equations, together with the zero equation, forms a subgroup of order 2^{n-m} of the elementary Abelian group of order 2^n. Hence the subgroup is generated by a set of $(n-m)$ independent, consistent equations and, because of the one-to-one correspondence between equations and effects, the set of confounded effects is generated by $(n-m)$ independent effects. The points in each block satisfy the $(n-m)$ independent equations so each block is an m-flat of $AG(n, 2)$. □

As we pointed out in Chapter 8 the m-flat containing $\mathbf{0}$ is a subspace of the vector space and the m-flats parallel to it are its cosets. When used as a block of a factorial experiment the m-flat containing $\mathbf{0}$ is called the *principal block*. Now $\mathbf{0}$ (or, equivalently, (1)) intersects each main effect or interaction effect in an even number of letters, namely zero. Thus, by Lemma 3, in a replicate in which no effects are partly confounded, all the elements in the principal block must intersect each of the confounded effects in an even number of letters, and any element which intersects each of the confounded effects in an even number of letters is in the principal block.

Example 7. Suppose we conduct a 2^3 factorial experiment in $2^{3-1}=4$ blocks of size 2 so that three effects are completely confounded with blocks. Suppose two of these are $\mathscr{A}$ and $\mathscr{ABC}$. Then

$$\begin{aligned}
W_0 &= \{(x_1, x_2, x_3) | x_1 = 0\} = \{000, 001, 010, 011\}, \\
W_1 &= \{(x_1, x_2, x_3) | x_1 = 1\} = \{100, 101, 110, 111\}, \\
V_0 &= \{(x_1, x_2, x_3) | x_1 + x_2 + x_3 = 0\} = \{000, 011, 101, 110\}, \\
V_1 &= \{(x_1, x_2, x_3) | x_1 + x_2 + x_3 = 1\} = \{001, 010, 100, 111\},
\end{aligned}$$

and the blocks are

$$W_0 \cap V_0 = \{000, 011\},\ W_0 \cap V_1 = \{001, 010\},\ W_1 \cap V_0 = \{101, 110\}$$
$$\text{and } W_1 \cap V_1 = \{100, 111\}.$$

The other confounded effect is $\mathscr{BC}$.

Alternatively, we can write the blocks as

$$\{(1), bc\}, \{c, b\}, \{ac, ab\}, \{a, abc\}.$$

The principal block is $\{(1), bc\}$. Observe that (1) intersects each of the confounded effects $\mathscr{A}$, $\mathscr{BC}$, and $\mathscr{ABC}$ in 0 letters whereas bc intersects them in 0, 2 and 2 letters respectively. Every other treatment combination in the experiment intersects at least one of the confounded effects in an odd number of letters. □

At this point several questions spring to mind. How should we choose the set of $n-m$ effects which are to be confounded? Having chosen them, how do we determine the other effects which are confounded? What is the best way to determine the actual layout of the design?

The choice of the set of confounded effects is usually made on the assumption that the higher-order interactions can be regarded as negligible, or at least that the experimenter is prepared to forgo information about the higher-order interactions in return for more accurate information about lower-order interactions. Thus, a set of $n-m$ independent higher-order interactions is chosen and all other effects which are linear combinations of these are calculated. As long as this set does not contain any effect we are not prepared to confound then we have an acceptable confounding scheme. There is, unfortunately, no short-cut to determine all the vectors generated by a set of $n-m$ independent vectors.

The easiest way to determine the layout of the design is to calculate the m-flats determined by the $(n-m)$ independent effects. These flats are the blocks of the design.

Example 8. Suppose we want to construct a 2^3 design in four blocks of size 2. We list the possible systems of confounding and the blocks in Table 9.6. □

Table 9.6. Systems of confounding for a 2^3 design in blocks of size 2

Confounded effects	Blocks
$\mathcal{A}, \mathcal{B}, \mathcal{AB}$	{000, 001}, {100, 101}, {010, 011}, {110, 111}
$\mathcal{A}, \mathcal{C}, \mathcal{AC}$	{000, 010}, {100, 110}, {001, 011}, {101, 111}
$\mathcal{B}, \mathcal{C}, \mathcal{BC}$	{000, 100}, {010, 110}, {001, 101}, {011, 111}
$\mathcal{A}, \mathcal{BC}, \mathcal{ABC}$	{000, 011}, {100, 111}, {010, 001}, {110, 101}
$\mathcal{B}, \mathcal{AC}, \mathcal{ABC}$	{000, 101}, {100, 001}, {010, 111}, {110, 011}
$\mathcal{C}, \mathcal{AB}, \mathcal{ABC}$	{000, 110}, {100, 010}, {001, 111}, {101, 011}
$\mathcal{AB}, \mathcal{AC}, \mathcal{BC}$	{000, 111}, {100, 011}, {010, 101}, {001, 110}

Note that the set of seven possible systems of confounding form a $B[3, 1; 7]$ design on the seven varieties $\mathcal{A}, \mathcal{B}, \mathcal{C}, \mathcal{AB}, \mathcal{AC}, \mathcal{BC}, \mathcal{ABC}$. This is not chance, but follows from the results on the number of m-flats passing through a point given in Chapter 8.

9.5. Incomplete blocks and the q^n factorial design

As we said earlier, we will represent the treatment combinations in a q^n factorial experiment either as an n-tuple with elements from $GF[q]$ or as a sequence of n lower-case letters with superscripts from $\{0, 1, \ldots, q-1\}$.

Example 9. Suppose that $n = 2$ and $q = 3$. Then the two representations for the treatment combinations appear in Table 9.7. □

Table 9.7

Sequence	(1)	a	a^2	b	ab	a^2b	b^2	ab^2	a^2b^2
Pairs from $GF[3]$	00	10	20	01	11	21	02	12	22

If we are interested in the effect of factor $\mathscr{I}$ then we are interested in comparing the yields amongst the sets $\{(x_1, \ldots, x_n)|x_i = \theta\}$, $\theta \in GF[q]$. Thus, we are comparing the yields in the flats of $\mathscr{P}(e_i)$. There are $q-1$ independent contrasts among these flats and so in a q^n factorial experiment the main effect of a factor is represented by $q-1$ independent, orthogonal contrasts.

If we are interested in the interaction effect of factors $\mathscr{I}$ and $\mathscr{J}$ then we are interested in comparing the yields in the flats of $\mathscr{P}(e_i+\theta e_j)$, $\theta \in GF[q]$, $\theta \neq 0$. Hence, the interaction of factors $\mathscr{I}$ and $\mathscr{J}$ is represented by $(q-1)^2$ independent, orthogonal contrasts. Higher-order interactions are defined similarly; an interaction involving t factors is represented by $(q-1)^t$ independent, orthogonal contrasts.

Given a pencil $\mathscr{P}(a_1, a_2, \ldots, a_n)$ we will use $\mathscr{A}_1^{a_1} \mathscr{A}_2^{a_2} \ldots \mathscr{A}_n^{a_n}$ to represent a set of $(q-1)$ independent, orthogonal contrasts among the flats of the pencil $\mathscr{P}(a_1, a_2, \ldots, a_n)$. Thus, the main effect of $\mathscr{I}$ is represented by $\mathscr{I}$ and the interaction of factors $\mathscr{I}$ and $\mathscr{J}$ by the $(q-1)$ terms $\mathscr{I}\mathscr{J}^\theta$, $\theta \in GF\ [q]$, $\theta \neq 0$. (We do not bother to include factors which correspond to $a_i = 0$.) We will call $\mathscr{I}$, $\mathscr{I}\mathscr{J}^\theta$ and so on *effects*.

Example 10. Consider a 3^2 factorial experiment. The points and pencils of $AG(2, 3)$ are given in Table 9.8. The main effect of $\mathscr{A}$ is given by two orthogonal contrasts between the flats of $\mathscr{P}(1, 0)$. We need two vectors (a, b, c) and (d, e, f) such that

$$a+b+c = d+e+f = ad+be+cf = 0$$

over Q. We use $a = -1, b = 0, c = d = f = 1, e = -2$. Thus the two contrasts which give the main effect of $\mathscr{A}$ are

$$(-1)(00+01+02)+(0)(10+11+12)+(1)(20+21+22)$$
$$= 20+21+22-(00+01+02)$$

and

$$(1)(00+01+02)-2(10+11+12)+(1)(20+21+22).$$

We can get a similar expression for the two contrasts which give the main effect of $\mathscr{B}$ from the flats of $\mathscr{P}(0, 1)$.

To get the contrasts in $\mathscr{A}\mathscr{B}$ and $\mathscr{A}\mathscr{B}^2$ we use the same coefficients and the flats of $\mathscr{P}(1, 1)$ and $\mathscr{P}(1, 2)$ respectively.

In Table 9.9 we give the matrix of coefficients of the eight contrasts. From this it is easy to see that they are all orthogonal and so form a basis, with j_9^T, for the nine-dimensional inner product space over Q. □

It is important to stress that the definition of the interaction of two (or more) factors only specifies that the contrasts are between the flats of a particular pencil and not what the contrasts are. However, we show that any contrast involving the flats in one pencil is orthogonal to one involving the flats of another pencil.

Table 9.8. Points and pencils of $AG(2, 3)$

Points:

00, 01, 02, 10, 11, 12, 20, 21, 22

Pencils:

$\mathscr{P}(01) = \{\{00, 10, 20\}, \{01, 11, 21\}, \{02, 12, 22\}\}$
$\mathscr{P}(10) = \{\{00, 01, 02\}, \{10, 11, 12\}, \{20, 21, 22\}\}$
$\mathscr{P}(11) = \{\{00, 12, 21\}, \{01, 10, 22\}, \{02, 11, 20\}\}$
$\mathscr{P}(12) = \{\{00, 11, 22\}, \{02, 10, 21\}, \{01, 12, 20\}\}$

Table 9.9. Coefficients of contrasts for a 3^2 design

00	01	02	10	11	12	20	21	22		
−1	−1	−1	0	0	0	1	1	1	$\mathscr{A}$	main effect
1	1	1	−2	−2	−2	1	1	1		
−1	0	1	−1	0	1	−1	0	1	$\mathscr{B}$	main effect
1	−2	1	1	−2	1	1	−2	1		
−1	0	1	0	1	−1	1	−1	0	$\mathscr{A}\mathscr{B}$	interaction effect
1	−2	1	−2	1	1	1	1	−2		
−1	1	0	0	−1	1	1	0	−1	$\mathscr{A}\mathscr{B}^2$	
1	1	−2	−2	1	1	1	−2	1		

Lemma 11. *Consider a q^n factorial experiment. Let **c** be any contrast in the set of contrasts representing some particular t-factor interaction ($t \geq 1$) and let **d** be any contrast in the set of contrasts representing some particular s-factor interaction ($s \geq 1$). Then **c** and **d** are orthogonal contrasts.*

Proof. If $\boldsymbol{c}$ and $\boldsymbol{d}$ are contrasts associated with the flats of the same pencil then $\boldsymbol{c}$ and $\boldsymbol{d}$ are orthogonal by definition. If not, suppose that $\boldsymbol{c}$ is a contrast among the flats $S_1, S_2, \ldots, S_q$ and that $\boldsymbol{d}$ is a contrast among the flats $T_1, T_2, \ldots, T_q$. Now

$$\sum_i c_i = \sum_i d_i = 0,$$

since $\boldsymbol{c}$ and $\boldsymbol{d}$ are contrasts, and

$$|S_i \cap T_j| = q^{n-2}$$

since the points in $S_i \cap T_j$ satisfy two independent equations. All the elements in $S_i \cap T_j$ have coefficient c_i in $\boldsymbol{c}$ and d_j in $\boldsymbol{d}$. Thus, the inner product of the vectors of the contrasts is

$$\sum_i \sum_j c_i d_j |S_i \cap T_j| = q^{n-2} \sum_i c_i \left(\sum_j d_j \right) = 0,$$

showing they are orthogonal, as required. □

The matrix of Table 9.9 gives an example of this result.

As in the 2^n case we will construct the blocks so that the contrasts between block totals are also either main effects or interaction effects. Thus the blocks are flats of $AG(n, q)$. From the independent equations defining these flats we can determine which effects are confounded.

Example 12. Suppose we want to conduct a 3^3 experiment in blocks of size 9. Then each block is a 2-flat of $AG(3, 3)$ and there is one effect, representing two contrasts, confounded. That effect is either $\mathscr{A}$, $\mathscr{B}$, $\mathscr{C}$, $\mathscr{A}\mathscr{B}$, $\mathscr{A}\mathscr{B}^2$, $\mathscr{A}\mathscr{C}$, $\mathscr{A}\mathscr{C}^2$, $\mathscr{B}\mathscr{C}$, $\mathscr{B}\mathscr{C}^2$, $\mathscr{A}\mathscr{B}\mathscr{C}$, $\mathscr{A}\mathscr{B}\mathscr{C}^2$, $\mathscr{A}\mathscr{B}^2\mathscr{C}$, or $\mathscr{A}\mathscr{B}^2\mathscr{C}^2$. The flats corresponding to $\mathscr{A}^a\mathscr{B}^b\mathscr{C}^c$ are given by $ax_1+bx_2+cx_3=\theta$, $\theta \in GF[3]$. □

Example 13. Suppose we want to construct a 3^3 experiment in blocks of size 3. Then eight contrasts are confounded with blocks, or equivalently, four sets of two contrasts. The blocks are 1-flats of $AG(3, 3)$. If we decide to confound $\mathscr{A}$ and $\mathscr{B}$ then the blocks are the simultaneous solutions of

$$x_1 = \theta \text{ and } x_2 = \rho,$$

θ, $\rho \in GF[3]$. The blocks are given in Table 9.10.

We can get two more equations from the two listed (recalling that the first non-zero coefficient must be 1), namely $x_1+x_2=\nu$ and $x_1+2x_2=\psi$; ν, $\psi \in GF[3]$. Now any two of these equations are independent so the same blocks will be obtained using any of the pairs of equations listed in Table 9.11. In all cases we see that the confounded effects are $\mathscr{A}$, $\mathscr{B}$, $\mathscr{A}\mathscr{B}$, and $\mathscr{A}\mathscr{B}^2$. □

Table 9.10. Blocks of size 3 for a 3^3 design

	$x_1=0$	$x_1=1$	$x_1=2$
$x_2=0$	000, 001, 002	100, 101, 102	200, 201, 202
$x_2=1$	010, 011, 012	110, 111, 112	210, 211, 212
$x_2=2$	020, 021, 022	120, 121, 122	220, 221, 222

Table 9.11

$x_1=\theta$	$x_1=\theta$	$x_1=\theta$	$x_2=\rho$	$x_2=\rho$	$x_1+x_2=\nu$
$x_2=\rho$	$x_1+x_2=\nu$	$x_1+2x_2=\psi$	$x_1+x_2=\nu$	$x_1+2x_2=\psi$	$x_1+2x_2=\psi$

Lemma 14. *Consider a q^n factorial experiment conducted in q^{n-m} blocks, where each block is an m-flat of $AG(n, q)$. Then there are $(q^{n-m}-1)/(q-1)$ effects confounded with blocks.*

Proof. Any m-flat of $AG(n, q)$ is determined by a set of $(n-m)$ independent, consistent equations and every point on the m-flat is a solution of these $(n-m)$

equations and, of course, of any linear combination of these equations. There are $(q^{n-m}-1)/(q-1)$ equations with first non-zero coefficient equal to 1 satisfied by the points of an m-flat. Hence there are $(q^{n-m}-1)/(q-1)$ $(n-1)$-flats which contain a given m-flat. Now the blocks are parallel m-flats and so each of the $(n-1)$-flats which contains a given m-flat is, in fact, expressible as a union of blocks. Thus, all of the flats in the pencils determined by the $(q^{n-m}-1)/(q-1)$ equations are expressible as unions of blocks and so contrasts between the flats of these pencils are also contrasts between block totals. The number of contrasts between block totals is $q^{n-m}-1$ and the number accounted for by contrasts between the flats of the pencils is $(q-1)(q^{n-m}-1)/(q-1)$. The result follows. □

Example 15. Let $q = 2, n = 3$ and $m = 1$. Each block is a line of $AG(3, 2)$ (see Table 7.12). If we use as blocks the 1-flats given by $x_1 = \nu, x_2 = \theta, \nu, \theta \in GF[2]$, then the blocks are {000, 001}, {010, 011}, {100, 101} and {110, 111}. It is straightforward to check that the flats of the pencils $\mathscr{P}(100)$, $\mathscr{P}(010)$, and $\mathscr{P}(110)$, but no others, can be expressed as the unions of blocks. □

9.6. Pseudo-factorials

The constructions given so far show how to arrange q^n factorial experiments in blocks of size q^m. If $q = p^a$, for some prime p and integer $a \geq 2$, then it is possible to arrange the experiment in blocks of size p^l, $l \not\equiv 0 \pmod{a}$. This increases the range of available block sizes.

We associate the levels of each factor with an a-tuple over $GF[p]$. Hence the original n factors with q levels each become an factors with p levels each. The design is now constructed using these an factors, remembering that a main effect of one of the original factors is now represented by the main effects and interaction effects of the corresponding a pseudo-factors.

Example 16. Consider a 4^2 experiment which is to be laid out in two blocks each of eight plots. Let the original two factors be $\mathscr{A}$ and $\mathscr{B}$ and represent the levels of each by ordered pairs over $GF[2]$. Thus, the levels are (0, 0), (0, 1), (1, 0), and (1, 1). Now let the levels for $\mathscr{A}$ represent the levels of $\mathscr{W}$ and $\mathscr{X}$ and those for $\mathscr{B}$ the levels of $\mathscr{Y}$ and $\mathscr{Z}$. Thus, if $\mathscr{A}$ is at level (0, 1) then $\mathscr{W}$ is at level 0 and $\mathscr{X}$ is at level 1.

The main effect of $\mathscr{A}$ corresponds to three contrasts between the flats of the pencil $\mathscr{P}(10)$ in $AG[2, 4]$. In the new representation, the three contrasts are the main effects of $\mathscr{W}$ and $\mathscr{X}$ and the interaction effect $\mathscr{W}\mathscr{X}$. Similarly, the three contrasts corresponding to the interaction effect $\mathscr{A}\mathscr{B}^{\alpha}$ (remember that $GF[4] = \{0, 1, \alpha, \alpha+1\}$) in the original experiment are those corresponding to $\mathscr{X}\mathscr{Y}$, $\mathscr{W}\mathscr{X}\mathscr{Z}$ and $\mathscr{W}\mathscr{Y}\mathscr{Z}$ in the new representation. The full correspondence is given in Table 9.12.

Table 9.12. Effects of a 2^4 and of a 4^2 design

Treatment combinations	0	0	0	0	0	0	0	0	1	1	1	1	1	1	1	1	$\mathscr{W}$	$\mathscr{A}$
	0	0	0	0	1	1	1	1	0	0	0	0	1	1	1	1	$\mathscr{X}$	
	0	0	1	1	0	0	1	1	0	0	1	1	0	0	1	1	$\mathscr{Y}$	$\mathscr{B}$
	0	1	0	1	0	1	0	1	0	1	0	1	0	1	0	1	$\mathscr{Z}$	
	−	−	−	−	−	−	−	−	1	1	1	1	1	1	1	1	$\mathscr{W}$	$\mathscr{A}$
	−	−	−	−	1	1	1	1	−	−	−	−	1	1	1	1	$\mathscr{X}$	
	1	1	1	1	−	−	−	−	−	−	−	−	1	1	1	1	$\mathscr{W}\mathscr{X}$	
	−	−	1	1	−	−	1	1	−	−	1	1	−	−	1	1	$\mathscr{Y}$	$\mathscr{B}$
	−	1	−	1	−	1	−	1	−	1	−	1	−	1	−	1	$\mathscr{Z}$	
	1	−	−	1	1	−	−	1	1	−	−	1	1	−	−	1	$\mathscr{Y}\mathscr{Z}$	
	1	−	1	−	−	1	−	1	1	−	1	−	−	1	−	1	$\mathscr{X}\mathscr{Z}$	$\mathscr{A}\mathscr{B}$
	1	1	−	−	1	1	−	−	−	−	1	1	−	−	1	1	$\mathscr{W}\mathscr{Y}$	
	1	−	−	1	−	1	1	−	−	1	1	−	1	−	−	1	$\mathscr{W}\mathscr{X}\mathscr{Y}\mathscr{Z}$	
	1	1	−	−	−	−	1	1	1	1	−	−	−	−	1	1	$\mathscr{X}\mathscr{Y}$	$\mathscr{A}\mathscr{B}^{\alpha}$
	−	1	−	1	1	−	1	−	1	−	1	−	−	1	−	1	$\mathscr{W}\mathscr{X}\mathscr{Z}$	
	−	1	1	−	−	1	1	−	1	−	−	1	1	−	−	1	$\mathscr{W}\mathscr{Y}\mathscr{Z}$	
	1	−	1	−	1	−	1	−	−	1	−	1	−	1	−	1	$\mathscr{W}\mathscr{Z}$	$\mathscr{A}\mathscr{B}^{(\alpha+1)}$
	−	−	1	1	1	1	−	−	1	1	−	−	−	−	1	1	$\mathscr{W}\mathscr{X}\mathscr{Y}$	
	−	1	1	−	1	−	−	1	−	1	1	−	1	−	−	1	$\mathscr{X}\mathscr{Y}\mathscr{Z}$	

Hence, to construct the design we would choose to confound one of the nine effects of the 2^4 experiment associated with the effects $\mathscr{A}\mathscr{B}$, $\mathscr{A}\mathscr{B}^{\alpha}$, or $\mathscr{A}\mathscr{B}^{\alpha+1}$ of the 4^2 experiment. For instance, if we confound $\mathscr{W}\mathscr{X}\mathscr{Y}\mathscr{Z}$ then the principal block is

$$(1),\ wx,\ wy,\ xy,\ wz,\ xz,\ yz,\ wxyz$$

or, in the original notation,

$$(1),\ a^{\alpha+1},\ a^{\alpha}b^{\alpha},\ ab^{\alpha},\ a^{\alpha}b,\ ab,\ b^{\alpha+1},\ a^{\alpha+1}b^{\alpha+1}.$$ □

Clearly, the same method can be used to construct blocks for an experiment in which the numbers of levels for each factor are various powers of the same prime.

Again, if we can conduct more than one replicate it is preferable to confound different effects in the different replicates.

Finally, we briefly consider the relationship between pseudo-factorials and lattice designs. For two-dimensional lattice designs we represent the k^2 treatments by two factors, each with k levels. The blocks in each replicate can be determined by confounding a particular effect and using the associated flats. The restriction that any two treatments appear together at most once is satisfied if a different effect is confounded for each resolution class (replicate). An example of this is given in the exercises.

For three-dimensional lattices the k^3 treatments are represented by three factors, each with k levels, and the resolution classes are obtained by confounding two effects (and, of course, all the effects which are determined by linear combinations of the equations associated with these two effects).

9.7. Orthogonal arrays

An *orthogonal array* $OA[N, k, s, t]$ is a $k \times N$ matrix with elements from a set of s symbols such that any $t \times N$ submatrix of A has each t-tuple appearing as a column N/s^t times. Often N/s^t is called the *index* of the array, t the *strength* of the array, k the number of *constraints* and s the number of *levels*.

An $OA[s^2, k, s, 2]$ is a transversal design $T[k, 1; s]$. (Transversal designs were introduced in Chapter 6.)

Example 17. Consider the two arrays in Table 9.13. Table 9.13(a) is an $OA[8, 5, 2, 2]$. Table 9.13(b) is an $OA[8, 5, 2, 2]$ except for rows 1 and 2 where $(0, 0)^T$ and $(1, 1)^T$ appear four times each. Note that if we regard the elements in the arrays as elements of $GF[2]$ and the columns of 5-tuples as representing the treatments in a 2^5 experiment then Table 9.13(a) could be written as

$$(1),\ ab,\ de,\ abde,\ ace,\ bce,\ acd,\ bcd$$

and Table 9.13(b) could be written as

$$(1),\ ab,\ cd,\ abcd,\ de,\ abde,\ ce,\ abce.$$

Thus, using Lemma 3, we see that (a) is the principal block of the design in which $\mathcal{ABC}$, $\mathcal{CDE}$ and $\mathcal{ABDE}$ are confounded and that (b) is the principal block of the design in which $\mathcal{AB}$, $\mathcal{CDE}$ and $\mathcal{ABCDE}$ are confounded. □

Table 9.13

(a)

0	1	0	1	1	0	1	0
0	1	0	1	0	1	0	1
0	0	0	0	1	1	1	1
0	0	1	1	0	0	1	1
0	0	1	1	1	1	0	0

(b)

0	1	0	1	0	1	0	1
0	1	0	1	0	1	0	1
0	0	1	1	0	0	1	1
0	0	1	1	1	1	0	0
0	0	0	0	1	1	1	1

Theorem 18. *The principal block in a q^n factorial experiment, with q^{n-m} blocks, in which no interactions involving t or fewer factors are confounded, is equivalent to an $OA[q^m, n, q, t]$.*

Proof. Consider a q^n design in q^{n-m} blocks of size q^m where these blocks are a set of parallel m-flats of $AG[n, q]$. Suppose that the design is such that no interaction involving t or fewer factors is confounded. The confounded effects are determined by all possible linear combinations of the $(n-m)$ independent consistent defining expressions. As no interaction involving t or fewer factors is confounded, all these linear combinations must have at least $t+1$ non-zero coefficients.

Now consider the principal block of this design. The elements in it satisfy $n-m$ independent consistent equations

$$\sum_{j=1}^{n} a_{ij}x_j = 0, \quad i = 1, 2, \ldots, n-m. \tag{9.1}$$

Suppose that we consider t of the x_js and that we give these t x_js some particular values, say $x_{r_s} = \theta_{r_s}, s = 1, \ldots, t; \theta_{r_s} \in GF[q]$. Then the equations in (9.1), with these values substituted, still have rank $n-m$. For suppose not. Then there exist values v_i such that

$$\sum_{i=1}^{n-m} v_i \left(\sum_{s=t+1}^{n} a_{ir_s} x_{r_s} \right) = 0.$$

Consider the same linear combination of the original equations. We have

$$\sum_{i=1}^{n-m} v_i \left(\sum_{s=1}^{n} a_{ir_s} x_{r_s} \right) = \sum_{i=1}^{n-m} v_i \left(\sum_{s=1}^{t} a_{ir_s} x_{r_s} \right) + \sum_{i=1}^{n-m} v_i \left(\sum_{s=t+1}^{n} a_{ir_s} x_{r_s} \right)$$
$$= \sum_{i=1}^{n-m} v_i \left(\sum_{s=1}^{t} a_{ir_s} x_{r_s} \right)$$

and so at most t of the x_js have non-zero coefficients. This is a contradiction. Hence specifying the values of any t of the x_js still leaves $n-m$ independent equations to determine the other x_js. Thus there are $n-m+t=n-(m-t)$ equations and q^{m-t} solutions. The result follows. □

9.8. Fractional replication

Often, when conducting a factorial experiment, it is not feasible to include every possible treatment combination because the total number of such combinations makes the cost prohibitive. In such cases much useful information can still be obtained by conducting an experiment which includes only a subset of all possible treatment combinations. Such an experiment is called a *fractional replicate*. Indeed, a $1/q^m$ *replicate* of a q^n factorial experiment is a design in which q^{n-m} treatment combinations are included.

Example 19. Suppose that only four treatment combinations of a 2^3 experiment can be included. Suppose that the four combinations applied are (*1*), *ab*, *ac*, *bc*. Then, using the coefficients in Table 9.4, we can see that the estimates of the various effects are as follows:

$$\mathscr{A} = ab + ac - (bc + (1)) = -\mathscr{BC}$$
$$\mathscr{B} = ab + bc - (ac + (1)) = -\mathscr{AC}$$
$$\mathscr{C} = ac + bc - (ab + (1)) = -\mathscr{AB}.$$

$\mathscr{ABC}$ can not be estimated at all. $\mathscr{A}$ and $\mathscr{BC}$, $\mathscr{B}$ and $\mathscr{AC}$, and $\mathscr{C}$ and $\mathscr{AB}$ are said to be *aliased*, and $\mathscr{ABC}$ is called the *defining contrast*. □

From the example we can see that if we want to estimate an effect easily from a fractional replicate then the treatment combinations applied must include the same fraction from each of the flats in the pencil corresponding to the effect. Thus, for a $1/q^m$ replicate the design is an m-flat of $AG(n, q)$. The defining contrast is chosen so that main effects will be confounded with higher-order interactions which are then assumed to be negligible.

9.9. References and comments

The first formal exposition of factorial designs was given by Yates (1935). Fisher (1926) gave a very strong recommendation for factorial designs and Yates described experiments in which all possible treatment combinations were used dating back to 1843–4 and 1852. References to this early work appear in Raktoe, Hedayat, and Federer (1981) as well as an extensive survey of the subsequent literature. The usefulness of affine geometries in constructing confounded designs was pointed out by Bose and Kishen (1940) and exploited by Bose (1947). Rao (1946) pointed out the relationship between orthogonal arrays and principal blocks of confounded designs. The construction in Exercise 9.12 is due to Bush (1952).

Kempthorne (1952) and Raghavarao (1971) provide accounts of various topics not covered in this chapter or the next, including bounds for the strength of an orthogonal array. Tables of designs are available; see McLean and Anderson (1984), Kempthorne (1952), and Cochran and G. M. Cox (1957). Bose (1980) gives a survey of the combinatorial problems in factorial designs.

Exercises

9.1. Calculate the effect $\mathscr{ABC}$ as half the difference between $\mathscr{AC}$ at high $\mathscr{B}$ and $\mathscr{AC}$ at low $\mathscr{B}$ and also as half the difference between $\mathscr{BC}$ at high $\mathscr{A}$ and $\mathscr{BC}$ at low $\mathscr{A}$. Verify that these two are the same.

9.2. For a 2^4 design what signs do (i) ab (ii) abc (iii) $abcd$ have in the effects $\mathscr{A}$, $\mathscr{B}$, $\mathscr{AB}$, and $\mathscr{ABC}$?

9.3. Express $(1, 1, 1, -1, -1, -1, 1, -1)$ as a linear combination of the basis vectors of Q^8 given in Example 2.

9.4. (i) Construct a 2^6 design in blocks of size 8 in which no main effects nor two-factor interactions are confounded. Show that it is not possible to construct such a design in which no interaction involving three or fewer factors is confounded.

(ii) If a factorial experiment is laid out in a rectangular array then some effects are confounded with the rows of the array and some with the columns of the array. For a 2^5 experiment to be laid out in four rows and eight columns, find a layout in which only two two-factor interactions (and, of course, interactions of higher order) are confounded.

9.5. Suppose that a 2^4 experiment is conducted in four blocks of size 4. Suppose that the blocks are flats of $AG(4, 2)$ and that one block is (1), abc, cd, abd. What are the other blocks of the design? What effects are confounded with blocks?

9.6. Give the equations defining the pencils of $AG(4, 3)$. Construct a 3^4 design with nine blocks. What effects are confounded in the design?

9.7. Verify that the same design will be obtained from any of the pairs of equations in Table 9.11.

9.8. Consider a factorial experiment in which two factors have four levels each and one factor has two levels. Construct a design in four blocks each of size 8. What effects are confounded in the design?

9.9. Consider a three-factor factorial experiment in which $\mathscr{A}$ has three levels and $\mathscr{B}$ and $\mathscr{C}$ have two levels each.

(i) Using the definition of the $\mathscr{B}$, $\mathscr{C}$, and $\mathscr{BC}$ effects given in Section 9.2 and the coefficients given in Example 10 for the $\mathscr{A}$ effect, construct a set of eleven contrasts which represent the effects $\mathscr{A}$, $\mathscr{B}$, $\mathscr{C}$, $\mathscr{AB}$, $\mathscr{AC}$, $\mathscr{BC}$, and $\mathscr{ABC}$.

(ii) Verify that if the design is conducted in two blocks of size 6 and these are (1), bc, ab, ac, a^2b, a^2c and b, c, a, abc, a^2, a^2bc, then the effects $\mathscr{BC}$ and $\mathscr{ABC}$ are partly confounded with blocks.

9.10. Let $v = 16$ and regard the sixteen varieties as being the combinations of a 4^2 factorial design. Construct a lattice design with three replicates by confounding three different effects, one for each replicate.

9.11. Construct an $OA[9, 4, 3, 2]$. Arrange the entries in one row of the orthogonal array to be three zeros, followed by three ones and then three twos. In another row, order the entries as 0, 1, 2, 0, 1, 2, 0, 1, 2. Now use the entries in these two rows to define the positions in a 3×3 array and construct two arrays, one from each of the other two rows. Verify that these two arrays are a pair of orthogonal Latin squares.

9.12. Let $\alpha_0 = 0, \alpha_1, \ldots, \alpha_{q-1}$, be the elements of $GF[q]$, let $t < q$, and consider the q^t polynomials

$$y_j(x) = a_{t-1}x^{t-1} + a_{t-2}x^{t-2} + \ldots + a_1 x + a_0, a_i \in GF[q],\ i = 0, 1, \ldots, t-1.$$

Consider a $q \times q^t$ array with position (i, j) occupied by u, where

$$y_j(\alpha_i) = \alpha_u.$$

We adjoin a $(q+1)$st row which has a u in position j if and only if y_j has leading coefficient α_u. Verify that this array is an $OA[q^t, q+1, q, t]$. Hence construct an $OA[27, 4, 3, 3]$.

10 Single replicate factorial designs

10.1. The designs

A *single replicate* factorial design is a factorial design in which every treatment combination appears precisely once in the design. We look at how to subdivide the treatment combinations into incomplete blocks, each of size k, and at how to calculate the effects which have been confounded by such an allocation.

Single replicate designs are the first step beyond the designs constructed using finite geometries in the previous chapter and we adopt a very different approach to them. To begin with, however, we just introduce some notation and establish some useful results.

Suppose we have n factors and that the ith factor has s_i levels. There are then $v = \prod_{i=1}^{n} s_i$ possible treatment combinations and we represent these by n-tuples in which the ith position is occupied by an element of Z_{s_i}.

Lemma 1. *Let A be the $v \times b$ incidence matrix of a single replicate factorial design. Then*

$$AA^TAA^T = kAA^T.$$

Proof. We can reorder the v treatment combinations so that the first k appear in the first block, the next k in the second block and so on. Thus AA^T can be written as $I_{v/k} \times J_k$. The result follows. □

Example 2. Consider a 2×3 factorial experiment conducted in two blocks each of size three. Suppose that the first block has treatment combinations 00, 01, and 11 and that the second block has treatment combinations 02, 10, and 12. The matrices A and AA^T for this design are given in Table 10.1. □

Table 10.1

$$A = \begin{bmatrix} 1 & 0 \\ 1 & 0 \\ 0 & 1 \\ 0 & 1 \\ 1 & 0 \\ 0 & 1 \end{bmatrix} \qquad AA^T = \begin{bmatrix} 1 & 1 & 0 & 0 & 1 & 0 \\ 1 & 1 & 0 & 0 & 1 & 0 \\ 0 & 0 & 1 & 1 & 0 & 1 \\ 0 & 0 & 1 & 1 & 0 & 1 \\ 1 & 1 & 0 & 0 & 1 & 0 \\ 0 & 0 & 1 & 1 & 0 & 1 \end{bmatrix}$$

In the next section we look at a convenient method of representing interactions in any factorial design. The representation does not depend on the pencils of affine geometries. We go on to look at the linear model for factorial designs in incomplete blocks, at how tests are performed and at how to put this information to use when constructing designs.

10.2. Interaction effects

Let $v = \prod_i s_i$ and assume that the v treatment combinations have been ordered lexicographically. Let $\boldsymbol{\tau}^T = (\tau_1, \tau_2, \ldots, \tau_v)$, where τ_i is the expected yield (or result or gain in weight or whatever) when the ith treatment combination is applied.

The *generalized interaction* of factors $i_1, i_2, \ldots, i_r$ is represented by a binary vector $\mathbf{x}$, of length n and with non-zero entries in positions $i_1, \ldots, i_r$, and is defined as the vector space of contrasts spanned by the rows of

$$M^{\mathbf{x}} = \mathop{\times}_{i=1}^{n} M^{x_i}, \quad \text{where} \quad M^{x_i} = \begin{cases} s_i I - J, & x_i = 1, \\ j^T \quad , & x_i = 0. \end{cases}$$

Example 3. In Table 10.2 we give the matrices $M^{\mathbf{x}}$ for a 2^3 and a 2×3 factorial design. In the case of the 2^3 design, note that a row of any of the $M^{\mathbf{x}}$ is either a row, or the negative of a row, of Table 9.4. Also, recall that we are now listing the treatments as (level of C, level of B, level of A) (that is, in the reverse order to that used in Chapter 9). □

A generalized interaction involving only one factor is often called a *main effect*.

The dimension of the vector space associated with the generalised interaction represented by $\mathbf{x}$ is $\prod_{i=1}^{n} (s_i - 1)^{x_i}$.

Recall that the aim of a factorial experiment is to be able to test hypotheses of the form ‘factor A has no effect’, ‘there is no interaction between factors A and B’, and so on. In the remainder of this section we introduce notation which will facilitate the statement of these hypotheses. We let

$$\boldsymbol{a}_i^T = (a_i(0), a_i(1), \ldots, a_i(s_i - 1)),$$

$$\boldsymbol{a}_{ij}^T = (a_{ij}(0, 0), a_{ij}(0, 1), \ldots, a_{ij}(0, s_j - 1), a_{ij}(1, 0), \ldots, a_{ij}(1, s_j - 1), \ldots, a_{ij}(s_i - 1, 0), \ldots, a_{ij}(s_i - 1, s_j - 1)), \quad i \neq j,$$

and so on, where $a_i(m)$ is the effect of the mth level of the ith factor, $a_{ij}(m, t)$ is the effect of the interaction of factors i and j at levels m and t respectively, and so on. Clearly these terms cannot all be independent, and we assume they satisfy

Table 10.2

x	M^x
001	$(1\quad 1)\times(1\quad 1)\times\begin{bmatrix}1&-1\\-1&1\end{bmatrix}=\begin{bmatrix}1&-1&1&-1&1&-1&1&-1\\-1&1&-1&1&-1&1&-1&1\end{bmatrix}$
010	$(1\quad 1)\times\begin{bmatrix}1&-1\\-1&1\end{bmatrix}\times(1\quad 1)=\begin{bmatrix}1&1&-1&-1&1&1&-1&-1\\-1&-1&1&1&-1&-1&1&1\end{bmatrix}$
011	$(1\quad 1)\times\begin{bmatrix}1&-1\\-1&1\end{bmatrix}\times\begin{bmatrix}1&-1\\-1&1\end{bmatrix}=\begin{bmatrix}1&-1&-1&1&1&-1&-1&1\\-1&1&1&-1&-1&1&1&-1\\-1&1&1&-1&-1&1&1&-1\\1&-1&-1&1&1&-1&-1&1\end{bmatrix}$
100	$\begin{bmatrix}1&-1\\-1&1\end{bmatrix}\times(1\quad 1)\times(1\quad 1)=\begin{bmatrix}1&1&1&1&-1&-1&-1&-1\\-1&-1&-1&-1&1&1&1&1\end{bmatrix}$
101	$\begin{bmatrix}1&-1\\-1&1\end{bmatrix}\times(1\quad 1)\times\begin{bmatrix}1&-1\\-1&1\end{bmatrix}=\begin{bmatrix}1&-1&1&-1&-1&1&-1&1\\-1&1&-1&1&1&-1&1&-1\\-1&1&-1&1&1&-1&1&-1\\1&-1&1&-1&-1&1&-1&1\end{bmatrix}$
110	$\begin{bmatrix}1&-1\\-1&1\end{bmatrix}\times\begin{bmatrix}1&-1\\-1&1\end{bmatrix}\times(1\quad 1)=\begin{bmatrix}1&1&-1&-1&-1&-1&1&1\\-1&-1&1&1&1&1&-1&-1\\-1&-1&1&1&1&1&-1&-1\\1&1&-1&-1&-1&-1&1&1\end{bmatrix}$
111	$\begin{bmatrix}1&-1\\-1&1\end{bmatrix}\times\begin{bmatrix}1&-1\\-1&1\end{bmatrix}\times\begin{bmatrix}1&-1\\-1&1\end{bmatrix}=\begin{bmatrix}1&-1&-1&1&-1&1&1&-1\\-1&1&1&-1&1&-1&-1&1\\-1&1&1&-1&1&-1&-1&1\\1&-1&-1&1&-1&1&1&-1\\-1&1&1&-1&1&-1&-1&1\\1&-1&-1&1&-1&1&1&-1\\1&-1&-1&1&-1&1&1&-1\\-1&1&1&-1&1&-1&-1&1\end{bmatrix}$

M^x for a 2^3 design

x	M^x
01	$(1\quad 1)\times\begin{bmatrix}2&-1&-1\\-1&2&-1\\-1&-1&2\end{bmatrix}=\begin{bmatrix}2&-1&-1&2&-1&-1\\-1&2&-1&-1&2&-1\\-1&-1&2&-1&-1&2\end{bmatrix}$
10	$\begin{bmatrix}1&-1\\-1&1\end{bmatrix}\times(1\quad 1\quad 1)=\begin{bmatrix}1&1&1&-1&-1&-1\\-1&-1&-1&1&1&1\end{bmatrix}$
11	$\begin{bmatrix}1&-1\\-1&1\end{bmatrix}\times\begin{bmatrix}2&-1&-1\\-1&2&-1\\-1&-1&2\end{bmatrix}=\begin{bmatrix}2&-1&-1&-2&1&1\\-1&2&-1&1&-2&1\\-1&-1&2&1&1&-2\\-2&1&1&2&-1&-1\\1&-2&1&-1&2&-1\\1&1&-2&-1&-1&2\end{bmatrix}$

M^x for a 2×3 design

$$\sum_{m=0}^{s_i-1} a_i(m) = 0, \quad i = 1, 2, \ldots, n, \tag{10.1}$$

$$\sum_{m=0}^{s_i-1} a_{ij}(m, t) = 0, \quad 1 \le i < j \le n, \quad t = 0, 1, \ldots, s_j - 1, \tag{10.2}$$

$$\sum_{t=0}^{s_j-1} a_{ij}(m, t) = 0, \quad 1 \le i < j \le n, \quad m = 0, 1, \ldots, s_i - 1, \tag{10.3}$$

and so forth.

If the ith treatment combination is $(j_1, j_2, \ldots, j_n)$ then

$$\tau_i = \sum_{m=1}^{n} a_m(j_m) + \sum_{1 \le m < t \le n} \sum a_{mt}(j_m, j_t) + \ldots + a_{12\ldots n}(j_1, \ldots, j_n).$$

(For those familiar with the parametric, or 'Greek letter', notation we are saying that

$$\tau = \alpha_{j_1} + \beta_{j_2} + \ldots + \nu_{j_n} + (\alpha\beta)_{j_1 j_2} + (\alpha\gamma)_{j_1 j_3} + \ldots + (\alpha\beta \ldots \nu)_{j_1 \ldots j_n}.)$$

We define the binary operation '$*$' by $a_i * a_j = a_{ij}$. This is an operation on symbolic quantities, not on numerical quantities.

If $\boldsymbol{u}^T = (u_0, \ldots, u_{s_i-1})$ and $\boldsymbol{w}^T = (w_0, \ldots, w_{s_j-1})$ are vectors of scalar quantities then

$$\begin{aligned}(\boldsymbol{u}^T \boldsymbol{a}_i) * (\boldsymbol{w}^T \boldsymbol{a}_j) &= \sum_{m=0}^{s_i-1} u_m a_i(m) * \sum_{t=0}^{s_j-1} w_t a_j(t) \\ &= \sum_m \sum_t u_m w_t a_{ij}(m, t) \\ &= (\boldsymbol{u}^T \times \boldsymbol{w}^T)(\boldsymbol{a}_i * \boldsymbol{a}_j).\end{aligned}$$

We can extend this notation to matrices in a natural way. For example,

$$B\boldsymbol{a}_i * C\boldsymbol{a}_j = (B \times C)(\boldsymbol{a}_i * \boldsymbol{a}_j).$$

Finally, we say $\boldsymbol{a}_i * \boldsymbol{j} = \boldsymbol{a}_i \times \boldsymbol{j}$ and $\boldsymbol{j} * \boldsymbol{a}_i = \boldsymbol{j} \times \boldsymbol{a}_i$.

The conditions in eqns (10.1), (10.2), and (10.3) can be represented by

$$\boldsymbol{j}^T \boldsymbol{a}_i = 0, \qquad i = 1, 2, \ldots, n, \tag{10.1'}$$

$$(\boldsymbol{j}^T \times I)(\boldsymbol{a}_i * \boldsymbol{a}_j) = \boldsymbol{0}, \qquad 1 \le i < j \le n, \tag{10.2'}$$

$$(I \times \boldsymbol{j}^T)(\boldsymbol{a}_i * \boldsymbol{a}_j) = \boldsymbol{0}, \qquad 1 \le i < j \le n, \tag{10.3'}$$

respectively.

Example 2 (continued). In a 2×3 factorial design,

$$a_1^T = (a_1(0), a_1(1)),\ a_2^T = (a_2(0), a_2(1), a_2(2)) \text{ and}$$

$$a_{12}^T = (a_{12}(0, 0), a_{12}(0, 1), a_{12}(0, 2), a_{12}(1, 0), a_{12}(1, 1), a_{12}(1, 2)).$$

Also

$$a_1(0)+a_1(1)=0=a_2(0)+a_2(1)+a_2(2)$$
$$=a_{12}(0,0)+a_{12}(0,1)+a_{12}(0,2)$$
$$=a_{12}(0,0)+a_{12}(1,0),$$

and so on.

Finally, let $i=2$, say. The second treatment combination is (0, 1) so

$$\tau_2=a_1(0)+a_2(1)+a_{12}(0,1). \qquad \square$$

Let

$$a_i^{x_i}=\begin{cases}a_i, & x_i=1,\\ j, & x_i=0.\end{cases}$$

The next result is immediate.

Lemma 4. *The vector τ can be written as*

$$\tau=\sum_{k=1}^{n}\left(\sum_{\{x\mid \sum_i x_i=k\}}(a_1^{x_1}*a_2^{x_2}*\ldots*a_n^{x_n})\right). \qquad \square$$

Corollary 4.1. *Let B_i, $i=1,\ldots,n$, be matrices such that B_i is of size $p_i\times s_i$. Then*

$$(B_1\times\ldots\times B_n)\tau=\sum_{k=1}^{n}\left(\sum_{\{x\mid \sum_i x_i=k\}}(B_1a_1^{x_1}*\ldots*B_na_n^{x_n})\right). \qquad \square$$

Example 2 (continued).

$$\tau^T=\left(\sum_{k=1}^{2}\left(\sum_{\{x\mid \sum_i x_i=k\}}(a_1^{x_1}*a_2^{x_1})\right)\right)^T$$
$$=(a_1^0*a_2^1+a_1^1*a_2^0+a_1^1*a_2^1)^T$$
$$=(j*a_2+a_1*j+a_1*a_2)^T$$
$$=(a_2(0),a_2(1),a_2(2),a_2(0),a_2(1),a_2(2))$$
$$+(a_1(0),a_1(0),a_1(0),a_1(1),a_1(1),a_1(1))$$
$$+(a_{12}(0,0),a_{12}(0,1)\,a_{12}(0,2),a_{12}(1,0),a_{12}(1,1),a_{12}(1,2)). \qquad \square$$

We say that C is a *contrast matrix* if $Cj^T=0$. Thus M^1 and M^x are contrast matrices.

Lemma 5.

(i) $Ia_i^{x_i}=a_i^{x_i}$.

(ii) $j^Ta_i^{x_i}=\begin{cases}s_i, & x_i=0,\\ 0, & x_i=1.\end{cases}$

(iii) *If C is a contrast matrix* $Ca_i^{x_i} = \begin{cases} \mathbf{0}, & x_i = 0, \\ Ca_i, & x_i = 1. \end{cases}$

Proof. The results are immediate from the definition of $a_i^{x_i}$. □

Lemma 6. *Let B_{i_j}, $j = 1, \ldots, m$, be contrast matrices and let the remaining B_i be equal to j^T. Then*

$$(B_1 \times \ldots \times B_n)\tau = \frac{v}{\prod_{j=1}^{m} s_{i_j}} (B_{i_1} \times \ldots \times B_{i_m})(a_{i_1} * \ldots * a_{i_m}).$$

Proof.

$$(B_1 \times \ldots \times B_n)\tau = \sum_{k=1}^{n} \sum_{\{x \mid \sum_i x_i = k\}} (B_1 a_1^{x_1} * B_2 a_2^{x_2} * \ldots * B_n a_n^{x_n}).$$

Now $B_i a_i^{x_i}$ is equal to 0 if $x_i = 1$ and $i \neq \{i_1, \ldots, i_m\}$ or if $x_i = 0$. Hence

$$(B_1 \times \ldots \times B_n)\tau = \prod_{\substack{j \\ j \neq i_1, i_2, \ldots, i_m}} s_j (B_{i_1} a_{i_1} * \ldots * B_{i_m} a_{i_m})$$

$$= \frac{v}{\prod_{j=1}^{m} s_{i_j}} (B_{i_1} \times \ldots \times B_{i_m})(a_{i_1} * \ldots * a_{i_m}).$$ □

Corollary 6.1. *Let $B_{i_j} = M_{i_j}^1$. Then*

$$a_{i_1} * \ldots * a_{i_m} = \frac{1}{v}(M_1^{x_1} \times \ldots \times M_n^{x_n})\tau.$$ □

We set $C^x = \frac{1}{v} M^x$.

The hypotheses we want to test can now be written as '$a^x = C^x\tau = 0$'. In the next section we see how to test these hypotheses. But first we look at an example.

Example 2 (continued). Let $x = (0, 1)$. The evaluation of $M^x\tau$ is given in Table 10.3. (Remember that $\sum_j a_2(j) = 0 = \sum_i a_{12}(i, j)$, $j = 0, 1, 2$.) □

10.3. Factorial designs and the linear model

As in Section 1.4 we let Y_{ij} denote the yield of the plot in block j which receives the ith treatment combination (if such a plot exists). Then

$$Y_{ij} = \tau_i + \beta_j + E_{ij},$$

Table 10.3

$$M^x\tau = \begin{bmatrix} 2 & -1 & -1 & 2 & -1 & -1 \\ -1 & 2 & -1 & -1 & 2 & -1 \\ -1 & -1 & 2 & -1 & -1 & 2 \end{bmatrix} \begin{bmatrix} a_1(0)+a_2(0)+a_{12}(0,0) \\ a_1(0)+a_2(1)+a_{12}(0,1) \\ a_1(0)+a_2(2)+a_{12}(0,2) \\ a_1(1)+a_2(0)+a_{12}(1,0) \\ a_1(1)+a_2(1)+a_{12}(1,1) \\ a_1(1)+a_2(2)+a_{12}(1,2) \end{bmatrix}$$

$$= \begin{bmatrix} 2(2a_2(0)-a_2(1)-a_2(2))+(2(a_{12}(0,0)+a_{12}(1,0))-(a_{12}(0,1)+a_{12}(1,1))-(a_{12}(0,2)+a_{12}(1,2)) \\ 2(-a_2(0)+2a_2(1)-a_2(2))+(2(a_{12}(0,1)+a_{12}(1,1))-(a_{12}(0,0)+a_{12}(1,0))-(a_{12}(0,2)+a_{12}(1,2)) \\ 2(-a_2(0)-a_2(1)+2a_2(2))+2(a_{12}(0,2)+a_{12}(1,2))-(a_{12}(0,0)+a_{12}(1,0))-(a_{12}(0,1)+a_{12}(1,1)) \end{bmatrix}$$

$$= 2\begin{bmatrix} 3a_2(0) \\ 3a_2(1) \\ 3a_2(2) \end{bmatrix} = 6\boldsymbol{a}_2.$$

where β_j is the effect of block j and E_{ij} is a random error term having the same properties as before. Also

$$\boldsymbol{Y} = T\boldsymbol{\tau} + B\boldsymbol{\beta} + \boldsymbol{E} = X\boldsymbol{\gamma} + \boldsymbol{E} = \boldsymbol{\eta} + \boldsymbol{E}$$

and

$$(rI - (1/k)AA^T)\hat{\boldsymbol{\tau}} = (T^T - (1/k)AB^T)\boldsymbol{Y}.$$

Recall that an estimable function is one which has the same value for all elements of $\boldsymbol{\gamma} + \kappa(X)$. Equivalently, $\boldsymbol{v}^T\boldsymbol{\tau}$ is estimable if there is a vector $\boldsymbol{t}$ such that $\boldsymbol{v}^T\boldsymbol{\tau} = \boldsymbol{t}^T\mathscr{E}(\boldsymbol{Y})$. If $\boldsymbol{v}^T\boldsymbol{\tau}$ is estimable then:

(i) the expected value of $\boldsymbol{v}^T\hat{\boldsymbol{\tau}}$ is $\boldsymbol{v}^T\boldsymbol{\tau}$;
(ii) the variance of $\boldsymbol{v}^T\hat{\boldsymbol{\tau}}$ is smaller than that for any other $\boldsymbol{t}^T\boldsymbol{Y}$ such that $\mathscr{E}(\boldsymbol{t}^T\boldsymbol{Y}) = \boldsymbol{t}^T\mathscr{E}(\boldsymbol{Y}) = \boldsymbol{v}^T\boldsymbol{\tau}$;
(iii) $\boldsymbol{v}^T\hat{\boldsymbol{\tau}}$ is invariant for all possible choices of $\hat{\boldsymbol{\tau}}$.

A *testable hypothesis* is one which can be expressed in terms of estimable functions.

Example 2 (continued). The function $\tau_1 - \tau_2$ is estimable since $\tau_1 - \tau_2 = \tau_1 + \beta_1 - (\tau_2 + \beta_1) = \mathscr{E}(Y_{11}) - \mathscr{E}(Y_{21})$. Hence the hypothesis '$\tau_1 - \tau_2 = 0$' is testable. □

How do we test a testable hypothesis? We begin by summarizing the properties of the error terms, E_{ij}, in matrix notation. They are:

(i) $\mathrm{Var}(E_{ij}) = \sigma^2$, independent of i and j;
(ii) $\mathrm{Cov}(E_{ij}, E_{pq}) = \mathscr{E}((E_{ij} - \mathscr{E}(E_{ij}))(E_{pq} - \mathscr{E}(E_{pq})))$
$= \mathscr{E}(E_{ij}E_{pq}) - \mathscr{E}(E_{ij})\mathscr{E}(E_{pq}) = 0, \quad \text{for } (i,j) \neq (p,q).$

Since $\mathrm{Cov}(E_{ij}, E_{ij}) = \mathrm{Var}(E_{ij})$, if we define

$$\mathrm{Cov}(\boldsymbol{E}) = \mathscr{E}((\boldsymbol{E} - \mathscr{E}(\boldsymbol{E}))(\boldsymbol{E} - \mathscr{E}(\boldsymbol{E}))^T)$$

then

$$\text{Cov}(E) = \sigma^2 I.$$

Adding a constant to any random variable, such as E_{ij}, changes its expected value but not its variance. Hence $\text{Cov}(Y) = \sigma^2 I$. From the definition of Cov it follows that $\text{Cov}(AZ) = A\text{Cov}(Z)A^T$ for any random vector Z.

If V is positive definite we say that the random vector Z, of length p, has a *multivariate normal distribution* with mean vector μ and covariance matrix V, written $Z \sim N_p(\mu, V)$, if

$$\Pr(Z_1 \leq u_1, Z_2 \leq u_2, \ldots, Z_p \leq u_p) = \int_{-\infty}^{u_p} \cdots \int_{-\infty}^{u_1} f(z_1, \ldots, z_p)\, dz_1 \ldots dz_p,$$

where

$$f(z_1, \ldots, z_p) = \frac{1}{(\sqrt{2\pi})^p \sqrt{|V|}} \exp\left(-(z-\mu)^T V^{-1}(z-\mu)/2\right).$$

The function $f(z_1, \ldots, z_p)$ is called the *probability density function* of Z.

From now on we assume that Y has a multivariate normal distribution.

A linear combination of normal random variables has a normal distribution so $W = LZ$, say, has a normal distribution with mean vector $L\mu$ and covariance matrix LVL^T. If LVL^T is singular then the probability density function of W cannot be written down explicitly, although it does still exist. (This is guaranteed by the continuity theorem for characteristic functions – see Section 10.7.) We say that W has a *singular* multivariate normal distribution, written $W \sim SN_p(L\mu, LVL^T)$.

Let F be any matrix. A matrix F^- such that $FF^-F = F$ and $F^-FF^- = F^-$ is called the *reflexive generalized inverse* of F. Then

$$(rI - (1/k)AA^T)\hat{\tau} = (T^T - (1/k)AB^T)Y$$

so

$$\hat{\tau} = (rI - (1/k)AA^T)^-(T^T - (1/k)AB^T)Y.$$

Since $Y \sim N_{bk}(\eta, \sigma^2 I)$,

$$\hat{\tau} \sim SN_v((rI - (1/k)AA^T)^-(T^T - (1/k)AB^T)\eta, \sigma^2(rI - (1/k)AA^T)^-)$$

and

$$(rI - (1/k)AA^T)^-(T^T - (1/k)AB^T)\eta = (rI - (1/k)AA^T)^-(rI - (1/k)AA^T)\tau$$

(see Exercise 10.2).

The next step in deriving a test of a testable hypothesis is to obtain the distributions of some quadratic forms.

The random variable S is said to have a *non-central* $\chi^2(n, \lambda)$ *distribution* if the probability density function of S is given by

$$e^{-\lambda} \sum_{i=0}^{\infty} \frac{\lambda^i}{i!} \frac{s^{(n/2)+i-1} e^{-s/2}}{2^{(n/2)+i}\Gamma((n/2)+i)},$$

where $\Gamma(x)$ is the gamma function with argument x. The parameters n and λ are usually called the *degrees of freedom* and the *non-centrality parameter*, respectively. If $\lambda = 0$, then s is said to have a *central* χ^2 distribution, written $\chi^2(n)$.

A *quadratic form* is a function of the form $z^T Hz$. Any quadratic form can be written uniquely in terms of a symmetric matrix: $z^T Hz = z^T\{\frac{1}{2}(H + H^T)\}z = z^T Gz$, say. Because of this uniqueness, we will assume, in the sequel, that the matrix of any quadratic form is symmetric.

Theorem 7. *If $Z \sim N_p(\mu, V)$, whether V is singular or non-singular, then the quadratic form $Z^T GZ$ has a χ^2 distribution with degrees of freedom equal to $tr(GV)$ and non-centrality parameter $\frac{1}{2}\mu^T G\mu$ if and only if:*

(i) $VGVGV = VGV$;

(ii) $\mu^T GV = \mu^T GVGV$;

(iii) $\mu^T G\mu = \mu^T GVG\mu$. □

Theorem 8. *If $Z \sim SN_p(\mu, V)$, then the quadratic forms $Z^T GZ$ and $Z^T HZ$ are independent if and only if:*

(i) $VGVHV = 0$;

(ii) $VGVH\mu = VHVG\mu = 0$;

(iii) $\mu^T GVH\mu = 0$. □

We now apply these results. Let K^T be a $t \times v$ matrix of full row rank. Suppose that $K^T\tau = \mathbf{0}$ is a testable hypothesis. Then one estimate of $K^T\tau$ is $K^T\hat{\tau}$ and

$$K^T\hat{\tau} \sim N_t(K^T\tau, \sigma^2 K^T(rI - (1/k)AA^T)^- K).$$

Let

$$Q = (K^T\hat{\tau})^T(K^T(rI - (1/k)AA^T)^- K)^{-1} K^T\hat{\tau}.$$

Then, applying Theorem 7 with $Z = K^T\hat{\tau}$ and

$$G = (K^T(rI - (1/k)AA^T)^- K)^{-1},$$

we see that Q/σ^2 has a non-central χ^2 distribution with t degrees of freedom and non-centrality parameter

$$(K^T\tau)^T(K^T(rI - (1/k)AA^T)^- K)^{-1} K^T\tau/2\sigma^2.$$

This is zero under the hypothesis $K^T\tau = 0$, and Q is the appropriate statistic for testing this hypothesis.

If $K^T\tau$ is not estimable, then

$$K^T\hat{\tau} \sim N_t(K^T(rI - (1/k)AA^T)^-(rI - (1/k)AA^T)\tau, \sigma^2 K^T(rI - (1/k)AA^T)^- K),$$

and Q is the appropriate statistic for testing the hypothesis $K^T(rI - (1/k)AA^T)^-(rI - (1/k)AA^T)\tau = 0$.

If K^T is not of full row rank but $K^T\tau = \mathbf{0}$ is a testable hypothesis then the appropriate statistic for testing this hypothesis is

$$Q = (K^T\hat{\tau})^T(K^T(rI - (1/k)AA^T)^- K)^- K^T\hat{\tau}.$$

(In Exercise 10.3 we see that Q is invariant for all choices of generalized inverse.) Again, by Theorem 7 we see that Q/σ^2 has a χ^2 distribution with degrees of freedom equal to the trace of

$$(K^T(rI-(1/k)AA^T)^-K)^-(K^T(rI-(1/k)AA^T)^-K)$$

and non-centrality parameter $(K^T\tau)^T(K^T(rI-(1/k)AA^T)^-K)^-K^T\tau/2\sigma^2$.

If K^T is not of full row rank and $K^T\tau$ is not estimable then Q is the appropriate statistic for testing $K^T(rI-(1/k)AA^T)^-(rI-(1/k)AA^T)\tau=0$.

In the case of factorial experiments we are interested in hypotheses in which $K^T=C^{\mathbf{x}}$. Theorem 8 lets us say when the tests for different interactions are independent. If $C^{\mathbf{x}}\tau$ is not a testable hypothesis then we say that $\mathbf{x}$ has been confounded with blocks.

Example 2 (concluded). Here $r=1$ and $k=3$ and $(I-(1/3)AA^T)$ is idempotent. Hence, one choice for $(I-(1/3)AA^T)^-$ is $(I-(1/3)AA^T)$. Also, $T=I$ and $B=A$ so

$$\hat{\tau}=(I-(1/3)AA^T)Y$$

and

$$\hat{\tau}\sim SN_6((I-(1/3)AA^T)\tau,\sigma^2(I-(1/3)AA^T)).$$

Let $K^T=C^{\mathbf{x}}$. Then

$$Q=\hat{\tau}^TC^{\mathbf{x}^T}(C^{\mathbf{x}}(I-(1/3)AA^T)C^{\mathbf{x}^T})^-C^{\mathbf{x}}\hat{\tau}=\hat{\tau}^TG^{\mathbf{x}}\hat{\tau},\text{ say.}$$

The values of $G^{\mathbf{x}}$ are given in Table 10.4. Also,

$$(I-(1/3)AA^T)G^{\mathbf{x}}(I-(1/3)AA^T)G^{\mathbf{y}}(I-(1/3)AA^T)$$

does not equal zero for any choices of $\mathbf{x}$ and $\mathbf{y}$, $\mathbf{x}\neq\mathbf{y}$, so the three quadratic forms are not independent. None of the hypotheses $C^{\mathbf{x}}\tau$ is testable, since $C^{\mathbf{x}}(I-(1/3)AA^T)\tau\neq C^{\mathbf{x}}\tau$, as can be seen from Table 10.4 where the values of $C^{\mathbf{x}}(I-(1/3)AA^T)$ are given. □

Lemma 9. *In a single replicate factorial design, $T=I$, $B=A$, and $(I-(1/k)AA^T)$ is idempotent.*

Proof. This follows from the definition of T and B and from Lemma 1. □

Hence, in a single replicate factorial design $\hat{\tau}=(I-(1/k)AA^T)Y$ and $Q=\hat{\tau}^TC^{\mathbf{x}^T}(C^{\mathbf{x}}(I-(1/k)AA^T)C^{\mathbf{x}^T})^-C^{\mathbf{x}}\hat{\tau}$. The degrees of freedom of Q equal the trace of $(C^{\mathbf{x}}(I-(1/k)AA^T)C^{\mathbf{x}^T})^-(C^{\mathbf{x}}(I-(1/k)AA^T)C^{\mathbf{x}^T})$. The hypothesis $C^{\mathbf{x}}\tau$ is testable if and only if $C^{\mathbf{x}}AA^T=\mathbf{0}$. AA^T is often called the *design matrix*.

In the next two sections we look at constructions where the matrix $C^{\mathbf{x}}(I-(1/k)AA^T)C^{\mathbf{x}^T}$ is a scalar multiple of an idempotent matrix.

Table 10.4

x	G^x	$C^x(I-(1/3)AA^T)$
01	$\frac{1}{6}\begin{bmatrix} 2 & -1 & -1 & 2 & -1 & -1 \\ -1 & 5 & -4 & -1 & 5 & -4 \\ -1 & -4 & 5 & -1 & -4 & 5 \\ 2 & -1 & -1 & 2 & -1 & -1 \\ -1 & 5 & -4 & -1 & 5 & -4 \\ -1 & -4 & 5 & -1 & -4 & 5 \end{bmatrix}$	$\frac{1}{6}\begin{bmatrix} 2 & -1 & -1 & 2 & -1 & -1 \\ -2 & 1 & 0 & 0 & 1 & 0 \\ 0 & 0 & 1 & -2 & 0 & 1 \end{bmatrix}$
10	$\frac{1}{16}\begin{bmatrix} 3 & 3 & 3 & -3 & -3 & -3 \\ 3 & 3 & 3 & -3 & -3 & -3 \\ 3 & 3 & 3 & -3 & -3 & -3 \\ -3 & -3 & -3 & 3 & 3 & 3 \\ -3 & -3 & -3 & 3 & 3 & 3 \\ -3 & -3 & -3 & 3 & 3 & 3 \end{bmatrix}$	$\frac{1}{18}\begin{bmatrix} 2 & 2 & 4 & -2 & -4 & -2 \\ -2 & -2 & -4 & 2 & 4 & 2 \end{bmatrix}$
11	$\frac{1}{14}\begin{bmatrix} 6 & -3 & -3 & -6 & 3 & 3 \\ -3 & 5 & -2 & 3 & -5 & 2 \\ -3 & -2 & 5 & 3 & 2 & -5 \\ -6 & 3 & 3 & 6 & -3 & -3 \\ 3 & -5 & 2 & -3 & 5 & -2 \\ 3 & 2 & -5 & -3 & -2 & 5 \end{bmatrix}$	$\frac{1}{18}\begin{bmatrix} 4 & -5 & -1 & -4 & 1 & 5 \\ -2 & 7 & -4 & 2 & -5 & 2 \\ -2 & -2 & 5 & 2 & 4 & -7 \\ -4 & 5 & 1 & 4 & -1 & -5 \\ 2 & -7 & 4 & -2 & 5 & -2 \\ 2 & 2 & -5 & -2 & -4 & 7 \end{bmatrix}$

10.4. A construction based on orthogonal arrays

Let E_i be a circulant matrix with first row equal to $\boldsymbol{e}_i$, where $\boldsymbol{e}_i$ is of length s. We state the next result for ease of referral.

Lemma 10.

(i) $\sum_i E_i = J$; (ii) $JE_i = E_iJ = J$; (iii) $E_i^T = E_{s-i+2}$; (iv) $E_iE_j = E_{i+j-1}$. □

Lemma 11. *Suppose that* $s_1 = s_2 = \ldots = s_n = s$, *say. Let the matrix* D *be given by*

$$D = \sum_{i_1=1}^{s} \sum_{i_2=1}^{s} \ldots \sum_{i_m=1}^{s} (E_{i_1} \times \ldots \times E_{i_m} \times E_{y_1} \times \ldots \times E_{y_{n-m}}),$$

where

$$y_f = \sum_{j=1}^{m} a_{jf}\, i_j - \sum_{j=1}^{m} a_{jf} + 1 \pmod{s}$$

and the a_{jf} *are constants giving suitable values of* y_f. *Then* D *is the design matrix of a single replicate* s^n *factorial design in blocks of size* s^m.

Proof. $D = D^T$ since $(E_{i_1} \times \ldots \times E_{y_{n-m}})^T$

$$= E_{s-i_1+2} \times \ldots \times E_{s-i_m+2} \times E_{s-y_1+2} \times \ldots \times E_{s-y_{n-m}+2}$$

and

$$s-y_f+2=\sum_{j=1}^{m} a_{jf}(s-i_j+2)-\sum_{j=1}^{m} a_{jf}+1 \pmod s.$$

$DJ = s^m J$ since $E_i J = J$ and there are s^m summands in D. Also,

$$(E_{i_1}\times \ldots \times E_{y_{n-m}})(E_{j_1}\times \ldots \times E_{w_{n-m}}) = E_{i_1+j_1-1}\times \ldots \times E_{y_{n-m}+w_{n-m}-1}.$$

But

$$y_f+w_f-1=\sum_{k=1}^{m}(a_{kf}i_k+a_{kf}j_k)-2\sum_k a_{kf}+2-1 \qquad \pmod s$$

$$=\sum_k a_{kf}(i_k+j_k-1)-\sum_k a_{kf}+1 \qquad \pmod s$$

and this is also a summand in D. As there are s^m such pairs, $DD = s^m D$. The result follows. □

Example 12. Let $s = n = 3$, $m = 2$ and define $y_1 = i_1 + i_2 - 1 \pmod 3$. Then

$$D=\sum_{i_1=1}^{3}\sum_{i_2=1}^{3}(E_{i_1}\times E_{i_2}\times E_{i_1+i_2-1})$$
$$=E_1\times E_1\times E_1+E_1\times E_2\times E_2+\ldots+E_3\times E_3\times E_2.$$

D is given in Table 10.5. The array in Table 10.6 is an OA [9, 3, 3, 2]; it has column vectors $(i_1, i_2, y_1)^T$. It is possible to check that $M^x D = \mathbf{0}$ if $\mathbf{x} \neq (111)$. For instance, if $\mathbf{x} = (100)$ then

$$M^x D = ((3I-J)\times j^T\times j^T)\Big(\sum_{i_1}\sum_{i_2} E_{i_1}\times E_{i_2}\times E_{i_1+i_2-1}\Big)$$

$$=\sum_{i_1}\sum_{i_2}((3I-J)E_{i_1}\times j^T\times j^T)$$

$$=\sum_{i_1}(3I-J)E_{i_1}\times\sum_{i_2}(j^T\times j^T)$$

$$=0.$$

Hence only 111 is confounded with blocks. □

The next result extends Theorem 9.19.

Theorem 13. *If the array with column vectors* $(i_1, i_2, \ldots, i_m, y_1, \ldots, y_{n-m})^T$, $1 \leq i_j \leq s$, $1 \leq j \leq m$, *is an orthogonal array, H say, of strength g then all interactions involving fewer than g factors are unconfounded in the design with design matrix D.*

Table 10.5

1 0 0	0 1 0	0 0 1	0 1 0	0 0 1	1 0 0	0 0 1	1 0 0	0 1 0
0 1 0	0 0 1	1 0 0	0 0 1	1 0 0	0 1 0	1 0 0	0 1 0	0 0 1
0 0 1	1 0 0	0 1 0	1 0 0	0 1 0	0 0 1	0 1 0	0 0 1	1 0 0
0 0 1	1 0 0	0 1 0	1 0 0	0 1 0	0 0 1	0 1 0	0 0 1	1 0 0
1 0 0	0 1 0	0 0 1	0 1 0	0 0 1	1 0 0	0 0 1	1 0 0	0 1 0
0 1 0	0 0 1	1 0 0	0 0 1	1 0 0	0 1 0	1 0 0	0 1 0	0 0 1
0 1 0	0 0 1	1 0 0	0 0 1	1 0 0	0 1 0	1 0 0	0 1 0	0 0 1
0 0 1	1 0 0	0 1 0	1 0 0	0 1 0	0 0 1	0 1 0	0 0 1	1 0 0
1 0 0	0 1 0	0 0 1	0 1 0	0 0 1	1 0 0	0 0 1	1 0 0	0 1 0
0 0 1	1 0 0	0 1 0	1 0 0	0 1 0	0 0 1	0 1 0	0 0 1	1 0 0
1 0 0	0 1 0	0 0 1	0 1 0	0 0 1	1 0 0	0 0 1	1 0 0	0 1 0
0 1 0	0 0 1	1 0 0	0 0 1	1 0 0	0 1 0	1 0 0	0 1 0	0 0 1
0 1 0	0 0 1	1 0 0	0 0 1	1 0 0	0 1 0	1 0 0	0 1 0	0 0 1
0 0 1	1 0 0	0 1 0	1 0 0	0 1 0	0 0 1	0 1 0	0 0 1	1 0 0
1 0 0	0 1 0	0 0 1	0 1 0	0 0 1	1 0 0	0 0 1	1 0 0	0 1 0
1 0 0	0 1 0	0 0 1	0 1 0	0 0 1	1 0 0	0 0 1	1 0 0	0 1 0
0 1 0	0 0 1	1 0 0	0 0 1	1 0 0	0 1 0	1 0 0	0 1 0	0 0 1
0 0 1	1 0 0	0 1 0	1 0 0	0 1 0	0 0 1	0 1 0	0 0 1	1 0 0
0 1 0	0 0 1	1 0 0	0 0 1	1 0 0	0 1 0	1 0 0	0 1 0	0 0 1
0 0 1	1 0 0	0 1 0	1 0 0	0 1 0	0 0 1	0 1 0	0 0 1	1 0 0
1 0 0	0 1 0	0 0 1	0 1 0	0 0 1	1 0 0	0 0 1	1 0 0	0 1 0
1 0 0	0 1 0	0 0 1	0 1 0	0 0 1	1 0 0	0 0 1	1 0 0	0 1 0
0 1 0	0 0 1	1 0 0	0 0 1	1 0 0	0 1 0	1 0 0	0 1 0	0 0 1
0 0 1	1 0 0	0 1 0	1 0 0	0 1 0	0 0 1	0 1 0	0 0 1	1 0 0
0 0 1	1 0 0	0 1 0	1 0 0	0 1 0	0 0 1	0 1 0	0 0 1	1 0 0
1 0 0	0 1 0	0 0 1	0 1 0	0 0 1	1 0 0	0 0 1	1 0 0	0 1 0
0 1 0	0 0 1	1 0 0	0 0 1	1 0 0	0 1 0	1 0 0	0 1 0	0 0 1

Table 10.6

1	1	1	2	2	2	3	3	3
1	2	3	1	2	3	1	2	3
1	2	3	2	3	1	3	1	2

Proof. Let $\mathbf{x}$ be a binary vector of length n with $r \le g$ non-zero entries. We need to show that $M^{\mathbf{x}}D = 0$. We know that

$$M^{\mathbf{x}}D = \sum_{i_1} \ldots \sum_{i_m} (M^{x_1} E_{i_1} \times \ \ldots \ \times M^{x_n} E_{y_{n-m}})$$

and that $j^T E_i = j^T$ and $\sum_i (sI - J)\, E_i = 0$.

The rows of H which correspond to the non-zero elements of x form a submatrix in which each possible s^r-tuple appears as a column equally often. We may need to change the variables over which we sum, but we can always write $M^x D$ as a sum of Kronecker products which involve factors of the form $\sum_a (sI-J)E_a = 0$. The result follows. □

Lemma 14. *Let $u = s^{\Sigma x_i}$. Then the following hold:*

(i) $C^x DC^{x^T} C^x = \frac{u}{v} C^x D$;

(ii) $\frac{v}{u} C^x (I-(1/k)D)C^{x^T}$ *is idempotent;*

(iii) *Let z be a binary vector of length n. Then, if $x \neq z$,*

$$C^{x^T} C^x (I-(1/k)D)C^{z^T} C^z = 0.$$

Proof. (i) Now

$$C^x DC^{x^T} C^x = C^x D(M^{x_1^T} M^{x_1} \times \ldots \times M^{x_n^T} M^{x_n})$$

and, by definition,

$$M^{0^T} M^0 = jj^T = J,\ M^{1^T} M^1 = sM^1 = s(sI-J),$$

and, by Lemma 10,

$$JE_i = E_i J,\ M^1 E_i = (sI-J)E_i = E_i M^1.$$

Thus

$$C^x DC^{x^T} C^x = C^x C^{x^T} C^x D.$$

The factors in $C^x C^{x^T} C^x$ are either $j^T J = sj^T$ or $M^1 sM^1 = s^2 M^1$, so $C^x C^{x^T} C^x = \frac{u}{v} C^x$, as required.

(ii)

$$\frac{v^2}{u^2} C^x (I-(1/k)D)C^{x^T} C^x (I-(1/k)D)C^{x^T}$$

$$= \frac{v^2}{u^2} C^x C^{x^T} C^x (I-(1/k)D)C^{x^T}$$

$$= \frac{v}{u} C^x (I-(1/k)D)C^{x^T}, \text{ as required.}$$

(iii) $C^{x^T} C^x (I-(1/k)D)C^{z^T} C^z = C^{x^T} C^x C^{z^T} C^z (I-(1/k)D)$. Since $x \neq z$, there is at least one factor of the form $J(sI-J) = 0$ or $(sI-J)J = 0$ and the result follows. □

Hence, by Theorem 8, the tests of the hypotheses '$C^x(I-(1/k)D)\tau = 0$' and '$C^z(I-(1/k)D)\tau = 0$' are independent.

A generalized inverse of $C^x(I-(1/k)D)C^{x^T}$ is $\frac{v^2}{u^2}C^x(I-(1/k)D)C^{x^T}$ so the number of degrees of freedom of Q equals the trace of $\frac{v}{u}C^x(I-(1/k)D)C^{x^T}$. If x is not confounded with blocks, this is the trace of $\frac{v}{u}C^xC^{x^T}$ which equals $\prod_i (s-1)^{x_i}$; if x is confounded with blocks, then the number of degrees of freedom confounded with blocks is given by the trace of $\frac{v}{uk}C^x DC^{x^T}$.

Lemma 15. *The trace of* $C^{x^T}C^xD$ is $\frac{u}{v}\sum_{i_1}\ldots\sum_{i_m}(s-1)^t(-1)^w$, *where t is the number of* 1s *in* $(i_1, i_2, \ldots, i_m, y_1, \ldots, y_{n-m})$ *in the positions where x has a* 1, *and* $w = \sum_i x_i - t$.

Proof.

$$C^{x^T}C^xD = \frac{1}{s^{2n}}\left(\mathop{\times}_{j=1}^{n}(M^{x_j})^T M^{x_j}\right)\left(\sum_{i_1}\ldots\sum_{i_m}(E_{i_1}\times\ldots\times E_{y_{n-m}})\right)$$

$$= \frac{1}{s^{2n}}\sum_{i_1}\ldots\sum_{i_m}(M^{x_1^T}M^{x_1}E_{i_1}\times\ldots\times M^{x_n^T}M^{x_n}E_{y_{n-m}}).$$

Now $M^{0^T}M^0E_i = JE_i = J$, $M^{1^T}M^1E_1 = s(sI-J)$ and $M^{1^T}M^1E_i = s(sE_i - J)$, $i \neq 1$, and these matrices have $1, s(s-1)$ and $-s$ on the diagonals, respectively. Hence the diagonal elements of $C^{x^T}C^xD$ are $u(s-1)^t(-1)^w$, where there are t terms $M^{1^T}M^1E_1$ and w terms $M^{1^T}M^1E_i$, $i\neq 1$, in a summand. Each such term appears s^n times, as that is the size of the submatrices in $C^{x^T}C^xD$. □

Example 12 (continued). We know that the corresponding orthogonal array has strength 2 so the main effects and the two-factor interaction effects are not confounded with blocks. The number of degrees of freedom of the three-factor interaction which is confounded with blocks is

$$\frac{27}{27\times 27}\times\frac{27}{27\times 9}(2^3+6(2)(-1)^2+2(-1)^3)27 = 2.$$

Since $b = 3$, $b-1 = 2$ degrees of freedom are confounded with blocks and this is a check. □

The blocks of the design with design matrix given in Lemma 11 can be constructed from the array with columns

$(i_1, \ldots, i_m, y_1, \ldots, y_{n-m})^T$ by adding the elements $(0, \ldots, 0, a_1, \ldots, a_{n-m})^T$, $a_j \in Z_s$, in turn to the array.

Example 12 (concluded). Table 10.7 lists the three blocks of this design. □

Table 10.7

1 1 1 2 2 2 3 3 3	1 1 1 2 2 2 3 3 3	1 1 1 2 2 2 3 3 3
1 2 3 1 2 3 1 2 3	1 2 3 1 2 3 1 2 3	1 2 3 1 2 3 1 2 3
1 2 3 2 3 1 3 1 2	2 3 1 3 1 2 1 2 3	3 1 2 1 2 3 2 3 1

Let H be an orthogonal array with h rows, $r_1, r_2, \ldots, r_h$, say. Let H_A be an array with $h+a$ rows, $r_1, r_2, \ldots, r_h, r_h, \ldots, r_h$, where the row r_h appears $a+1$ times altogether. Then H_A can be used to construct a factorial design but some two-factor interactions will be confounded.

Example 16. Let an orthogonal array H have column vectors $(i_1, i_2, i_1+i_2-1)^T$, $1 \le i_1,\ i_2 \le 6$, and use H, with its final row repeated, to construct a factorial design. It has 36 blocks of size 36. The interactions confounded are (0011), (1110), (1101), and (1111) and the number of degrees of freedom of each that are confounded is 5, 5, 5, and 25 respectively. □

10.5. A construction using generalized cyclic designs

In this section and the next we regard the v treatment combinations as elements of the additive Abelian group $Z_{s_1} \oplus Z_{s_2} \oplus \ldots \oplus Z_{s_n}$. Suppose that B is a set of k treatment combinations. Then the *generalized cyclic design* developed from B consists of the v blocks obtained by adding each treatment combination in turn to B. As before, we call B the *starter block* and we assume B contains no repeated elements.

Theorem 17. *Cyclic development of a starter block B, of order k, gives a generalized cyclic design with exactly v/k distinct blocks if and only if the blocks of the design consist of a subgroup of order k and its cosets.*

Proof. Suppose that the generalized cyclic design obtained by developing the starter block B has v/k distinct blocks. Thus any two blocks are either disjoint or identical, since each of the v treatment combinations must appear in one of the v/k blocks of size k. Without loss of generality suppose that $0 \in B$. Then, for $b \in B$, $|(b+B) \cap B| > 0$ so $B+b = B$, showing that B is closed and so a subgroup.

The converse is obvious. □

By Lemma 1, the design matrix of a single replicate generalized cyclic design can be written as

$$D = \sum_{i_1=1}^{s_1} \cdots \sum_{i_n=1}^{s_n} w_{i_1 \ldots i_n} (E_{i_1} \times \ldots \times E_{i_n}), \tag{10.4}$$

where $w_{i_1 \ldots i_n}$ is 1 if $(i_1 - 1, \ldots, i_n - 1)$ is in the subgroup and is 0 otherwise.

Example 18. Let $s_1 = 2$, $s_2 = 3$, $s_3 = 6$ and let the subgroup of order 12 be $\{000, 100, 011, 111, 022, 122, 003, 103, 014, 114, 025, 125\} = B$, say. The other two blocks are $B + (010)$ and $B + (202)$. Also,

$$D = AA^T = (I \times I \times I + E_2 \times I \times I + \ldots + E_2 \times E_3 \times E_6). \qquad \square$$

The next result is proved in the same way as Lemma 15.

Lemma 19. *The number of degrees of freedom of the generalized interaction represented by* $\mathbf{x}$, *that are confounded in the design with D given by* (10.4), *is*

$$\frac{v}{uk} tr\,(C^{\mathbf{x}} D C^{\mathbf{x}^T}) = \frac{1}{k} \sum_{i_1} \cdots \sum_{i_n} w_{i_1 \ldots i_n} \left(\prod_j z_{i_j}^{x_j} \right),$$

where

$$z_{i_j}^{x_j} = \begin{cases} s_j - 1, & i_j = 1, \quad x_j = 1, \\ -1, & i_j \neq 1, \quad x_j = 1, \\ 1, & \qquad\quad\; x_j = 0. \end{cases} \qquad \square$$

Example 18 (concluded). The number of degrees of freedom confounded is: for $\mathbf{x} = 100$

$$\frac{1}{12} \sum_{i_1=1}^{2} \sum_{i_2=1}^{3} \sum_{i_3=1}^{6} w_{i_1 i_2 i_3} z_{i_1}^1 = \frac{1}{12} \{6(2-1) + 6(-1)\} = 0;$$

for $\mathbf{x} = 010$ is

$$\frac{1}{12} \sum_{i_1} \sum_{i_2} \sum_{i_3} w_{i_1 i_2 i_3} z_{i_2}^1 = \frac{1}{12} \{4(3-1) + 4(-1) + 4(-1)\} = 0;$$

and for $\mathbf{x} = 011$ is

$$\begin{aligned} \frac{1}{12} \sum_{i_1} \sum_{i_2} \sum_{i_3} w_{i_1 i_2 i_3} z_{i_2}^1 z_{i_3}^1 &= \frac{1}{12} \{2(3-1)(6-1) + 8(-1)^2 + 2(3-1)(-1)\} \\ &= 2. \end{aligned} \qquad \square$$

10.6. The DSIGN method

This is a general method of factorial design construction; we present it here only for incomplete block designs.

Each plot in the design has associated with it a p-tuple of plot factors. In an incomplete block design this means that each plot has a pair of plot factors associated with it. The first records the block containing the plot and the second records which plot it is within that block. The ordering of both blocks and 'plots-within-blocks' is arbitrary, but fixed. Thus the first plot factor has b levels and the second plot factor has k levels. In fact, we can represent the 'plots-within-blocks' factor by one factor with k levels or by several pseudo-factors, the products of whose levels is k. We let the number of levels of the jth plot factor be p_j.

The treatment combination applied to the plot $\boldsymbol{u} = (u_1, u_2, \ldots, u_p)^T$ is given by $z = K^T\boldsymbol{u}$. The matrix K^T is called the *key matrix*. Let U be the $p \times v$ array with plot factor levels as columns and let $Q = K^T U$. Then in a single replicate design Q is an $n \times v$ array which contains each treatment combination precisely once in its columns.

Example 20. Let $s_1 = 3, s_2 = 2, s_3 = 6$ and construct a design with three blocks of size 12. We represent the 'plots-within-blocks' factor by two factors, one with two levels and the other with six levels. Let

$$K = \begin{bmatrix} 1 & 0 & 0 \\ 0 & 1 & 0 \\ 1 & 0 & 1 \end{bmatrix}.$$

Then the first block of the design is

$$\{K^T u \mid u = (1, u_2, u_3)^T;\ u_2 = 0, 1;\ u_3 = 0, \ldots, 5\};$$

that is,

$$\{(1 + u_3, u_2, u_3)^T \mid u_2 = 0, 1;\ u_3 = 0, \ldots, 5\},$$

$$= \{100, 201, 002, 103, 204, 005, 110, 211, 012, 113, 214, 015\}.$$

After interchanging factors 1 and 2, this becomes one of the blocks of the design given in Example 18. □

How do we use this method of construction to determine which effects have been confounded with blocks? We could construct C^x and AA^T and proceed as before but another technique, which will work for a larger class of designs generated by the DSIGN method, is available.

Let $\lambda_s = lcm[s_1, s_2, \ldots, s_n]$, and let Δ_s be an $n \times n$ diagonal matrix with the ith diagonal position occupied by λ_s/s_i. We associate with each treatment combination, $\boldsymbol{x}$ say, a set, $\mathrm{T}(z)$, of all contrasts between treatment combinations with different values of

$$[\boldsymbol{w}, z] = \boldsymbol{w}^T \Delta_s z \pmod{\lambda_s}.$$

Example 20 (continued). Here $\lambda_s = lcm[3, 2, 6] = 6$ so $\Delta_s = \mathrm{diag}(2, 3, 1)$. The values of $[\boldsymbol{w}, (215)^T]$ are given in Table 10.8.

Table 10.8. Values of $[w, (215)^T]$

Level of		Level of F_3					
F_1	F_2	0	1	2	3	4	5
0	0	0	5	4	3	2	1
0	1	3	2	1	0	5	4
1	0	4	3	2	1	0	5
1	1	1	0	5	4	3	2
2	0	2	1	0	5	4	3
2	1	5	4	3	2	1	0

We see there are six distinct values in the table and so $T((215)^T)$ represents five degrees of freedom. The contrast between the w for which $[w, (215)^T]$ is even and those for which it is odd is in the interaction (011). The contrasts between the sets $\{w|[w, (215)^T] \equiv 0 \pmod 3\}$, $\{w|[w, (215)^T] \equiv 1 \pmod 3\}$ and $\{w|[w, (215)^T] \equiv 2 \pmod 3\}$ represents two degrees of freedom and is in the interaction (101). The remaining two degrees of freedom are in the interaction (111).

The values of $[w, (013)^T]$ are given in Table 10.9. The contrast in $\mathrm{T}((013)^T)$ is in the interaction (011) and is also in $\mathrm{T}((215)^T)$. □

Table 10.9. Values of $[w, (013)^T]$

Level of		Level of F_3					
F_1	F_2	0	1	2	3	4	5
0	0	0	3	0	3	0	3
0	1	3	0	3	0	3	0
1	0	0	3	0	3	0	3
1	1	3	0	3	0	3	0
2	0	0	3	0	3	0	3
2	1	3	0	3	0	3	0

The aim is to find sets $\mathrm{T}_*(z)$ such that all the contrasts in $\mathrm{T}_*(z)$ are in the same generalized interaction x and such that each contrast is in exactly one $\mathrm{T}_*(z)$.

For any treatment combination z, let $\langle z \rangle$ be the subgroup generated by z (in the Abelian group of treatment combinations). If $|\langle z \rangle| = m$ then $\langle z \rangle = \langle rz \rangle$ for every integer r coprime to m. There are $\phi(m)$ such r, where ϕ is Euler's ϕ-function. We subdivide the treatment combinations into *families*, where the members of each family generate the same subgroup.

Now $[w, rz] = r[w, z] \pmod{\lambda_s}$, so $[w, z]$ takes m different values as w varies over the v treatment combinations. Hence, the dimension of the vector space of contrasts $\mathrm{T}(z)$ is $m-1$. Also, $\mathrm{T}(rz)$ is a subset of $\mathrm{T}(z)$ and $\mathrm{T}(z) = \mathrm{T}(rz)$

for r coprime to m. Suppose that $\langle z_1 \rangle \cap \langle z_2 \rangle = \langle z_3 \rangle$. Then the contrasts common to $\mathrm{T}(z_1)$ and $\mathrm{T}(z_2)$ are those in $\mathrm{T}(z_3)$.

$\mathrm{T}_*(z)$ is defined to be the set of contrasts in $\mathrm{T}(z)$ which are orthogonal to those in $\mathrm{T}(m_1 z), \ldots, \mathrm{T}(m_h z)$, where $m_1, m_2, \ldots, m_h$ are the proper prime divisors of m. We need to select one z from each family; we make this the earliest in the lexicographic ordering.

Example 20 (continued). In Table 10.10 we give the families, and the interaction in which each $\mathrm{T}_*(z)$ lies, for the $3 \times 2 \times 6$ factorial design. □

Table 10.10

z	Order, m, of z	The rest of the family	Generators of proper subgroups of $\langle z \rangle$	Interactions in $\mathrm{T}_*(z)$ (dimension)
010	2	—	—	010 (1)
003	2	—	—	001 (1)
013	2	—	—	011 (1)
100	3	200	—	100 (2)
002	3	004	—	001 (2)
102	3	204	—	101 (2)
104	3	202	—	101 (2)
001	6	005	002, 003	001 (2)
011	6	015	002, 013	011 (2)
012	6	014	002, 010	011 (2)
101	6	205	104, 003	101 (2)
103	6	203	100, 003	101 (2)
105	6	201	102, 003	101 (2)
110	6	210	100, 010	110 (2)
111	6	215	104, 013	111 (2)
112	6	214	102, 010	111 (2)
113	6	213	100, 013	111 (2)
114	6	212	104, 010	111 (2)
115	6	211	102, 013	111 (2)

Lemma 21. *The sets* $\mathrm{T}_*(z)$, *where there is one* z *from each family, are such that all contrasts in each* $\mathrm{T}_*(z)$ *are in the same interaction and such that each contrast is in precisely one* $\mathrm{T}_*(z)$.

Proof. The contrasts common to $\mathrm{T}(z_1)$ and $\mathrm{T}(z_2)$ have been removed in the construction of the $\mathrm{T}_*(z)$. Hence, the sets $\mathrm{T}_*(z)$ have no common contrasts.

The contrasts in $\mathrm{T}_*(e_i)$ are those between treatment combinations with different values of

$$[w, e_i] = \mathbf{w}^T \Delta_s e_i = w_i \lambda_s / s_i \pmod{\lambda_s};$$

that is, between the elements of the s_i sets

$$\{(w_1, \ldots, w_n) | w_i = \theta\}, \quad 0 \le \theta \le s_i - 1.$$

Any such contrast is a linear combination of the rows of M^{e_i} so all the elements of $\mathrm{T}_*(e_i)$ are elements of the main effect of the ith factor.

The contrasts in $\mathrm{T}_*(e_i + \delta e_j)$ are those between treatment combinations with different values of

$$[w, e_i + \delta e_j] = w_i \lambda_s / s_i + \delta w_j \lambda_s / s_i \pmod{\lambda_s};$$

that is, between the elements of the sets

$$\{(w_1, \ldots, w_n) | w_i = \theta, \quad w_j = v\}, \quad 0 \le \theta \le s_i - 1, \quad 0 \le v \le s_j - 1.$$

Consider the sets $\{w | w_i = \theta_1, w_j = v\}$ and $\{w | w_i = \theta_2, w_j = v\}$. Contrasts between the elements of these sets appear in $\mathrm{T}_*(e_i)$. Contrasts between the elements of the sets $\{w | w_i = \theta, w_j = v_1\}$ and $\{w | w_i = \theta, w_j = v_2\}$ appear in $\mathrm{T}_*(e_j)$. Hence, the elements in $\mathrm{T}_*(e_i + \delta e_j)$ are linear combinations of the rows of $M^{e_i + e_j}$ and so are elements of the interaction of factors i and j.

By induction the contrasts in $\mathrm{T}_*(z)$ are elements of the interaction between the factors corresponding to the positions in z with non-zero entries.

The dimension of $\mathrm{T}_*(z)$ is $\phi(m)$, where $m = |\langle z \rangle|$. Hence, the $\mathrm{T}_*(z)$ account for $v - 1$ independent contrasts, the total number of independent contrasts possible in an experiment with v treatments (or the dimension of $\langle j_v \rangle^{\perp}$). □

Plot effects $\mathrm{P}(u)$ and $\mathrm{P}_*(u)$ can be defined similarly. If the contrasts in $\mathrm{T}_*(z)$ are the same as those in $\mathrm{P}_*(u)$ then we say that $\mathrm{T}_*(z)$ is *aliased* with $\mathrm{P}_*(u)$. In a single replicate design if $\mathrm{T}_*(z)$ is aliased with $\mathrm{P}_*(100 \ldots 0)^T$ then $\mathrm{T}_*(z)$ is completely confounded with blocks. Let $\alpha(z) = Q^T \Delta_s z \pmod{\lambda_s}$ and $\beta(u) = U^T \Delta_p u \pmod{\lambda_p}$. If $\mathrm{P}_*(u)$ and $\mathrm{T}_*(z)$ are aliased then $\alpha(z) = \gamma \beta(u)$ for some constant γ.

Let $K_* = \Delta_p^{-1} K \Delta_s$. If $K_* z$ is a valid plot vector, then $\mathrm{T}_*(z)$ is estimated by $\mathrm{P}_*(K_* z)$. $K_* z$ will be a valid plot vector if $p_j k_{ij}$ is divisible by s_i, for instance. Suppose that there is a one-to-one correspondence between plot and treatment factors in a generalized cyclic design. If K has as its rows $(10 \ldots 0)$ and the generators of the subgroup, that is of the starter block, then K^T is a key matrix in which $p_j k_{ij}$ is divisible by s_i. The designs in Examples 18 and 20 illustrate this.

Example 20 (concluded). Here

$$K_* = \begin{bmatrix} 1 & 0 & 0 \\ 0 & 1 & 0 \\ 2 & 0 & 1 \end{bmatrix}, \qquad K_*^{-1} = \begin{bmatrix} 1 & 0 & 0 \\ 0 & 1 & 0 \\ -2 & 0 & 1 \end{bmatrix}$$

and $\mathrm{P}_*((100)^T) = \mathrm{P}_*(K_* z) = \mathrm{T}_*(z) = \mathrm{T}_*(K^{-1}(100)^T) = \mathrm{T}_*((104)^T)$. In the case of a single replicate incomplete block design we have now identified the two degrees of freedom confounded with blocks.

If the first factor was the rows of the design, the second factor the columns of the design and the third factor the plots, then we might want to know what is confounded with columns and with the interaction of rows and columns. These are given by $\mathrm{P}_*((010)^T) = \mathrm{T}_*(K_*^{-1}(010)^T) = \mathrm{T}_*((010)^T)$ and $\mathrm{P}_*((110)^T) = \mathrm{T}_*(K_*^{-1}(110)^T) = \mathrm{T}_*((114)^T)$, respectively. □

10.7. References and comments

The results in Section 10.2 are due to Kurkjian and Zelen (1962, 1963).

The results in Section 10.3 are given in Searle (1971). In particular, Theorem 7 is Corollary 2s.1 of Chapter 2, Theorem 8 is Theorem 4s of Chapter 2, and the results on testable hypotheses are to be found in Chapter 5. Generalized inverses are discussed by Searle (1971) in Chapter 1; the definition we have used ensures that the rank of F^- equals the rank of F (Rohde (1966)).

The results in Section 10.4 are due to Cotter (1974), who also gives values of a_{jf} for $s = 4$, 5 or 7 and $m = 3$, 4, 5, or 6. Some of these values are given in Exercise 10.4.

The results in Section 10.5 are due to John and Dean (1975) and Dean and John (1975) except Theorem 17, which is due to Dean and Lewis (1980). Dean and John (1975) give a table which lists, for various values of $n \leq 5$, $v \leq 50$, $k \leq 30$, and $s_i \leq 7$, about 50 suitable subgroups and the interactions confounded for each. Lewis (1982) has extended these to $n \leq 7$ and $v \leq 200$. Exercise 10.5 is from Lewis (1982).

The DSIGN method is a general method of factorial design construction that was introduced by Patterson (1965) and has been studied recently by Patterson (1976), Bailey (1977), Bailey, Gilchrist, and Patterson (1977), and Patterson and Bailey (1978). We gave it here in a simplified form for incomplete block designs; as Bailey notes, it can be used to construct designs with any of the simple block structures defined by Nelder (1965). One extension, to row and column designs, forms Exercise 10.7. This is an example in Patterson and Bailey (1978). Another systematic approach to the construction of confounding schemes is described by Collings (1984).

A design is said to have *orthogonal factorial structure* if the treatment sum of squares, adjusted for blocks, can be partitioned into interaction sums of squares which are independent. The designs described in Sections 10.4, 10.5, and 10.6 all have this property.

Cotter, John, and Smith (1973) showed that a block design has orthogonal factorial structure if

$$(rI - (1/k)AA^T) = \sum_{i_1} \dots \sum_{i_{n-1}} (E_{i_1} \times \dots \times E_{i_{n-1}} \times G_{i_1 \dots i_{n-1}}),$$

where $G_{i_1 \ldots i_{n-1}}$ is an $s_n \times s_n$ matrix with row and column sums equal.

Mukerjee (1979) has shown that a proper and equi-replicate block design will have orthogonal factorial structure if and only if AA^T may be written as the sum of matrices of the form $\alpha(N_1 \times \ldots \times N_n)$, where N_j is an $s_j \times s_j$ matrix with constant row and column sums and α is some real number.

Constructions for other designs having this structure are given by John (1981), Mukerjee (1981), and Gupta (1983).

Lewis and Dean (1980) describe an experiment in which a generalized cyclic design was used.

Exercises

10.1. For the 2×3 factorial design given in Example 10.2, let $x = (10)$ and evaluate $M^x\tau$.

10.2. Show that

$$(T^T - (1/k)AB^T)\eta = (rI - (1/k)AA^T)\tau$$

and that

$$(rI - (1/k)AA^T)^- (T^T - (1/k)AB^T)(T - (1/k)BA^T)(rI - (1/k)AA^T)^- = (rI - (1/k)AA^T)^-.$$

10.3. Suppose that K^T is an $s \times v$ matrix of row rank t and that $K^T = \begin{bmatrix} D^T \\ E^T \end{bmatrix}$, where D^T is a $t \times v$ matrix of full row rank. Let

$$Q_K = (K^T\hat{\tau})^T (K^T V^- K)^- K^T\hat{\tau}$$

and

$$Q_D = (D^T\hat{\tau})^T (D^T V^- D)^{-1} D^T\hat{\tau}.$$

(i) Show that the value of Q_D is unaffected by permutations of the rows of D.

(ii) Verify that

$$\begin{bmatrix} (D^T V^- D)^{-1} & 0 \\ 0 & 0 \end{bmatrix}$$

is a generalized inverse of $K^T V^- K$. (Since D is of full row rank we can write $E^T = F^T D^T$ for some F^T.)

(iii) Using the generalized inverse in (ii), show that $Q_K = Q_D$.

10.4. Show that:

(i) if $s = 5$, $m = 3$, $y_1 = i_1 + i_2 + i_3 + 3$, $y_2 = i_1 + 3i_2 + 2i_3$, and $y_3 = i_1 + 4i_2 + 3i_2 + 3$, then the array with column vectors $(i_1, i_2, i_3, y_1, y_2, y_3)^T$ has strength 3;

(ii) if $s = 5, m = 4, y_1 = i_1 + i_2 + i_3 + i_4 + 2$ and $y_2 = i_1 + 2i_2 + 3i_3 + 4i_4 + 1$, then the array with column vectors $(i_1, i_2, i_3, i_4, y_1, y_2)$ has strength 2.

In the designs generated from these arrays, how many degrees of freedom of each interaction are confounded?

10.5. (i) Let $s_1 = s_2 = 2$, $s_3 = 4$. Consider the subgroup of $Z_2 \oplus Z_2 \oplus Z_4$ generated by 101 and 110. Let this subgroup be a starter block. Write down the distinct blocks of the generalized cyclic design developed from this starter block. How many degrees of freedom of each interaction are confounded with blocks?

(ii) Let $s_1 = 2$, $s_2 = 4$, $s_3 = 6$. Consider the subgroup of $Z_2 \oplus Z_4 \oplus Z_6$ generated by 011 and 123. Let this subgroup be a starter block. Write down the distinct blocks of the generalized cyclic design developed from this starter block. How many degrees of freedom of each interaction are confounded with blocks?

10.6. What is the key matrix which gives each of the designs in the previous question? Use the results in Section 10.6 to calculate what has been confounded in each design.

10.7. Consider a 5×5 factorial design conducted in a 5×5 array. There are two plot factors, the rows and the columns of the array. Suppose that

$$K^T = \begin{bmatrix} 1 & 1 \\ 1 & 2 \end{bmatrix}.$$

(i) Verify that the resulting layout is equivalent to a pair of MOLS.

(ii) Verify that $P_*((10)^T)$ and $T_*((12)^T)$, and $P_*((01)^T)$ and $T_*((14)^T)$ are aliased. Thus 8 degrees of freedom of the interaction have been confounded: 4 with the rows of the array and 4 with the columns of the array.

(iii) Give a key matrix which would result in the two main effects being confounded with rows and with columns.

11 Designs with partial balance

11.1. Partially balanced incomplete block designs

In previous chapters we have looked at some of the existence criteria for BIBDs; in many cases a BIBD cannot exist and so we look for designs with most, but not all, of the properties of a BIBD. We have already seen transversal designs (Chapters 6 and 7) and α-designs (Section 8.7), for both of which the property of balance is relaxed, and briefly considered pairwise balanced designs (Section 1.2), for which the property of constant block size is relaxed. In this chapter we look at partially balanced incomplete block designs (PBIBDs), which generalize transversal designs.

We begin by imposing a structure on the underlying set, although this structure may well be artificial (unlike the structure of the set of treatments in a factorial design). An *association scheme with m associate classes* on a v-set X is a family of m symmetric anti-reflexive binary relations on X such that:

(i) any two distinct elements of X are ith associates for exactly one value of i, where $1 \leq i \leq m$;

(ii) each element of X has n_i ith associates, $1 \leq i \leq m$;

(iii) for each i, $1 \leq i \leq m$, if x and y are ith associates, then there are p^i_{jl} elements of X which are both jth associates of x and lth associates of y. The numbers v, $n_i (1 \leq i \leq m)$, and p^i_{jl} $(1 \leq i, j, l \leq m)$ are called the *parameters of the association scheme.* We see that $p^i_{jl} = p^i_{lj}$. Often we write $P_i = (p^i_{jl})$, $i = 1, 2, \ldots, m$.

A *partially balanced incomplete block design with m associate classes* (PBIBD(m)) is a design based on a v-set X, with b blocks each of size k and with replication number r, such that there is an association scheme with m classes defined on X where, if elements x and y are ith associates, $1 \leq i \leq m$, then they occur together in precisely λ_i blocks. The numbers v, b, r, k, λ_i $(1 \leq i \leq m)$ are called the *parameters* of the PBIBD(m). We denote such a design by $PB[k, \lambda_1, \ldots, \lambda_m; v]$. A PBIBD (m) determines an association scheme but the converse is false.

Example 1. Let $v = 6, b = 8, r = 4, k = 3, \lambda_1 = \lambda_2 = 2, \lambda_3 = 1$. We place the six varieties in a 2×3 array. For any given variety, the first associates are those varieties in the same row, the second associates those in the same column, and the third associates the remaining varieties; see Table 11.1. Here $n_1 = n_3 = 2$, $n_2 = 1$.

Table 11.1

	Element	Associates: First	Second	Third
123 456 Array	1	2, 3	4	5, 6
	2	1, 3	5	4, 6
	3	1, 2	6	4, 5
	4	5, 6	1	2, 3
	5	4, 6	2	1, 3
	6	4, 5	3	1, 2

$$P_1 = \begin{bmatrix} 1 & 0 & 0 \\ 0 & 0 & 1 \\ 0 & 1 & 1 \end{bmatrix}, \quad P_2 = \begin{bmatrix} 0 & 0 & 2 \\ 0 & 0 & 0 \\ 2 & 0 & 0 \end{bmatrix}, \quad P_3 = \begin{bmatrix} 0 & 1 & 1 \\ 1 & 0 & 0 \\ 1 & 0 & 0 \end{bmatrix}$$

124 134 235 456
125 136 236 456

Design

Notice that even though the design based on the association scheme has $\lambda_1 = \lambda_2$, we cannot combine the first and second associate classes into one class. For suppose that we could, giving 1, 2, and 4 the first associates $\{2, 3, 4\}$, $\{1, 3, 5\}$ and $\{1, 5, 6\}$ respectively. Then $p_{11}^1 = 1$, for first associates 1 and 2, but $p_{11}^1 = 0$ for first associates 1 and 4. This contradicts requirement (iii) of the definition of an association scheme. □

If, however, $m \le 2$ and $\lambda_1 = \lambda_2$, then the design is a BIBD (see Exercise 11.1).

We define the $v \times b$ incidence matrix, A, of a PBIBD in the same way as we did for a BIBD. For a PBIBD we also define *association matrices.*

The association matrices $B_i = (b_{jl}^i)$, $1 \le i \le m$, $1 \le j, l \le v$, of a PBIBD(m) are $v \times v$ (0, 1) matrices given by

$$b_{jl}^i = \begin{cases} 1, & \text{if } j \text{ and } l \text{ are } i\text{th associates,} \\ 0, & \text{otherwise.} \end{cases}$$

Example 1 (continued). In Table 11.2 we give the incidence matrix and association matrices for the design in Table 11.1. □

For convenience in the following proof we say that each variety is the 0th associate of itself, and write $B_0 = I_v$. If δ_{ij} (the Kronecker delta) is defined by

$$\delta_{ij} = \begin{cases} 1, & i = j, \\ 0, & \text{otherwise,} \end{cases}$$

then $p_{ij}^0 = n_i\delta_{ij}$, $p_{0h}^i = \delta_{ih}$.

Table 11.2

$$B_1 = \begin{bmatrix} & 1 & 1 & & & \\ 1 & & 1 & & & \\ 1 & 1 & & & & \\ & & & & 1 & 1 \\ & & & 1 & & 1 \\ & & & 1 & 1 & \end{bmatrix} \qquad B_2 = \begin{bmatrix} & & & 1 & & \\ & & & & 1 & \\ & & & & & 1 \\ 1 & & & & & \\ & 1 & & & & \\ & & 1 & & & \end{bmatrix}$$

$$B_3 = \begin{bmatrix} & & & & 1 & 1 \\ & & & 1 & & 1 \\ & & & 1 & 1 & \\ & 1 & 1 & & & \\ 1 & & 1 & & & \\ 1 & 1 & & & & \end{bmatrix} \qquad A = \begin{bmatrix} 1 & 1 & 1 & 1 & & & & \\ 1 & 1 & & & 1 & 1 & & \\ & & 1 & 1 & 1 & 1 & & \\ 1 & & 1 & & & & 1 & 1 \\ & 1 & & & 1 & & 1 & 1 \\ & & & 1 & & 1 & 1 & 1 \end{bmatrix}$$

Lemma 2. *Let $B_0, B_1, \ldots, B_m$ be the association matrices for an m-associate class association scheme. Then:*

(i) $\sum_{i=0}^{m} B_i = J_v$; (11.1)

(ii) $B_i J_v = n_i J_v$; (11.2)

(iii) $B_i B_j = \sum_{h=0}^{m} p_{ij}^{h} B_h$; (11.3)

(iv) $\sum_u \sum_t p_{jh}^{u} p_{iu}^{t} B_t = B_i(B_j B_h) = (B_i B_j) B_h = \sum_u \sum_t p_{ij}^{u} p_{uh}^{t} B_t$. (11.4)

Proof. (i) Every variety is an ith associate of every other, for exactly one value of i.

(ii) Every variety has n_i ith associates.

(iii) The (x, y) element of $B_i B_j$ is the number of symbols which are ith associates of x and jth associates of y. Of course, x and y are themselves hth associates for precisely one h, so the result follows.

(iv) From (iii),

$$\begin{aligned} B_i(B_j B_h) &= B_i\left(\sum_{u=0}^{m} p_{jh}^{u} B_u \right) \\ &= \sum_{u=0}^{m} p_{jh}^{u} B_i B_u \\ &= \sum_u p_{jh}^{u} \sum_t p_{iu}^{t} B_t \\ &= \sum_u \sum_t p_{jh}^{u} p_{iu}^{t} B_t \end{aligned}$$

and

$$(B_i B_j)B_h = \left(\sum_u p^u_{ij} B_u\right) B_h$$

$$= \sum_u \sum_t p^u_{ij} p^t_{uh} B_t, \text{ as required.}$$ □

Lemma 3. *Let A be the incidence matrix of a $PB[k, \lambda_1, \ldots, \lambda_m; v]$ design and let $B_0, B_1, \ldots, B_m$ be the corresponding association matrices. Then*

$$AA^T = rI + \sum_{u=1}^{m} \lambda_u B_u \tag{11.5}$$

and

$$J_v A = kJ_{v \times b}. \tag{11.6}$$

Conversely, if A is a $v \times b$ (0, 1) matrix which satisfies eqns (11.5) *and* (11.6) *where $B_0, B_1, B_2, \ldots, B_m$ satisfy the conditions of Lemma* 2, *then, provided $k < v$, A is the incidence matrix of a $PB[k, \lambda_1, \lambda_2, \ldots, \lambda_m; v]$ design.*

Proof. (i) Suppose that A is the incidence matrix of a $PB[k, \lambda_1, \ldots, \lambda_m; v]$ design. Equation (11.6) holds, since the design has k varieties per block, and hence A has k ones per column (and all other entries are zeros).

The (i, j) entry of AA^T is the dot product of rows i and j of A, and so counts the number of times that varieties i and j appear in the same block. This is r, if $i = j$, and λ_u if i and j are uth associates. Hence eqn (11.5) is true.

(ii) Now suppose that A is a $v \times b$ (0, 1) matrix which satisfies eqns (11.5) and (11.6), where $B_0, B_1, \ldots, B_m$ satisfy the conditions of Lemma 2. Define a block design D with varieties $c_1, c_2, \ldots, c_v$ and blocks $E_1, \ldots, E_b$, where

$$c_i \in E_j \text{ if and only if } a_{ij} = 1.$$

A straightforward check shows that D is a PBIBD with the required parameters. □

Also note that $AJ_b = rJ_{v \times b}$ since each variety appears in r blocks; so A has r ones per row (and all other entries are zeros).

Lemma 4. *The parameters of a PBIBD(m) satisfy:*

(i) $vr = bk;$ (11.7)

(ii) $\sum_{i=1}^{m} n_i = v - 1;$ (11.8)

(iii) $\sum_{i=1}^{m} n_i \lambda_i = r(k-1);$ (11.9)

(iv) $\sum_{u=0}^{m} p^h_{ju} = n_j;$ (11.10)

(v) $n_i p^i_{jh} = n_j p^j_{ih}.$ (11.11)

Proof. (i) Now

$$J_v(AJ_b) = J_v r J_{v\times b} = rvJ_{v\times b}$$

and

$$(J_v A)J_b = kJ_{v\times b}J_b = kbJ_{v\times b}$$

so $vr = bk$.

(ii) Here

$$(J_v - I_v)J_v = vJ_v - J_v = (v-1)J_v$$

and

$$\left(\sum_{u=1}^{m} B_u\right)J_v = \sum_{u=1}^{m} B_u J_v = \sum_{u=1}^{m} n_u J_v$$

so the result follows.

(iii) Again

$$AA^T J_v = AkJ_{b\times v} = krJ_v$$

and

$$\left(rI + \sum_{u=1}^{m} \lambda_u B_u\right)J_v = rJ_v + \sum_{u=1}^{m} \lambda_u n_u J_v.$$

Hence

$$(kr - r)J_v = \sum_{u=1}^{m} \lambda_u n_u J_v$$

so the result follows.

(iv) Here note that

$$B_j J_v = n_j J_v = \sum_{h=0}^{m} n_j B_h$$

and that

$$B_j J_v = B_j \sum_{u=0}^{m} B_u = \sum_u \sum_h p^h_{ju} B_h.$$

Equating coefficients we get

$$n_j = \sum_u p^h_{ju}.$$

(v) Let x be any element of X, let G_i be the set of ith associates of x and let G_j be the set of jth associates of x. Let y be any element of G_i. There are p^i_{jh} elements of G_j which are hth associates of y and there are n_i possible choices for y. Any element in G_j has p^j_{ih} hth associates in G_i. There are n_j elements in G_j. Thus $n_i p^i_{jh} = n_j p^j_{ih}$, counting hth associate pairs. □

Example 1 (concluded). Consider the association scheme. Let $x = 1$. Then $G_1 = \{2, 3\}$, $G_3 = \{5, 6\}$. If $y = 2$, then 5 is a second associate of 2, and 6 is a third associate of 2. If $y = 3$ then 5 is a third associate and 6 is a second associate. Thus

$$n_1 p^1_{32} = 2 \times 1 = 2 \quad \text{and} \quad n_3 p^3_{12} = 2 \times 1 = 2. \qquad \square$$

11.2. Two-associate-class PBIBDs

Of all the PBIBDs, those with two associate classes have been studied the most, because they are both the simplest and the most useful kind. If a BIBD with parameters appropriate for a given experiment does not exist, then usually a suitable PBIBD(2) can be found.

PBIBD(2)s are usually classified, according to properties of their association schemes, into six types as follows:

(a) group divisible;
(b) triangular;
(c) Latin-square-type;
(d) cyclic;
(e) partial geometry type;
(f) miscellaneous.

We look in some detail at group divisible and triangular designs, which are of both theoretical and practical interest. Group divisible designs with $\lambda_1 = 0$ have been used in proving the existence of various classes of BIBDs; see Chapter 13. Triangular designs occur as duals of the residual designs of SBIBDs with $\lambda = 2$; see Section 11.6. We also look briefly at Latin-square-type and cyclic designs. The other two classes are listed for completeness; no general methods are known for constructing designs in these classes.

As part of the classification depends on the eigenvalues of AA^T we begin by evaluating them.

Theorem 5. *Let A be the incidence matrix of a* $PB[k, \lambda_1, \lambda_2; v]$ *design. Let*

$$\gamma = p^2_{12} - p^1_{12}, \quad \beta = p^2_{12} + p^1_{12}, \quad \Delta = \gamma^2 + 2\beta + 1.$$

Then the eigenvalues of AA^T *are*

$$rk \text{ and } \theta_i = r + \tfrac{1}{2}[(\lambda_1 - \lambda_2)(\gamma + (-1)^i \sqrt{\Delta}) - (\lambda_1 + \lambda_2)], \quad i = 1, 2,$$

with multiplicities

$$1 \text{ and } \alpha_i = (n_1 + n_2)/2 + (-1)^i [(n_2 - n_1) - \gamma(n_1 + n_2)]/2\sqrt{\Delta}, \quad i = 1, 2,$$

respectively.

Proof. We know that $AA^T = rI + \lambda_1 B_1 + \lambda_2 B_2$ and $n_1\lambda_1 + n_2\lambda_2 = r(k-1)$. Hence

$$AA^T j = (r + r(k-1))j = rkj$$

so rk is an eigenvalue of AA^T.

By eqn (11.1) $B_2 = J - I - B_1$; thus $AA^T = (r-\lambda_2)I + (\lambda_1 - \lambda_2)B_1 + \lambda_2 J$. Also, if x is any other eigenvector of AA^T then $x^T j = 0$, since AA^T is symmetric. Now, let x be an eigenvector of AA^T with corresponding eigenvalue θ. Then

$$AA^T x = \theta x = (r-\lambda_2)x + (\lambda_1 - \lambda_2)B_1 x \tag{11.12}$$

and

$$\begin{aligned} AA^T AA^T x = \theta^2 x &= ((r-\lambda_2)I + \lambda_2 J + (\lambda_1 - \lambda_2)B_1)^2 x \\ &= \{(r-\lambda_2)^2 I + 2\lambda_2(r-\lambda_2)J + \lambda_2^2 vJ + 2(\lambda_1-\lambda_2)(r-\lambda_2)B_1 \\ &\quad + 2(\lambda_1-\lambda_2)\lambda_2 JB_1 + (\lambda_1-\lambda_2)^2 B_1^2\}x \\ &= I((r-\lambda_2)^2 + (n_1 - p_{11}^2)(\lambda_1-\lambda_2)^2)x \\ &\quad + (2(\lambda_1-\lambda_2)(r-\lambda_2) + (p_{11}^1 - p_{11}^2)(\lambda_1-\lambda_2)^2)B_1 x \end{aligned} \tag{11.13}$$

by eqns (11.2) and (11.3). By (11.12),

$$(\lambda_1 - \lambda_2)B_1 x = \theta x - (r-\lambda_2)x.$$

Substituting into (11.13) gives

$$\begin{aligned} \theta^2 x = ((r-\lambda_2)^2 + p_{12}^2(\lambda_1-\lambda_2)^2)x \\ + (2(r-\lambda_2) + (p_{11}^1 - p_{11}^2)(\lambda_1-\lambda_2))(\theta - (r-\lambda_2))x, \end{aligned}$$

since $p_{12}^2 = n_1 - p_{11}^2$ by eqn (11.10). Hence θ satisfies the quadratic equation

$$\begin{aligned} \theta^2 - \theta(2(r-\lambda_2) + (p_{11}^1 - p_{11}^2)(\lambda_1-\lambda_2)) + (p_{11}^1 - p_{11}^2)(\lambda_1-\lambda_2)(r-\lambda_2) \\ + (r-\lambda_2)^2 - p_{12}^2(\lambda_1-\lambda_2)^2 = 0. \end{aligned}$$

The discriminant of this quadratic is

$$\begin{aligned} \{2(r-\lambda_2) + (p_{11}^1 - p_{11}^2)(\lambda_1-\lambda_2)\}^2 - 4\{(p_{11}^1 - p_{11}^2)(\lambda_1-\lambda_2)(r-\lambda_2) \\ + (r-\lambda_2)^2 - p_{12}^2(\lambda_1-\lambda_2)^2\} = (\lambda_1-\lambda_2)^2((p_{11}^1 - p_{11}^2)^2 + 4p_{12}^2). \end{aligned}$$

By eqn (11.3), $p_{11}^1 - p_{11}^2 = p_{12}^2 - p_{12}^1 - 1 = \gamma - 1$ and the discriminant becomes

$(\lambda_1-\lambda_2)^2((\gamma-1)^2 + 2(\beta+\gamma)) = (\lambda_1-\lambda_2)^2\Delta$. Also,

$$\begin{aligned} 2(r-\lambda_2) + (\lambda_1-\lambda_2)(p_{11}^1 - p_{11}^2) &= 2(r-\lambda_2) + (\lambda_1-\lambda_2)(\gamma-1) \\ &= 2r + (\lambda_1-\lambda_2)\gamma - (\lambda_1+\lambda_2) \end{aligned}$$

and we have that

$$\theta_i = r + \tfrac{1}{2}[(\lambda_1-\lambda_2)(\gamma + (-1)^i\sqrt{\Delta}) - (\lambda_1+\lambda_2)], \quad i = 1, 2, \text{ as required.}$$

Let α_i be the multiplicity of the root θ_i, $i = 1, 2$, of AA^T. Then

$$\alpha_1 + \alpha_2 = v - 1$$

and

$$\operatorname{tr}(AA^T) = vr = rk + \alpha_1\theta_1 + \alpha_2\theta_2.$$

Hence

$$\alpha_2 = \frac{rv - rk - (v-1)\theta_1}{\theta_2 - \theta_1}.$$

Now

$$\theta_2 - \theta_1 = (\lambda_1 - \lambda_2)\sqrt{\Delta}$$

so

$$\alpha_2 = \frac{2r - 2rk - (v-1)(\lambda_1 - \lambda_2)(\gamma - \sqrt{\Delta}) + (v-1)(\lambda_1 + \lambda_2)}{2(\lambda_1 - \lambda_2)\sqrt{\Delta}}$$

$$= \frac{-2(n_1\lambda_1 + n_2\lambda_2) - (n_1+n_2)(\lambda_1 - \lambda_2)(\gamma - \sqrt{\Delta}) + (n_1+n_2)(\lambda_1+\lambda_2)}{2(\lambda_1 - \lambda_2)\sqrt{\Delta}}$$

from (11.9),

$$= \frac{n_2 - n_1 - \gamma(n_1+n_2) + \sqrt{\Delta}(n_1+n_2)}{2\sqrt{\Delta}}$$

$$= \frac{n_1+n_2}{2} + \frac{n_2 - n_1 - \gamma(n_1+n_2)}{2\sqrt{\Delta}},$$

and

$$\begin{aligned}\alpha_1 &= v - 1 - \alpha_2\\ &= n_1 + n_2 - \alpha_2\\ &= \frac{n_1+n_2}{2} - \frac{n_2 - n_1 - \gamma(n_1+n_2)}{2\sqrt{\Delta}}.\end{aligned}$$ □

Example 6. Consider the design of Table 11.3, where the first associates of any variety are the varieties in the same row of the array and all other varieties are the second associates. The parameters of the association scheme and of the design are

$$v = 6, b = 14, r = 7, k = 3, \lambda_1 = 2, \lambda_2 = 3, n_1 = 1, n_2 = 4,$$
$$p_{11}^1 = 0, p_{12}^1 = 0, p_{22}^1 = 4, p_{11}^2 = 0, p_{12}^2 = 1, p_{22}^2 = 2.$$

Hence

$$\gamma = p_{12}^2 - p_{12}^1 = 1 - 0 = 1, \ \beta = p_{12}^1 + p_{12}^2 = 0 + 1 = 1,$$
$$\Delta = \gamma^2 + 2\beta + 1 = 1 + 2 + 1 = 4,$$
$$\theta_1 = 7 + \tfrac{1}{2}\{(2-3)(1-2) - (2+3)\} = 5,$$
$$\theta_2 = 7 + \tfrac{1}{2}\{(2-3)(1+2) - (2+3)\} = 3,$$

$$\alpha_1 = \frac{5}{2} - \frac{4-1-5}{4} = 3,$$

$$\alpha_2 = 2.$$

Table 11.3 also contains the expansion of the determinant of $AA^T - \theta I$. The general principle is that if no row or column has all entries but one equal to zero, then all rows are added to the first, and then the first column is subtracted from all the others. $\square$

Table 11.3

Array		Blocks				
1 4	1 2 3	3 4 5	1 2 3	3 4 6	2 3 6
2 5	1 4 6	1 3 5	1 5 6	1 2 4	3 4 5
3 6	2 5 6	2 4 6	2 4 5	1 5 6	

$$A = \begin{bmatrix} 1 & 1 & & & 1 & & 1 & 1 & & & 1 & 1 & & \\ 1 & & 1 & & & 1 & 1 & & 1 & & 1 & & 1 & \\ 1 & & & 1 & 1 & & 1 & & & 1 & & & 1 & 1 \\ & 1 & & 1 & & 1 & & & 1 & 1 & 1 & & & 1 \\ & & 1 & 1 & 1 & & & 1 & 1 & & & 1 & & 1 \\ & 1 & 1 & & & 1 & & 1 & & 1 & & 1 & 1 & \end{bmatrix} \qquad AA^T = \begin{bmatrix} 7 & 3 & 3 & 2 & 3 & 3 \\ 3 & 7 & 3 & 3 & 2 & 3 \\ 3 & 3 & 7 & 3 & 3 & 2 \\ 2 & 3 & 3 & 7 & 3 & 3 \\ 3 & 2 & 3 & 3 & 7 & 3 \\ 3 & 3 & 2 & 3 & 3 & 7 \end{bmatrix}$$

$$|AA^T - \theta I| = (21-\theta) \begin{vmatrix} 4-\theta & 0 & 0 & -1 & 0 \\ 0 & 4-\theta & 0 & 0 & -1 \\ 1 & 1 & 5-\theta & 1 & 1 \\ -1 & 0 & 0 & 4-\theta & 0 \\ 0 & -1 & 0 & 0 & 4-\theta \end{vmatrix}$$

$$= (21-\theta)(5-\theta) \begin{vmatrix} 3-\theta & 3-\theta & 3-\theta & 3-\theta \\ 0 & 4-\theta & 0 & -1 \\ -1 & 0 & 4-\theta & 0 \\ 0 & -1 & 0 & 4-\theta \end{vmatrix}$$

$$= (21-\theta)(5-\theta)^2(3-\theta) \begin{vmatrix} 4-\theta & -1 \\ -1 & 4-\theta \end{vmatrix}$$

$$= (21-\theta)(5-\theta)^3(3-\theta)^2$$

11.3. Group divisible PBIBDs

Let X be a set of v varieties such that

$$X = \bigcup_{i=1}^{m} G_i, \; |G_i| = n \quad \text{for} \quad 1 \le i \le m, \; G_i \cap G_j = \varnothing \quad \text{for } i \ne j.$$

The G_i's are called *groups* (though they are not groups in the usual algebraic sense) and an association scheme defined on X is said to be *group divisible* if the varieties in the same group are first associates and those in different groups are second associates.

Designs in which the underlying association scheme is group divisible are called *group divisible* (*GD*) *designs*. Thus the transversal designs of Chapter 6 are group divisible designs with $\lambda_1 = 0$ and $\lambda_2 = 1$.

The parameters of the association scheme are $n_1 = n-1, n_2 = n(m-1)$ and

$$P_1 = \begin{bmatrix} n-2 & 0 \\ 0 & n(m-1) \end{bmatrix}, \quad P_2 = \begin{bmatrix} 0 & n-1 \\ n-1 & n(m-2) \end{bmatrix}.$$

The eigenvalues of AA^T are $rk, r-\lambda_1, rk-v\lambda_2$ with multiplicities $1, m(n-1)$ and $m-1$ respectively. The class of group divisible designs is further subdivided on the basis of the values of $r-\lambda_1$ and $rk-v\lambda_2$, as follows:

(i) singular (S) if $r = \lambda_1$;
(ii) semiregular (SR) if $r > \lambda_1, rk - v\lambda_2 = 0$;
(iii) regular (R) if $r > \lambda_1, rk > v\lambda_2$.

Hence the design of Example 6 is a regular group divisible design.

If $\lambda_2 = 0$ then varieties in different groups never occur together and the design consists of unions of designs, each constructed on the varieties of a group. The design is said to be *disconnected* and the difference of effects of two varieties in different groups is not estimable (that is, not all elementary contrasts are estimable).

Theorem 7. *The parameters of a group divisible association scheme uniquely determine the association scheme.*

Proof. Let x be any variety and let $x_1, x_2, \ldots, x_{n-1}$ be the first associates of x. Since $p_{11}^1 = n-2$, the first associates of x_1, other than x, are $x_2, \ldots, x_{n-1}$. Now $p_{12}^1 = 0$; so we can subdivide the set of mn varieties into m groups, where each group contains varieties which are first associates. Varieties in different groups are second associates. □

Theorem 8. *The existence of a singular $PB[kn, r, \lambda; mn]$ design is equivalent to the existence of a $B[k, \lambda; m]$ design.*

Proof. Consider a $B[k, \lambda; m]$ design constructed on the set $X = \{1, 2, \ldots, m\}$. In each block of the $B[k, \lambda; m]$ design replace all occurrences of i by $i_1, i_2, \ldots, i_n$. The resulting design is a group divisible $PB[kn, r, \lambda; mn]$ design, and so is singular by definition.

Conversely, given a singular $PB[kn, r, \lambda; mn]$ design, we know that $\lambda_1 = r$ and so any block which contains at least one variety of a group must contain all the varieties of that group. Replacing each group by a single symbol gives a $B[k, \lambda; m]$ design. □

Corollary 8.1. *In a singular $PB[kn, r, \lambda; mn]$ design, $b \geq m$.*

Proof. This is just Fisher's inequality for the equivalent $B[k, \lambda; m]$ design. □

Example 9. In Table 11.4, we give a $B[3, 1; 7]$ design and a singular $PB[6, 3, 1; 14]$ design constructed from it. The groups are $\{i, i+7\}$ mod 14. □

Table 11.4

$B[3,1;7]$			$PB[6,3,1;14]$					
1	2	4	1	8	2	9	4	11
2	3	5	2	9	3	10	5	12
3	4	6	3	10	4	11	6	13
4	5	7	4	11	5	12	7	14
5	6	1	5	12	6	13	1	8
6	7	2	6	13	7	14	2	9
7	1	3	7	14	1	8	3	10

Theorem 10. *In a semi-regular group divisible design, $m|k$. Every block has k/m varieties from each group.*

Proof. Let x_{ij} be the number of varieties from the ith group in the jth block of the design. Then, counting occurrences of varieties gives

$$\sum_j x_{ij} = nr \tag{11.14}$$

and counting occurrences of pairs of varieties gives $\sum_j \binom{x_{ij}}{2} = \binom{n}{2}\lambda_1$; so

$$\sum_j x_{ij}(x_{ij}-1) = n(n-1)\lambda_1. \tag{11.15}$$

Now define

$$\bar{x}_i = \left(\sum_j x_{ij}\right)/b = nr/b = mnr/mb = vr/mb = k/m. \tag{11.16}$$

Then

$$\begin{aligned}\sum_j (x_{ij}-\bar{x}_i)^2 &= \sum_j x_{ij}(x_{ij}-1) + (1-2\bar{x}_i)\sum_j x_{ij} + \sum_j \bar{x}_i^2 \\ &= n(n-1)\lambda_1 + \left(1-\frac{2k}{m}\right)nr + b\frac{k^2}{m^2}\end{aligned}$$

by (11.14), (11.15), and (11.16). But

$$\frac{bk^2}{m^2} = \frac{vrk}{m^2} = \frac{mnrk}{m^2} = \frac{nrk}{m};$$

so

$$\sum_j (x_{ij} - \bar{x}_i)^2 = n(n-1)\lambda_1 + nr - (nrk/m). \tag{11.17}$$

By eqn (11.9), we have

$$n_1\lambda_1 + n_2\lambda_2 + r = rk.$$

Since $n_1 = n - 1$ and $n_2 = n(m-1)$,

$$\begin{aligned}(n-1)\lambda_1 + r &= rk - n(m-1)\lambda_2\\ &= rk - v\lambda_2 + n\lambda_2\\ &= n\lambda_2, \quad \text{since } rk = v\lambda_2.\end{aligned}$$

Substituting into eqn (11.17) gives

$$\begin{aligned}\sum_j (x_{ij} - \bar{x}_i)^2 &= n^2\lambda_2 - (nrk/m)\\ &= \frac{n}{m}(mn\lambda_2 - rk) = \frac{n}{m}(v\lambda_2 - rk) = 0.\end{aligned}$$

Hence $x_{ij} = \bar{x}_i$ for $j = 1, 2, \ldots, b$. The same argument holds for each of the n groups. Thus k/m is an integer. □

Theorem 11. *The dual of an affine resolvable $B[\beta(\gamma(\beta-1)+1), \beta\gamma+1; \beta k]$ design is a semi-regular group divisible design with*

$$v = \beta(\gamma\beta^2 + \beta + 1), \quad b = \beta^2(\gamma(\beta-1)+1), \quad r = \beta(\gamma(\beta-1)+1),$$
$$k = \beta^2\gamma + \beta + 1, \quad m = \beta^2\gamma + \beta + 1, \quad n, \quad \lambda_1 = 0, \quad \lambda_2 = \gamma(\beta-1)+1.$$

Proof. This follows directly from the proof of Theorem 8.2.11. □

Example 12. In Table 11.5 we give a $B[3, 1; 9]$ design and its dual. □

Table 11.5

1: 1 2 3	4: 1 4 7	7: 1 5 9	A: 1 6 8
2: 4 5 6	5: 2 5 8	8: 2 6 7	B: 2 4 9
3: 7 8 9	6: 3 6 9	9: 3 4 8	C: 3 5 7

$B[3, 1; 9]$

1 4 7 A	2 4 9 B	3 4 8 C
1 5 8 B	2 5 7 C	3 5 9 A
1 6 9 C	2 6 8 A	3 6 7 B

$SR\ PB[4, 0, 1; 12]$

Theorem 13. *The existence of a $B[k, 1; v]$ design implies the existence of a group divisible $PB[k, 0, 1; v-1]$ design.*

Proof. Consider the r blocks containing the variety x. Delete x from each of these r blocks, giving $(k-1)$-sets which form the groups of the design. The remaining $b-r$ blocks are the blocks of the design and the varieties are those of the original design, other than x. □

Example 14. Consider the design of Table 11.5. Let $x = 1$. Then the groups are $\{2, 3\}, \{4, 7\}, \{5, 9\}$, and $\{6, 8\}$ and the blocks numbered 2, 3, 5, 6, 8, 9, B, C form a $PB[3,0,1;8]$ design. It is regular since $r>\lambda_1$ and $rk=9>8=v\lambda_2$. □

The technique of Theorem 13 is sometimes called 'block cutting'.

The next result is a straightforward generalization of Theorem 3.2.4.

Theorem 15. *Let G be an abelian group of order m, and for each element $x \in G$, let there be defined n symbols $x_1, x_2, \ldots, x_n$, giving a set X of mn symbols altogether; symbols with the same subscript are said to form a class. Suppose that s k-subsets $S_1, \ldots, S_s$ and one n-subset T of X are chosen so that:*

(i) *the $n(n-1)$ differences arising from T are all distinct;*

(ii) *among the ks symbols occurring in the s k-subsets, exactly r' symbols belong to each of the n classes;*

(iii) *the differences from the s subsets are symmetrically repeated, each occurring λ_2 times, except those arising from T which occur λ_1 times.*

Then, by adding each element of G in turn to each of the sets $S_1, \ldots, S_s$, we develop the blocks of a group divisible design with parameters

$$v = mn, b = ms, r = r', k, \lambda_1, \lambda_2;$$

the groups are obtained by developing T. □

Example 16. Let $G = Z_7, S_1 = \{0_1, 1_2, 2_2, 4_2\}, S_2 = \{0_2, 1_1, 2_1, 4_1\}$, $T = \{0_1, 0_2\}$. The fourteen blocks of the $PB[4, 0, 1; 14]$ design are given in Table 11.6. □

Table 11.6

0_1	1_2	2_2	4_2	0_2	1_1	2_1	4_1
1_1	2_2	3_2	5_2	1_2	2_1	3_1	5_1
2_1	3_2	4_2	6_2	2_2	3_1	4_1	6_1
3_1	4_2	5_2	0_2	3_2	4_1	5_1	0_1
4_1	5_2	6_2	1_2	4_2	5_1	6_1	1_1
5_1	6_2	0_2	2_2	5_2	6_1	0_1	2_1
6_1	0_2	1_2	3_2	6_2	0_1	1_1	3_1

The next result shows one way of using finite geometries to construct *GD* designs.

Theorem 17. *If q is a prime or a prime power, a semi-regular GD design exists with*

$$v = mq, \quad b = q^3, \quad r = q^2, \quad k = m, \quad m, \quad n = q, \quad \lambda_1 = 0, \quad \lambda_2 = q$$

where $m \leq q^2 + q + 1$.

Proof. Take the geometry $PG(3, q)$ and let x be one of its points. Using the results of Theorem 8.6.29 we know there are $q^2 + q + 1$ lines through x, each containing q other points. Choose any m of these lines and let the points on these lines, other than x, correspond to the varieties of the design. The groups are determined by these lines. The planes not passing through x are the blocks of the design. □

Example 18. Let $q = 2$. $PG(3, 2)$ is given in Table 8.6.13. Let $x = 0001$ and let $m = 5$. The groups and blocks are given in Table 11.7. □

Table 11.7. *GD PB*[5, 0, 2; 10]

Groups	Blocks
0010, 0011	0010, 0100, 0110, 1000, 1010
0100, 0101	0010, 0100, 0110, 1001, 1011
0110, 0111	0010, 0101, 0111, 1000, 1010
1000, 1001	0010, 0101, 0111, 1001, 1011
1010, 1011	0011, 0100, 0111, 1000, 1011
	0011, 0100, 0111, 1001, 1010
	0011, 0101, 0110, 1000, 1011
	0011, 0101, 0110, 1001, 1010

Many regular *GD* designs have duals which are regular *GD* designs. If the dual of a regular *GD* design is isomorphic to the original design then the design is said to be *truly self-dual.*

Example 19. In Table 11.8 we give a $PB[4, 3, 1; 9]$ design and its dual. The groups are $\{i, i+3, i+6\}$ (mod 9). The isomorphism (23) (56) (89) maps the dual design onto the original design. □

Table 11.8

Original			Dual		
1 4 7 2	4 7 1 5	7 1 4 8	1 4 7 9	1 3 4 7	1 4 6 7
2 5 8 3	5 8 2 6	8 2 5 9	1 2 5 8	2 4 5 8	2 5 7 8
3 6 9 4	6 9 3 7	9 3 6 1	2 3 6 9	3 5 6 9	3 6 8 9

11.4. Triangular PBIBDs

Let X be a set of $v = n(n-1)/2$ varieties, $n \geq 5$. Arrange the varieties in a symmetrical $n \times n$ array with the diagonal entries blank. Then the association scheme is said to be *triangular* if the first associates of any variety are the varieties in the same row or column of the array, and all the other varieties are its second associates. The parameters of the association scheme are

$$n_1 = 2(n-2), \quad n_2 = (n-2)(n-3)/2$$

and

$$P_1 = \begin{bmatrix} n-2 & n-3 \\ n-3 & (n-3)(n-4)/2 \end{bmatrix}, \quad P_2 = \begin{bmatrix} 4 & 2n-8 \\ 2n-8 & (n-4)(n-5)/2 \end{bmatrix}.$$

Example 20. Let $n = 5$. The array for X, and a triangular $PB[5, 1, 2; 10]$ design, are given in Table 11.9. Note that by placing the variety A in the diagonal positions of the array and having as blocks the original six blocks, together with the five rows of the array, we obtain an $SB[5, 2; 11]$ design. □

Table 11.9. Triangular $PB[5, 1, 2; 10]$

Array

*	1	2	3	4
1	*	5	6	7
2	5	*	8	9
3	6	8	*	0
4	7	9	0	*

Blocks

0	1	2	7	8
0	1	3	5	9
0	2	4	5	6
1	4	6	8	9
2	3	6	7	9
3	4	5	7	8

The next result we prove is that, except for $n = 8$, the parameters of the triangular association scheme uniquely determine the association scheme. We begin by considering the form of possible submatrices of the symmetric matrix B_1. Recall that if M is a principal submatrix of B_1, then the least eigenvalue of M must be at least as large as the least eigenvalue of B_1, and that the corresponding eigenvectors must be orthogonal in the subspace corresponding to M.

Lemma 21. *Let B_1 be the first association matrix for an association scheme with the parameters of a triangular association scheme. Then*

$$B_1 B_1^T = B_1^2 = 2(n-2)I + (n-2)B_1 + 4(J - I - B_1),$$

and the eigenvalues of B_1 are $2n-4$, $n-4$ and -2 with multiplicities 1, $n-1$ and $v-n$ respectively.

Proof. The expression for B_1^2 follows directly from (11.3) and the parameters of the triangular association scheme. Now

$$B_1^2 = (2n-8)I + (n-6)B_1 + 4J \tag{11.18}$$

and the matrices in (11.18) can be simultaneously diagonalized. Also

$$B_1 j = 2(n-2)j \quad \text{and} \quad Jj = vj;$$

so $2n-4$ is an eigenvalue of B_1 and v is the only non-zero eigenvalue of J. Thus, if θ is any other eigenvalue of B_1, then θ corresponds to a zero eigenvalue of J. So

$$\theta^2 = (2n-8) + (n-6)\theta.$$

Hence

$$\theta = n-4 \text{ or } -2.$$

Let y be the multiplicity of $n-4$ and z be the multiplicity of -2. Then

$$\operatorname{tr}(B_1) = 0 = 2n-4+(n-4)y+(-2)z$$

and

$$y+z = v-1.$$

Hence

$$y = n-1 \text{ and } z = v-n.$$ □

The matrices referred to in the next five lemmata appear in Table 11.10.

Lemma 22. *For $n \neq 8$, B_1 does not contain A_1 as a principal submatrix.*

Table 11.10

$$A_1 = \begin{bmatrix} 0&0&0&1\\ 0&0&0&1\\ 0&0&0&1\\ 1&1&1&0 \end{bmatrix} \quad A_2 = \begin{bmatrix} 0&0&0&1&1\\ 0&0&0&1&1\\ 0&0&0&1&1\\ 1&1&1&0&0\\ 1&1&1&0&0 \end{bmatrix} \quad A_3 = \begin{bmatrix} 0&0&0&1&1\\ 0&0&0&1&1\\ 0&0&0&1&1\\ 1&1&1&0&1\\ 1&1&1&1&0 \end{bmatrix} \quad A_4 = \begin{bmatrix} 0&0&0&1&1\\ 0&0&0&1&1\\ 0&0&0&1&0\\ 1&1&1&0&0\\ 1&1&0&0&0 \end{bmatrix}$$

$$A_5 = \begin{bmatrix} 0&0&1&1&1&1\\ 0&0&1&1&1&1\\ 1&1&0&0&1&1\\ 1&1&0&0&1&1\\ 1&1&1&1&0&0\\ 1&1&1&1&0&0 \end{bmatrix} \quad A_6 = \begin{bmatrix} 0&0&1&1&1&1&1&1\\ 0&0&1&1&1&1&0&0\\ 1&1&0&0&1&1&1&1\\ 1&1&0&0&1&1&0&0\\ 1&1&1&1&0&0&1&0\\ 1&1&1&1&0&0&0&1\\ 1&0&1&0&1&0&0&x\\ 1&0&1&0&0&1&x&0 \end{bmatrix} \quad A_7 = \begin{bmatrix} 0&0&1&1&1&1&1&1\\ 0&0&1&1&1&1&0&0\\ 1&1&0&0&1&1&1&1\\ 1&1&0&0&1&1&0&0\\ 1&1&1&1&0&0&1&1\\ 1&1&1&1&0&0&0&0\\ 1&0&1&0&1&0&0&0\\ 1&0&1&0&1&0&0&0 \end{bmatrix}$$

Proof. Suppose that B_1 does contain A_1 as a principal submatrix. Label the rows and columns of B_1 which correspond to A_1 by $1, 2, 3, 4$, so that $1, 2$, and 3 are all first associates of 4, and second associates of each other. Suppose that some other treatment is a first associate of each of $1, 2$, and 3. Then it is either a first or a second associate of 4, and so B_1 has a principal submatrix equal to A_2 or A_3. But A_2 has an eigenvalue equal to $-\sqrt{6}$, which is less than -2, and hence A_2 cannot be a principal submatrix of B_1. A_3 does have an eigenvalue equal to -2, but with corresponding eigenvector equal to $(1, 1, 1, -1, -1)^T$, which is not orthogonal to j_5. Hence A_3 cannot be a principal submatrix of B_1, either. This means that 4 is the only variety which is a first associate of $1, 2$, and 3.

Now 1 and 2 are second associates, and since the association scheme is triangular they must have four first associates in common. One of these is variety 4, so there are three other varieties which are first associates of 1 and 2, but not 3. Similar statements are true for 1 and 3, and for 2 and 3. Thus there are nine additional varieties. If any of these nine were a second associate of 4, then B_1 would have a submatrix equal to A_4. But A_4 has an eigenvalue equal to -2.1358, and hence cannot occur as a submatrix of B_1. Thus 4 has at least twelve first associates, so $12 \leq 2n-4$, which rules out $n \leq 7$.

If $n \geq 9$, then there are $n-2$ varieties which are first associates of both variety 1 and variety 4. Three of these are also first associates of 2, and three of 3. Hence there are $n-8$ new first associates. Similar statements are true for 2 and 4, and for 3 and 4. Hence $2(n-2) \geq 12 + 3(n-8)$, which is impossible for $n \geq 9$. □

Lemma 23. *Let $n \neq 8$. If 1 and 2 are second associates, and if 3, 4, 5, 6 are first associates of both 1 and 2, then (after renumbering, if necessary) the principal submatrix of B_1 corresponding to varieties 1, 2, 3, 4, 5, 6, is A_5.*

Proof. Variety 3 has $2(n-2)$ first associates, two of which are 1 and 2. Suppose that x is a first associate of 3 and a second associate of both 1 and 2. Then the matrix of $1, 2, x, 3$ is just the matrix A_1, giving a contradiction. Hence a first associate of 3 must be a first associate of 1, or of 2, or of both. Let t be the number of first associates of 3 which are first associates of both 1 and 2. Then

$$t + (n-2-t) + (n-2-t) = 2(n-2) - 2,$$

so $t = 2$. These two must be among varieties 4, 5, 6; without loss of generality, we take them to be 5 and 6. Then 4 is a second associate of 3, and the Lemma follows. □

Lemma 24. *A_6 is not a principal submatrix of B_1, for $x = 0$ or 1, for $n \neq 8$.*

Proof. Suppose that A_6 is a principal submatrix of B_1, with $x = 0$ or 1. If $x = 0$, then 2, 7, 8 are pairwise second associates, and 3 is a first associate of each of them. This contradicts Lemma 22. If $x = 1$, then 6 and 7 are second associates, and 1, 3, 8 are first associates of both of them. This contradicts Lemma 23. □

Lemma 25. *A_7 is not a principal submatrix of B_1, for $n \neq 8$.*

Proof. Here 7 and 8 are second associates and 1, 3, and 5 are first associates of both, contradicting Lemma 23. □

Lemma 26. *The $2(n-2)$ first associates of any variety fall into two classes of $n-2$ varieties each, such that any two varieties in the same class are first associates of each other, provided that $n \neq 8$.*

Proof. Suppose that 1, 2, 3, 4, 5, 6 correspond to the submatrix A_5, so that both 5 and 6 are first associates of both 1 and 3. Since $p_{11}^1 = n-2$, there must be $n-4$ other varieties which are also first associates of both 1 and 3; without loss of generality let variety 7 be one of them. If 7 is a second associate of both 5 and 6, then the matrix of 5, 6, 7, 1 is just A_1, contradicting Lemma 22. Hence each of these $n-4$ varieties is a first associate of at least one of 5 and 6; without loss of generality, suppose that 7 and 8 are among these varieties, that 7 is a first associate of 5 and that 8 is a first associate of 6. But now 1, 2, 3, 4, 5, 6, 7, 8 correspond to a submatrix A_6 of B_1, contradicting Lemma 24. Hence, we may suppose that 7 and 8 are both first associates of 5, say. If 7 and 8 are second associates of each other, then 1, 2, 3, 4, 5, 6, 7, 8 correspond to a submatrix A_7 of B_1, contradicting Lemma 25. Hence 7 and 8 are first associates of each other. Altogether, 3, 5, and $n-4$ other varieties are first associates of 1, and are all mutually first associates.

Of the $p_{11}^1 = n-2$ first associates of both 1 and 4, variety 5 is in the class described above but 6 is not, and there are $n-4$ other varieties. By the same reasoning as above, these $n-4$ other varieties are mutually first associates. Further, since they are second associates of 4 they form a class disjoint from the previous one.

Hence the set of first associates of 1 (or of any variety) can be partitioned as required. □

We now show that the result of Lemma 26 is sufficient to characterize the triangular association scheme.

Theorem 27. *Let D be a partially balanced incomplete block design with two associate classes, and with $v = n(n-1)/2$ and $p_{11}^1 = n-2$, for $n \neq 8$. Then D has a triangular association scheme if and only if the $2(n-2)$ first associates of any variety fall into two classes so that the $n-2$ varieties of each class are mutual first associates.*

Proof. The necessity of the condition is obvious.

To prove sufficiency, consider variety 1. Let the two classes of first associates of variety 1 be $X = \{3, 5, x_1, \ldots, x_{n-4}\}$ and $Y = \{4, 6, y_1, \ldots, y_{n-4}\}$. As in Lemma 23, 3 and 6 are first associates, and similarly 4 and 5.

Consider 1 and x_1. Since $p_{11}^1 = n-2$, and there are $n-3$ elements of X which are first associates of both 1 and x_1, there is precisely one element of Y which is a first associate of x_1, say y_1. We may assume, without loss of generality, that x_i and y_i are first associates for $i = 1, 2, \ldots, n-4$. Hence we can write an array (Table 11.11(a)) where elements in the same row and column are first associates.

Table 11.11

$$\begin{array}{cccccccc} * & 1 & 3 & 5 & x_1 & x_2 & \cdots & x_{n-4} \\ 1 & * & 6 & 4 & y_1 & y_2 & \cdots & y_{n-4} \end{array}$$

(a)

$$\begin{array}{cccccccc} * & 1 & 3 & 5 & x_1 & x_2 & \cdots & x_{n-4} \\ 1 & * & 6 & 4 & y_1 & y_2 & \cdots & y_{n-4} \\ 3 & 6 & * & 2 & z_1 & z_2 & \cdots & z_{n-4} \end{array}$$

(b)

$$\begin{array}{cccccccc} * & 1 & 3 & 5 & x_1 & x_2 & \cdots & x_{n-4} \\ 1 & * & 6 & 4 & y_1 & y_2 & \cdots & y_{n-4} \\ 3 & 6 & * & 2 & z_1 & z_2 & \cdots & z_{n-4} \\ 5 & 4 & 2 & * & w_1 & w_2 & \cdots & w_{n-4} \end{array}$$

(c)

Now consider the first associates of 3, which fall into the two sets $\{1, 5, x_1, \ldots, x_{n-4}\}$ and $\{2, 6, z_1, \ldots, z_{n-4}\}$. Hence 1 and 6 are first associates, 2 and 5 are first associates, and again without loss of generality we assume that x_i and z_i are first associates for $i = 1, 2, \ldots, n-4$. Thus the array of varieties is now that of Table 11.11(b).

Next consider 5. One set of first associates is $\{1, 3, x_1, \ldots, x_{n-4}\}$ and the other is $\{2, 4, w_1, \ldots, w_{n-4}\}$. As before we get the array of Table 11.11(c).

Proceeding in this way we obtain a symmetric $n \times n$ array with diagonal positions empty, where the first associates are varieties in the same row or column of the array. □

We have therefore shown the following.

Theorem 28. *The parameters of the triangular association scheme uniquely determine the scheme for $n \neq 8$.* □

Where $n = 8$, there are three possible association schemes which are pseudo-triangular, that is, non-triangular but with the same parameters as the triangular scheme. They are listed in Table 11.12.

We give one construction for triangular PBIBDs.

Theorem 29. *If an $SB[n+2, 2; (n^2+3n+4)/2]$ design exists, then so does a triangular PBIBD with parameters*

$$v = (n+1)(n+2)/2, b = n(n+1)/2, r = n, k = n+2, \lambda_1 = 1, \lambda_2 = 2.$$

Proof. Choose some fixed variety, x, in the $SB[n+2, 2; (n^2+3n+4)/2]$ design; it occurs in $n+2$ blocks.

Table 11.12. The three pseudo-triangular association schemes when $n = 8$

Variety	First associates											
1	2	3	4	5	6	7	8	9	10	11	12	13
2	1	3	4	5	6	7	8	14	15	16	17	18
3	1	2	4	5	9	10	11	14	15	16	19	20
4	1	2	3	5	6	9	12	14	15	21	22	23
5	1	2	3	4	7	10	13	14	15	21	24	25
6	1	2	4	7	8	9	12	17	18	22	23	26
7	1	2	5	6	8	10	13	17	18	24	25	26
8	1	2	6	7	11	12	13	16	17	18	27	28
9	1	3	4	6	10	11	12	19	20	22	23	26
10	1	3	5	7	9	11	13	19	20	24	25	26
11	1	3	8	9	10	12	13	16	19	20	27	28
12	1	4	6	8	9	11	13	21	22	23	27	28
13	1	5	7	8	10	11	12	21	24	25	27	28
14	2	3	4	5	15	16	17	19	21	22	24	27
15	2	3	4	5	14	16	18	20	21	23	25	28
16	2	3	8	11	14	15	17	18	19	20	27	28
17	2	6	7	8	14	16	18	19	22	24	26	27
18	2	6	7	8	15	16	17	20	23	25	26	28
19	3	9	10	11	14	16	17	20	22	24	26	27
20	3	9	10	11	15	16	18	19	23	25	26	28
21	4	5	12	13	14	15	22	23	24	25	27	28
22	4	6	9	12	14	17	19	21	23	24	26	27
23	4	6	9	12	15	18	20	21	22	25	26	28
24	5	7	10	13	14	17	19	21	22	25	26	27
25	5	7	10	13	15	18	20	21	23	24	26	28
26	6	7	9	10	17	18	19	20	22	23	24	25
27	8	11	12	13	14	16	17	19	21	22	24	28
28	8	11	12	13	15	16	18	20	21	23	25	27

(i)

Variety	First associates											
1	2	3	4	5	6	7	8	9	10	11	12	13
2	1	3	4	5	6	7	8	14	15	16	17	18
3	1	2	4	5	9	10	11	14	15	16	19	20
4	1	2	3	6	9	12	13	14	15	16	21	22
5	1	2	3	7	8	10	11	14	17	19	23	24
6	1	2	4	7	8	12	13	15	18	21	25	26
7	1	2	5	6	8	10	12	17	18	23	25	27
8	1	2	5	6	7	11	13	17	18	24	26	28
9	1	3	4	10	11	12	13	16	20	22	27	28
10	1	3	5	7	9	11	12	19	20	23	25	27
11	1	3	5	8	9	10	13	19	20	24	26	28
12	1	4	6	7	9	10	13	21	22	23	25	27
13	1	4	6	8	9	11	12	21	22	24	26	28
14	2	3	4	5	15	16	17	19	21	22	23	24
15	2	3	4	6	14	16	18	19	20	21	25	26
16	2	3	4	9	14	15	17	18	20	22	27	28
17	2	5	7	8	14	16	18	22	23	24	27	28
18	2	6	7	8	15	16	17	20	25	26	27	28
19	3	5	10	11	14	15	20	21	23	24	25	26
20	3	9	10	11	15	16	18	19	25	26	27	28
21	4	6	12	13	14	15	19	22	23	24	25	26
22	4	9	12	13	14	16	17	21	23	24	27	28
23	5	7	10	12	14	17	19	21	22	24	25	27
24	5	8	11	13	14	17	19	21	22	23	26	28
25	6	7	10	12	15	18	19	20	21	23	26	27
26	6	8	11	13	15	18	19	20	21	24	25	28
27	7	9	10	12	16	17	18	20	22	23	25	28
28	8	9	11	13	16	17	18	20	22	24	26	27

(ii)

Variety	First associates											
1	2	3	4	5	6	7	8	9	10	11	12	13
2	1	3	4	5	6	7	8	14	15	16	17	18
3	1	2	4	5	6	9	10	14	15	19	20	21
4	1	2	3	5	6	9	11	14	16	22	23	24
5	1	2	3	4	7	9	11	15	17	19	22	25
6	1	2	3	4	8	10	12	14	16	20	23	26
7	1	2	5	8	11	12	13	15	17	18	25	27
8	1	2	6	7	10	12	13	16	17	18	26	28
9	1	3	4	5	10	11	13	19	21	22	24	28
10	1	3	6	8	9	12	13	19	20	21	26	28
11	1	4	5	7	9	12	13	22	23	24	25	27
12	1	6	7	8	10	11	13	20	23	25	26	27
13	1	7	8	9	10	11	12	18	21	24	27	28
14	2	3	4	6	15	16	18	20	21	23	24	27
15	2	3	5	7	14	17	18	19	20	21	25	27
16	2	4	6	8	14	17	18	22	23	24	26	28
17	2	5	7	8	15	16	18	19	22	25	26	28
18	2	7	8	13	14	15	16	17	21	24	27	28
19	3	5	9	10	15	17	20	21	22	25	26	28
20	3	6	10	12	14	15	19	21	23	25	26	27
21	3	9	10	13	14	15	18	19	20	24	27	28
22	4	5	9	11	16	17	19	23	24	25	26	28
23	4	6	11	12	14	16	20	22	24	25	26	27
24	4	9	11	13	14	16	18	21	22	23	27	28
25	5	7	11	12	15	17	19	20	22	23	26	27
26	6	8	10	12	16	17	19	20	22	23	25	28
27	7	11	12	13	14	15	18	20	21	23	24	25
28	8	9	10	13	16	17	18	19	21	22	24	26

(iii)

The varieties and blocks of the triangular design are respectively the varieties of the original design, other than x, and the blocks of the original design which do not contain x. Any two varieties which occur together in a block with x are first associates in the triangular design. The parameters of both the association scheme and the design follow immediately. □

This is the construction used to obtain Example 20.

11.5. Other two-associate-class association schemes

We define two other association schemes.

(a) Consider a set of $v = n^2$ varieties arranged in an $n \times n$ array onto which a set of $i-2$ mutually orthogonal Latin squares of side n has been superimposed. The first associates of a variety are precisely the varieties in the same row or column of the array or associated with the same symbols in one of the Latin squares. This is an *L_i-type association scheme.* The parameters are

$$n_1 = i(n-1), \quad n_2 = (n-1)(n-i+1)$$

and

$$P_1 = \begin{bmatrix} i^2-3i+n & (i-1)(n-i+1) \\ (i-1)(n-i+1) & (n-i)(n-i+1) \end{bmatrix}, \quad P_2 = \begin{bmatrix} i(i-1) & i(n-i) \\ i(n-i) & (n-i)^2+i-2 \end{bmatrix},$$

where $i \leq n-1$. If $i = n$, then the scheme is a group divisible one; if $i = n+1$, then all varieties are first associates of each other.

The lattice designs of Chapter 8, with $t = k-1$, are examples of L_2 designs.

(b) Let $\{1, 2, \ldots, v-1\} = \{d_1, \ldots, d_{n_1}\} \cup \{e_1, \ldots, e_{n_2}\} = D \cup E$, where $n_1 + n_2 = v-1$. Then a non-group divisible association scheme defined on Z_v is said to be *cyclic* if $D = -D$ and amongst the $n_1(n_1-1)$ differences of distinct elements of D, each element of D appears p^1_{11} times and each element of E appears p^2_{11} times. The first associates of i are $i + D$. The parameters of the only known cyclic association schemes are

$$n_1 = n_2 = 2t, \quad P_1 = \begin{bmatrix} t-1 & t \\ t & t \end{bmatrix}, \quad P_2 = \begin{bmatrix} t & t \\ t & t-1 \end{bmatrix}.$$

Example 30. Let $D = \{3, 5, 6, 7, 10, 11, 12, 14\}$, $v = 17$. Then there are:
(i) a $PB[8, 4, 3; 17]$ design with starter block $\{1, 2, 4, 8, 9, 13, 15, 16\}$;
(ii) a $PB[4, 1, 2; 17]$ design with starter blocks $\{0, 1, 4, 5\}$ and $\{1, 8, 10, 16\}$. □

11.6. Residual designs of biplanes

We know from Section 8.4 that the residual design of any projective plane is an affine plane, and that any affine plane can be extended to a projective plane. In other words, a $B[n, 1; n^2]$ design must be the residual of an $SB[n+1, 1; n^2+n+1]$ design if it exists, giving an existence criterion for a BIBD which is not symmetric.

For the symmetric designs with $\lambda = 2$ (or *biplanes*), we can obtain a similar result, by considering the duals of their residual designs; these dual designs happen to be triangular PBIBDs, and Theorem 27 allows us to prove that a BIBD with the parameters of the residual of a biplane is in fact the residual of a biplane.

A biplane is an SBIBD with parameters

$$v' = b' = (n^2+3n+4)/2, \quad r' = k' = n+2, \quad \lambda' = 2,$$

and its residual design has parameters

$$v = n(n+1)/2, \quad b = (n+1)(n+2)/2, \quad r = n+2, \quad k = n, \quad \lambda = 2.$$

We consider the structure of the dual of the residual design.

Theorem 31. *Let D be a $B[n, 2; n(n+1)/2]$ design. Then D^*, the dual of D, is a PBIBD with parameters*

$$v^* = (n+1)(n+2)/2, \quad b^* = n(n+1)/2, \quad r^* = n, \quad k^* = n+2,$$

$$n_1 = 2n, \quad n_2 = n(n-1)/2, \quad \lambda_1 = 1, \quad \lambda_2 = 2, \quad p^1_{11} = n, \quad p^2_{11} = 4.$$

Proof. Certainly the values of v^*, b^*, r^*, k^* follow immediately from the definition of dual design.

Suppose first that $n = 2$. Then D is the $B[2, 2; 3]$ design, with blocks 12, 12, 13, 13, 23, 23. Its dual, D^*, is the design with blocks 1234, 1256, 3456. Thus D^* is a triangular design with parameters as shown in Table 11.13.

Table 11.13

*	1	3	5
1	*	6	4
3	6	*	2
5	4	2	*

Array

$v^* = 6 \quad r^* = 2 \quad n_1 = 4 \quad \lambda_1 = 1$

$b^* = 3 \quad k^* = 4 \quad n_2 = 1 \quad \lambda_2 = 2 \quad P_1 = \begin{bmatrix} 2 & 2 \\ 2 & 0 \end{bmatrix}, \quad P_2 = \begin{bmatrix} 4 & 0 \\ 0 & 0 \end{bmatrix}$

We now assume that $n \geq 3$, and consider the intersection pattern of blocks of D. We let x_i denote the number of blocks which intersect some particular block of D in precisely i elements, for $i = 0, 1, \ldots, k$, and apply Theorems 2.1.1 and 2.1.2. This gives

$$\sum_{i=0}^{k} x_i = b - 1 = (n^2+3n)/2, \tag{11.19}$$

$$\sum_{i=0}^{k} ix_i = k(r-1) = n^2+n, \tag{11.20}$$

$$\sum_{i=0}^{k} i(i-1)x_i = k(k-1)(\lambda-1) = n^2-n. \tag{11.21}$$

From these three equations we obtain

$$\sum_{i=0}^{k} (i(i-1)-2i+2)x_i = n^2-n-2(n^2+n)+(n^2+3n) = 0;$$

that is,

$$\sum_{i=0}^{k} (i-1)(i-2)x_i = 0.$$

For $i = 0$ or $i > 2$, we have $(i-1)(i-2) > 0$, so $x_i = 0$ for all $i \neq 1, 2$. From eqn (11.21) $x_2 = (n^2-n)/2$; then from (11.19), $x_1 = 2n$.

Since $x_i = 0$ for all $i \neq 1$, 2, any pair of blocks intersect in one or two elements. Since the blocks of D correspond to the varieties of D^*, we shall say that two blocks of D are *first associates* if they intersect in one element, and *second associates* if they intersect in two.

Suppose first that blocks B_1 and B_2 are first associates. Without loss of generality we assume that $B_1 = \{1, 2, \ldots, n\}$, $B_2 = \{1, n+1, \ldots, 2n-1\}$. The remaining blocks of the design fall into two families: the n other blocks which contain 1, and the $(n+1)(n+2)/2-(n+2) = (n+2)(n-1)/2$ blocks which do not contain 1. Each of these families can be further subdivided into four classes.

First we consider the n remaining blocks which contain 1. These blocks include α_{ij} which are ith associates of B_1 and jth associates of B_2, for $i, j = 1$, 2, so there are:

α_{11} of the form $\{1, a_2, \ldots, a_k\}$, $a_i \geq 2n$;
α_{21} of the form $\{1, b_2, a_3, \ldots, a_k\}$, $2 \leq b_2 \leq n$, $a_i \geq 2n$;
α_{12} of the form $\{1, c_2, a_3, \ldots, a_k\}$, $n+1 \leq c_2 \leq 2n-1$, $a_i \geq 2n$;
α_{22} of the form $\{1, b_2, c_3, a_4, \ldots, a_k\}$, $2 \leq b_2 \leq n$, $n+1 \leq c_3 \leq 2n-1$, $a_i \geq 2n$.

Thus

$$\alpha_{11}+\alpha_{21}+\alpha_{12}+\alpha_{22} = n. \tag{11.22}$$

Counting the number of occurrences of pairs $\{1, i\}$, for $2 \leq i \leq n$, gives

$$\alpha_{21}+\alpha_{22} = n-1 \tag{11.23}$$

and similarly counting pairs $\{1, i\}$, for $n+1 \leq i \leq 2n-1$, gives

$$\alpha_{12}+\alpha_{22} = n-1; \tag{11.24}$$

thus

$$\alpha_{21} = \alpha_{12}$$

and

$$\alpha_{11}+\alpha_{12}=1.$$

Since each α_{ij} is a non-negative integer, we have two possibilities: either

$$\alpha_{11}=1, \alpha_{12}=\alpha_{21}=0, \alpha_{22}=n-1 \tag{11.25}$$

or

$$\alpha_{11}=0, \alpha_{12}=\alpha_{21}=1, \alpha_{22}=n-2. \tag{11.26}$$

We now consider the $(n+1)(n+2)/2-(n+2)=(n+2)(n-1)/2$ blocks which do not contain the variety 1; there are β_{ij} of them which are ith associates of B_1 and jth associates of B_2, for $i, j=1, 2$. Thus there are:

β_{11} of the form $\{b_1, c_2, a_3, \ldots, a_n\}$, $2 \leq b_1 \leq n$, $n+1 \leq c_2 \leq 2n-1$, $a_i \geq 2n$;
β_{21} of the form $\{b_1, b_2, c_3, a_4, \ldots, a_n\}$, $2 \leq b_i \leq n$, $n+1 \leq c_3 \leq 2n-1$, $a_i \geq 2n$;
β_{12} of the form $\{b_1, c_2, c_3, a_4, \ldots, a_n\}$, $2 \leq b_1 \leq n$, $n+1 \leq c_i \leq 2n-1$, $a_i \geq 2n$;
β_{22} of the form $\{b_1, b_2, c_3, c_4, a_5, \ldots, a_n\}$, $2 \leq b_i \leq n$, $n+1 \leq c_i \leq 2n-1$, $a_i \geq 2n$.

Thus

$$\beta_{11}+\beta_{21}+\beta_{12}+\beta_{22}=(n+2)(n-1)/2. \tag{11.27}$$

Counting the number of occurrences of pairs $\{i, j\}$ for $2 \leq i, j \leq n$, gives

$$\beta_{21}+\beta_{22}=(n-1)(n-2)/2 \tag{11.28}$$

and similarly counting pairs $\{i, j\}$ for $n+1 \leq i, j \leq 2n-1$, gives

$$\beta_{12}+\beta_{22}=(n-1)(n-2)/2; \tag{11.29}$$

so

$$\beta_{12}=\beta_{21} \tag{11.30}$$

and

$$\beta_{11}+\beta_{12}=2(n-1). \tag{11.31}$$

Finally, counting occurrences of pairs $\{i, j\}$ for $2 \leq i \leq n$ and $n+1 \leq j \leq 2n-1$ gives

$$\alpha_{22}+\beta_{11}+2\beta_{21}+2\beta_{12}+4\beta_{22}=2(n-1)^2. \tag{11.32}$$

By (11.30), we have

$$\alpha_{22}+\beta_{11}+4(\beta_{12}+\beta_{22})=2(n-1)^2$$

and hence, by (11.29),

$$\alpha_{22}+\beta_{11}=2(n-1).$$

If eqn (11.25) holds, then $\beta_{11}=n-1$ and, by (11.29) and (11.31), we have

$$\beta_{12}=\beta_{21}=n-1.$$

Finally, by (11.28),

$$\beta_{22}=(n-1)(n-4)/2.$$

On the other hand, if eqn (11.26) holds, then $\beta_{11} = n$ and, by (11.29) and (11.31), we have

$$\beta_{12} = \beta_{21} = n-2.$$

Finally, by (11.28),

$$\beta_{22} = (n-2)(n-3)/2.$$

Table 11.14 shows the possible values of α_{ij}, β_{ij}.

Table 11.14

α_{11}	1	0
$\alpha_{12} = \alpha_{21}$	0	1
$\alpha_{22} = \beta_{12} = \beta_{21}$	$n-1$	$n-2$
β_{11}	$n-1$	n
β_{22}	$(n-1)(n-4)/2$	$(n-2)(n-3)/2$

For the varieties of D^* which correspond to B_1 and B_2 (that is, for first associates) we have

$$p_{11}^1 = \alpha_{11} + \beta_{11} = n, \quad p_{12}^1 = p_{21}^1 = \alpha_{12} + \beta_{12} = n-1,$$
$$p_{22}^1 = \alpha_{22} + \beta_{22} = (n-1)(n-2)/2.$$

These values are independent of the particular choice of blocks.

If we consider instead the second associates $B_1 = \{1, 2, 3, \ldots, n\}$, $B_2 = \{1, 2, n+1, \ldots, 2n-2\}$, then a similar argument shows that

$$p_{11}^2 = 4, \quad p_{12}^2 = p_{21}^2 = 2n-4, \quad p_{22}^2 = (n-2)(n-3)/2.$$

Hence D^* is a PBIBD with the required parameters. □

Corollary 31.1. *If $n \neq 8$, then D^* is a triangular PBIBD.* □

Theorem 32. *Let D be a $B[n, 2; n(n+1)/2]$ design, $n \neq 8$. Then there exists an $SB[n+2, 2; (n^2+3n+4)/2]$ design, D', of which D is a residual design.*

Proof. Since the design D exists, so also does its dual, D^*, which is triangular by Corollary 31.1. Hence the blocks $B_1, B_2, \ldots, B_b$ of D can be written in the array of Table 11.15 (where i stands for B_i), so that any two blocks have precisely one variety in common if they appear in the same row or column of the array, and precisely two varieties in common otherwise; here $b = (n+1)(n+2)/2$.

Now let B' be a new block consisting of new varieties $u_1, u_2, \ldots, u_{n+2}$. Let the $(n+1)(n+2)/2$ pairs of these varieties be listed as in Table 11.16. We now adjoin the pairs of varieties in Table 11.16 to the blocks in the corresponding positions of Table 11.15, to give blocks B_i', $i = 1, 2, \ldots, b$. Thus B_1' consists of all the varieties in the block B_1 together with u_1 and u_2, B'_2 of all the varieties in B_2 together with u_1 and u_3, and so on.

This gives a design D' with $v = n(n+1)/2$ original varieties together with $n+2$ new varieties, and blocks $B'_1, \ldots, B'_b, B'$, each containing $n+2$ varieties. Thus D' has parameters

$$v' = b' = (n^2+3n+4)/2, \quad r' = k' = n+2.$$

Further, if i and j belong to the same row or column of Table 11.15, then blocks B_i and B_j have one element in common, the pairs of varieties in corresponding positions in Table 11.16 have one element in common, and thus $|B'_i \cap B'_j| = 2$. On the other hand, if i and j do not belong to the same row or column of Table 11.15, then blocks B_i and B_j have two elements in common, the pairs of varieties in corresponding positions in Table 11.16 have no element in common, and again $|B'_i \cap B'_j| = 2$. Thus we have a linked symmetric design which must be balanced, with $\lambda = 2$. This is the required design. □

Table 11.15

*	1	2	3	...		$n+1$
1	*	$n+2$	$n+3$	...		$2n+1$
2	$n+2$	*	$2n+2$	...	⋮	⋮
3	$n+3$	$2n+2$	*			
⋮	⋮	⋮	.		...	$\frac{(n+1)(n+2)}{2}$
$n+1$	$2n+1$			...	$\frac{(n+1)(n+2)}{2}$	*

Table 11.16

*	u_1u_2	u_1u_3	...		u_1u_{n+2}
u_2u_1	*	u_2u_3	...		u_2u_{n+2}
u_3u_1	u_3u_2	*	...		u_3u_{n+2}
⋮	⋮	⋮			⋮
⋮	⋮	⋮			$u_{n+1}u_{n+2}$
$u_{n+2}u_1$	$u_{n+2}u_2$	$u_{n+2}u_3$	...	$u_{n+2}u_{n+1}$	*

Finally, we show that a $B[8,2;36]$ design does not exist.

Lemma 33. *Let M be a $(v+t)\times(v+t)$ matrix, where*

$$M = \begin{bmatrix} W & X \\ Y & Z \end{bmatrix}$$

and $W = (\alpha-\beta)I_v + \beta J_v$. Then $\det M = (\alpha+(v-1)\beta)^{-t+1}(\alpha-\beta)^{v-t-1}\det N$, *where $N = (\alpha+(v-1)\beta)(\alpha-\beta)Z - (\alpha+(v-1)\beta)YX + \beta YJX$.*

Proof. See Exercise 11.13. □

Choose t blocks of the design and permute the columns of A so that these t blocks correspond to the first t columns of A. Let A_0 be this $v \times t$ submatrix of A. Then

$$S_t = A_0^T A_0$$

is called the *structural matrix* of the t blocks. Let

$$A_1 = \begin{bmatrix} A \\ I_t \quad 0 \end{bmatrix}.$$

Then

$$A_1 A_1^T = \begin{bmatrix} AA^T & A_0 \\ A_0^T & I_t \end{bmatrix}$$

and, by Lemma 33,

$$\det(A_1 A_1^T) = kr^{-t+1}(r-\lambda)^{v-t-1} \det(C_t),$$

where

$$\begin{aligned} C_t &= (1/k)((r+(v-1)\lambda)(r-\lambda)I - (r+(v-1)\lambda)A_0^T A_0 + \lambda A_0^T J A_0) \\ &= r(r-\lambda)I + \lambda k J - rS_t. \end{aligned}$$

The matrix C_t is called the *characteristic matrix* of the t chosen blocks and C_b is the characteristic matrix of the design.

Theorem 34. *If C_t is the characteristic matrix of any set of t blocks chosen from a $B[k, \lambda; v]$ design, then:*

(i) $\det(C_t) \geq 0$ *if* $t < b - v$;
(ii) $\det(C_t) < 0$ *if* $t > b - v$;
(iii) $kr^{-t+1}(r-\lambda)^{v-t-1} \det(C_t)$ *is a perfect square if* $t = b - v$.

Proof. If P is a matrix with real entries then $\det(PP^T) \geq 0$. Thus, if $b > v + t$ then $\det(A_1 A_1^T) \geq 0$. Since the elements of A_1 are integers, if $b = v + t$ then $\det(A_1 A_1^T)$ is a perfect square. If $b < v + t$ then $\det(A_1 A_1^T) = 0$. □

Lemma 35. *The matrices T_1, T_2, and T_3, given in Table* 11.17, *cannot be submatrices of the structural matrix of a $B[8, 2; 36]$ design; that is, $S_5 \neq T_1, T_2$ or T_3.*

Table 11.17

$$T_1 = \begin{bmatrix} 8 & 1 & 1 & 2 & 2 \\ 1 & 8 & 1 & 2 & 2 \\ 1 & 1 & 8 & 1 & 1 \\ 2 & 2 & 1 & 8 & 2 \\ 2 & 2 & 1 & 2 & 8 \end{bmatrix} \quad T_2 = \begin{bmatrix} 8 & 1 & 1 & 2 & 2 \\ 1 & 8 & 1 & 1 & 1 \\ 1 & 1 & 8 & 1 & 1 \\ 2 & 1 & 1 & 8 & 2 \\ 2 & 1 & 1 & 2 & 8 \end{bmatrix} \quad T_3 = \begin{bmatrix} 8 & 1 & 1 & 1 & 2 \\ 1 & 8 & 1 & 2 & 1 \\ 1 & 1 & 8 & 2 & 1 \\ 1 & 2 & 2 & 8 & 1 \\ 2 & 1 & 1 & 1 & 8 \end{bmatrix}$$

Proof. The determinants of the characteristic matrices corresponding to $S_5 = T_1, T_2$ or T_3 are $-80{,}000$, $-160{,}000$, and $-240{,}000$ respectively. The result follows by Theorem 34. □

Theorem 36. *There is no $B[8, 2; 36]$ design.*

Proof. Let B_1, B_2 and B_3 be blocks of a $B[8, 2; 36]$ design which are mutually first associates. By the calculations in the proof of Theorem 31, with $n = 8$, we have

$$x_1 = 16,\ x_2 = 28,\ p^1_{11} = 8.$$

Hence the first three rows of S_{45}, the structural matrix of the design, can be arranged as in Table 11.18.

Table 11.18

$$\begin{matrix} 8 & 1 & 1 & 1\dots1 & 1\dots1 & 1\dots1 & 1\dots1 & 2\dots2 & 2\dots2 & 2\dots2 & 2\dots2 \\ 1 & 8 & 1 & 1\dots1 & 1\dots1 & 2\dots2 & 2\dots2 & 1\dots1 & 1\dots1 & 2\dots2 & 2\dots2 \\ 1 & 1 & 8 & \underbrace{1\dots1}_{j} & \underbrace{2\dots2}_{7-j} & \underbrace{1\dots1}_{7-j} & \underbrace{2\dots2}_{j} & \underbrace{1\dots1}_{7-j} & \underbrace{2\dots2}_{j} & \underbrace{1\dots1}_{j} & \underbrace{2\dots2}_{21-j} \end{matrix}$$

Now consider the structural matrix S_{10} given in Table 11.19. We know that F and H are symmetric, and that each entry on the main diagonal equals 8. Also each off-diagonal entry of S_{10} is either 1 or 2. Without loss of generality, consider the submatrices of S_{10} which contain rows and columns 1, 2, 3, 4, 5 or 1, 2, 3, $4+j$, $5+j$ or 1, 2, 3, 4, $4+j$; that is, of the form shown in Table 11.20.

To avoid the submatrices T_1, T_2, T_3 respectively of Lemma 35, we must have $x = y = 1, z = 2$, and thus $F = 7I + J, H = 7I + J, G = 2J$, for appropriately-sized I and J.

Table 11.19

$$S_{10} = \begin{bmatrix} 8 & 1 & 1 & 1\dots1 & 2\dots2 \\ 1 & 8 & 1 & 2\dots2 & 1\dots1 \\ 1 & 1 & 8 & 2\dots2 & 1\dots1 \\ 1 & 2 & 2 & & \\ \vdots & & & F & G \\ 1 & 2 & 2 & & \\ 2 & 1 & 1 & & \\ \vdots & & & G^T & H \\ 2 & 1 & 1 & & \end{bmatrix} \begin{matrix} \\ \\ \\ \left.\begin{matrix} \\ \\ \\ \end{matrix}\right\} j \\ \left.\begin{matrix} \\ \\ \\ \end{matrix}\right\} 7-j \end{matrix}$$

Table 11.20

$$\begin{matrix} 8 & 1 & 1 & 1 & 1 \\ 1 & 8 & 1 & 2 & 2 \\ 1 & 1 & 8 & 2 & 2 \\ 1 & 2 & 2 & 8 & x \\ 1 & 2 & 2 & x & 8 \end{matrix} \qquad \begin{matrix} 8 & 1 & 1 & 2 & 2 \\ 1 & 8 & 1 & 1 & 1 \\ 1 & 1 & 8 & 1 & 1 \\ 2 & 1 & 1 & 8 & y \\ 2 & 1 & 1 & y & 8 \end{matrix} \qquad \begin{matrix} 8 & 1 & 1 & 1 & 2 \\ 1 & 8 & 1 & 2 & 1 \\ 1 & 1 & 8 & 2 & 1 \\ 1 & 2 & 2 & 8 & z \\ 2 & 1 & 1 & z & 8 \end{matrix}$$

Let $C_{10} = 80I + 16J - 10S_{10}$. Then, applying Lemma 34 three times, first with $v = 3$ and $t = 7$, next with $v = j$ and $t = 7 - j$ and finally with $v = 7 - j$ and $t = 0$, we get that $\det(C_{10}) = j(j-6)2(10)^9$. Thus, by Theorem 34, either $j = 0$ or $j = 6$.

If $j = 0$ let S_9 be the structural matrix from the first 9 rows and columns of S_{10}. Then, by (iii) of Theorem 34,

$$8(10)^{-8}(8)^{26} \det(C_9) = 2^{83}$$

should be a perfect square.

Similarly, if $j = 6$, 2^{85} should be a perfect square. Thus the design does not exist. □

11.7. References and comments

Partially balanced incomplete block designs were introduced by Bose and Nair (1939). They have since been studied extensively and Raghavarao (1971) provides a comprehensive summary and list of references. The known two-associate-class designs have been tabulated by Clatworthy (1973). The classification used here is his.

Kageyama (1974) has determined conditions under which a PBIBD(m) with $\lambda_i = \lambda_j$ becomes, in fact, a PBIBD($m-1$) with classes i and j combined, in contrast to the situation in Example 1. Theorem 5 was originally proved by Connor and Clatworthy (1954). Theorems 8 and 10 are due to Bose and Connor (1952). Theorems 13 and 15 and Exercise 11.5 are due to Bose, Shrikhande, and Bhattacharya (1953). For the linear algebra used in the proof of Lemma 21, see Hohn (1973, §9.9). The proof given here that the triangular association scheme is determined by its parameters, except for $n = 8$, is due to A. J. Hoffman (1960) and Shrikhande (1959). The pseudo-triangular schemes are due to Chang (1960), as reproduced in Raghavarao (1971). Theorem 29 is due to Shrikhande (1960). The designs of Example 30 and Exercise 11.11 are listed in Clatworthy (1973). The construction of Exercise 11.10 is due to Chang, Liu and Liu (1965). The results of Exercises 11.1 and 11.4 seem to be part of the folklore. For the relationship between association schemes and graph theory, see Biggs (1974).

Theorems 31 and 32 were originally proved by Hall and Connor (1953). The constructive proof given here is that of Shrikhande (1960). Lemmas 33 and 35 and Theorems 34 and 36 are from Connor (1952).

Exercises

11.1. If $m \leq 2$ and $\lambda_1 = \lambda_2$ show that a $PB[k, \lambda_1, \lambda_1; v]$ design is, in fact, a $B[k, \lambda_1; v]$ design.
11.2. Use Theorem 5 to verify that the eigenvalues for a group divisible design are rk, $r - \lambda_1$ and $rk - v\lambda_2$ and to check the multiplicity of each.
11.3. Show that if $\lambda_2 = 0$ in a group divisible design then the difference of two varieties in different groups is not estimable.
11.4. By considering the rank of AA^T, show that in a semiregular group divisible design $b \geq v - m + 1$.
11.5. Using Theorem 10, and writing the blocks of the designs as columns, show that the existence of a semiregular $PB[m, 0, \lambda_2; mn]$ design is equivalent to the existence of an $OA[\lambda_2 n^2, m, n, 2]$.
11.6. Construct a $PB[6, 0, 1; 30]$ design.
11.7. Construct a $PB[2(2t-1), 0, t-1; 2(4t-1)]$ design for $4t-1$ a prime or a prime power.
11.8. Show that the design obtained by developing the initial block $\{1, 2, 4\}$ mod 6 is truly self-dual. What are the groups of the design?
11.9 Construct a $PB[4, 1, 2; 6]$ design.
11.10. By considering the sets of mutual first associates, show that the existence of a $B[k, \lambda; n-1]$ design implies the existence of a $PB[k, \lambda, 0; n(n-1)/2]$ design.
11.11. Let $D = \{1, 3, 4, 9, 10, 12\}$ and $v = 13$. Show that $\{0, 1, 2, 3, 6, 10\}$ is the starter block of a $PB[6, 3, 2; 13]$ design with a cyclic association scheme.
11.12. Construct an $SB[5, 2; 11]$ design from a $B[3, 2; 6]$ design by the method of Theorem 32.
11.13. To prove Lemma 33 carry out the following operations on M:

(i) multiply each of the last t columns by $(\alpha + (v-1)\beta)(\alpha - \beta)$ and divide by the appropriate scalar;
(ii) add rows $1, 2, \ldots, v-1$, to row v;
(iii) factor row v by $(\alpha + (v-1)\beta)$;
(iv) multiply row v by β and subtract from each of rows $1, 2, \ldots, v-1$;
(v) factor each of rows $1, 2, \ldots, v-1$, by $(\alpha - \beta)$;
(vi) subtract each of rows $1, 2, \ldots, v-1$, from row v;
(vii) subtract suitable multiples of columns $1, \ldots, v$ from columns $v+1, \ldots, v+t$ so that all the entries in the 'X' submatrix are now zero.

Evaluate the determinant of this matrix.

12 Existence results: symmetric balanced designs

12.1. Statement of results

We saw earlier that if a block design has parameters v, b, r, k, then those parameters satisfy

$$vr = bk \tag{12.1}$$

and that if the design is balanced with index λ, then the parameters satisfy both

$$\lambda(v-1) = r(k-1) \tag{12.2}$$

and

$$v \leq b \tag{12.3}$$

as well. If $k = 2$, so that these constraints become

$$vr = 2b \text{ and } \lambda(v-1) = r,$$

then the corresponding $B[2, \lambda; v]$ design must exist (Example 2.3.6). Similarly, if $k = 3$, giving

$$vr = 3b \text{ and } \lambda(v-1) = 2r,$$

then the constructions of Chapter 3 show that the corresponding $B[3, \lambda; v]$ design must exist. But no $SB[7, 2; 22]$ design exists (Example 2.4.11), even though the parameters $v = b = 22$, $r = k = 7$, $\lambda = 2$, satisfy eqns (12.1), (12.2), and (12.3). These examples raise several questions.

If parameters v, b, r, k, λ satisfy constraints (12.1) and (12.2), and if $k \geq 4$, what can we say about the existence of the (v, b, r, k, λ) BIBD? Might there be just a finite number of exceptional sets of parameters which satisfy (12.1) and (12.2) but for which no design exists? Might there be some additional constraints which must be satisfied by the parameters of a design? In this chapter we begin to consider such problems.

The first important result that we prove is a necessary condition for the existence of an $SB[k, \lambda; v]$ design. It is convenient here to work in terms of $k - \lambda$ which we denote by n. Further, we let $n = l^2h$, where h is square-free, and denote by π any prime which divides h. In this notation, $\lambda = 1$ implies that $k = n+1$ and $v = n^2+n+1$, $\lambda = 2$ implies that $k = n+2$ and $v = (n^2+3n+4)/2$, and so on.

Theorem 1. *Suppose that there exists an $SB[k, \lambda; v]$ design.*

(i) *If v is even, then $h = 1$; so n is a perfect square.*
(ii) *If v is odd, then the equation*

$$x^2 = ny^2 + (-1)^{(v-1)/2}\lambda z^2 \tag{12.4}$$

has a solution in integers x, y, z, not all zero. □

Corollary 1.1. *Suppose that there exists an $SB[n+1, 1; n^2+n+1]$ design. Then:*

$$n \equiv 1 \pmod 4 \text{ implies } \pi \equiv 1 \pmod 4;$$
$$n \equiv 2 \pmod 4 \text{ implies } \pi = 2 \text{ or } \pi \equiv 1 \pmod 4.$$ □

Corollary 1.2. *Suppose that there exists an $SB[n+2, 2; (n^2+3n+4)/2]$ design. If $n \equiv 0$ or $1 \pmod 4$, then n is a perfect square. If $n \equiv 2$ or $3 \pmod 4$, then:*

$$n \equiv 2 \pmod 8 \text{ implies } \pi = 2 \text{ or } \pi \equiv 1 \text{ or } 3 \pmod 8;$$
$$n \equiv 3 \pmod 8 \text{ implies } \pi \equiv 1 \text{ or } 3 \pmod 8;$$
$$n \equiv 6 \pmod 8 \text{ implies } \pi = 2 \text{ or } \pi \equiv 1 \text{ or } 7 \pmod 8;$$
$$n \equiv 7 \pmod 8 \text{ implies } \pi \equiv 1 \text{ or } 7 \pmod 8.$$ □

By Theorem 2.5.15, the residual design of an $SB[n+1, 1; n^2+n+1]$ design (or projective plane of order n) is a $B[n, 1; n^2]$ design (or affine plane of order n), and similarly the residual design of an $SB[n+2, 2; (n^2+3n+4)/2]$ design (or biplane of order n) is a $B[n, 2; n(n+1)/2]$ design. Conversely, any affine plane may be embedded in a projective plane of the same order (Section 8.2) and any $B[n, 2; n(n+1)/2]$ design in a biplane of order n (Theorem 11.6.33). Thus we could rephrase Corollaries 1.1 and 1.2 in terms of the residual designs; see Exercise 12.1. However, if $\lambda \geq 3$, then a *quasi-residual* design (one with the parameters of a residual design) may not in fact be a residual design; see Exercise 12.6.

The other main result of this chapter concerns the existence of *Hadamard* designs; that is, $SB[2h-1, h-1; 4h-1]$ designs. We have already seen two designs with these parameters, namely the $SB[3, 1; 7]$ and $SB[5, 2; 11]$ designs (see Examples 2.3.9 and 2.4.10). These designs are conjectured to exist for every positive integer h. If $[x]$ denotes the integer part of x, then the best result in this direction can be stated as

Theorem 2. *If α is any positive integer, then there exists an Hadamard design with $h = 2^{s-2}\alpha$ for every $s \geq [2\log_2(\alpha-3)]$.* □

In view of Theorem 2.6.18, we can rephrase Theorem 2 in the following way.

Theorem 2′. *If α is any positive integer, then there exists an Hadamard matrix of side $2^s\alpha$ for every $s \geq [2\log_2(\alpha-3)]$.* □

This is the form in which we shall prove the result.

The proofs of all these results are number theoretic. We begin by developing the necessary background. As usual, (x, y) denotes the greatest common divisor of the integers x and y.

12.2. The quadratic character

We have already seen in Section 3.3 that the even powers of a primitive element θ in the field $GF[p^n]$ are known as the *quadratic residues* of the field, and that -1 is a quadratic residue if and only if $p^n \equiv 1 \pmod 4$. We now consider the properties of quadratic residues in a little more detail.

The *quadratic character* on a field is a map χ from the field to the integers defined by

$$\chi(y) = \begin{cases} 1 & \text{if } y = \theta^{2a} \quad \text{for some } a, \\ -1 & \text{if } y = \theta^{2a+1} \quad \text{for some } a, \\ 0 & \text{if } y = 0. \end{cases}$$

It is easily checked that $\chi(x^{-1}) = \chi(x)$, for any $x \neq 0$, and that for all elements x and y in the field,

$$\chi(xy) = \chi(x) . \chi(y).$$

If the field F has odd order, then it has equal numbers of quadratic and non-quadratic residues; so

$$\sum_{y \in F} \chi(y) = 0. \tag{12.5}$$

If the field has prime order we often speak of the 'quadratic character modulo p'.

Lemma 3. *If $p = 2P + 1$ is an odd prime and $p \nmid a$, then $\chi(a) \equiv a^P \pmod p$.*

Proof. Let θ be a primitive element (modulo p), and let $a \equiv \theta^r \pmod p$. Then $a^P \equiv \theta^{rP} = x$, say and $x^2 \equiv \theta^{2rP} \equiv 1^r \equiv 1 \pmod p$. Hence

$$x \equiv 1 \quad \text{or} \quad x \equiv -1 \pmod p.$$

Now $a^P \equiv 1 \pmod p$ if and only if $\theta^{rP} \equiv 1 \pmod p$, that is, if and only if $2P | rP$, or in other words, if and only if r is even. But this is equivalent to the statement that a is a quadratic residue, $\chi(a) = 1$, and the Lemma is proved. □

Corollary 3.1. (i) *2 is a quadratic residue* (mod p) *if and only if $p \equiv 1$ or 7* (mod 8).

(ii) -2 *is a quadratic residue* (mod p) *if and only if $p \equiv 1$ or 3* (mod 8).

Proof. Let $p = 2P + 1$ as before.

(i) By Lemma 3, it is sufficient to show that

$$2^P \equiv \begin{cases} 1 & \text{if } p \equiv 1 \text{ or } 7 \pmod 8, \\ -1 & \text{if } p \equiv 3 \text{ or } 5 \pmod 8. \end{cases}$$

Certainly

$$2^P.P! = 2^P.1.2.3.\cdots.P = 2.4.6.\cdots.2P. \tag{12.6}$$

If $p \equiv 1 \pmod 4$, so that $(p-1)/4 = P/2$ is an integer, then the right side of eqn (12.6) may be written as

$$\begin{aligned} &2.4.6.\cdots.P.(P+2).\cdots.(2P-4).(2P-2).2P \\ &\equiv 2.4.6.\cdots.P(-(P-1)).\cdots.(-5).(-3).(-1) \pmod p \\ &= P!(-1)^{P/2}. \end{aligned}$$

Comparing this with the left side of eqn (12.6) we have

$$2^P \equiv (-1)^{P/2} \pmod p.$$

If $p \equiv 1 \pmod 8$, then $P/2$ is even, and $2^P \equiv 1 \pmod p$;
if $p \equiv 5 \pmod 8$, then $P/2$ is odd, and $2^P \equiv -1 \pmod p$.

If $p \equiv 3 \pmod 4$, so that $(p-3)/4 = (P-1)/2$ is an integer, then a similar argument shows that

$$2^P \equiv (-1)^{(P+1)/2} \pmod p.$$

If $p \equiv 3 \pmod 8$, then $(P+1)/2$ is odd, and

$$2^P \equiv -1 \pmod p;$$

if $p \equiv 7 \pmod 8$, then $(P+1)/2$ is even, and

$$2^P \equiv 1 \pmod p.$$

(ii) Now -2 is a quadratic residue if both -1 and 2 are quadratic residues, that is, if $p \equiv 1 \pmod 8$, or if neither -1 nor 2 is a quadratic residue, that is, if $p \equiv 3 \pmod 8$. □

12.3. Sums of squares

Lemma 4. (i) *No integer of the form* $4A+3$ *is a sum of two squares.*
(ii) *No integer of the form* $8A+7$ *is a sum of three squares.*

Proof. Both statements follow immediately on checking that any integer square is congruent to 0 or 1 (mod 4) and to 0, 1 or 4 (mod 8). □

Lemma 5. (i) *If* $q = a^2 + Nb^2$ *and* $r = c^2 + Nd^2$, *then*

$$\begin{aligned} qr &= (ac + Nbd)^2 + N(ad - bc)^2 \\ &= (ac - Nbd)^2 + N(ad + bc)^2. \end{aligned}$$

(ii) *If* $q = a^2+b^2+c^2+d^2$ *and* $r = e^2+f^2+g^2+h^2$, *then*

$$qr = A^2+B^2+C^2+D^2,$$

where

$$A = ae+bf+cg+dh, \qquad B = af-be+ch-dg,$$
$$C = ag-bh-ce+df, \qquad D = ah+bg-cf-de. \qquad \square$$

Corollary 5.1. *If each prime factor of the positive integer n can be written in the form* a^2+Nb^2, *then so can n.* □

Corollary 5.2. *If every prime can be written as the sum of four squares, then so can every positive integer.* □

Because of these corollaries, we temporarily direct our attention to primes. Several approaches are available; our next proof depends on the pigeonhole principle.

Theorem 6. *If* $n > 1$, $(a, n) = 1$ *and m is the least integer greater than* $\sqrt{n}$, *then there exist x and y such that*

$$0 < x < m, \quad 0 < y < m,$$

and

$$ay \equiv x \pmod{n} \text{ or } ay \equiv -x \pmod{n}.$$

Proof. Consider $ay-x$ in the m^2 cases where $x, y \in \{0, 1, 2, \ldots, m-1\}$. Since $m^2 > n$, there must exist at least two pairs (x_1, y_1), (x_2, y_2) such that

$$ay_1 - x_1 \equiv ay_2 - x_2 \pmod{n},$$

and without loss of generality $y_1 \geq y_2$.

If $y_1 = y_2$, then $x_1 \equiv x_2 \pmod{n}$ and thus $x_1 = x_2$. Hence $y_1 > y_2$. Since $(a, n) = 1$, we also have $x_1 \neq x_2$. Now let

$$y = y_1 - y_2 > 0 \quad \text{and} \quad x = \pm(x_1 - x_2) > 0$$

and we have

$$ay \equiv x \quad \text{or} \quad ay \equiv -x \pmod{n}$$

proving the Theorem. □

Lemma 7. *If* $-N$ *is a quadratic residue modulo p, then there exist integers x and y such that* $0 < x < \sqrt{p}$, $0 < y < \sqrt{p}$ *and*

$$x^2 + Ny^2 \equiv 0 \pmod{p}.$$

Proof. If $-N$ is a quadratic residue modulo p, then there exists s, such that $p \nmid s$, and

$$s^2 + N \equiv 0 \pmod{p}.$$

By Theorem 6, there exist positive integers x and y such that $x < \sqrt{p}$, $y < \sqrt{p}$, and

$$ys \equiv \pm x \pmod{p}.$$

Since $(y, p) = 1$, there exists an integer z such that $yz \equiv 1 \pmod{p}$ and hence

$$(yzs)^2 + N \equiv x^2z^2 + N \equiv s^2 + N \equiv 0 \pmod{p}.$$

Multiplying by y^2 gives

$$x^2 + Ny^2 \equiv 0 \pmod{p}.$$

□

Corollary 7.1. *If $p \equiv 1 \pmod 4$, then p is the sum of two integer squares.*

Proof. If $p \equiv 1 \pmod 4$, then -1 is a quadratic residue modulo p. If we let $N = 1$ in the Lemma, then there exist integers x and y such that $0 < x^2 < p$, $0 < y^2 < p$, and $x^2 + y^2 \equiv 0 \pmod p$. But now $0 < x^2 + y^2 < 2p$, and thus

$$x^2 + y^2 = p.$$

□

Corollary 7.2. *If $p \equiv 1$ or $7 \pmod 8$, then $p = a^2 - 2b^2$ for some integers a and b.*

Proof. If $p \equiv 1$ or $7 \pmod 8$, then 2 is a quadratic residue modulo p by Corollary 3.1. If we let $N = -2$ in the Lemma, then as before there exist integers x and y such that

$$x^2 - 2y^2 \equiv 0 \pmod{p}$$

and

$$-2p < x^2 - 2y^2 < p.$$

Since $\sqrt{2}$ is irrational, we must have

$$x^2 - 2y^2 = -p,$$

and thus

$$(x + 2y)^2 - 2(x + y)^2 = p.$$

So $a = x + 2y$, $b = x + y$, as required. □

Corollary 7.3. *If $p \equiv 1$ or $3 \pmod 8$, then $p = a^2 + 2b^2$ for some integers a and b.*

Proof. If $p \equiv 1$ or $3 \pmod 8$, then -2 is a quadratic residue modulo p by Corollary 3.1. If we let $N = 2$ in the Lemma, then as before there exist integers x and y such that

$$x^2 + 2y^2 \equiv 0 \pmod{p}$$

and

$$0 < x^2 + 2y^2 < 3p.$$

If $x^2+2y^2=p$, then we just choose $a=x, b=y$; if $x^2+2y^2=2p$, then x must be even, say $x=2w$, so that $y^2+2w^2=p$ and we choose $a=y, b=w$. □

If an integer n is represented as a sum of two squares, say

$$n=x^2+y^2,$$

then the representation is said to be *proper* if x and y are relatively prime, that is, their greatest common divisor, (x, y), equals 1.

Lemma 8. *Let p be a prime congruent to* 3 (modulo 4) *and let n be a positive multiple of p. Then n has no proper representation as the sum of two squares.*

Proof. Suppose that $n=x^2+y^2$, where x and y are integers such that $(x, y)=1$. Now $p|n$; if $p|x$, then $p|y$ also so that $p|(x, y)$ which is a contradiction. Hence p does not divide x and there exists an inverse for x (modulo p). Call this inverse u, so that $ux \equiv 1 \pmod{p}$. Then $uxy \equiv y \pmod{p}$, so

$$n=x^2+y^2 \equiv x^2+u^2x^2y^2 \equiv x^2(1+u^2y^2) \pmod{p}.$$

Since $p|n$, we have $x^2(1+u^2y^2) \equiv 0 \pmod{p}$. Since $p \nmid x$, we have $1+u^2y^2 \equiv 0 \pmod{p}$, so that $-1 \equiv u^2y^2=(uy)^2 \pmod{p}$. But this means that -1 is a quadratic residue modulo p, and since $p \equiv 3 \pmod{4}$, this is impossible, proving the Lemma. □

Lemma 9. *If $n=p^cm$, where p is a prime congruent to* 3 (modulo 4), *c is odd and p does not divide m, then n has no representation as the sum of two squares.*

Proof. Suppose that $n=x^2+y^2$. Let $(x, y)=d$, so that $x=dX$, $y=dY$ and $(X, Y)=1$. Then $n=Nd^2$, where $N=X^2+Y^2$. Now p^c divides n and c is odd, so p must divide N. But N cannot have the proper representation X^2+Y^2 by Lemma 8 so we have a contradiction which proves the Lemma. □

Theorem 10. *A positive integer n is the sum of two squares if and only if each of its prime factors congruent to* 3 (modulo 4) *appears to an even power.*

Proof. If n has any prime factor congruent to 3 (modulo 4) which appears to an odd power then, by Lemma 9, n cannot be represented as a sum of two squares.

So suppose any such prime factor appears to an even power and consider the possible factors of n. Let p be a prime which divides n. If $p \equiv 1 \pmod{4}$, then by Corollary 8.1, it is the sum of two squares. If $p \equiv 3 \pmod{4}$, then p^{2s} divides n for some $s \geq 0$ and $p^{2s}=(p^s)^2+0^2$, so p^{2s} is the sum of two squares. If $p=2$, then $2=1^2+1^2$, again a sum of two squares. Now n itself can be written as the sum of two squares. This covers every case except $n=1$, and $1=1^2+0^2$. □

We now turn our attention to sums of four squares.

Lemma 11. *For every prime p, there exist integers a, b, c, d, and m, with* $1 \le m < p$, *such that*

$$mp = a^2 + b^2 + c^2 + d^2.$$

Proof. For $p = 2$, the Lemma is obvious. Now assume that $p = 2P + 1$, and consider the $p+1$ values of

$$x^2 \text{ and } -(x^2+1),$$

with $x \in \{0, 1, \ldots, P\}$. Again by the pigeonhole principle, at least two of these $p+1$ values must be congruent modulo p. If we have $0 \le x_1 < x_2 \le P$, then $x_1^2 \equiv x_2^2 \pmod{p}$ implies that

$$p \mid (x_1 + x_2)(x_1 - x_2),$$

which is impossible since $1 \le x_1 + x_2 \le p - 1$ and $1 \le x_1 - x_2 \le P$. Similarly, $-(x^2+1)$ runs through $P+1$ different residue classes modulo p, so we must have

$$x_i^2 \equiv -1 - x_j^2 \pmod{p}$$

for some $i \ne j$. Thus

$$mp = x_i^2 + x_j^2 + 1^2 + 0^2$$

and since $1 \le mp < 2P^2 + 1 < Pp$, we have $1 \le m < P$. □

Lemma 12. *If p is an odd prime and if there exist integers, w, x, y, z, and m, with* $1 < m < p$, *such that*

$$w^2 + x^2 + y^2 + z^2 = mp, \tag{12.7}$$

then there exist integers W, X, Y, Z, and M, with $1 \le M < m$, *such that*

$$W^2 + X^2 + Y^2 + Z^2 = Mp.$$

Proof. (i) If m is even, then w, x, y, and z are all odd, or all even, or two are odd and two are even. In any case we can assume that $w \equiv x \pmod{2}$ and $y \equiv z \pmod{2}$, without loss of generality. Then

$$\left(\frac{w+x}{2}\right)^2 + \left(\frac{w-x}{2}\right)^2 + \left(\frac{y+z}{2}\right)^2 + \left(\frac{y-z}{2}\right)^2 = \frac{m}{2} \cdot p,$$

so $W = \dfrac{w+x}{2}$, $X = \dfrac{w-x}{2}$, $Y = \dfrac{y+z}{2}$, $Z = \dfrac{y-z}{2}$, and $M = \dfrac{m}{2}$ satisfy the conditions of the Lemma.

(ii) If m is odd, we apply the division algorithm to show that

$$w = am + e,\ x = bm + f,\ y = cm + g,\ z = dm + h,$$

where $|e| < m/2$, $|f| < m/2$, $|g| < m/2$, $|h| < m/2$. Substituting into (12.7) we find that (in the notation of Lemma 5 (ii))

$$e^2+f^2+g^2+h^2+2Am+(a^2+b^2+c^2+d^2)m^2 = mp, \qquad (12.8)$$

which implies that m divides $e^2+f^2+g^2+h^2$. Define M by letting $Mm = e^2+f^2+g^2+h^2$, where $M \geq 0$.

If $M=0$, then $e=f=g=h=0$, so that $m^2|(w^2+x^2+y^2+z^2)$, which implies that $m^2|mp$ and hence that $m|p$. But $1<m<p$ and p is prime, giving a contradiction. Hence $M \geq 1$. Also $e^2+f^2+g^2+h^2 < 4.\left(\frac{m^2}{4}\right) = m^2$, so $M<m$, and altogether $1 \leq M < m$.

Substituting into (12.8) we have

$$Mm+2Am+(a^2+b^2+c^2+d^2)m^2 = mp;$$

multiplying by M/m gives

$$M^2+2AM+A^2+B^2+C^2+D^2 = Mp,$$

again in the notation of Lemma 5 (ii). But now

$$(M+A)^2+B^2+C^2+D^2 = Mp,$$

so the integers $W=M+A$, $X=B$, $Y=C$, $Z=D$ and M satisfy the conditions of the Lemma. □

Theorem 13. *Any positive integer can be represented as the sum of four squares.*

Proof. By Lemma 11, for any odd prime p, there exist integers a, b, c, d, m, with $1 \leq m < p$, such that

$$mp = a^2+b^2+c^2+d^2.$$

If $m=1$, then p is the sum of four squares. If $m>1$, then by Lemma 12 there exist integers W, X, Y, Z, M, with $1 \leq M < m$, such that

$$Mp = W^2+X^2+Y^2+Z^2.$$

If $M=1$, then p is the sum of four squares. If $M>1$, then apply Lemma 12 again. This process of 'descent' cannot continue for ever because there are only a finite number of integers between 1 and p, but it can stop only if at some stage $M=1$. Hence p can be written as the sum of four squares.

Since $2 = 1^2+1^2+0^2+0^2$, the result follows by Lemma 5. □

12.4. Necessary conditions for existence of an SBIBD

We now have all the information that we need in order to prove Theorem 1 and its corollaries; we proceed to do so.

Proof of Theorem 1.

(i) Suppose there exists an $SB[k, \lambda; v]$ design with incidence matrix A. From Theorem 2.2.4,

$$\det(AA^T) = k^2(k-\lambda)^{v-1} = (n+\lambda)^2 l^{2(v-1)} h^{v-1}$$

and since A is square

$$\det(AA^T) = (\det A)^2.$$

Hence

$$\det A = \pm(n+\lambda) l^{v-1} h^{(v-1)/2}$$

and since v is even, and the determinant of an integer matrix must be an integer, we see that $h = 1$ and so n is a square.

(ii) By Theorem 13, we may write

$$n = a^2 + b^2 + c^2 + d^2,$$

where a, b, c, and d are non-negative integers.

Now consider the matrix

$$B = \begin{bmatrix} a & -b & -c & -d \\ b & a & -d & c \\ c & d & a & -b \\ d & -c & b & a \end{bmatrix},$$

where $BB^T = nI_4$, $\det(BB^T) = n^4$, $\det B = n^2 \neq 0$ and B is non-singular.

Now let $W = [w_1, w_2, w_3, w_4]^T$ where each w_i is a rational number and let

$$U = BW,$$

so that

$$\begin{aligned} u_1 &= aw_1 - bw_2 - cw_3 - dw_4, \\ u_2 &= bw_1 + aw_2 - dw_3 + cw_4, \\ u_3 &= cw_1 + dw_2 + aw_3 - bw_4, \\ u_4 &= dw_1 - cw_2 + bw_3 + aw_4. \end{aligned}$$

Then

$$U^T U = (BW)^T . BW = W^T B^T BW,$$

or equivalently,

$$\begin{aligned} u_1^2 + u_2^2 + u_3^2 + u_4^2 &= (a^2+b^2+c^2+d^2)(w_1^2+w_2^2+w_3^2+w_4^2) \\ &= n(w_1^2+w_2^2+w_3^2+w_4^2). \end{aligned} \tag{12.9}$$

All the numbers above are rational, and it should be observed that all the calculations in this proof take place over the rational numbers.

Assume there is an $SB[k, \lambda; v]$ design with incidence matrix $A = [a_{ij}]$. Assume $x_1, x_2, \ldots, x_v$ are variables, and write A_j for the linear homogeneous function

$$A_j = \sum_{i=1}^{v} x_i a_{ij}.$$

If X is the row-vector $(x_1, x_2, \ldots, x_v)$, then

$$XAA^TX^T = \sum_{j=1}^{v} A_j^2.$$

But, from Theorem 2.2.3,

$$XAA^TX^T = nXIX^T + \lambda XJX^T;$$

it is easy to see that $XIX^T = \sum_{j=1}^{v} x_j^2$ and $XJX^T = \left(\sum_{j=1}^{v} x_j\right)^2$, so

$$\sum_{j=1}^{v} A_j^2 = n\sum_{j=1}^{v} x_j^2 + \lambda\left(\sum_{j=1}^{v} x_j\right)^2. \tag{12.10}$$

Case (a) Now assume $v \equiv 1 \pmod 4$, so that $(-1)^{(v-1)/2} = 1$. Define a new set of variables $y_1, y_2, \ldots, y_v$ by

$$\begin{pmatrix} y_1 \\ y_2 \\ y_3 \\ y_4 \end{pmatrix} = B\begin{pmatrix} x_1 \\ x_2 \\ x_3 \\ x_4 \end{pmatrix}, \begin{pmatrix} y_5 \\ y_6 \\ y_7 \\ y_8 \end{pmatrix} = B\begin{pmatrix} x_5 \\ x_6 \\ x_7 \\ x_8 \end{pmatrix}, \ldots, \begin{pmatrix} y_{v-4} \\ y_{v-3} \\ y_{v-2} \\ y_{v-1} \end{pmatrix} = B\begin{pmatrix} x_{v-4} \\ x_{v-3} \\ x_{v-2} \\ x_{v-1} \end{pmatrix}, y_v = x_v,$$

and write ξ for Σx_j. Then, from eqn (12.9),

$$n\sum_{j=1}^{v-1} x_j^2 = \sum_{j=1}^{v-1} y_j^2,$$

so (12.10) becomes

$$\sum_{j=1}^{v} A_j^2 = \left(\sum_{j=1}^{v-1} y_j^2\right) + ny_v^2 + \lambda\xi^2. \tag{12.11}$$

Now we use the non-singularity of B in two different ways. First, it shows that each x_i is a linear function of the y_i with rational coefficients, so that A_1 (and in fact each of the A_j) is a linear rational function of the y_i. We need not compute the coefficients, but simply write

$$A_1 = \sum_{i=1}^{v} e_i y_i.$$

Secondly, the non-singularity of B also shows that, by proper choice of the variables x_j, we can give any values we wish to the variables y_j, so that eqn (12.11) remains true whatever values the y_j may take. In particular, it will be true if we restrict ourselves to values such that

$$y_1 = (1 - e_1)^{-1} \sum_{i=2}^{v} e_i y_i$$

– or, in the particular case where $e_1 = 1$ and the above inverse does not exist,

$$y_1 = (-1 - e_1)^{-1} \sum_{i=2}^{v} e_i y_i.$$

In these cases $A_1 = y_1$ or $A_1 = -y_1$ respectively; in any event $A_1^2 = y_1^2$, so eqn (12.11) becomes

$$\sum_{j=2}^{v} A_j^2 = \left(\sum_{j=2}^{v-1} y_j^2\right) + ny_v^2 + \lambda\xi^2. \qquad (12.12)$$

Each of the A_j, and also ξ, are linear homogeneous functions of the variables $y_1, y_2, \ldots, y_v$. But y_1 has been made into a linear homogeneous function of the remaining y_j. So the A_j, and ξ, can be considered to be linear functions of $y_2, y_3, \ldots$, and y_v, and they will all have zero constant terms.

Now operate on (12.12) just as we did on (12.11). This continues until we reach

$$A_v^2 = ny_v^2 + \lambda\xi^2. \qquad (12.13)$$

The value of y_v is not yet specified. The process of eliminating $y_1, y_2, \ldots, y_{v-1}$ will result in A_v and ξ being multiples of y_v; since all our operations have taken place over the rational field, and since they have been linear, A_v and ξ are rational multiples of y_v: we have

$$\left(\frac{\alpha y_v}{\beta}\right)^2 = ny_v^2 + \lambda\left(\frac{\gamma y_v}{\delta}\right)^2$$

for some integers α, β, γ, and δ and for any value of y_v. In particular, put $y_v = \beta\delta$. Then

$$(\alpha\delta)^2 = n(\beta\delta)^2 + \lambda(\beta\gamma)^2.$$

Recalling that $(-1)^{(v-1)/2} = 1$ in the case under consideration, we see that we have an integer solution

$$x = \alpha\delta,\ y = \beta\delta,\ z = \beta\gamma$$

to eqn (12.4); since β and δ are denominators of rational numbers, y is not zero.

Case (b) In the case $v \equiv 3 \pmod 4$ we introduce a new variable x_{v+1}, and continue the change of variables as far as

$$\begin{pmatrix} y_{v-2} \\ y_{v-1} \\ y_v \\ y_{v+1} \end{pmatrix} = B \begin{pmatrix} x_{v-2} \\ x_{v-1} \\ x_v \\ x_{v+1} \end{pmatrix}.$$

Then eqn (12.12) becomes

$$\left(\sum_{j=1}^{v} A_j^2\right) + nx_{v+1}^2 = \left(\sum_{j=1}^{v} y_j^2\right) + y_{v+1}^2 + \lambda\xi^2$$

and rather than (12.13) we obtain

$$nx_{v+1}^2 = y_{v+1}^2 + \lambda\xi^2.$$

The proof continues similarly to that of the previous case. □

Example 2.3.11 is a particular case of Theorem 1(i): $v = 22$, $n = k - \lambda = 7 - 2$ which is not a square, and hence no $SB[7, 2; 22]$ design exists. Applying Theorem 1(ii) for $v = 53$, $k = 13$, $\lambda = 3$, we have $n = 10$. Thus the equation $x^2 = 10y^2 + 3z^2$ must have a solution in integers. Now $x^2 \equiv 3z^2 \pmod 5$ forcing $x \equiv z \equiv 0 \pmod 5$. Thus $x^2 - 3z^2$ is divisible by an even power of 5, but $10y^2$ is divisible by an odd power of 5. Hence no $SB[13, 3; 53]$ design exists.

Proof of Corollary 1.1. If $\lambda = 1$, then $k = n+1$ and $v = n^2 + n + 1$, so that

$$(v-1)/2 = n(n+1)/2,$$

which is even if $n \equiv 0$ or $3 \pmod 4$, odd otherwise. Thus eqn (12.4) becomes

$$x^2 = ny^2 + z^2$$

when $n \equiv 0$ or $3 \pmod 4$, which tells us nothing, because it has the solution $x = z = 1$, $y = 0$. But when $n \equiv 1$ or $2 \pmod 4$, eqn (12.4) becomes

$$x^2 = ny^2 - z^2$$

and more conveniently,

$$ny^2 = x^2 + z^2.$$

By Theorem 10, any prime factor congruent to 3 (modulo 4) which divides ny^2 must divide it to an even power, and hence must divide n to an even power. This proves the Corollary. □

Example 14. Neither 6 nor 14 can be written as the sum of two squares, but both are congruent to 2 (mod 4). Hence by Corollary 1.1, there are no $SB[7, 1; 43]$ or $SB[15, 1; 211]$ designs, that is, no projective planes of orders 6 or 14. On the other hand, $10 \equiv 2 \pmod 4$ but $10 = 3^2 + 1^2$, $12 \equiv 0 \pmod 4$ and $15 \equiv 3 \pmod 4$; so we have no information concerning the existence of projective planes of orders 10, 12, 15. □

No example has yet been constructed of an $SB[n+1, 1; n^2+n+1]$ design for which n is not a prime power, but as we saw in Chapter 8, such designs are known whenever n is a prime power. Maybe n must always be a prime power for the design to exist; maybe every design not forbidden by Corollary 1.1 exists; maybe there are further necessary conditions yet to be discovered but the design exists in some cases where n is not a prime power. Any of these possibilities is consistent with the known theory.

Proof of Corollary 1.2. If $\lambda = 2$, we have $k = n+2$ and $v = (n^2 + 3n + 4)/2$. Now $n \equiv 0$ or $1 \pmod 4$ implies that v is even; so by Theorem 1.1, n is a perfect square. We now consider the cases where $n \equiv 2$ or $3 \pmod 4$. If $n \equiv 2$ or $3 \pmod 8$, then $(v-1)/2$ is odd; if $n \equiv 6$ or $7 \pmod 8$, then $(v-1)/2$ is even. Thus, eqn (12.4) becomes

$$x^2 = ny^2 - 2z^2 \text{ for } n \equiv 2 \text{ or } 3 \pmod 8 \tag{12.14}$$

and

$$x^2 = ny^2 + 2z^2 \text{ for } n \equiv 6 \text{ or } 7 \pmod 8. \tag{12.15}$$

Since $n = l^2h$, where h is square-free, these equations can be written as

$$x^2 - h(ly)^2 \pm 2z^2 = 0,$$

so we may as well deal with

$$x^2 - hy^2 \pm 2z^2 = 0,$$

and assume that the coefficient of y^2 is square-free. Now suppose that $(x, z) = t > 1$, and let $x = tX$, $z = tZ$. Then

$$t^2X^2 - hy^2 \pm 2t^2Z^2 = 0$$

and hence $t^2 | hy^2$. But h is square-free, so $t | y$, and we may cancel any common factor. Hence, without loss of generality, we assume $(x, z) = 1$.

For any prime π dividing h, we have

$$x^2 \pm 2z^2 \equiv 0 \pmod \pi.$$

If $\pi | z$, then $\pi | x$ also, contradicting our assumption that $(x, z) = 1$. Hence $\pi \nmid z$, and there exists an integer m such that

$$mz \equiv x \pmod \pi.$$

This means that

$$m^2z^2 \equiv x^2 \pmod \pi$$

and hence that $m^2 \equiv \pm 2 \pmod \pi$.

Now we apply Corollary 3.1. If $n \equiv 2$ or $3 \pmod 8$, eqn (12.14) holds and we must have $m^2 \equiv -2 \pmod \pi$, so $\pi \equiv 1$ or $3 \pmod 8$ or $\pi = 2$. If $n \equiv 6$ or $7 \pmod 8$, eqn (12.15) holds and we must have $m^2 \equiv 2 \pmod \pi$, so $\pi \equiv 1$ or $7 \pmod 8$ or $\pi = 2$. This proves the Corollary. □

12.5. Matrices defined in terms of the quadratic character

Some of the matrices needed for the proof of Theorem 2 are defined in terms of the quadratic character. We need one further property of this map itself first.

Lemma 15. *Let F be the field $GF[p^n]$ where p is odd and let $F^* = F \backslash \{0\}$. If c is any element of F^*, then*

$$\sum_{x \in F^*} \chi(x).\chi(x + c) = -1.$$

Proof. Choose an element $c \in F^*$. Provided $x \neq 0$, $\chi(x^{-1}) = \pm 1$; so $\chi(x^{-1}).\chi(x^{-1}) = 1$. Hence

$$\begin{aligned}\chi(x).\chi(x+c) &= \chi(x).1.\chi(x+c)\\ &= \{\chi(x).\chi(x^{-1})\}.\{\chi(x^{-1}).\chi(x+c)\}\\ &= \chi(xx^{-1}).\chi(x^{-1}(x+c))\\ &= 1.\chi(1+x^{-1}c)\\ &= \chi(1+x^{-1}c).\end{aligned}$$

Since $c \neq 0$, $x^{-1}c \neq 0$. If $x \neq y$, then $x^{-1}c \neq y^{-1}c$. Thus, as x runs through F^*, the product $x^{-1}c$ also runs through F^*. So

$$\begin{aligned}\sum_{x\in F^*} \chi(1+x^{-1}c) &= \sum_{x^{-1}c\in F^*} \chi(1+x^{-1}c)\\ &= \sum_{\substack{y\in F\\ y\neq 1}} \chi(y), \text{ where } y = 1+x^{-1}c,\\ &= \left(\sum_{y\in F} \chi(y)\right) - \chi(1).\end{aligned}$$

By eqn (12.5), $\sum_{y\in F} \chi(y) = 0$; since $\chi(1) = 1$, the result follows. □

For any odd prime p, we define two $p \times p$ matrices, $M = [m_{ij}]$ and $N = [n_{ij}]$. M is circulant, with

$$m_{ij} = \chi(j-i) \pmod p$$

and N is back-circulant, with

$$n_{ij} = \begin{cases}\chi(i+j), & i+j \not\equiv 0 \pmod p,\\ 1, & i+j \equiv 0 \pmod p.\end{cases}$$

Lemma 16. *For p odd, we have:*
(i) $MJ = JM = 0$;
(ii) $MM^T = pI - J$;
(iii) $M^T = (-1)^{(p-1)/2} M$.

Proof.
(i) Since $\sum_{i=0}^{p-1} \chi(i) = 0$, we have $MJ = JM = 0$.
(ii) If $MM^T = [l_{ij}]$, then

$$l_{ij} = \sum_{h=0}^{p-1} m_{ih}m_{jh} = \sum_{h=0}^{p-1} \chi(h-i).\chi(h-j).$$

If $i \neq j$, then

$$\begin{aligned}l_{ij} &= \sum_{h=0}^{p-1} \chi(x).\chi(x+c) \text{ where } x = h-i \text{ and } c = i-j\\ &= -1, \text{ by Lemma 15.}\end{aligned}$$

If $i = j$, then

$$l_{ij} = \sum_{h=0}^{p-1} \chi^2(x) \text{ where } x = h-i = h-j$$

$$= p-1.$$

Now (ii) follows.

(iii) Since $m_{ij} = \chi(j-i)$ and $m_{ji} = \chi(i-j)$, we have $m_{ij} = m_{ji}$ for $p \equiv 1 \pmod 4$ and $m_{ij} = -m_{ji}$ for $p \equiv 3 \pmod 4$. This proves (iii). □

Corollary 16.1. *If p is a prime, $p \equiv 1 \pmod 4$, let the matrices X and Y be defined by*

$$X = I + M \quad \text{and} \quad Y = I - M.$$

Then (i) $X^T = X$, $Y^T = Y$;
(ii) $XY^T = YX^T$;
(iii) $XX^T + YY^T = 2(p+1)I - 2J$;
(iv) $XJ = YJ = J$;
(v) $X(J-2I) = (J-2I)X$, $\quad Y(J-2I) = (J-2I)Y$. □

Corollary 16.2. *If p is a prime, $p \equiv 3 \pmod 4$, then:*
(i) $NJ = JN = J$;
(ii) $NN^T = (p+1)I - J$;
(iii) $N^T = N$;
(iv) $N(J-2I) = (J-2I)N$. □

12.6. Orthogonal designs

An orthogonal design is a generalization of an Hadamard matrix, and we use it in proving Theorem 2.

An *orthogonal design*, D, of side n and type $(u_1, u_2, \ldots, u_t)$ is a square matrix with the following properties:

(i) its entries are chosen from the set $\{0, \pm x_1, \pm x_2, \ldots, \pm x_t\}$, where the x_i's are commuting variables;

(ii) $\pm x_i$ occurs u_i times in each row and each column, for $i = 1, \ldots, t$;

(iii) the distinct rows of D are formally orthogonal.

Thus

$$DD^T = \sum_{i=1}^{n} u_i x_i{}^2 I_n.$$

We shall often refer to D as an $(n; u_1, u_2, \ldots, u_t)$ design.

An Hadamard matrix is thus an orthogonal design of side h and type (h); it has no zero entries, and just one variable which is assigned the value one.

Example 17. Some small orthogonal designs are shown in Table 12.1. Notice that (ii) is the matrix B which occurs in the proof of Theorem 1(ii). □

Table 12.1. Orthogonal designs

$$\begin{bmatrix} x & y \\ y & -x \end{bmatrix} \quad \begin{bmatrix} a & -b & -c & -d \\ b & a & -d & c \\ c & d & a & -b \\ d & -c & b & a \end{bmatrix} \quad \begin{bmatrix} a & b & b & d \\ -b & a & d & -b \\ -b & -d & a & b \\ -d & b & -b & a \end{bmatrix} \quad \begin{bmatrix} a & 0 & -c & 0 \\ 0 & a & 0 & c \\ c & 0 & a & 0 \\ 0 & -c & 0 & a \end{bmatrix}$$

(i) (2; 1, 1); (ii) (4; 1, 1, 1, 1); (iii) (4; 1, 1, 2); (iv) (4; 1, 1).

Lemma 18. *Let D be an orthogonal design of side n and type $(u_1, u_2, \ldots, u_t)$ on the commuting variables $x_1, x_2, \ldots, x_t$. Then D can be written as*

$$D = x_1A_1 + x_2A_2 + \ldots + x_tA_t,$$

where for each $i, j \in \{1, \ldots, t\}$:
 (i) *A_i is an $n \times n$ matrix with entries $0, \pm 1$;*
 (ii) *$A_iA_i^T = u_iI_n$;*
 (iii) *$A_iA_j^T + A_jA_i^T = 0, \quad i \neq j$.*

Proof. By definition, $DD^T = \sum_{i=1}^{t} u_ix_i^2I_n$. $D = x_1A_1 + x_2A_2 + \ldots + x_tA_t$, and since $\pm x_i$ occurs u_i times in each row and each column, A_i is a $(0, \pm 1)$-matrix with u_i non-zero entries in each row and each column. This proves (i). Now

$$\begin{aligned} DD^T &= (x_1A_1 + \ldots + x_tA_t)(x_1A_1^T + \ldots + x_tA_t^T) \\ &= x_1^2A_1A_1^T + x_2^2A_2A_2^T + \ldots + x_t^2A_tA_t^T \\ &\quad + x_1x_2(A_1A_2^T + A_2A_1^T) + \ldots + x_{t-1}x_t(A_{t-1}A_t^T + A_tA_{t-1}^T). \end{aligned}$$

Let $D_{(r)}$ denote the matrix D, with $x_i = 0$ for all $i \neq r$. Then

$$D_{(r)}D_{(r)}^T = u_rx_r^2I_n = x_r^2A_rA_r^T$$

proving (ii). Now let $D_{(r,s)}$ denote the matrix D, with $x_i = 0$ for all $i \neq r, i \neq s$. Then

$$\begin{aligned} D_{(r,s)}D_{(r,s)}^T &= u_rx_r^2I_n + u_sx_s^2I_n \\ &= x_r^2A_rA_r^T + x_s^2A_sA_s^T + x_rx_s(A_rA_s^T + A_sA_r^T); \end{aligned}$$

with (ii), this proves (iii). □

Lemma 19. *Let D be an $(n;\ u_1, u_2, \ldots, u_t)$ orthogonal design, on the t commuting variables $x_1, x_2, \ldots, x_t$. Then the following orthogonal designs exist:*
 (i) *$(n;\ u_1, u_2, \ldots, u_i + u_j, \ldots, u_t)$ on $t-1$ variables;*
 (ii) *$(n;\ u_1, \ldots, u_{i-1}, u_{i+1}, \ldots, u_t)$ on $t-1$ variables;*
 (iii) *$(2n;\ u_1, u_2, \ldots, u_t)$ on t variables;*

(iv) $(2n; 2u_1, 2u_2, \ldots, 2u_t)$ *on* t *variables;*
(v) $(2n; u_1, u_1, u_2, \ldots, u_t)$ *on* $t+1$ *variables;*
(vi) $(2n; u_1, u_1, 2u_2, \ldots, 2u_t)$ *on* $t+1$ *variables.*

Proof. In the design D make the following changes.

(i) Set $x_i = x_j$.
(ii) Set $x_i = 0$.
(iii) Replace x_i by $\begin{bmatrix} x_i & 0 \\ 0 & x_i \end{bmatrix}$ for each $i = 1, 2, \ldots, t$.
(iv) Replace x_i by $\begin{bmatrix} x_i & x_i \\ x_i & -x_i \end{bmatrix}$ for each $i = 1, 2, \ldots, t$.
(v) Replace x_1 by $\begin{bmatrix} x_0 & x_1 \\ -x_1 & x_0 \end{bmatrix}$; replace x_i by $\begin{bmatrix} 0 & x_i \\ x_i & 0 \end{bmatrix}$ for each $i = 2, 3, \ldots, t$.
(vi) Replace x_1 by $\begin{bmatrix} x_0 & x_1 \\ -x_1 & x_0 \end{bmatrix}$; replace x_i by $\begin{bmatrix} x_i & x_i \\ x_i & -x_i \end{bmatrix}$ for each $i = 2, 3, \ldots, t$. □

Example 20. Let D_1 and D_2 be the designs of Table 12.1(ii) and (i) respectively. Applying methods of Lemma 19 to these designs gives examples as follows. D_1 is a (4; 1, 1, 1, 1) design; letting $b = c$ as in Lemma 19(i) gives the (4; 1, 1, 2) design in Table 12.1 (iii); letting $d = 0$ as in Lemma 19(ii) gives the (4; 1, 1, 1) design in Table 12.2(i). D_2 is a (2; 1, 1) design; replacing variables by 2×2 matrices as in Lemma 19(iii), (iv), (v) gives the (4; 1, 1), (4; 2, 2), (4; 1, 1, 1) designs in Table 12.2(ii), (iii), (iv) respectively. □

Table 12.2

$$\begin{bmatrix} a & -b & -c & 0 \\ b & a & 0 & c \\ c & 0 & a & -b \\ 0 & -c & b & a \end{bmatrix} \quad \begin{bmatrix} x & 0 & y & 0 \\ 0 & x & 0 & y \\ y & 0 & -x & 0 \\ 0 & y & 0 & -x \end{bmatrix} \quad \begin{bmatrix} x & x & y & y \\ x & -x & y & -y \\ y & y & -x & -x \\ y & -y & -x & x \end{bmatrix} \quad \begin{bmatrix} z & x & 0 & y \\ -x & z & y & 0 \\ 0 & y & -z & -x \\ y & 0 & x & -z \end{bmatrix}$$

(i) (4; 1, 1, 1) (ii) (4; 1, 1) (iii) (4; 2, 2) (iv) (4; 1, 1, 1)

Lemma 21. *Let* a, b, c, x, y, z *be non-negative integers such that* $a+b+c = n$ *and* $x+y+z = 2n$. *Suppose that for all possible choices of* a, b, c, *an* $(n; a, b, c)$ *orthogonal design exists. Then for all possible choices of* x, y, z, *a* $(2n; x, y, z)$ *orthogonal design exists.*

Proof. Notice first that we make the convention that an $(n; a, b)$ orthogonal design may also be considered as an $(n; a, b, 0)$ orthogonal design, and so on.

Let x, y, z, be non-negative integers such that $x+y+z = 2n$, and assume that $0 \leq x \leq y \leq z$, so that $y \leq n$. Four cases arise.

(i) Both x and y are even, so we may write $x = 2a$, $y = 2b$, and $a+b < n$. By hypothesis an $(n; a, b, c)$ orthogonal design exists, where $c = n-a-b$. Hence by Lemma 19(vi), a $(2n; a, a, 2b, 2c)$ orthogonal design exists, and, by Lemma 19(i), a $(2n; 2a, 2b, 2c)$ orthogonal design also. This is the design we want.

(ii) Next, let x be even and y odd, so we may take $x = 2a$, $y = 2a+l$. Now $a+y = 3a+l$, and $z = 2n-4a-l$. Since $y \leq z$, we have $3a+l \leq n$. Thus an $(n; y, a, n-a-y)$ orthogonal design exists, and as before this means that a $(2n; y, y, 2a, 2n-2a-2y)$ orthogonal design also exists. Setting $x_1 = x_4$, we get a $(2n; y, 2a, 2n-2a-y)$ orthogonal design. Since $2a = x$ and $2n-2a-y = z$, the last design is the required one.

(iii) If x is odd and y is even, we may take $x = 2a+1$, $y = 2b$, and $z = 2t+1$. Since $x+y+z = 2n$, we have $a+b+t+1 = n$. Now by assumption $a < t$, so $x+b = 2a+b+1 < n$. Hence we have the following orthogonal designs: $(n; x, b, n-x-b)$, $(2n; x, x, 2b, 2n-2x-2b)$ and $(2n; x, 2b, 2n-x-2b)$. Since $y = 2b$, and $z = 2n-x = y$, we have the required design.

(iv) Finally, if x and y are both odd, we let $y = x+2b$, where $b \geq 0$. Since $x+b \leq n$, we have orthogonal designs

$$(n;\ x,\ b,\ n-x-b),\ (2n;\ x,\ x,\ 2b,\ 2n-2x-2b),$$

and finally $(2n; x, x+2b, 2n-2x-2b)$, as required. □

Corollary 21.1. *If x, y, z are non-negative integers such that $x+y+z = 2^m$, then a $(2^m; x, y, z)$ orthogonal design exists.*

Proof. From Example 17 and Lemma 19, the statement is true for $m = 2$. It then follows from Lemma 21 for all $m > 2$. □

Corollary 21.2. *If x, y are non-negative integers such that $x+y = 2^m$, then a $(2^m; x, y)$ orthogonal design exists.*

Proof. Apply Lemma 19(i) to the $(2^m; x, y, z)$ orthogonal design obtained from the previous corollary. □

12.7. An existence theorem for Hadamard designs

We need one further result from number theory.

Theorem 22. *Let x and y be positive integers such that $(x, y) = 1$. Then every integer $N \geq (x-1)(y-1)$ can be written as a linear combination $N = ax+by$, where a and b are non-negative integers.*

Proof. Let $x > y$, and consider the set, S, of the first x multiples of y: $S = \{0, y, 2y, \ldots, (x-1)y\}$. Since $(x, y) = 1$, one of these multiples of y

appears in each congruence class, modulo x. Suppose that $by \equiv i \pmod{x}$, where $by \in S$. If $j \equiv i \pmod{x}$, and $j \geq by$, then

$$j = ax + by,$$

where a and b are non-negative integers.

The largest element of S is $M = (x-1)y$, and hence the largest integer which cannot be written as a non-negative linear combination of x and y is $(x-1)y - x$, that is, the predecessor of M in its congruence class modulo x. But $(x-1)y - x = (x-1)(y-1) - 1$, and hence every integer $N \geq (x-1)(y-1)$ can be written as required. □

Corollary 22.1. *Let $z \geq 9$ be an odd integer, let $x = z+1$ and let $y = z-3$. Then there exist non-negative integers a and b such that*

$$a(z+1) + b(z-3) = n = 2^t$$

for some t.

Proof. Let $d = (z+1, z-3) = \begin{cases} 2 & \text{if } z \equiv 1 \pmod{4}, \\ 4 & \text{if } z \equiv 3 \pmod{4}. \end{cases}$

Let $N = \left(\frac{z+1}{d} - 1\right)\left(\frac{z-3}{d} - 1\right)$ and choose m so that $2^{m-1} < N \leq 2^m$. By Theorem 22 there exist non-negative integers a and b such that

$$a(z+1)/d + b(z-3)/d = 2^m$$

and thus

$$a(z+1) + b(z-3) = 2^{m+s},$$

where $s = \begin{cases} 1 & \text{if } z \equiv 1 \pmod{4}, \\ 2 & \text{if } z \equiv 3 \pmod{4}, \end{cases}$

and $t = m + s$. □

Lemma 23. *Let p be a prime, $p \geq 11$. Then there exists a positive integer t such that an Hadamard matrix of side $2^s p$ exists for every $s \geq t$.*

Proof. Let $x = p+1$ and $y = p-3$. By Corollary 22.1 there exist non-negative integers a and b such that $ax + by = 2^t = n$ for some t. By Corollary 21.1, there exists an $(n; a, b, n-a-b)$ orthogonal design, D, on the variables x_1, x_2, x_3.

The proof now divides into two cases.

Case 1: $p \equiv 3 \pmod{4}$. We replace each variable in D by a $p \times p$ $(1, -1)$ matrix: x_1 by J_p, x_2 by $J_p - 2I_p$, and x_3 by the back-circulant matrix N of Section 12.5. This gives an $np \times np$ $(1, -1)$ matrix E.

By the definition of D,

$$DD^T = (ax_1^2 + bx_2^2 + (n-a-b)x_3^2)I_n,$$

and by Corollary 16.2,

$$N^T = N,\ NJ_p = J_p,\ N(J_p - 2I_p) = (J_p - 2I_p)N,\ NN^T = (p+1)I_p - J_p.$$

Hence

$$\begin{aligned} EE^T &= (aJ_p^2 + b(J_p - 2I_p)^2 + (n-a-b)\,NN^T) \times I_n \\ &= (apJ_p + 4bI_p + b(p-4)J_p + (n-a-b)\,(p+1)I_p - (n-a-b)J_p) \times I_n \\ &= ((n(p+1) - a(p+1) - b(p-3))I_p + (a(p+1) + b(p-3) - n)J_p) \times I_n. \end{aligned}$$

But a and b were chosen so that $a(p+1)+b(p-3) = n$, which means that

$$EE^T = npI_p \times I_n = npI_{np}.$$

Hence E is an Hadamard matrix of side $np = 2^t p$, and by Corollary 2.6.23.2, the Lemma follows for $p \equiv 3 \pmod 4$.

Case 2: $p \equiv 1 \pmod 4$. There exists a $(2n;\ 2a,\ 2b,\ n-a-b,\ n-a-b)$ orthogonal design F on the variables $x_1,\ x_2,\ x_3,\ x_4$ by Lemma 19(vi). We replace each variable in F by a $p \times p$ $(1, -1)$ matrix: x_1 by J_p, x_2 by $J_p - 2I_p$, x_3, and x_4 respectively, by the circulant matrices X and Y of Corollary 16.1. This gives an $np \times np$ $(1, -1)$ matrix G.

By the definition of F,

$$FF^T = (2ax_1^2 + 2bx_2^2 + (n-a-b)x_3^2 + (n-a-b)x_4^2)I_{2n},$$

and by Corollary 16.1

$$X^T = X,\ Y^T = Y,\ XY^T = YX^T,\ XX^T + YY^T = 2(p+1)I_p - 2J_p,$$
$$XJ_p = YJ_p = J_p,\ X(J_p - 2I_p) = (J_p - 2I_p)X,\ Y(J_p - 2I_p) = (J_p - 2I_p)Y.$$

Hence GG^T

$$\begin{aligned} &= (2aJ_p^2 + 2b(J_p - 2I_p)^2 + (n-a-b)\,(XX^T + YY^T)) \times I_{2n} \\ &= (2apJ_p + 8bI_p + 2b(p-4)J_p + (n-a-b)\,(2(p+1)I_p - 2J_p)) \times I_{2n} \\ &= ((2n(p+1) - 2a(p+1) - 2b(p-3))I_p + (2a(p+1) + 2b(p-3) - 2n)J_p \times I_{2n}. \end{aligned}$$

But, as before, a and b were chosen so that $a(p+1)+b(p-3) = n$, which means that

$$GG^T = 2npI_p \times I_{2n} = 2npI_{2np}.$$

Hence, G is an Hadamard matrix of side $2np = 2^{t+1}p$, and by Corollary 2.6.23.2, the Lemma also follows for $p \equiv 1 \pmod 4$.

This completes the proof for all primes, except 2, 3, 5, and 7. □

Lemma 24. *There exist Hadamard matrices of sides 2^t for all $t \geq 1$, and $2^t p$ for all $t \geq 2$ and $p = 3, 5, 7$.*

Proof. There exists an Hadamard matrix of side 2^t for $t \geq 1$, by Corollary 2.6.23.1.

By Corollary 2.6.23.2, if there exist Hadamard matrices of sides 12, 20, and 28, then there exist Hadamard matrices of sides $2^t p$ for all $t \geq 2$ and $p = 3, 5, 7$.

By Theorem 2.6.18, the existence of an Hadamard matrix of side 12 is equivalent to the existence of an $SB[5, 2; 11]$ design. Such a design was constructed in Example 2.4.10. Similarly, the difference sets of quadratic residues in $GF[19]$ and $GF[27]$ respectively lead to $SB[9, 4; 19]$ and $SB[13, 6; 27]$ designs (by Theorem 3.4.7) and hence to Hadamard matrices of sides 20 and 28. □

Theorem 25. *Let α be any positive integer. Then there exists $t = t(\alpha)$ such that an Hadamard matrix of side $2^s\alpha$ exists for every $s \geq t$.*

Proof. We apply Lemma 23 and/or Lemma 24 to each prime factor of α. Since a Kronecker product of Hadamard matrices is an Hadamard matrix, the result follows. □

Proof of Theorem 2. By Corollary 22.1, we can choose t so that

$$2^t \geq \left(\frac{z+1}{d} - 1\right)\left(\frac{z-3}{d} - 1\right),$$

where z is odd, and $d = (z+1, z-3)$.

If $z \equiv 1 \pmod 4$, then $d = 2$ and we must have $2^t \geq (z-1)(z-5)/4$. Since $(z-3)^2 > (z-1)(z-5)$, it is sufficient to ensure that

$$2^{t+2} > (z-3)^2;$$

that is,

$$t + 2 > 2\log_2(z-3).$$

Since t is an integer, we may choose

$$t = [2\log_2(z-3)] - 1.$$

Similarly, if $z \equiv 3 \pmod 4$, then $d = 4$ and we may choose

$$t = [2\log_2(z-5)] - 3.$$

By Lemma 23 and Lemma 24, these choices of t ensure the existence of an Hadamard matrix of side $2^t z$.

If $z = pq$ where p and q are primes, $p \equiv 1 \pmod 4$, $q \equiv 1 \pmod 4$, then there exists an Hadamard matrix of side $2^r pq$, where

$$r = [2\log_2(p-3)] + [2\log_2(q-3)] < [2\log_2(pq-3)].$$

Since an integer z which is a product of primes congruent to 1 (modulo 4) gives the greatest lower bound on the value of t for which we know an Hadamard matrix of side $2^t z$ exists, we have proved Theorem 2′. But now by Theorem 2.6.18, this is equivalent to Theorem 2. □

12.8. References and comments

Theorem 1 is usually known as the Bruck–Ryser–Chowla Theorem. Condition (i) was first proved by Schutzenberger (1949) and independently by Shrikhande (1950); condition (ii) with $\lambda = 1$ appeared in Bruck and Ryser (1949) and for arbitrary λ in Chowla and Ryser (1950). Related results are due to Hall and Ryser (1951) for the cyclic case and to Hughes (1957) for finite geometries.

Hadamard matrices were first described in Hadamard (1893). The proof of Theorem 2 is due to J. S. Wallis (1976). For accounts of these matrices and related designs see W. D. Wallis, A. P. Street, and J. S. Wallis (1972); for the orthogonal designs which generalize Hadamard matrices, see also Geramita and Seberry (1979).

The number theory needed in the proofs is described in various books. In particular, Theorem 6 is due to Axel Thue and Theorem 13 to Lagrange; see, for example, Shanks (1978) or Griffin (1954). For a fuller discussion of Theorem 22 see Chapter 13 of Honsberger (1976).

The quasi-residual design which is not residual, in Exercise 12.6, is due to Bhattacharya (1944); Lawless (1970) constructed other such designs. A quasi-residual $B[k, \lambda; v]$ design, with $\lambda \geq 3$, is a residual design if k is greater than some function of λ; see Bose, Shrikhande, and Singhi (1976) and Neumaier (1982). Kelly (1982) discusses the number of non-isomorphic embeddings possible. Results on block intersection sizes in quasi-residual designs are given by Stanton, Mullin and Lawless (1969). Some recent results on quasi-residual designs are given by Baartmans, Danhof, and Tan (1980), Mavron, Mullin, and Rosa (1984), and van Lint and Tonchev (1984).

Exercises

12.1. What does Theorem 1 imply about the possible parameters of an Hadamard design? What do Corollaries 1.1 and 1.2 say about the parameters of $B[n, 1; n^2]$ and $B[n, 2; n(n+1)/2]$ designs, respectively?

12.2. Which of the parameters in Table 12.3 do not correspond to an SBIBD with $\lambda = 2$?

Table 12.3

v	k	v	k	v	k
29	8	56	11	92	14
37	9	67	12	106	15
46	10	79	13	121	16

12.3. For $p = 5$, construct the matrices M of Lemma 16 and X, Y of Corollary 16.1. For $p = 7$, construct the matrices M of Lemma 16 and N of Corollary 16.2.

12.4. Let D denote the orthogonal design of Table 12.1(ii). Find A_i, $i = 1, \ldots, 4$, as defined in Lemma 18.

12.5. Let D denote the orthogonal design of Table 12.1(ii); that is, a (4; 1, 1, 1, 1) design. Construct from D designs with the following parameters:

$$(8; 1, 1, 1, 1);\ (8; 2, 2, 2, 2);\ (8; 1, 1, 1, 1, 1);\ (8; 1, 1, 2, 2, 2).$$

12.6. Table 12.4 shows a $B[6, 3; 16]$ design, that is, a quasi-residual of an $SB[9, 3; 25]$ design. By considering block intersection sizes, show that the design in Table 12.4 is not the residual of an $SB[9, 3; 25]$ design.

Table 12.4. A quasi-residual design which is not residual

1	2	3	4	5	6	2	5	8	11	13	15
1	2	3	4	7	8	2	6	7	11	12	14
1	2	9	10	12	13	2	7	8	10	14	16
1	3	10	11	12	15	3	4	11	14	15	16
1	4	9	13	14	16	3	5	6	10	13	14
1	5	7	10	14	15	3	5	8	9	12	14
1	5	7	11	13	16	3	6	7	12	13	16
1	6	8	9	11	14	3	7	8	9	13	15
1	6	8	12	15	16	4	5	7	9	11	12
2	3	9	10	11	16	4	5	8	10	12	16
2	4	12	13	14	15	4	6	7	9	10	15
2	5	6	9	15	16	4	6	8	10	11	13

13 Existence results: designs with index 1 and given block size

13.1. Statement of results

If there exists a BIBD with parameters v, b, r, k, λ, then by Theorem 1.2.1 we have $vr = bk$ and $\lambda(v-1) = r(k-1)$. This means that $b = \lambda v(v-1)/k(k-1)$ and $r = \lambda(v-1)/(k-1)$. Also by Theorem 2.1.2, we must have $v \leq b$. In other words,

$$\lambda(v-1) \equiv 0 \pmod{k-1}, \tag{13.1}$$

$$\lambda v(v-1) \equiv 0 \pmod{k(k-1)} \tag{13.2}$$

and

$$v \leq b \tag{13.3}$$

are necessary conditions for the existence of a $B[k, \lambda; v]$ design. By the constructions of Chapter 5, they are also sufficient in the cases $k = 2$ and $k = 3$. By Example 2.4.11 and Theorem 12.1.1, they are not always sufficient. In this chapter, we apply recursive methods to prove three results concerning the existence of designs with $\lambda = 1$.

Theorem 1. *A $B[4, 1; v]$ design exists if and only if $v \equiv 1$ or $4 \pmod{12}$.* □

Theorem 2. *Let $v \equiv 3 \pmod 6$. Then there exists a resolvable $B[3, 1; v]$ design, that is, a Kirkman triple system.* □

Finally, we prove an asymptotic existence result.

Theorem 3. *Let k be a fixed positive integer. Then there exists a $B[k, 1; v]$ design for all sufficiently large v such that*

$$v-1 \equiv 0 \pmod{(k-1)} \quad \textit{and} \quad v(v-1) \equiv 0 \pmod{k(k-1)}. \qquad \square$$

Although the final results of this chapter concern BIBDs, the proofs involve several more general kinds of designs, and we begin by describing these designs.

13.2. Designs related to BIBDs

As with a BIBD, we consider a design $(X, \mathscr{B})$, based on a v-set X, and having a collection $\mathscr{B}$ of blocks (or subsets of X). If all the block-sizes are chosen from the set K, and if every unordered pair of elements of X occurs in precisely λ of the blocks, then $(X, \mathscr{B})$ is said to be a *pairwise balanced* design (PBD) of index λ, based on X, or a $B[K,\lambda; v]$ design. For example, the design $(D3)$ of Section 1.2 is based on $X = Z_7$, and has blocks

$$\mathscr{B} = \{123456, 01, 02, 03, 04, 05, 06\}.$$

Thus it is a $B[\{2, 6\}, 1; 7]$ design. Notice that K may contain positive integers which do not actually occur as sizes of blocks: we may consider $(D3)$ to be a $B[\{2, 3, 4, 5, 6\}, 1; 7]$ design, or even a $B[\mathrm{N}, 1; 7]$ design.

Suppose that all blocks of the design are the same size, so that we may choose $K = \{k\}$. If the variety x appears in precisely r_x blocks, then the other $r_x(k-1)$ positions in these blocks must be occupied by the other $v-1$ varieties, λ times each. Thus

$$\lambda(v-1) = r_x(k-1),$$

so that each variety occurs in $r = \lambda(v-1)/(k-1)$ blocks. Hence, a $B[\{k\}, \lambda; v]$ design is a BIBD, and we write, as before, $B[k, \lambda; v]$.

If S is a set of integers, then we write a multiple or shift of S as $aS = \{as | s \in S\}$ or $S+a = \{s+a | s \in S\}$ respectively, as we did in Chapter 4.

The set of integers v for which a $B[K,\lambda; v]$ design exists will be denoted by $B(K,\lambda)$, and similarly, the set of integers v for which a $B[k,\lambda; v]$ exists will be denoted by $B(k, \lambda)$. Certainly,

$$K \subseteq B(K, \lambda), \tag{13.4}$$

for if $k \in K$, we can simply choose λ blocks, each consisting of $\{1, 2, \ldots, k\}$. Also, if there is a $B[K,\lambda; v]$, and $K \subseteq H$, then the same design is a $B[H,\lambda; v]$, so

$$B(K, \lambda) \subseteq B(H, \lambda). \tag{13.5}$$

Further, m copies of each block of a $B[K,\lambda; v]$ form a $B[K,m\lambda; v]$, so

$$B(K,\lambda) \subseteq B(K,m\lambda); \tag{13.6}$$

this is simply our earlier construction for multiple BIBDs applied to pairwise balanced designs.

Theorem 4. *If $v \in B(K',\lambda')$, and $K' \subseteq B(K,\lambda)$, then $v \in B(K,\lambda\lambda')$.*

Proof. If $v \in B(K', \lambda')$, then choose a $B[K', \lambda'; v]$ design and consider each of its blocks in turn. If the block A has k elements, then $k \in K'$ so $k \in B(K, \lambda)$. Hence on the set of k elements of A, we construct a $B[K, \lambda; k]$ design, and we replace the block A by the set of all blocks of this new design. (Notice that the design we choose might simply consist of λ copies of the block A itself.)

If $x, y \in A$, $x \neq y$, then $\{x, y\}$ is a subset of precisely λ of the blocks replacing A. Since $\{x, y\}$ was a subset of precisely λ' blocks of the original design, it is now a subset of precisely $\lambda\lambda'$ blocks of the new design.

The size of any new block is a member of K; hence we have a $B[K, \lambda\lambda'; v]$ design, as required. □

Corollary 4.1. *If $v \in B(K, 1)$ and $K \subseteq B(k, \lambda)$ then $v \in B(k, \lambda)$.* □

The procedure used in the proof of Theorem 4 is known as 'breaking up blocks'.

Next we need a special kind of *group divisible* design. This design is a triple $(X, \mathcal{G}, \mathcal{B})$, where X is a v-set of elements and $\mathcal{G}$ and $\mathcal{B}$ are two families of subsets of X with the following properties:

$\mathcal{G} = \{G_1, G_2, \ldots, G_h\}$ is a family of h n-subsets (or *groups*) of X which partition X;
$\mathcal{B} = \{B_1, B_2, \ldots, B_b\}$ is a family of b k-subsets (or *blocks*) of X;
$|G_i \cap B_j| \leq 1$, for $i = 1, \ldots, h$, for $j = 1, \ldots, b$;
each unordered pair of elements of X which belongs to different groups of the design occurs in precisely λ blocks of $\mathcal{B}$.

Notice that the 'groups' of $\mathcal{G}$ are not groups in the usual algebraic sense, and that these designs are both a generalization of the transversal designs of Chapters 6 and 7, and a special case of the group divisible designs of Chapter 11, in which we have $\lambda_1 = 0$, $\lambda_2 = \lambda$. A group divisible design with these parameters is a $GD[k, \lambda, n; v]$ design, and the set of positive integers v for which a $GD[k, \lambda, n; v]$ design can exist is denoted by $GD(k, \lambda, n)$. Certainly, since n-sets partition a v-set, we have $n \mid v$. If $n = 1$, so that every element forms a group, then we have a BIBD; that is, a $GD[k, \lambda, 1; v]$ design is a $B[k, \lambda; v]$ design. If $n = v$, then no block could contain more than one element, so a $GD[k, \lambda, v; v]$ design is trivial. We must always have $k \leq h$, the number of groups.

As before, we have multiple designs: if we have a $GD[k, \lambda, n; v]$ design, then writing down l copies of each block gives a $GD[k, l\lambda, n; v]$ design or in other words

$$GD(k, \lambda, n) \subseteq GD(k, l\lambda, n). \tag{13.7}$$

Also, if we take the groups as additional blocks, using λ copies of each group, then a $GD[k, \lambda, n; v]$ leads to a $B[\{k, n\}, \lambda; v]$, and

$$GD(k, \lambda, n) \subseteq B(\{k, n\}, \lambda). \tag{13.8}$$

Similarly, if we take a point, say ∞, where $\infty \notin X$, adjoin it to each group G_i, and then take λ copies of each $G_i \cup \{\infty\}$, $i = 1, \ldots, n$, as blocks, we again have a PBD, this time a $B[\{k, n+1\}, \lambda; v+1]$ design, so

$$GD(k, \lambda, n) + 1 \subseteq B(\{k, n+1\}, \lambda). \tag{13.9}$$

The most commonly used cases are those where $k = n$ in eqn (13.8), giving

$$GD(k, \lambda, k) \subseteq B(k, \lambda) \qquad (13.8')$$

and where $n = k - 1$ in (13.9) giving

$$GD(k, \lambda, k-1)+1 \subseteq B(k, \lambda). \qquad (13.9')$$

If $\lambda = 1$, we can say more.

Lemma 5. $GD(k, 1, k-1)+1 = B(k, 1)$.

Proof. We already know that $GD(k, 1, k-1)+1 \subseteq B(k, 1)$. To show equality, we must take a $B[k, 1; v]$ design and construct from it a $GD[k, 1, k-1; v-1]$ design. We choose any fixed point, say θ, of the $B[k, 1; v]$ design and delete it from each of the blocks it belongs to. The $(k-1)$-sets so formed must partition the remaining $v-1$ points and we take them to be the groups of our new design. The remaining blocks of the original design become the blocks of the new design. □

Example 6. Suppose we take the $B[3,1;9]$ design given in Table 13.1, and let $\theta = 9$. The groups come from blocks 789, 369, 159, 249, leaving the remaining eight blocks to form a $GD[3, 1, 2; 8]$ design. □

Table 13.1

123	147	159	168
456	258	267	249
789	369	348	357

Blocks of a $B[3,1;9]$ design

G_1: 15	B_1: 123	B_5: 267
G_2: 24	B_2: 456	B_6: 348
G_3: 36	B_3: 147	B_7: 168
G_4: 78	B_4: 258	B_8: 357

Groups and blocks of a $GD[3, 1, 2; 8]$ design

Corollary 5.1. *If $k-1$ is a prime power, then $(k-1)k \in GD(k, 1, k-1)$.*

Proof. By Theorems 8.2.12 and 8.4.18,

$$(k-1)^2+(k-1)+1 = (k-1)k+1 \in B(k, 1),$$

and hence by Lemma 5, $(k-1)k \in GD(k, 1, k-1)$. □

Corollary 5.2. *If k is a prime power, then $(k-1)(k+1) \in GD(k, 1, k-1)$.*

Proof. By Theorem 8.2.12, $(k-1)(k+1)+1 = k^2 \in B(k, 1)$, and hence by Lemma 5, $(k-1)(k+1) \in GD(k, 1, k-1)$. □

Example 7. If $k = 4$, then by Corollary 5.1 we have $12 \in GD(4, 1, 3)$, and by Corollary 5.2 we have $15 \in GD(4, 1, 3)$; see Exercise 13.1. □

Suppose that $v \in GD(k, \lambda, n)$. In view of eqn (13.8), if $n \in B(k, \lambda)$, then the blocks (groups) with n points can be broken up as in the proof of Theorem 4. That is, if $n \in B(k, \lambda)$, then

$$GD(k, \lambda, n) \subseteq B(k, \lambda). \tag{13.8''}$$

Similarly, in view of (13.9), if $n+1 \in B(k, \lambda)$, then

$$GD(k, \lambda, n)+1 \subseteq B(k, \lambda). \tag{13.9''}$$

In proving (13.9), we adjoined one point to the design. Sometimes we may adjoin a whole subdesign. For suppose that we have a $GD[k, \lambda, n; v]$ design, and a $B[k, 1; n+k]$ design. We choose k fixed points and adjoin them to the v-set of the $GD[k, \lambda, n; v]$ design. Now on the $(n+k)$-set consisting of the k new points together with the points of one group, we construct a $B[k, 1; n+k]$ design, chosen so that the set of k new points forms a block. We repeat this process with each group in turn, and take as the blocks of the new design both the blocks of the original $GD[k, \lambda, n; v]$ design and λ copies of each of the distinct blocks arising from the $B[k, 1; n+k]$ designs. Thus, if $n+k \in B(k, 1)$, then

$$GD(k, \lambda, n)+k \subseteq B(k, \lambda). \tag{13.10}$$

Example 8. Table 13.2 shows the three groups and 16 blocks of a $GD[3, 1, 4; 12]$ design, with $X = \{1, 2, \ldots, 9, \mathrm{A}, \mathrm{B}, \mathrm{C}\}$. We adjoin points D, E, F, and construct a $B[3, 1; 7]$ design on $G_i \cup \{\mathrm{D}, \mathrm{E}, \mathrm{F}\}$, $i = 1, 2, 3$, with $\{\mathrm{D}, \mathrm{E}, \mathrm{F}\}$ a block of each design. This gives 19 new blocks, as shown in Table 13.2, which together with the 16 original blocks form a $B[3, 1; 15]$ design. □

Table 13.2

G_1: 1234	159, 16A, 17B, 18C,
G_2: 5678	25A, 26B, 27C, 289,
G_3: 9ABC	35B, 36C, 379, 38A,
	45C, 469, 47A, 48B.

Group and blocks: $GD[3, 1, 4; 12]$ design

	D12,	D34,	E13,	E24,	F14,	F23,
DEF,	D56,	D78,	E57,	E68,	F58,	F67,
	D9A,	DBC,	E9B,	EAC,	F9C,	FAB.

New blocks, forming a $B[3, 1; 15]$ design

Just as we generalized from BIBDs, with all block sizes the same, to PBDs with different block sizes allowed, so we may generalize from group divisible to pairwise group divisible designs, where both groups and blocks may have different sizes. More precisely, if $K, M \subseteq \mathrm{N}$, then the triple $(X, \mathscr{G}, \mathscr{B})$ is a $GD[K, \lambda, M; v]$ design, *pairwise group divisible*, if X is a v-set of elements, and $\mathscr{G}$ and $\mathscr{B}$ are two families of subsets of X with the following properties:

$\mathscr{G} = \{G_1, G_2, \ldots, G_h\}$ is a family of h subsets (or groups) of X which partition X, and $|G_i| \in M$ for $i = 1, \ldots, h$;
$\mathscr{B} = \{B_1, B_2, \ldots, B_b\}$ is a family of b subsets (or blocks of X) and $|B_i| \in K$ for $i = 1, \ldots, b$;
$|G_i \cap B_j| \leq 1$ for $i = 1, \ldots, h$, and $j = 1, \ldots, b$;
each unordered pair of elements of X which belongs to different groups of the design occurs in precisely λ blocks of $\mathscr{B}$.

Notice that if $K = \{k\}$ and $M = \{n\}$, then a $GD[K, \lambda, M; v]$ design is just a $GD[k, \lambda, n; v]$ design. We denote by $GD(K, \lambda, M)$ the set of integers v for which $GD[K, \lambda, M; v]$ designs exist. Properties of $GD[k, \lambda, n; v]$ designs can be generalized; thus

$$M \subseteq GD(K, \lambda, M), \tag{13.11}$$

and

$$GD(K, \lambda, M) \subseteq GD(K, l\lambda, M) \tag{13.12}$$

where $l \in \mathrm{N}$. Again, we may generalize Theorem 4 and Corollary 4.1 in the following way; the proofs are left as exercises.

Theorem 9. *If* $v \in GD(K, \lambda', M')$ *and* $M' \subseteq GD(K, \lambda, M)$, *then* $v \in GD(K, \lambda\lambda', M)$. □

Corollary 9.1. *If* $\lambda = \lambda' = 1$, *and if* $v \in GD(K, 1, M')$ *and* $M' \subseteq GD(K, 1, M)$, *then* $v \in GD(K, 1, M)$. □

Similarly, eqns (13.8) and (13.9) respectively generalize to

$$GD(K, \lambda, M) \subseteq B(K \cup M, \lambda) \tag{13.13}$$

and

$$GD(K, \lambda, M) + 1 \subseteq B(K \cup (M+1), \lambda). \tag{13.14}$$

Theorem 10. *If* $v \in GD(S, 1, R)$, $mR \subseteq B(k, \lambda)$ *and* $mS \subseteq GD(k, \lambda, m)$, *then* $m \in B(k, \lambda)$.

Proof. Construct a $GD[S, 1, R; v]$ design based on Z_v. If G_i is a group of the original design, then $|G_i| \in R$, and we construct a $B[k, \lambda; m|G_i|]$ design based on $Z_m \times G_i$. If B_j is a block of the original design, then $|B_j| \in S$, and we construct a $GD[k, \lambda, m; m|B_j|]$ design based on $Z_m \times B_j$. The collection of all the blocks of these designs, for each G_i and B_j, is a $B[k, \lambda; mv]$ design based on $Z_m \times Z_v$. □

A similar proof (again left as an exercise) leads to

Theorem 11. *If* $v \in GD(S, 1, R)$, $mR+1 \subseteq B(k, \lambda)$ *and* $mS \subseteq GD(k, \lambda, m)$, *then* $mv+1 \in B(k, \lambda)$. □

We now generalize the technique of breaking up blocks introduced in the proof of Theorem 4. If a design is based on a set X, we may assign a *weight*, w_x, to each individual element of X. This means that we take a set W_x consisting of w_x copies of x, thus

$$W_x = \{x_1, x_2, \ldots, x_{w_x}\}.$$

We assume that the sets $\{W_x | x \in X\}$ are pairwise disjoint, and let

$$W_X = X^* = \bigcup_{x \in X} W_x.$$

Similarly, if $(X, \mathscr{G}, \mathscr{A})$ is a group divisible design, then assigning weights to the elements of X induces a new family $\mathscr{G}^*$ of groups, in the following way: if $G \in \mathscr{G}$, then

$$W_G = G^* = \bigcup_{x \in G} W_x$$

and

$$\mathscr{G}^* = \{G^* | G \in \mathscr{G}\}.$$

Also, each block $A \in \mathscr{A}$ corresponds to the set

$$W_A = \bigcup_{x \in A} W_x$$

which has a natural partition π_A into the sets W_x; that is,

$$\pi_A = (W_A, \{W_x | x \in A\}).$$

We are now ready to state

Theorem 12. *Suppose that* $(X, \mathscr{G}, \mathscr{A})$ *is a* $GD[K, 1, M; v]$ *design, the 'master' design, and that a weight,* w_x, *is assigned to each element* $x \in X$. *Suppose that, for each block* $A \in \mathscr{A}$, $(W_A, \{W_x | x \in A\}, \mathscr{B}_A)$ *is a* $GD[K_A, 1, \{w_x | x \in A\}; |W_A|]$ *design. Let*

$$\mathscr{A}^* = \bigcup_{A \in \mathscr{A}} \mathscr{B}_A.$$

Then $(X^*, \mathscr{G}^*, \mathscr{A}^*)$ *is a* $GD[\bigcup_{A \in \mathscr{A}} K_A, 1, \{|G^*| \,|\, G^* \in \mathscr{G}^*\}; |X^*|]$ *design.* □

Verifying the properties of $(X^*, \mathscr{G}^*, \mathscr{A}^*)$ is left as an exercise, but we give a small case in detail, in the next example.

Example 13. Let $X = \{1, 2, 3, 4, 5, 6\}$, let $K = \{2, 3\}$, let $M = \{1, 2, 3\}$, and let the groups and blocks of $(X, \mathscr{G}, \mathscr{A})$ be those listed in Table 13.3. Now choose weights $w_1 = w_4 = w_6 = 3$, $w_5 = 2$, $w_2 = w_3 = 1$. The sets W_G, for $G \in \mathscr{G}$, and W_A, for $A \in \mathscr{A}$, are listed in Table 13.4; the listing of W_A indicates its natural partition. We construct a $GD[\{3,2\}, 1, \{3,2,1\}; 6]$ design on W_{A_2}, a $GD[3, 1, 3; 9]$ design on W_{A_1}, a $GD[2, 1, \{3, 2\}; 5]$ design on W_{A_3},

$GD[2, 1, \{3, 1\}; 4]$ designs on each of W_{A_4}, W_{A_5}, W_{A_7}, and a $GD[2, 1, \{2, 1\}; 3]$ design on W_{A_6}. The blocks of all these designs are listed in Table 13.4. Taken together they form a $GD[\{3, 2\}, 1, \{5, 3\}; 13]$ design with 33 blocks. □

Table 13.3. The master design of Example 13

Groups		Blocks			
G_1:	1 2 3	A_1:	1 4 6	A_5:	3 4
G_2:	4 5	A_2:	2 5 6	A_6:	3 5
G_3:	6	A_3:	1 5	A_7:	3 6
		A_4:	2 4		

Groups and blocks of $(X, \mathscr{G}, \mathscr{A})$.

Table 13.4. Groups and blocks of a $GD[\{3, 2\}, 1, \{5, 3\}; 13]$ design

Groups, G_i^*		W_{A_i}, showing partitions		Blocks of $\mathscr{A}^*$		
G_1^*: $1_1\ 1_2\ 1_3\ 2\ 3$	W_{A_1}:	$1_1\ 1_2\ 1_3$	$\mathscr{B}_{A_1}$:	$1_1\ 4_1\ 6_1$	$1_1\ 4_2\ 6_3$	$1_1\ 4_3\ 6_2$
G_2^*: $4_1\ 4_2\ 4_3\ 5_1\ 5_2$		$4_1\ 4_2\ 4_3$		$1_2\ 4_2\ 6_2$	$1_2\ 4_3\ 6_1$	$1_2\ 4_1\ 6_3$
G_3^*: $6_1\ 6_2\ 6_3$		$6_1\ 6_2\ 6_3$		$1_3\ 4_3\ 6_3$	$1_3\ 4_1\ 6_2$	$1_3\ 4_2\ 6_1$
	W_{A_2}:	2	$\mathscr{B}_{A_2}$:	$2\ 5_1\ 6_1$	$5_1\ 6_2$	$5_2\ 6_1$
		$5_1\ 5_2$		$2\ 5_2\ 6_2$	$5_1\ 6_3$	$5_2\ 6_3$
		$6_1\ 6_2\ 6_3$		$2\ 6_3$		
	W_{A_3}:	$1_1\ 1_2\ 1_3$	$\mathscr{B}_{A_3}$:	$1_1\ 5_1$	$1_2\ 5_1$	$1_3\ 5_1$
		$5_1\ 5_2$		$1_1\ 5_2$	$1_2\ 5_2$	$1_3\ 5_2$
	W_{A_4}:	2	$\mathscr{B}_{A_4}$:	$2\ 4_1$	$2\ 4_2$	$2\ 4_3$
		$4_1\ 4_2\ 4_3$				
	W_{A_5}:	3	$\mathscr{B}_{A_5}$:	$3\ 4_1$	$3\ 4_2$	$3\ 4_3$
		$4_1\ 4_2\ 4_3$				
	W_{A_6}:	3	$\mathscr{B}_{A_6}$:	$3\ 5_1$	$3\ 5_2$	
		$5_1\ 5_2$				
	W_{A_7}:	3	$\mathscr{B}_{A_7}$:	$3\ 6_1$	$3\ 6_2$	$3\ 6_3$
		$6_1\ 6_2\ 6_3$				

We also need some further properties of *transversal* designs, the special class of group divisible designs in which $|G_i \cap B_j| = 1$, for each group G_i and each block B_j. Thus h, the number of groups, equals k, the block-size, so that $v = kn$ and $b = \lambda n^2$ (Lemma 6.3.9). In the notation of Chapters 6 and 7, a $T[k, \lambda; n]$ design is a $GD[k, \lambda, n; kn]$ design, and the set of integers n for which a $T[k, \lambda; n]$ design exists is denoted by $T(k, \lambda)$. Theorem 7.3.4 shows that $n \in T(5, 1)$, provided $n \neq 2, 3, 6$, or 10.

We also use *truncated* transversal designs, similar to those used in the proof of Theorem 7.1.1. Suppose that we have a $T[k+1, \lambda; n]$ design, and that we delete some points from at most one of its groups, leaving a truncated group of size m, $0 \leq m \leq n$. These points are also deleted from all the blocks in which they occur, giving a $GD[\{k,k+1\},\lambda,\{m,n\};kn+m]$ design which is pairwise group divisible. Thus if $n \in T(k+1,1)$ and $0 \le m \le n$, then

$$kn+m \in GD(\{k,k+1\},\lambda,\{m,n\}). \tag{13.15}$$

From eqns (13.13) and (13.15), and Theorem 4, we have also (see Exercise 13.4)

Lemma 14. *If* $n \in T(k+1,1)$, $0 \le m \le n$, *and* $\{k,k+1,m,n\} \subseteq B(K,\lambda)$, *then*

$$kn+m \in B(K,\lambda).$$ □

13.3. Existence of BIBDs with block-size 4

Lemma 15. *Let* $M_4 = \{1,4,5,8,9,12,13,28,29\}$. *If* $u \equiv 0$ *or* 1 (mod 4), *then* $u \in GD(\{4,5\},1,M_4)$.

Proof. If $u \in M_4$, then certainly $u \in GD(\{4,5\},1,M_4)$. If $u \notin M_4$, then we may write $u = 4n+m$, where $m,n \equiv 0$ or 1 (mod 4), with $0 \le m \le n$, and $n \in T(5,1)$. We may truncate a $T[4+1,1;n]$ design to obtain D, a $GD[\{4,5\},1,\{m,n\};u]$ design; see eqn (13.15). If $m,n \in M_4$, then D is the design we need. If either group-size does not belong to M_4, then break up the appropriate group(s) by repeating the procedure, until we arrive at a $GD[\{4,5\},1,M_4;u]$ design. The blocks of D, and of all designs created in breaking up groups, are taken as the blocks of the required design. □

Example 16. Suppose that $u = 153$. Then we may choose $m = 5, n = 37$ or $m = 9, n = 36$ or $m = 21, n = 33$ or $m = 25, n = 32$. If we choose $m = 9, n = 36$, then we immediately construct a $GD[\{4,5\},1,\{9,36\};153]$ design, D, with a family, $\mathscr{B}$, of blocks. Since $9 \in M_4$, we keep the group of size 9. Since $36 \notin M_4$, we must break up the remaining four groups.

Now $u' = 36$, and we may choose $m' = 0, n' = 9$ or $m' = 4, n' = 8$. If we choose $m' = 4, n' = 8$, for each group G_i, $i = 1,2,3,4$, then we construct a $GD[\{4,5\},1,\{4,8\};36]$ design, D_i, based on G_i, with a family, $\mathscr{B}_i$, of blocks. Our final design has one group of size 9 (from D), four groups of size 4 (one from each D_i), sixteen groups of size 8 (four from each D_i), and the blocks of all five designs D, D_1, D_2, D_3, D_4.

Notice that there is no special reason to choose the same values of m', n' for each D_i. □

Lemma 17. $3M_4+1 \subseteq B(4,1)$.

Proof. We check each case, as shown in Table 13.5. □

Table 13.5

u	$v = 3u+1$	$B[4,1; v]$
1	4	Trivial
4	13	Theorem 8.3.14
5	16	Theorem 6.2.12
8	25	Exercise 3.1(a)
9	28	Example 3.2.6
12	37	Exercise 3.1(b)
13	40	Exercise 3.1(c)
28	85	Exercise 13.5
29	88	Example 3.4.12

Lemma 18. *If* $v \equiv 1$ *or* 4 (mod 12), *then* $v \in B(4, 1)$.

Proof. Let $v = 3u+1$, where $u \equiv 0$ or 1 (mod 4). By Lemma 15, $u \in GD(\{4, 5\}, 1, M_4)$. By Lemma 17, $3M_4+1 \subseteq B(4, 1)$. By Example 7, $\{12, 15\} = 3\{4, 5\} \subseteq GD(4, 1, 3)$. Hence by Theorem 11, $v = 3u+1 \in B(4, 1)$. □

Proof of Theorem 1. By Lemma 18, if $v \equiv 1$ or 4 (mod 12), then a $B[4, 1; v]$ design exists. Conversely, if a $B[4, 1; v]$ design exists, then eqn (13.1) implies that $v \equiv 1$ (mod 3), and eqn (13.2) that $v(v-1) \equiv 0$ (mod 12). Hence we must have $v \equiv 1$ or 4 (mod 12). □

13.4. Existence of Kirkman triple systems

We give a recursive proof of Theorem 2 based essentially on Theorem 1 and Theorem 12. A Kirkman triple system of order v is a resolvable $B[3, 1; v]$ or $RB[3, 1; v]$ design, and $RB(3, 1)$ denotes the set of values of v for which an $RB[3, 1; v]$ design exists.

Lemma 19. *Let* $(X, \mathscr{B})$ *be a* $B[K, 1; v]$ *design such that* $2K+1 \subseteq RB(3, 1)$. *Then* $2v+1 \in RB(3, 1)$.

Proof. Let $Y = (X \times \{1, 2\}) \cup \{\infty\}$, where $\infty \notin X \times \{1, 2\}$. For every $B \in \mathscr{B}$, with $|B| = k$, an $RB[3, 1; 2k+1]$ design is defined on $(B \times \{1, 2\}) \cup \{\infty\}$, in such a way that $\{x_1, x_2, \infty\}$ is a block of this design for every $x \in B$. Now take the collection $\mathscr{C}$ of all the blocks of these designs, including just one copy of each block of the form $\{x_1, x_2, \infty\}$ for each $x \in X$.

This collection $\mathscr{C}$ of blocks clearly forms a $B[3, 1; 2v+1]$ design. We still need to check that this design is resolvable.

For each $x \in X$, let $B_{x1}, B_{x2}, \ldots, B_{xh}$ be the blocks of $\mathscr{B}$ which contain x. In the $RB[3,1;2k+1]$ design defined on $(B_{xi} \times \{1,2\}) \cup \{\infty\}$, consider the blocks of the resolution class which includes $\{x_1, x_2, \infty\}$, and denote the collection of blocks of this class, other than $\{x_1, x_2, \infty\}$, by $\mathscr{C}_{xi}$. Then $(\{x_1, x_2, \infty\}) \cup \mathscr{C}_{x1} \cup \mathscr{C}_{x2} \cup \ldots \cup \mathscr{C}_{xh} = \pi_x$ is a resolution class of $(Y, \mathscr{C})$ and $\{\pi_x | x \in X\}$ is a resolution of the design $(Y, \mathscr{C})$. □

Example 20. Let $K = \{4\}$ in Lemma 19, and let $(X, \mathscr{B})$ be the $B[4, 1; 13]$ design with $X = Z_{13}$, where $\mathscr{B}$ is the set of blocks developed from the starter block $B_0 = \{0, 1, 3, 9\}$ (mod 13). The $RB[3, 1; 9]$ design corresponding to B_0 is given by the rows of Table 13.6, where the columns indicate resolution classes.

The blocks of $\mathscr{B}$ containing 0 are

$$B_0 = \{0, 1, 3, 9\},\ B_4 = \{4, 5, 7, 0\},\ B_{10} = \{10, 11, 0, 6\},\ B_{12} = \{12, 0, 2, 8\},$$

where B_i denotes $B_0 + i$. Then the resolution class π_0 consists of the blocks listed in Table 13.7, and the remaining 12 resolution classes are developed (mod 13) from π_0. □

Table 13.6. Blocks of $\mathscr{C}$ from B_0

$\infty\ 0_1\ 0_2$	$\infty\ 1_1\ 1_2$	$\infty\ 3_1\ 3_2$	$\infty\ 9_1\ 9_2$
$1_1\ 3_1\ 9_1$	$0_1\ 3_1\ 9_2$	$0_1\ 9_1\ 1_2$	$0_1\ 1_1\ 3_2$
$1_2\ 9_2\ 3_2$	$0_2\ 9_1\ 3_2$	$0_2\ 1_1\ 9_2$	$0_2\ 3_1\ 1_2$

Table 13.7. The blocks of π_0

$\infty\ 0_1\ 0_2$	$1_1\ 3_1\ 9_1$	$4_1\ 5_1\ 7_2$	$10_1\ 6_1\ 11_2$	$12_1\ 2_1\ 8_2$
	$1_2\ 9_2\ 3_2$	$4_2\ 7_1\ 5_2$	$10_2\ 11_1\ 6_2$	$12_2\ 8_1\ 2_2$

Notice that in Example 20, we have used the cyclic nature of $(X, \mathscr{B})$ to simplify the description of $(Y, \mathscr{C})$; in general, this simplification is not available.

We already have Kirkman triple systems (or $RB[3, 1; v]$ designs) of order 9 (Table 13.6), order 15 (Example 8.1.2), order 21 (Table 13.8) and order 39 (Table 13.9). Let $K_3 = \{4, 7, 10, 19\}$, so that $2K_3 + 1 = \{9, 15, 21, 39\}$. If we can show that for every $v \equiv 1 \pmod 3$, there exists a $B[K_3, 1; v]$ design, then Theorem 2 will follow. To start the proof, we must deal individually with some small values of v.

Lemma 21. *Let* $V = \{v | v \equiv 1 \pmod 3), 4 \leq v \leq 46\} \cup \{79, 82\}$. *If* $v \in V$, *then there exists a* $B[K_3, 1; v]$ *design.*

Proof. For $v = 4, 7, 10$, or 19, the trivial design will do.

For $v = 13, 16, 25, 28, 37$, or 40, the existence of a $B[4, 1; v]$ design follows from Theorem 1.

For $v = 22$, take the Kirkman triple system of order 15 (Example 8.1.2) and let the resolution classes be denoted by $\pi_i, i = 1, \ldots, 7$. To the set $\{1, 2, \ldots, 15\}$ adjoin new elements $\{\infty_i, i = 1, \ldots, 7\}$; to each block of π_i, adjoin the element ∞_i; to the set of blocks, adjoin the new block $\{\infty_1, \infty_2, \infty_3, \infty_4, \infty_5, \infty_6, \infty_7\}$. This gives a $B[\{4,7\},1;22]$ design.

Table 13.8. Kirkman triple system of order 21

1	2	4	1	11	16	1	8	15	1	12	21	1	5	6
3	8	21	2	3	5	2	12	17	2	9	16	2	13	15
5	9	18	4	9	15	3	4	6	3	13	18	3	10	17
6	13	20	6	10	19	5	10	16	4	5	7	4	14	19
7	10	15	7	14	21	7	11	20	6	11	17	7	12	18
11	12	14	8	12	13	9	13	14	8	10	14	8	9	11
16	17	19	17	18	20	18	19	21	15	19	20	16	20	21

1	13	19	1	3	7	1	9	17	1	10	20	1	14	18
2	6	7	2	14	20	2	10	18	2	11	21	2	8	19
3	14	16	4	8	17	3	11	19	3	12	15	3	9	20
4	11	18	5	12	19	4	12	20	4	13	16	4	10	21
5	8	20	6	9	21	5	13	21	5	14	17	5	11	15
9	10	12	10	11	13	6	14	15	6	8	18	6	12	16
15	17	21	15	16	18	7	8	16	7	9	19	7	13	17

Table 13.9. 19 starter blocks for a Kirkman triple system of order 39; each block gives a resolution class when developed (mod 39) by adding $3i$, $i = 0, 1, \ldots, 12$

0	1	2	12	31	38	12	32	37
3	9	27	13	32	36	13	30	38
4	10	28	14	30	37	14	31	36
5	11	29	21	25	35	21	26	34
6	15	18	22	26	33	22	24	35
7	16	19	23	24	34	23	25	33
8	17	20						

For $v = 31$, an exactly analogous construction starting with the Kirkman triple system of order 21 given in Table 13.8 leads to a $B[\{4, 10\}, 1; 31]$ design.

For $v = 34$, take $X = (Z_9 \times Z_3) \cup \{\infty\}$ and take four starter blocks

$$B_1 = \{0_0, 3_0, 6_0, \infty\},\ B_2 = \{0_0, 2_1, 2_2, 3_2\},\ B_3 = \{0_0, 3_1, 5_1\},$$
$$B_4 = \{0_0, 4_1, 8_1\},$$

which can be developed to give a $B[\{3,4\}, 1; 28]$ design, where (i, x) is written as i_x. The blocks of size three fall into six resolution classes: if $k \in \{0, 1, 2\}$, then

$$\pi(k_3) = \{B_3 + i_x, \quad i - x \equiv k \pmod 3\}$$

and

$$\pi(k_4) = \{B_4 + i_x, \quad i \equiv k \pmod 3\}.$$

To the set X adjoin six new elements $\{\infty_{ab}; a = 3, 4; b = 0, 1, 2\}$; to each block of the class $\pi(k_3)$, adjoin ∞_{3k}; to each block of the class $\pi(k_4)$, adjoin ∞_{4k}; to the set of blocks adjoin $\{\infty, \infty_{30}, \infty_{31}, \infty_{32}, \infty_{40}, \infty_{41}, \infty_{42}\}$. This gives a $B[\{4, 7\}, 1; 34]$ design.

For the four remaining cases, we apply the method of Theorem 12. In each case, $v = 3u + 1$, and we take a $GD[4, 1, \{2, 3\}; u]$ design as a master design. We give each point a weight of 3 and, on the set of 12 points corresponding to each block, we construct a $GD[4, 1, 3; 12]$ design (as given in Example 6.3.7). The groups now correspond to sets of 6 or 9 points each, and we adjoin a new element, ∞, to each of them, giving sets of size 7 or 10. We now have $3u + 1 = v$ elements, all the size 4 blocks from the $GD[4, 1, 3; 12]$ designs, and all the size 7 or 10 blocks from the groups together with ∞. This gives a $B[\{4, 7, 10\}, 1; v]$ design, as required. The master designs for each case, $v = 43, 46, 79, 82$, are listed in Table 13.10.

This completes the proof. □

Table 13.10. Master designs for Lemma 21

$v = 43$: $X = Z_{14}, G_0 = \{0, 7\}, B_0 = \{2, 4, 7, 8\}$.
Seven groups, 14 blocks, are developed modulo 14, giving a $GD[4, 1, 2; 14]$ design.

$v = 46$: From $B[4, 1; 16]$ design, (Theorem 6.2.12), delete one point, leaving a $GD[4, 1, 3; 15]$ design.

$v = 79$: $X = Z_{26}, G_0 = \{0, 13\}, B_{01} = \{2, 6, 13, 18\}, B_{02} = \{4, 12, 10, 13\}$.
13 groups, 52 blocks are developed modulo 26, giving a $GD[4, 1, 2; 26]$ design.

$v = 82$: From $B[4, 1; 28]$ design (Example 3.2.6), delete one point, leaving a $GD[4, 1, 3; 27]$ design.

Lemma 22. *If* $v \equiv 1 \pmod 3$, *then* $v \in B(K_3, 1)$.

Proof. We have verified this statement in Lemma 21 for $v \leq 46$, $v = 79$, $v = 82$. We now assume that $v \equiv 1 \pmod 3$, $v \geq 49$, $v \neq 79$, $v \neq 82$, and that $u \in B(K_3, 1)$ for all $u \equiv 1 \pmod 3$, $u < v$. We can write

$$v = 12m + 3t + 1,$$

where $m \geq 4$, $0 \leq t \leq m$, $m \neq 6$, $m \neq 10$. Hence by Theorem 7.3.4, there exists a $T[5, 1; m]$ design, which is a $GD[5, 1, m; 5m]$ design. If we delete $m - t$ elements from one of the groups, and delete each of these elements from each block which contains it, we obtain a $GD[\{4, 5\}, 1, \{m, t\}; 4m + t]$ design.

Now we apply the method of Theorem 12 again, giving each element a weight of 3, and on the set of 12 or 15 points corresponding to each block, we construct a $GD[4, 1, 3; 12]$ or a $GD[4, 1, 3; 15]$ design. Collecting the blocks of all these designs gives a $GD[4, 1, \{3m, 3t\}; 12m+3t]$ design. Finally, we adjoin a new element, ∞, to each of the groups, and take these extended groups as blocks also. This gives a $B[\{4, 3t+1, 3m+1\}, 1; v]$ design.

Our choice of m and t ensures that $\{3t+1, 3m+1\} \subseteq B(K_3, 1)$. By Corollary 4.1, there exists a $B[K_3, 1; v]$ design. □

Proof of Theorem 2. If there exists an $RB[3,1;v]$ design, then certainly $v \equiv 3 \pmod 6$. Conversely suppose that $v = 6u+3$. By Lemma 22, there exists a $B[K_3, 1; 3u+1]$ design. Since $2K_3+1 \subseteq RB(3, 1)$, by Lemma 19, $2(3u+1)+1 = v \in RB(3, 1)$. □

13.5. Asymptotic existence results: initial designs

To start the recursion we need difference families in finite fields. Let p be a prime, $q = p^n = mf+1$, and consider $F = GF[q]$, the field of order q. Its multiplicative group $F^* = F\backslash\{0\}$ is cyclic, and has a unique subgroup C_0 of index m, consisting of the mth powers of the non-zero field elements. If θ is a primitive element of F, then the multiplicative cosets $C_0, C_1 = \theta C_0, \ldots, C_{m-1} = \theta^{m-1}C_0$ partition F^*. They are called the *cyclotomic classes of index m and order f.*

Lemma 23. *Let $q = 2mt+1$ be a prime power, where $m = \binom{k}{2}$. If there exists a k-tuple $B = (a_1, a_2, \ldots, a_k) \in F^k$ such that the m differences*

$$\{a_j - a_i \mid 1 \le i < j \le k\}$$

form a system of representatives for the cyclotomic classes $C_0, C_1, \ldots, C_{m-1}$ of index m in F, then there exists a difference family from which a $B[k, 1; q]$ design may be developed.

Proof. Let $B = (a_1, a_2, \ldots, a_k)$. Now $2m \mid (q-1)$, so $-1 \in C_0$. Thus, if x is an element of C_0, so is $-x$. Let S be a system of representatives of the cosets of $\{1, -1\}$ in C_0. In other words, S consists of half the elements of C_0, one from each pair $\{x, -x\} \subseteq C_0$.

Then $\{sB \mid s \in S\}$ is a difference family with $\lambda = 1$, for among the differences from sB we find

$$\pm(sa_j - sa_i) = \pm s(a_j - a_i) \qquad \text{for all } s \in S.$$

These cover precisely the cyclotomic class represented by $a_j - a_i$. □

Example 24. Let $v = q = 19$, $k = m = t = 3$. Then $C_0 = \pm\{1, 7, 8\}$, $C_1 = \pm\{2, 3, 5\}$, $C_2 = \pm\{4, 6, 9\}$. The block $B = \{0, 1, 6\}$ has differences $1, 5, 6$, one in each cyclotomic class. The set $S = \{1, -7, 8\}$ is a suitable set of representatives. Then the blocks $1B = B = \{0, 1, 6\}$, $-7B = \{0, 12, 15\}$, $8B = \{0, 8, 10\}$, developed (mod 19), give a $B[3, 1; 19]$ design. □

For q a prime power, $q \equiv 1 \pmod{k(k-1)}$, let $N(k, q)$ denote the number of k-tuples $(a_1, a_2, \ldots, a_k)$ with the property required in Lemma 23. We show that if q is sufficiently large relative to k, then $N(k, q) > 0$ and therefore such a k-tuple exists. Our argument depends on properties of the mean and variance of sets of real numbers, and sets of elements in a finite field.

Lemma 25. *Let $\alpha_1, \alpha_2, \ldots, \alpha_N$ be real numbers, with mean*

$$\mu = (\alpha_1 + \alpha_2 + \ldots + \alpha_N)/N$$

and variance

$$\sigma^2 = \frac{1}{N}\sum_{i=1}^{N}(\alpha_i - \mu)^2 = \frac{1}{N}\left(\sum_{i=1}^{N}\alpha_i^2\right) - \mu^2.$$

If $m \le N$, then

$$|(\alpha_1 + \ldots + \alpha_m) - m\mu|^2 \le m(N-m)\sigma^2 \le N^2\sigma^2/4.$$

Proof. Without loss of generality, we assume $\mu = 0$. Let $\boldsymbol{a}^T = (\alpha_1, \ldots, \alpha_N)$, $\boldsymbol{w}^T = (1, \ldots, 1, 0, \ldots, 0)$ with m ones, $N - m$ zeros, and $\boldsymbol{j}^T = (1, \ldots, 1)$ as usual. Then the dot product $\langle \boldsymbol{a}, \boldsymbol{j} \rangle = 0$ and by the Cauchy–Schwartz inequality we have

$$|\langle \boldsymbol{a}, \boldsymbol{w} \rangle| = |\langle \boldsymbol{a}, \boldsymbol{w} - \beta \boldsymbol{j} \rangle| \le \|\boldsymbol{a}\| . \|\boldsymbol{w} - \beta \boldsymbol{j}\|$$

for any real β. If we choose $\beta = (N-m)/N$, then we have

$$\begin{aligned}(\alpha_1 + \ldots + \alpha_m)^2 &\le (\alpha_1^2 + \ldots + \alpha_N^2)\{m(N-m) + (N-2m)^2\}/N \\ &\le (\alpha_1^2 + \ldots + \alpha_N^2)m(N-m)/N\end{aligned}$$

as required. □

Next consider a prime power $q = mf + 1$. Let F denote the field $GF[q]$, and $C_i, i = 0, 1, \ldots, m-1$, the cyclotomic classes of index m. For any set X, we let X^r denote the set of all r-tuples $(x_1, \ldots, x_r)$ of elements of X, and $X^{(r)}$ the subset of X^r consisting of r-tuples with distinct elements. If $|X| = n$, then $|X^r| = n^r$ and $|X^{(r)}| = n(n-1)(n-2) \ldots (n-r+1) = n^{(r)}$.

Let $\boldsymbol{i} = (i_1, i_2, \ldots, i_r) \in Z_m^r$, and let $\boldsymbol{a} = (a_1, a_2, \ldots, a_r) \in F^{(r)}$. Consider a field element $x \in F$ such that

$$x - a_1 \in C_{i_1}, x - a_2 \in C_{i_2}, \ldots, x - a_r \in C_{i_r}.$$

We denote the number of such elements, x, by $E_{\boldsymbol{i}}(\boldsymbol{a})$.

Lemma 26. *Over the $m^r q^{(r)}$ choices of $\boldsymbol{i}$ and $\boldsymbol{a}$, the mean value of $E_{\boldsymbol{i}}(\boldsymbol{a})$ is*

$$\mu_r = (q-r)/m^r \tag{13.16}$$

and the variance is

$$\sigma_r^2 = \frac{q(q-1)}{m^r q^r}\left(\frac{q-m-1}{m}\right)^{(r)} + \frac{q-r}{m^r} - \left(\frac{q-r}{m^r}\right)^2 \tag{13.17}$$

$$< (q-r)/m^r. \tag{13.18}$$

Proof. For a given $\boldsymbol{a} \in F^{(r)}$, any value of $x \in F \backslash \{a_1, \ldots, a_r\}$ is feasible for some $\boldsymbol{i} \in Z_m^r$, so

$$\sum_i E_i(\boldsymbol{a}) = q - r.$$

Hence

$$\sum_{\boldsymbol{a}} \sum_i E_i(\boldsymbol{a}) = (q-r)q^{(r)} = q^{(r+1)},$$

and (13.16) follows.

Again we let $E^{(2)} = E(E-1)$, and $[E_i(\boldsymbol{a})]^{(2)}$ denote the number of pairs $(x, y) \in F^{(2)}$ such that $x - a_j \in C_{i_j}$, $y - a_j \in C_{i_j}$, for $j = 1, 2, \ldots, r$. Suppose that $\boldsymbol{a} \in F^{(r)}$ is fixed. Then

$$\sum_i [E_i(\boldsymbol{a})]^{(2)}$$

counts the number of $(x, y) \in F^{(2)}$ such that $x - a_j$, $y - a_j$ belong to the same cyclotomic class for $j = 1, 2, \ldots, r$. Further,

$$\sum_{\boldsymbol{a}} \sum_i [E_i(\boldsymbol{a})]^{(2)}$$

counts the number of $(r+2)$-tuples

$$(a_1, a_2, \ldots, a_r; x, y) \in F^{(r+2)},$$

such that for each $j = 1, 2, \ldots, r$, the differences $x - a_j$ and $y - a_j$ are in the same cyclotomic class. Now for fixed $(x, y) \in F^{(2)}$, we find such an $(r+2)$-tuple by considering the set

$$S(x, y) = \{c \in F \mid x - c,\ y - c \text{ are in the same class}\},$$

and choosing $a_1, \ldots, a_r$ as distinct elements of $S(x, y)$.

The differences $x - c$ and $y - c$ are in the same class if and only if

$$x - c = g(y - c)$$

for some $g \in C_0$, $g \neq 1$. There are $(q-1)/m$ elements in C_0 and hence $(q-m-1)/m$ choices for g, each of which corresponds to a unique value of c; that is, for any choice of $(x, y) \in F^{(2)}$, we have

$$|S(x, y)| = (q-m-1)/m.$$

Hence

$$\sum_{\boldsymbol{a}} \sum_i [E_i(\boldsymbol{a})]^{(2)} = q(q-1)\left(\frac{q-m-1}{m}\right)^{(r)}. \tag{13.19}$$

Now $\sigma_r^2 = \dfrac{1}{m^r q^{(r)}} \sum_a \sum_i E_i(a)(E_i(a)-1) + \mu_r - \mu_r^2$; so eqns (13.16) and (13.19) imply (13.17). Since $q = mf + 1$, we have

$$\frac{q(q-1)}{q^{(r)}} \left(\frac{q-m-1}{m} \right)^{(r)} < \frac{(q-r)^2}{m^r},$$

and the inequality (13.18) follows. □

We now apply these results to prove that $N(k, q) > 0$ for q large relative to k.

Theorem 27. *Let* $m = \binom{k}{2}$, $q = 2mt + 1$, *where* q *is a prime power. Then*

$$\left| N(k, q) - \frac{m!}{m^m} q^{(k)} \right| < m^{(k-1)/2} q^{(2k-1)/2}.$$

Proof. Again let F denote $GF[q]$. For $0 \le r \le k$, let $\mathcal{M}_r$ be the set of $a \in F^{(r)}$ such that the differences

$$a_j - a_i, \quad 1 \le i < j \le r$$

belong to $\binom{r}{2}$ distinct cyclotomic classes of index m in F. Let $M_r = |\mathcal{M}_r|$, so that $M_0 = 1$, $M_1 = q$, $M_2 = q(q-1)$, and $M_k = N(k, q)$.

The first r entries of a member of $\mathcal{M}_{r+1}$ determine a member of $\mathcal{M}_r$. For $a = (a_1, \ldots, a_r) \in \mathcal{M}_r$, we let $E'(a)$ be the number of $x \in F$ such that $(a_1, \ldots, a_r, x) \in \mathcal{M}_{r+1}$. Then

$$M_{r+1} = \sum_{a \in \mathcal{M}_r} E'(a).$$

Now there are $m - \binom{r}{2}$ elements $i \in Z_m$ for which the class C_i is not represented by a difference from $(a_1, \ldots, a_r)$, and from these elements we may choose $\left(m - \binom{r}{2}\right)^{(r)}$ r-tuples of distinct $i_1, \ldots, i_r$. Then

$$E'(a) = \sum E_i(a),$$

where the sum is taken over these $\left(m - \binom{r}{2}\right)^{(r)}$ choices of i. Thus altogether M_{r+1} is a sum of $M_r\left(m - \binom{r}{2}\right)^{(r)}$ of the numbers $E_i(a)$.

By Lemma 25

$$\left| M_{r+1} - \left\{ \left(m - \binom{r}{2} \right)^{(r)} M_r \right\} \frac{q-r}{m^r} \right| \le (m^r . q^{(r)} \sigma_r^2 / 4)^{1/2}.$$

By the inequality (13.18),

$$q^{(r)}\sigma_r^2 < q^{2r+1},$$

so

$$|M_{r+1} - c_r M_r| < d_r, \quad \text{for } r = 0, 1, \ldots, k-1,$$

where

$$c_r = \frac{q-r}{m^r}\left(m - \binom{r}{2}\right)^{(r)}, d_r = (m^{r/2}q^{(2r+1)/2})/2.$$

Since

$$|M_1 - c_0 M_0| < d_0 \quad \text{and} \quad |M_2 - c_1 M_1| < d_1,$$

we have

$$\begin{aligned}|M_2 - c_1 c_0 M_0| &= |M_2 - c_1 M_1 + c_1(M_1 - c_0 M_0)| \\ &\leq |M_2 - c_1 M_1| + c_1 |M_1 - c_0 M_0| \\ &< d_1 + c_1 d_0.\end{aligned}$$

Since $M_0 = 1$, repetition of this argument leads to

$$|M_{r+1} - c_r c_{r-1} \ldots c_1 c_0| < d_r + c_r d_{r-1} + c_r c_{r-1} d_{r-2} + \ldots + c_r c_{r-1} \ldots c_1 d_0.$$

Since

$$\prod_{r=0}^{k-1}\left(m - \binom{r}{2}\right)^{(r)} = m!$$

we have

$$c_{k-1}c_{k-2} \ldots c_1 c_0 = \frac{q^{(k)}}{m^m} \cdot m!$$

Since $c_r < q$, and $d_{r+1} = m^{1/2} q d_r$, we have

$$d_{k-1} + c_{k-1}d_{k-2} + \ldots + c_{k-1}c_{k-2} \ldots c_1 d_0 < d_{k-1} + q d_{k-2} + \ldots + q^{k-1} d_0.$$

Also

$$q^{k-r} d_{r-1} = (q^{(2k-1)/2} m^{(r-1)/2})/2$$

so

$$\begin{aligned}\left|M_k - \frac{m!\,q^{(k)}}{m^m}\right| &< (q^{(2k-1)/2}(m^{(k-1)/2} + m^{(k-2)/2} + \ldots + 1))/2 \\ &< m^{(k-1)/2} q^{(2k-1)/2}.\end{aligned}$$

□

Corollary 27.1. *Let $q = 2mt + 1$ be a prime power, where $m = \binom{k}{2}$. Then $N(k, q) > 0$ whenever $q > e^{k^2} k^{2k}$.*

Proof. By Theorem 27,

$$N(k, q) > \frac{m!}{m^m} q^{(k)} - m^{(k-1)/2} q^{(2k-1)/2},$$

so

$$q^{-(2k-1)/2}N^{(k,q)} > \frac{m!}{m^m}\sqrt{q}\left(\frac{q^{(k)}}{q^k}\right) - m^{(k-1)/2}.$$

Now $e^m = \sum_{i=0}^{\infty} \frac{m^i}{i!} > \frac{m^m}{m!}$, so $\frac{m!}{m^m} > e^{-m}$. Also $\frac{q^{(k)}}{q^k} > e^{-k/2}$.

Thus for $q > e^{k^2}k^{2k}$, we have

$$q^{-(2k-1)/2}N(k,q) > e^{-k^2/2}\sqrt{q} - k^k,$$

from which the corollary follows. □

At this stage we have $B[k, 1; v]$ designs for some $v \equiv 1 \pmod{k(k-1)}$. Now suppose that we are given v_0, with $v_0 - 1 \equiv 0 \pmod{k-1}$ and $v_0(v_0-1) \equiv 0 \pmod{k(k-1)}$. We show next how to construct a $B[k, 1; v]$ design for some $v \equiv v_0 \pmod{k(k-1)}$.

If X is the v-set on which our design is based, we work in terms of the $\binom{v}{2} \times \binom{v}{k}$ incidence matrix, N, of pairs and k-sets where, for the pair P and the k-set K,

$$n_{PK} = \begin{cases} 1 & \text{if } P \subseteq K, \\ 0 & \text{otherwise.} \end{cases}$$

A $B[k, \lambda; v]$ design is thus a column vector s with $\binom{v}{k}$ entries, each a non-negative integer, such that

$$Ns = \lambda j. \tag{13.20}$$

Lemma 28. *Let y be a row vector of length $\binom{v}{2}$ with rational entries, such that every entry of yN is an integer. Then $yj \in Z$.*

Proof. By hypothesis, $\sum_P y_P n_{PK}$ is an integer; that is,

$$\sum_P y_P n_{PK} \equiv 0 \pmod 1$$

for every k-set $K \subseteq X$.

(i) Let $L \subseteq X$ be a $(k-2)$-set and let f, g, h, i be distinct elements of $X \backslash L$. Then

$$\begin{aligned} &y_{\{f,h\}} - y_{\{f,i\}} - y_{\{g,h\}} + y_{\{g,i\}} \\ &\quad = \sum_P y_P(n_{P,L\cup\{f,h\}} - n_{P,L\cup\{f,i\}} - n_{P,L\cup\{g,h\}} + n_{P,L\cup\{g,i\}}) \\ &\quad \equiv 0 \pmod 1. \end{aligned}$$

Now for fixed f, g, h, there are rationals z_f, z_g, z_h, such that

$$y_{\{f,g\}} = z_f + z_g,$$
$$y_{\{f,h\}} = z_f + z_h,$$
$$y_{\{g,h\}} = z_g + z_h.$$

We choose z_i for all $i \neq f$, so that

$$y_{\{g,i\}} = z_f + z_i.$$

Then

$$y_{\{g,i\}} \equiv y_{\{f,i\}} + y_{\{g,h\}} - y_{\{f,h\}} \equiv z_g + z_i \pmod 1$$

and

$$y_{\{i,j\}} \equiv y_{\{f,i\}} + y_{\{g,j\}} - y_{\{f,g\}} \equiv z_i + z_j \pmod 1.$$

(ii) Now, let $M \subseteq X$ be a $(k-1)$-set, and let f, g be distinct elements of $X \backslash M$. Then

$$(k-1)z_f - (k-1)z_g = \sum_{f \in P \subseteq M \cup \{f\}} y_P - \sum_{g \in P \subseteq M \cup \{g\}} y_P$$

$$= \sum_P y_P(n_{P,M \cup \{f\}} - n_{P,M \cup \{g\}}) \equiv 0 \pmod 1.$$

(iii) Next, let K be a k-set, $K \subseteq X$, and let $i \in X$. Then

$$\sum_{P \subseteq K} y_p = \sum_{\{i,j\} \subseteq K} y_{\{i,j\}} \equiv \sum_{\{i,j\} \subseteq K} (z_i + z_j) \quad \text{(by (i))}$$

$$= (k-1) \sum_{i \in K} z_i \equiv k(k-1)z_i \quad \text{(by (ii))}.$$

But $\sum_{P \subseteq K} y_P = \sum_P y_P n_{PK} \equiv 0 \pmod 1$; thus $k(k-1)z_i \equiv 0 \pmod 1$.

(iv) Finally, let K be a k-set, $K \subseteq X$. Then for each $i \in X$,

$$\lambda \sum_{P \subseteq X} y_P = \lambda \sum_{\{i,j\} \subseteq X} n_p \equiv \lambda \sum_{\{i,j\} \subseteq X} (z_i + z_j) = \lambda(v-1) \sum_{j \in X} z_j$$
$$\equiv \lambda(v-1)vz_i \qquad \text{(since } (k-1) \mid \lambda(v-1))$$
$$\equiv 0 \pmod 1 \qquad \text{(since } k(k-1) \mid \lambda v(v-1)). \qquad \square$$

Corollary 28.1. *$Ns = \lambda j$ has a solution s with integer entries.*

Proof. This follows from the fact that for every rational y such that yN is a vector of integers, we have yj an integer. $\square$

Theorem 29. *For some λ_0 depending only on v and k, whenever $\lambda \geq \lambda_0$, $\lambda(v-1) \equiv 0 \pmod{(k-1)}$, $\lambda v(v-1) \equiv 0 \pmod{k(k-1)}$, and $v \geq k+2$, then there exists a $B[k, \lambda; v]$ design.*

Proof. By Corollary 28.1, eqn (13.20) has a solution s with integer entries. Also

$$N(s+j) = Ns + \binom{v-2}{k-2} j,$$

so we can find a solution with non-negative integer entries by adding to s an appropriate multiple of j. This gives a design for $\lambda + c\lambda_1$ where $\lambda_1 = \binom{v-2}{k-2}$ and c is some constant, $c = c(\lambda)$.

The theorem now follows by choosing

$$\lambda_0 = \max\{\lambda + c(\lambda).\lambda_1 \mid \lambda < \lambda_1, \lambda(v-1) \equiv 0 \pmod{k-1}, \lambda v(v-1) \equiv 0 \pmod{k(k-1)}\}. \qquad \square$$

We now have a $B[k, \lambda; v]$ design with large λ, and we unfold it to give a design with $\lambda = 1$.

Theorem 30. *Suppose that q is a prime power and that there exists a $B[k, q; u]$ design. If $q \geq u+2$ and $d \geq \binom{u}{2}$, then there also exists a $GD[k, 1, q^d; uq^d]$ design.*

Proof. Let $(X, \mathscr{B})$ be the given $B[k, q; u]$ design and let V be a d-dimensional vector space over $F = GF[q]$. Let θ be a primitive element of F. We construct our new design on the set $V \times X$, writing each element as x_i, $x \in V$, $i \in X$. Each subset $V \times \{i\}$ is a group of the design.

Consider a pair $P = \{i, j\} \subseteq X$. There is a (multi)set of $\lambda = q$ blocks of $\mathscr{B}$, each of which contains P, and we define an arbitrary one-one mapping ζ_P which takes this collection of blocks onto F.

Further, since $d \geq \binom{u}{2}$, we assume that $X = \{1, 2, \ldots, u\}$, and order the family of unordered pairs of X lexicographically: thus

$$P_1 = \{1, 2\}, P_2 = \{1, 3\}, \ldots, P_u = \{2, 3\}, \ldots, P_{\binom{u}{2}} = \{u-1, u\}.$$

We choose a basis for V, and associate the lth basis vector with the pair P_l, for $1 \leq l \leq \binom{u}{2}$.

Let $H = \left\{v = (v_1, v_2, \ldots, v_d) \in V \mid \sum_{l=1}^{d} v_l = 0\right\}$, a hyperplane of V. For every $i \in X$, let T_i: $V \to V$ be the linear map defined by

$$T_i(y) = T_i((y_1, y_2, \ldots, y_d)) = (z_1, z_2, \ldots, z_d)$$

where

$$z_l = \begin{cases} y & \text{if } 1 \le l \le \binom{u}{2} \text{ and } i \in P_l, \\ y_l\theta^i & \text{otherwise.} \end{cases} \tag{13.21}$$

Note that if $P_l = \{i, j\}$ and $z \in V$, then there exists $y \in H$ with $T_j(y) - T_i(y) = z$ if and only if $z_l = 0$.

For each block $B \in \mathscr{B}$, we define a function f_B: $B \to V$, with the following property. If $f_B(i) = z = (z_1, \ldots, z_d)$, and if $P_l = \{i, j\}$, then

$$z_l = \begin{cases} 0 & \text{if } i < j, \\ \zeta_P(B) & \text{if } i > j. \end{cases} \tag{13.22}$$

If $l > \binom{u}{2}$, then z_l is chosen arbitrarily.

We can now finish the definition of our design. Choose $x \in V$, $y \in H$, $B \in \mathscr{B}$. This leads to a k-set

$$\{z \mid i \in B, z = x + T_i(y) + f_B(i)\},$$

and these k-sets are the blocks of the design. We must check that each pair $\{x_i, y_j\}$, $i \neq j$, is covered by exactly one block.

First, suppose that i and j are given. The pair $\{i, j\}$ determines q blocks of $\mathscr{B}$, and we have q^d choices for x, q^{d-1} choices for y. Hence there are q^d blocks of our new design from $\{i, j\}$. But given $\{i, j\}$, there are also q^{2d} pairs $\{x_i, y_j\}$, since $i \neq j$. Thus if we show that each pair $\{x_i, y_j\}$ is covered at least once, we have the required property of partial balance.

Now $\{x_i, y_j\}$ is covered if and only if $\{\mathbf{0}_i, (y-x)_j\}$ is covered; that is, if and only if the expression

$$T_j(y) - T_i(y) + f_B(j) - f_B(i)$$

runs through all vectors in V. But this follows from eqns (13.21) and (13.22). □

Corollary 30.1. *If q is a prime power and if there exist a $B[k, q; u]$ design and a $B[k, 1; q^d]$ design, and if $q \ge u + 2$ and $d \ge \binom{u}{2}$, then there exists a $B[k, 1; uq^d]$ design.*

Proof. On each group of the $GD[k, 1, q^d; uq^d]$ design of Theorem 30, construct a $B[k, 1; q^d]$ design. □

Theorem 31. *If $v_0 \equiv 1 \pmod{(k-1)}$ and $v_0(v_0 - 1) \equiv 0 \pmod{k(k-1)}$, then for any $M \ge 1$, there exists a $B[k, 1; v]$ design with $v \equiv v_0 \pmod{Mk(k-1)}$.*

Proof. Without loss of generality let $v_0 \ge k + 2$. By Theorem 29, there exists a $B[k, q; v_0]$ design where q is a prime power, $q \equiv 1 \pmod{Mk(k-1)}$. If d is large enough then by Corollary 27.1 there exists a $B[k, 1; q^d]$ design, and hence by Corollary 30.1, a $B[k, 1; v_0 q^d]$ design. □

Example 32. Let $q = 4, k = u = d = 2$. Then the construction of Theorem 30 enables us to derive a $GD[2, 1, 16; 32]$ design from a $B[2, 4; 2]$ design. Obviously a $GD[2, 1, 16; 32]$ design can be written down immediately; we are simply illustrating the method. Here $X = \{1, 2\}$, $\mathscr{B} = \{B_1, B_2, B_3, B_4\}$ where $B_i = \{1, 2\}, i = 1, 2, 3, 4$. $F = GF[4] = \{0, 1, \alpha, \beta\}$, where $\beta = \alpha + 1 = \alpha^2$. V is a two-dimensional vector space over F, so the groups of the design are G_1 and G_2, where

$$G_i = \left\{\begin{matrix} (0,0)_i, (0,1)_i, (0,\alpha)_i, (0,\beta)_i, (1,0)_i, (1,1)_i, (1,\alpha)_i, (1,\beta)_i \\ (\alpha,0)_i, (\alpha,1)_i, (\alpha,\alpha)_i, (\alpha,\beta)_i, (\beta,0)_i, (\beta,1)_i, (\beta,\alpha)_i, (\beta,\beta)_i \end{matrix}\right\},$$

for $i = 1, 2$. We choose

$$\{(1, 0) = v_1, (0, 1) = v_2\}$$

as a basis for V, where v_1 is associated with the pair $P_1 = \{1, 2\}$. The hyperplane H is just

$$\{(0, 0), (1, 1), (\alpha, \alpha), (\beta, \beta)\}.$$

We choose a primitive element, θ, of F to be $\theta = \alpha$, so

$$T_1(y) = T_1(y_1, y_2) = (y_1, y_2\alpha),$$
$$T_2(y) = T_2(y_1, y_2) = (y_1, y_2\alpha^2) = (y_1, y_2\beta).$$

We define $\zeta_1\colon \{B_1, B_2, B_3, B_4\} \to F$ by

$$\zeta_1(B_1) = 0, \quad \zeta_1(B_2) = 1, \quad \zeta_1(B_3) = \alpha, \quad \zeta_1(B_4) = \beta.$$

Hence

$$f_{B_1}(1) = (0, 0), \quad f_{B_1}(2) = (0, 1),$$
$$f_{B_2}(1) = (0, 0), \quad f_{B_2}(2) = (1, 0),$$
$$f_{B_3}(1) = (0, 0), \quad f_{B_3}(2) = (\alpha, \beta),$$
$$f_{B_4}(1) = (0, 0), \quad f_{B_4}(2) = (\beta, \alpha)$$

where the second entry has in each case been chosen arbitrarily.

Now choose $x = (1, 0)$, $y = (\alpha, \alpha)$, $B = B_2$. This leads to the 2-set

$$\{z \mid i \in B_2, z = (1, 0) + T_i(\alpha, \alpha) + f_{B_2}(i)\}.$$

Thus we find

$$z = (1, 0) + (\alpha, \beta) + (0, 0) = (\beta, \beta) \text{ for } i = 1,$$
$$z = (1, 0) + (\alpha, 1) + (1, 0) = (\alpha, 1) \text{ for } i = 2,$$

leading to the block $\{(\beta, \beta)_1, (\alpha, 1)_2\}$. Further blocks of the design are listed in Table 13.11.

Note that if $i = 1$, $j = 2$, then the expression $T_j(y) - T_i(y) + f_B(j) - f_B(i)$

Table 13.11. The 16 blocks corresponding to $y = (\alpha, \alpha)$, $B = B_2$

x	Block	x	Block
$(0, 0)$	$(\alpha, \beta)_1\,(\beta, 1)_2$	$(\alpha, 0)$	$(0, \beta)_1\,(1, 1)_2$
$(0, 1)$	$(\alpha, \alpha)_1\,(\beta, 0)_2$	$(\alpha, 1)$	$(0, \alpha)_1\,(1, 0)_2$
$(0, \alpha)$	$(\alpha, 1)_1\,(\beta, \beta)_2$	(α, α)	$(0, 1)_1\,(1, \beta)_2$
$(0, \beta)$	$(\alpha, 0)_1\,(\beta, \alpha)_2$	(α, β)	$(0, 0)_1\,(1, \alpha)_2$
$(1, 0)$	$(\beta, \beta)_1\,(\alpha, 1)_2$	$(\beta, 0)$	$(1, \beta)_1\,(0, 1)_2$
$(1, 1)$	$(\beta, \alpha)_1\,(\alpha, 0)_2$	$(\beta, 1)$	$(1, \alpha)_1\,(0, 0)_2$
$(1, \alpha)$	$(\beta, 1)_1\,(\alpha, \beta)_2$	(β, α)	$(1, 1)_1\,(0, \beta)_2$
$(1, \beta)$	$(\beta, 0)_1\,(\alpha, \alpha)_2$	(β, β)	$(1, 0)_1\,(0, \alpha)_2$

becomes $(0, y_2) + f_B(2)$, and as y_2 and B run through four values each, this expression does indeed run through the whole of V. □

13.6. Asymptotic existence results: recursive constructions

Now we have a design for one value of v in each possible residue class and we use recursive constructions to find designs for all v sufficiently large.

Let $R_k = \{r \in \mathrm{N} \mid \text{there exists a } B[k, 1; r(k-1)+1] \text{ design}\}$; that is, R_k is the set of all replication numbers which can occur in $B[k, 1; v]$ designs. For example, by Theorem 1, if $k = 4$ then there exists a $B[4, 1; v]$ design if and only if $v = 12t + 1$ or $12t + 4$ for some positive integer t. But now $r = 4t$ or $4t + 1$, respectively. Notice that if $v = 12t + 1$ or $12t + 4$, then v occurs as a replication number of a design on $36t + 4$ or $(36t + 12) + 1$ elements respectively. In other words, if there exists a $B[4, 1; v]$ design, then $v \in R_4$. A similar property is true in general.

Lemma 33. *If there exists a* $B[R_k, 1;\ v]$ *design, then* $v \in R_k$.

Proof. Let $(X, \mathscr{B})$ be a $B[R_k, 1; v]$ design, so that if $B \in \mathscr{B}$, then $|B| \in R_k$. Let $I = \{1, 2, \ldots, k-1\}$, and consider the set $Y = (X \times I) \cup \{\infty\}$, where ∞ is any element not belonging to $X \times I$. Now construct a $B[k, 1; v(k-1)+1]$ design on the set Y, in the following way. For each $B \in \mathscr{B}$, take the set $(B \times I) \cup \{\infty\}$, and on that set construct a $B[k, 1; |B|(k-1)+1]$ design which contains the blocks $\{x_1, x_2, \ldots, x_{k-1}, \infty\}$ for each $x \in B$. Now take the collection of all the blocks of these designs, including only one copy of each of the blocks $\{x_1, x_2, \ldots, x_{k-1}, \infty\}$. □

Note that this is an application of Theorem 12, where the master design is the original $B[R_k, 1; v]$ design considered as a $GD[R_k, 1, 1; v]$ design.

Corollary 33.1. $B(R_k, 1) = R_k$.

Proof. $R_k \subseteq B(R_k, 1)$ is immediate; $B(R_k, 1) \subseteq R_k$ is the content of Lemma 33. □

Lemma 34. *There exists $r \in R_k$ such that $r+1 \in R_k$.*

Proof. Consider $v \in B(k, 1)$. By Theorem 7.4.5, if v is large enough then, whenever $u \geq v-1$, $u \in T(k, 1)$; that is, there exists a $T[k, 1; u]$ design.

First, we construct a $B[k, 1; (v-1)k+1]$ design starting from a $T[k, 1; v-1]$ design, with groups $G_1, \ldots, G_k$, together with a new element ∞. We base a $B[k, 1; v]$ design on $G_i \cup \{\infty\}$, for each $i = 1, 2, \ldots, k$. The replication number of this design is $(v-1)k/(k-1) = r$, say.

Next, we construct a $B[k, 1; vk]$ design by taking a $T[k, 1; v]$ design and basing a $B[k, 1; v]$ design on each of its groups. The replication number of this design is $(vk-1)/(k-1) = r+1$. □

Lemma 35. *Suppose that $r, r+1, s, t \in R_k$, where $s \geq t$ and $s \in T(r+1, 1)$. Then $rs+t \in R_k$.*

Proof. We start from a $T[r+1, 1; s]$ design, and from one of its $r+1$ groups, we delete $s-t$ elements. This leaves r groups of size s, and one of size t, together with blocks of sizes r and $r+1$. Taking all the groups and blocks of this truncated transversal design gives a $B[\{s, t, r, r+1\}, 1; rs+t]$ design; that is, a $B[R_k, 1; rs+t]$ design. Hence by Lemma 33, $rs+t \in R_k$. □

Lemma 36. *Let M be a set of non-negative integers and let r be a fixed positive integer. Suppose that:*

(i) $0, 1 \in M$;

(ii) *if $s, t \in M$, $s \geq t$, then $rs+t \in M$.*

Then $n \in M$ if $n \equiv 0$ or 1 (mod r), provided that n is sufficiently large.

Proof. (a) M contains all values given by finite sums $\Sigma a_i r^i$, where $a_i = 0$ or 1.

(b) M contains all values $r^{l+2} + \sum_{i=0}^{l} b_i r^i$, where $0 \leq b_i \leq i+1$. To see this, observe that $r^3 + 2r + 1 = r(r^2+1) + (r+1)$,

$$r^4 + 3r^2 + 2r + 1 = r(r^3 + 2r + 1) + (r^2 + r + 1), \text{ and so on.}$$

(c) M contains all multiples ρ of r^{r-1}, where $\rho \geq r^{r^3+1}$. To see this, note that

$$\begin{aligned}\rho &= c_{l+2} r^{l+2} + \ldots + c_{r-1} r^{r-1} \\ &= r^{l+2} + ((c_{l+2} - 1)r^2 + c_{l+1} r + c_l) r^l + \ldots + c_{r-1} r^{r-1}\end{aligned}$$

where $0 \leq c_i < r$, $c_{l+2} \geq 1$, $l \geq r^3 - 1$, and $(c_{l+2} - 1)r^2 + c_{l+1} r + c_l \leq r^3$.

(d) Finally, we show that M contains representatives of each congruence class modulo $r^r\ (= r.r^{r-1})$ which is congruent to 0 or 1 (modulo r). Since

$$r(\ldots + r^{i-1} + \ldots) + (\ldots + d_i r^i + \ldots) = \ldots + (d_i + 1) r^i + \ldots,$$

the coefficient d_i of r^i can take any value (except when $i = 0$), and the result follows. □

Proof of Theorem 3. Since $v - 1 \equiv 0 \pmod{(k-1)}$, we write $v = r(k-1)+1$. Now we must prove that if $r(r-1) \equiv 0 \pmod{k}$, and if r is sufficiently large, then $r \in R_k$. By Lemma 34, we can choose $r_0 \in R_k$ such that $r_0 + 1 \in R_k$; in fact $r_0 = k(v-1)/(k-1)$ will do.

Now suppose that $t \in R_k$. By Lemma 35 and Lemma 36, the set R_k contains all sufficiently large r such that $r \equiv t \pmod{r_0}$. Since $k|r_0$, it is sufficient to show that for each r_1, with $r_1(r_1-1) \equiv 0 \pmod{k}$, there exists $r \in R_k$ such that $r \equiv r_1 \pmod{r_0}$. In other words, it is sufficient to show that for each v_1 such that $v_1 - 1 \equiv 0 \pmod{k-1}$ and $v_1(v_1-1) \equiv 0 \pmod{k(k-1)}$, there exists a $B[k, 1; v]$ design for some $v \equiv v_1 \pmod{r_0(k-1)}$.

Such a design is exactly what Theorem 31 provides. □

13.7. References and comments

Theorems 1, 2, and 3 of this chapter are special cases of far more general results. Theorem 1 and its generalizations, many of the constructions in Section 13.2, and Lemma 33, are due to Hanani (1975, 1961). Theorem 2 is due to Ray-Chaudhuri and R. M. Wilson (1971); for more general results, see also Hanani, Ray-Chaudhuri, and R. M. Wilson (1972), and Ray-Chaudhuri and R. M. Wilson (1973). The proof given here is that of Vanstone (1982), as reported by P. W. Robinson (1984). Theorem 3 is due to R. M. Wilson (1972*a*, *b*; 1975); we have made much use of other accounts of the work by R. M. Wilson (1974*c*) and Brouwer (1979). For the linear algebra assumed in the proof of Corollary 28.1, see, for instance, Section 108 of van der Waerden (1950).

Exercises

13.1. Construct the designs referred to in Example 7.

13.2. Prove Theorem 9 and Corollary 9.1.

13.3. Prove Theorem 11.

13.4. Prove Lemma 14.

13.5. Construct a $B[4, 1; 85]$ design, D, in the following way. Let the $B[4, 1; 28]$ design, E, referred to in Table 13.5 be based on the set $X = \{1, 2, \ldots, 28\}$. Let $S = \{a, b, c\}$. Then the underlying set of D is $(X \times S) \cup \{\infty\}$. For the 595 blocks of D, take:

28 blocks of the form $\{\infty, (i, a), (i, b), (i, c)\}$, for $i = 1, \ldots, 28$;

12 blocks constructed from each of the 63 blocks of E, as follows.

If $\{1, 2, 3, 4\}$ is a block of D, then construct the $GD[4, 1, 3; 12]$ on $\{1, 2, 3, 4\} \times S$, with groups $\{(i, a), (i, b), (i, c)\}$, $i = 1, 2, 3, 4$, and blocks listed in Table 13.12. Show that D is a $B[4, 1; 85]$ design.

Table 13.12

$(1, a), (2, a), (3, a), (4, a)$;	$(1, a), (2, b), (3, b), (4, b)$;	$(1, a), (2, c), (3, c), (4, c)$;
$(1, b), (2, a), (3, b), (4, c)$;	$(1, b), (2, b), (3, c), (4, a)$;	$(1, b), (2, c), (3, a), (4, b)$;
$(1, c), (2, a), (3, c), (4, b)$;	$(1, c), (2, b), (3, a), (4, c)$;	$(1, c), (2, c), (3, b), (4, a)$.

13.6. Show that Z_{11} may be partitioned into starter blocks of a $B[\{2, 3\}, 2; 11]$ design thus: $\{1, 2, 4\}, \{3, 6, 8\}, \{5, 9, 10\}, \{7, 0\}$. Now consider $Z_{11} \times Z_2$, where a_b denotes (a, b), and show that the blocks of Table 13.13, developed modulo 11, lead to a $B[4, 2; 22]$ design.

Table 13.13

$1_0\ 2_0\ 4_0\ 0_1$	$1_1\ 2_1\ 4_1\ 0_0$	
$3_0\ 6_0\ 8_0\ 0_1$	$3_1\ 6_1\ 8_1\ 0_0$	$7_0\ 0_0\ 7_1\ 0_1$
$5_0\ 9_0\ 10_0\ 0_1$	$5_1\ 9_1\ 10_1\ 0_0$	

13.7. Complete the construction of the design in Example 32.

13.8. Show directly that $B(R_3, 1) = R_3$.

14 Designs balanced for neighbouring varieties

14.1. Motivation

So far we have considered designs in which varieties (or treatments) have been allocated at random to plots within a block, and the order of allocation within a block has had no significance. In this chapter we consider designs in which the number of times that two varieties appear in adjacent plots is of interest. The designs may be block designs, in which case the neighbours of a given plot are those to its immediate left and right, or they may be arrays (such as Latin squares), in which case a given plot has four neighbours, namely those to its immediate left, right, top, and bottom.

There are several situations in which designs with some sort of 'neighbour balance' might be appropriate, and in this context 'neighbours' could mean neighbours in time or neighbours in space.

The first situation arises, for instance, in psychological experiments where several subjects are to be tested on successive occasions and the response to a particular treatment may be modified by the treatment applied on the preceding occasion. If possible we would like to have as many subjects as treatments, on each occasion to have one subject receiving each treatment, and to have each treatment preceded by every other. Hence if we let the subjects be the rows of the Latin square and the occasions be the columns, we want each ordered pair of distinct treatments to appear once in the rows of the square. If it is not possible to have as many subjects as treatments then each subject must receive more treatments and the number of occasions required will increase. But still it is often possible to construct suitable arrangements.

The second situation arises in field trials where we need to make allowances for local variations in soil fertility. Thus, we assume that results from neighbouring plots are correlated but that non-adjacent plots are still independent. For this situation we look at both block designs and Latin squares, and at some generalizations.

Sometimes circular blocks may arise, for instance in serology with Ouchterlony gel diffusion tests. Such a test is performed on a round plate of agar gel (see Fig. 14.1), with a drop of antiserum (S) developed in response to one virus placed in the centre of the plate and drops of the virus preparations or antigens (A, B, C, D, E, F here) placed around the plate. The antiserum and antigens diffuse through the gel. If the virus to which S is an antiserum has a component in common with a virus on the plate then a line of precipitate will form as the antiserum and the virus diffuse into each other. Thus in Fig. 14.1, S corresponds to a virus having components in common with A, B, C, and F, but not D or E.

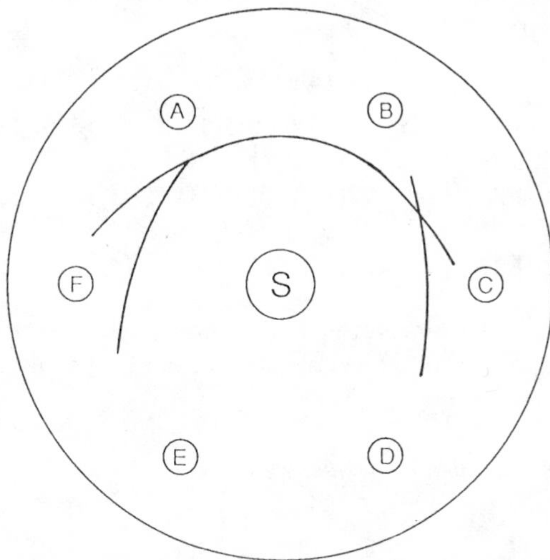

Fig. 14.1

Various types of precipitate lines form, three of which are shown in Fig. 14.1. The lines from viruses A and B merge, showing that the components of A and B reacting to S are also common to each other. The line from A continues behind that from F, showing that some component of S which does not react with F has diffused past the precipitate line from F and reacted with A. The lines from B and C cross each other, indicating that S has a component that reacts with B but not with C and one that reacts with C but not with B.

Hence, to get as much information as possible we need to have every virus next to every other virus in turn.

In the remainder of this chapter we consider methods of constructing various types of designs balanced for neighbouring treatments.

14.2. Circular block designs

A *neighbour design* (or a *circular block design*) for v treatments is a layout arranged in b blocks of size k such that each treatment appears r times and such that any two distinct treatments appear as *neighbours* (that is, in adjacent positions) λ' times. Note that we do not require that the treatments in a block be distinct, merely that a treatment never appear as its own neighbour. We will write a block of a neighbour design as $(a_1, a_2, \ldots, a_k)$ and say that the neighbours of a_i are a_{i-1} and a_{i+1}, where the subscripts are calculated modulo k. Thus the order of the treatments in a block is now important.

Example 1. The design in Table 14.1 is a neighbour design with $v = b = 15$, $r = k = 7$ and $\lambda' = 1$. □

Table 14.1

1	2	4	1	5	15	9	6	7	9	6	10	5	14	11	12	14	11	15	10	4
2	3	5	2	6	1	10	7	8	10	7	11	6	15	12	13	15	12	1	11	5
3	4	6	3	7	2	11	8	9	11	8	12	7	1	13	14	1	13	2	12	6
4	5	7	4	8	3	12	9	10	12	9	13	8	2	14	15	2	14	3	13	7
5	6	8	5	9	4	13	10	11	13	10	14	9	3	15	1	3	15	4	14	8

If $k = 2$ we adopt the convention that the pair of treatments in each block is counted twice as neighbours. Then we get the following relationships between the parameters.

Theorem 2. *In a neighbour design, we have:*
(i) $vr = bk$;
(ii) $\lambda'(v-1) = 2r$.

Proof. (i) Each of the v treatments appears r times and each block has k treatments in it.

(ii) Each of the r times that a treatment appears it is adjacent to two neighbours. Each of the $(v-1)$ other treatments appears as a neighbour λ' times. □

Hence from now on we shall usually refer to an $N[k, \lambda'; v]$ design.

Lemma 3. *If $k = 2$ or $v = 2$, then the necessary conditions are sufficient for the existence of a neighbour design.*

Proof. First, suppose $k = 2$. Then λ' is even, because of our convention of counting neighbours. Thus an $N[2, 2\mu; v]$ design consists of μ copies of each of the $\binom{v}{2}$ pairs as blocks.

Secondly, suppose $v = 2$. Then k must be even, otherwise a treatment is adjacent to itself. Thus an $N[2k_1, 2bk_1; 2]$ design has b blocks, each of size $2k_1$, and each of the form $(a_1, a_2, a_1, a_2, \ldots, a_1, a_2)$. □

In fact, the necessary conditions are sufficient for the existence of a neighbour design for all values of v and k. The proof is very long, but quite straightforward, and we look only at the essential ideas in the rest of this section.

Lemma 4. *An $N[k, \lambda'; v]$ design can be used to construct an $N[k, \lambda' t; v]$ design for every positive integer t.*

Proof. The second design is obtained by writing down t copies of the first design. □

An $N[k, \lambda'; v]$ design is said to have *property H* if the b blocks can be arranged into a sequence $B_1, B_2, \ldots, B_b$ such that, considering the blocks as sets of treatments,

$$|B_i \cap B_{i+1}| > 0 \quad \text{for } i = 1, 2, \ldots, b-1.$$

We denote a neighbour design with property H by $N_H[k, \lambda'; v]$. Note that there is no restriction on $B_1 \cap B_b$.

The design of Example 1, with the blocks in the order given by the left-most treatment in each block, can be seen to have property H.

We now allow designs to have blocks of varying sizes. Given any block $B = (a_1, a_2, \ldots, a_k)$, let $B(a_i) = (a_i, a_{i+1}, \ldots, a_k, a_1, \ldots, a_{i-1})$. The neighbours in the blocks B and $B(a_i)$ are the same. Let $|B_i| = k_i$, $i = 1, 2$, and let $j \in B_1 \cap B_2 \neq \emptyset$. Then we define $B_1(j) + B_2(j)$ to be a block of size $k_1 + k_2$ obtained by concatenating $B_1(j)$ and $B_2(j)$. Notice that if the element j occurs more than once in either of these blocks, then $B_1(j) + B_2(j)$ is not uniquely defined. For instance, if $B_1 = \{1, 2\}$ and $B_2 = \{1, 3, 1, 4\}$, we could interpret $B_1(1) + B_2(1)$ to be $\{1, 2, 1, 3, 1, 4\}$ or $\{1, 2, 1, 4, 1, 3\}$. Since this makes no difference to the neighbours in the block, either interpretation is valid.

Example 5. Let B_i be the ith block in Table 14.1, $i = 1, 2, \ldots, 15$. Then

$$B_1 \cap B_2 = \{1, 2, 5\}, B_2 \cap B_3 = \{2, 3, 6\},$$
$$B_1(2) + B_2(2) = \{2, 4, 1, 5, 15, 9, 1, 2, 3, 5, 2, 6, 1, 10\},$$

and

$$(B_1(2) + B_2(2)) + B_3(2) = \{2, 4, 1, 5, 15, 9, 1, 2, 3, 5, 2, 6, 1, 10, 2, 11, 3, 4, 6, 3, 7\}.$$

Table 14.2 gives the five blocks obtained by concatenating the blocks B_{3i+1}, B_{3i+2} and B_{3i+3}, $i = 0, 1, 2, 3, 4$. They are the blocks of an $N[21, 1; 15]$ design. □

Table 14.2

2	4	1	5	15	9	1	2	3	5	2	6	1	10	2	11	3	4	6	3	7
5	7	4	8	3	12	4	5	6	8	5	9	4	13	5	14	6	7	9	6	10
8	10	7	11	6	15	7	8	9	11	8	12	7	1	8	2	9	10	12	9	13
11	13	10	14	9	3	10	11	12	14	11	15	10	4	11	5	12	13	15	12	1
14	1	13	2	12	6	13	14	15	2	14	3	13	7	14	8	15	1	3	15	4

Lemma 6. *Neighbours in the following blocks or pairs of blocks are the same:*

(i) $\{B_1(j), B_2(j)\}$ *and* $B_1(j) + B_2(j)$;

(ii) $B_1(j)+B_2(j)$ *and* $B_2(j)+B_1(j)$;
(iii) $(B_1(j)+B_2(j))+B_3(j)$ *and* $B_1(j)+(B_2(j)+B_3(j))$;
(iv) $B_1(j)+B_2(j)$ *and* $B_1(j')+B_2(j')$ *for* $j, j' \in B_1 \cap B_2$.

Proof. See Exercise 14.1. □

Lemma 7. *Let v, k and λ' satisfy the necessary conditions of Theorem 2, and let $k = k_1k_2$. Then the existence of an $N_H[k_1, \lambda'; v]$ design implies the existence of an $N_H[k, \lambda'; v]$ design.*

Proof. Let $B_1, B_2, \ldots, B_b$ be the blocks of the $N_H[k_1, \lambda'; v]$ design arranged so that $|B_i \cap B_{i+1}| > 0, i = 1, 2, \ldots, b-1$. Then, choosing elements from the intersections as appropriate, the blocks

$$C_i = B_{ik_2+1} + B_{ik_2+2} + \ldots + B_{ik_2+k_2},$$

$i = 0, 1, \ldots, (b/k_2)-1$, are those of an $N_H[k, \lambda'; v]$ design. The parameters are quickly checked. Clearly v is still the number of treatments, $|C_i| = k_2k_1 = k$ and the neighbours in the C_i are precisely those of the B_i, so λ' is unchanged. The design has property H since

$$|B_{ik_2+k_2} \cap B_{(i+1)k_2+1}| > 0, i = 0, 1, \ldots, (b/k_2)-1.$$ □

Example 5 illustrates this construction.

We now look at three difference set constructions for neighbour designs. Because we are only interested in differences between adjacent elements in the starter block (or blocks), we define the sequence of forward differences of a starter block B as follows. Let $B = (b_1, b_2, \ldots, b_k)$ and let $f_i = b_{i+1} - b_i$ (mod v), $i = 1, 2, \ldots, k$, (where $b_{k+1} = b_1$). Then $F = (f_1, f_2, \ldots, f_k)$ is the *sequence of forward differences* of B. Note that the sum of the entries in F is zero and that, since B is a starter block, $F \cup (-F)$ contains all the non-zero elements modulo v precisely once.

Theorem 8. *There is an $N_H[k, 1; 2k+1]$ design for all $k \geq 3$.*

Proof. Let $F_k(1)$ denote the sequence of forward differences for the starter block of size k and define

$$F_k(1) \circ c = (f_1 \circ c, f_2 \circ c, \ldots, f_k \circ c),$$

where

$$f_i \circ c = \begin{cases} f_i + c \pmod{v}, & \text{if } f_i \geq 0, \\ f_i - c \pmod{v}, & \text{if } f_i < 0. \end{cases}$$

The proof is by induction. Let

$$F_3(1) = (1, 2, -3), \quad F_4(1) = (1, -2, -3, 4), \quad F_5(1) = (1, -2, 3, 4, -6),$$
$$F_6(1) = (1, -2, 3, -4, -5, 7).$$

It is clear that $F_3(1)$, $F_4(1)$, $F_5(1)$, and $F_6(1)$ are sequences of forward differences for $k = 3, 4, 5$, and 6 respectively. Suppose that $F_k(1)$ exists for $k = 3, 4, \ldots, K-1$ where $K \geq 7$. Construct $F_K(1)$ by

$$F_K(1) = \begin{cases} (1, -2, -3, 4, F_{K-4}(1) \circ 4), & \text{if } K \text{ is even,} \\ (1, 2, -3, F_{K-3}(1) \circ 3), & \text{if } K \text{ is odd.} \end{cases}$$

By construction, $F_{K-4}(1)$ and $F_{K-3}(1)$ have as many positive terms as negative terms; so the sum of the elements in $F_K(1)$ is zero. By the induction hypothesis $F_K(1) \cup (-F_K(1))$ contains all the non-zero elements modulo v precisely once. The fact that the design has property H follows from the fact that the first two elements in the starter block have a difference of 1. □

Example 9.

$$\begin{aligned} F_7(1) &= (1, 2, -3, F_4(1) \circ 3) \\ &= (1, 2, -3, 1 \circ 3, -2 \circ 3, -3 \circ 3, 4 \circ 3) \\ &= (1, 2, -3, 4, -5, -6, 7). \end{aligned}$$

The starter block corresponding to this sequence of forward differences is

$$\begin{gathered} (1, 1+1, 1+1+2, 1+1+2-3, 1+1+2-3+4, 1+1+2-3+4-5, \\ 1+1+2-3+4-5-6) = (1, 2, 4, 1, 5, 0, 9) \end{gathered}$$

and $f_7 = 7 = 1 - 9 = -8$. The design developed from this starter block is that given in Table 14.1. □

Theorem 10. *If $k \equiv 0 \pmod 2$ and $k > 2$, then there is an $N_H[k, 1; 2^i k + 1]$ design for all integers $i \geq 1$.*

Proof. We need 2^{i-1} starter blocks. Let the sequence of forward differences for the jth starter block, $j = 1, 2, \ldots, 2^{i-1}$, be $F_k(1) \circ 2(j-1)k$. It is straightforward to check that these starter blocks generate an $N_H[k, 1; 2^i k + 1]$ design. □

Example 11. Let $k = 4$, $i = 2$. Then an $N_H[4, 1; 17]$ design has two starter blocks with sequences of forward differences equal to $F_4(1)$ and $F_4(1) \circ 8$. The corresponding starter blocks are (1, 2, 0, 14) and (1, 10, 0, 6), respectively. □

Theorem 12. *If $k \equiv 0 \pmod 4$ there is an $N_H[k, 1; 2mk + 1]$ design for all integers $m \geq 1$.*

Proof. Let the sequence of forward differences for the jth starter block be

$$F_k(j) = \{\varepsilon_y(kj - k + y) \mid y = 1, 2, \ldots, k\}, \quad j = 1, 2, \ldots, m,$$

where

$$\varepsilon_y = \begin{cases} 1, & y \equiv 0, 1 \pmod 4), \\ -1, & y \equiv 2, 3 \pmod 4). \end{cases}$$

Again we leave the verification as an exercise. □

Example 13. Let $k = 4$ and $m = 2$. This time we get an $N[4, 1; 17]$ design with two starter blocks with sequences of forward differences equal to

$$F_4(1) = \{\varepsilon_y(y) \mid y = 1, 2, 3, 4\} = \{1, -2, -3, 4\}$$

and

$$F_4(2) = \{\varepsilon_y(4+y) \mid y = 1, 2, 3, 4\} = \{5, -6, -7, 8\}.$$

The corresponding starter blocks are (1, 2, 0, 14) and (1, 6, 0, 10) respectively. (Of course, (1, 6, 0, 10) and (1, 10, 0, 6) give rise to the same circular block: one is obtained by reading the block in a clockwise fashion and the other by reading it anticlockwise.) □

14.3. Complete and quasi-complete Latin squares and related designs

A Latin square is said to be *row-complete* if every ordered pair of distinct treatments appears in adjacent positions precisely once in the rows of the square. A *column-complete* square is defined similarly. A Latin square which is both row- and column-complete is said to be *complete.* A Latin square in which each unordered pair of treatments appears twice in rows and twice in columns is said to be *quasi-complete.*

Example 14. The squares in Table 14.3 are (a) a row-complete Latin square of order 4, (b) a quasi-complete Latin square of order 5, and (c) a complete Latin square of order 6. If we permute the rows of the square of order 4 so that the first column equals the first row, the resulting Latin square is complete. This cannot be done with every row-complete Latin square. For instance, the square of order 9 in Table 14.3(d) is row-complete but cannot be made column-complete as well; its properties were verified by computer search. □

Theorem 15. *There is a complete Latin square of order 2m for every integer $m \geq 1$.*

Proof. Let the Latin square be $L = (l_{ij})$, $1 \leq i, j \leq 2m$, and let the first row and column of L be

$$1 \;\; 2 \;\; 2m \;\; 3 \;\; 2m-1 \;\; 4 \; \ldots \; m \;\; m+2 \;\; m+1.$$

For $i > 1$, $j > 1$, let $l_{ij} \equiv l_{i1} + l_{1j} - 1 \pmod{2m}$. Clearly L is a Latin square. The

Table 14.3

(a)	(b)	(c)	(d)
1 4 2 3	1 2 5 3 4	1 2 6 3 5 4	1 4 7 2 6 8 5 9 3
2 1 3 4	2 3 1 4 5	2 3 1 4 6 5	2 5 8 3 4 9 6 7 1
3 2 4 1	5 1 4 2 3	6 1 5 2 4 3	3 6 9 1 5 7 4 8 2
4 3 1 2	3 4 2 5 1	3 4 2 5 1 6	4 2 1 8 7 3 9 5 6
	4 5 3 1 2	5 6 4 1 3 2	5 3 2 9 8 1 7 6 4
		4 5 3 6 2 1	6 1 3 7 9 2 8 4 5
			7 8 6 5 2 4 3 1 9
			8 9 4 6 3 5 1 2 7
			9 7 5 4 1 6 2 3 8

sequence of forward differences (not including $1-(m+1)$) for both the rows and the columns is

$$\{2-1, 3-2m, 4-(2m-1), \ldots, (m+1)-(m+2), 2m-2, 2m-1-3, \ldots,$$
$$m+2-m\} = \{1, 3, 5, \ldots, 2m-1, 2m-2, 2m-4, \ldots, 2\}$$

which contains all the non-zero elements modulo m precisely once. Hence every ordered pair of distinct treatments appears precisely once in rows and once in columns. □

The square of order 6 in Table 14.3 was constructed by this method. For odd orders the next result is the best general result known.

Theorem 16. *There is a quasi-complete Latin square of order $2m+1$ for every integer $m \geq 1$.*

Proof. Here we let the first row and column of the square be

$$1 \ 2 \ 2m+1 \ 3 \ 2m \ 4 \ldots m \ m+3 \ m+1 \ m+2.$$

Again let $l_{ij} \equiv l_{i1} + l_{1j} - 1 \pmod{2m+1}$ for $i > 1$, $j > 1$. It is easy to check that the sequence of forward differences (again not including $1-(m+2)$) contains each of the odd numbers modulo $2m+1$ twice (and so the negative of this sequence contains each of the even numbers modulo $2m+1$ twice). □

The square of order 5 in Table 14.3 was constructed by this method.

We can extend the concept of complete and quasi-complete Latin squares as follows. We consider $r \times c$ arrays of v treatments in which:

(i) each treatment appears equally often in the array, perhaps further restricted to appear equally often in rows or columns or both;

(ii) no treatment appears adjacent to itself in either rows or columns;

(iii) the array has either *directional balance* (each ordered pair of distinct treatments appears adjacent equally often in rows and in columns) or *non-*

directional balance (each pair of distinct treatments appears adjacent equally often in rows and columns considered together).

Thus a complete Latin square has directional balance and $v = r = c$. A quasi-complete Latin square has non-directional balance and $v = r = c$.

Example 17. The designs in Table 14.4 have the following properties: (a) $v = r = 4$, $c = 10$, non-directional balance; (b) $v = 4$, $r = c = 6$, non-directional balance; (c) $v = r = 4$, $c = 16$, directional balance. □

Table 14.4

1	4	2	3	2	1	3	4	2	3
4	3	1	2	1	4	2	3	1	2
2	1	3	4	3	2	4	1	3	4
3	2	4	1	4	3	1	2	4	1

(a)

1	2	1	4	3	2
3	1	2	3	4	1
1	3	1	4	2	4
4	2	4	1	3	2
2	4	3	2	1	3
3	1	4	3	2	4

(b)

1	4	2	3	2	1	3	4	2	1	3	4	1	4	2	3
4	3	1	2	1	4	2	3	1	4	2	3	4	3	1	2
2	1	3	4	3	2	4	1	3	2	4	1	2	1	3	4
3	2	4	1	4	3	1	2	4	3	1	2	3	2	4	1

(c)

Arrays with directional or non-directional balance can often be obtained by suitably juxtaposing known arrays with the same property.

Theorem 18. *If $v = 4$, and if $r = 6s$ and $c = 6d$, $s, d \geq 1$, then there exists an $r \times c$ array with non-directional balance.*

Proof. If $s = d = 1$ then Example 17(b) is the required array. The pairs obtained by placing two copies of this design side by side are $\{(2, 1), (3, 1), (4, 1), (4, 2), (3, 2), (4, 3)\}$ and each unordered pair appears once. Similarly, juxtaposing them vertically gives $\{(1, 3), (2, 1), (1, 4), (4, 3), (3, 2), (2, 4)\}$ and again each unordered pair appears once. Let F be this original array. Then $J_{s \times d} \times F$ is the required array. □

Example 19. Let C be the complete Latin square given in Table 14.5(a). Let $C\sigma$ be the array obtained from C by applying the permutation σ to each element of C. The array in Table 14.6(b) has $v = 4$, $r = c = 12$ and is obtained from C as indicated. □

For experiments where plants in border plots may exhibit different rates of growth from those in internal plots, a more appropriate design may be an array with (non-) directional balance but in which the border plots are neighbours for internal plots but are not otherwise part of the design.

Example 20. The design in Table 14.6 has seven treatments. The internal plots are those inside the line. The design has non-directional balance. Any treatments can be placed in the corner positions as those treatments are not the

Table 14.5

(a)

1	2	3	4
2	4	1	3
3	1	4	2
4	3	2	1

$$\begin{bmatrix} C & C(24) & C \\ C(24) & C(1432) & C(24) \\ C & C(24) & C \end{bmatrix} =$$

(b)

1	2	3	4	1	4	3	2	1	2	3	4
2	4	1	3	4	2	1	3	2	4	1	3
3	1	4	2	3	1	2	4	3	1	4	2
4	3	2	1	2	3	4	1	4	3	2	1
1	4	3	2	4	1	2	3	1	4	3	2
4	2	1	3	1	3	4	2	4	2	1	3
3	1	2	4	2	4	3	1	3	1	2	4
2	3	4	1	3	2	1	4	2	3	4	1
1	2	3	4	1	4	3	2	1	2	3	4
2	4	1	3	4	2	1	3	2	4	1	3
3	1	4	2	3	1	2	4	3	1	4	2
4	3	2	1	2	3	4	1	4	3	2	1

neighbours of any treatment and are included so that every interior plot has eight occupied plots around it. The neighbours of 1, for instance, are 5, 6, 7, 7, 2, 3, 4, 4, 2, 3, 5, 6. □

Table 14.6

6	1	7	3	5	2	4	6	5
7	2	4	6	1	7	3	5	4
4	1	3	5	7	6	2	4	1
5	4	6	1	3	2	5	7	2
1	3	5	2	6	7	1	4	3

However, no general methods of construction for such bordered designs appear to be known.

14.4. Directional seed orchard designs

Sometimes the end plants (or varieties or treatments) of a BIBD are treated as border plants. This is a useful approach when collecting seed from a wind-pollinated species, for instance. All plants are pollinators but seed is collected only from internal plants. The internal plants of each variety must have equal chances of being pollinated by plants of every other variety. We look at a construction for a $B[3, \lambda; v]$ design (together with some additional blocks) in which each block of the design is ordered and the blocks themselves are ordered. Each treatment (variety) must appear equally often in the middle position in the blocks of the design. Consider a particular treatment x and consider all the blocks containing x in the middle position. The set of treatments in the blocks immediately preceding these must contain every treatment, except x, equally often. These designs are appropriate for seed orchards in which the wind comes mainly from one direction and so are called *directional seed orchard designs.*

Example 21. The design in Table 14.7 shows a directional seed orchard design for seven varieties. Some plants are pollinators only. Plants from which seed is collected are indicated. □

Table 14.7. The 14 varieties shown in boxes are for seed collection; the remaining varieties are pollinators only

3	4	6		4	3	1
1	2	4		6	5	3
6	7	2		1	7	5
4	5	7		3	2	7
2	3	5		5	4	2
7	1	3		7	6	4
5	6	1		2	1	6
3	4	6		4	3	1

Theorem 22. *Let $v = 6t + 1 = p$, a prime, and let θ be a primitive element of $GF[p]$. Then the starter blocks*

$$\{\theta^i, \theta^{2t+i}, \theta^{4t+i}\} \quad \textit{and} \quad \{\theta^{3t+i}, \theta^{5t+i}, \theta^{t+i}\}, \quad i = 0, 1, \ldots, t-1,$$

can be developed to give a directional seed orchard design.

Proof. From Theorem 3.4.13, we know that the starter blocks given can be developed to form a $B[3, 2; v]$ design. We are interested in the varieties which appear in the blocks preceding those which have, in the centre, a given variety; thus the way in which each starter block is developed is also important. We develop the starter blocks $\{\theta^i, \theta^{2t+i}, \theta^{4t+i}\}$, $i = 0, 1, \ldots, t-1$, by subtracting θ^{2t+i} at each step. The number of distinct blocks obtained is $y-1$, where

$$\theta^{2t+i} - y\theta^{2t+i} = \theta^{2t+i}.$$

Thus $y \equiv 1 \pmod{6t+1}$ so $y = 6t+2$. We include each starter block twice however: once at the beginning and once at the end of the blocks developed from it. The first occurrence contains pollinators only. Similarly we develop the starter blocks $\{\theta^{3t+i}, \theta^{5t+i}, \theta^{t+i}\}$, $i = 0, 1, \ldots, t-1$, by subtracting θ^{5t+i} at each step. Again each starter block is included twice.

To check the varieties in the preceding blocks we proceed as follows. First, 0 is the only element of $GF[p]$ which does not appear in the starter blocks and it appears in the central position of every block developed at the first step from a starter block. So 0 has the required property.

Now, suppose that $\theta^a = \theta^{2t+i} - w\theta^{2t+i} = \theta^{2t+i}(1-w)$. Then the block preceding the block with θ^a in the centre (and developed from the initial block with θ^{2t+i} in the centre) is

$$\{\theta^i - (w-1)\theta^{2t+i}, \theta^{2t+i} - (w-1)\theta^{2t+i}, \theta^{4t+i} - (w-1)\theta^{2t+i}\}.$$

Thus the treatments in the blocks preceding θ^a (from blocks of the form $\{\theta^i, \theta^{2t+i}, \theta^{4t+i}\}$ are

$$\{\theta^i + \theta^a, \theta^{2t+i} + \theta^a, \theta^{4t+i} + \theta^a \mid i = 0, 1, \ldots, t-1\}.$$

Similarly, those from starter blocks of the form $\{\theta^{3t+i}, \theta^{5t+i}, \theta^{t+i}\}$ are

$$\{\theta^{3t+i} + \theta^a, \theta^{5t+i} + \theta^a, \theta^{t+i} + \theta^a \mid i = 0, 1, \ldots, t-1\}$$

and so each variety, except θ^a, appears exactly once. □

Example 23. Let $t = 2, v = 13$. Table 14.8 contains a directional seed orchard design constructed using the method of Theorem 22. □

Table 14.8

1	3	9	12	10	4	2	6	5	11	7	8
11	0	6	2	0	7	9	0	12	4	0	1
8	10	3	5	3	10	3	7	6	10	6	7
5	7	0	8	6	0	10	1	0	3	12	0
2	4	10	11	9	3	4	8	7	9	5	6
12	1	7	1	12	6	11	2	1	2	11	12
9	11	4	4	2	9	5	9	8	8	4	5
6	8	1	7	5	12	12	3	2	1	10	11
3	5	11	10	8	2	6	10	9	7	3	4
0	2	8	0	11	5	0	4	3	0	9	10
10	12	5	3	1	8	7	11	10	6	2	3
7	9	2	6	4	11	1	5	4	12	8	9
4	6	12	9	7	1	8	12	11	5	1	2
1	3	9	12	10	4	2	6	5	11	7	8

14.5. Serially balanced sequences

A *type* 1 *serially balanced sequence of order v and index λ* is a sequence of length $\lambda v^2 + 1$, which has the following properties:

(i) the first and last elements are the same;

(ii) the first element appears $\lambda v + 1$ times;

(iii) the remaining $v - 1$ elements appear λv times each;

(iv) each of the v^2 ordered pairs of elements appears λ times among the λv^2 pairs of consecutive elements;

(v) aside from the first element, each element appears precisely once in each of the λv successive sets of v elements.

We denote such a sequence by $SBS1(v, \lambda)$.

A *type* 2 *serially balanced sequence of order v and index* λ is a sequence of length $\lambda v(v-1)+1$ which has the following properties:

(i) the first and last elements are the same;

(ii) the first element appears $\lambda(v-1)+1$ times;

(iii) the remaining $v-1$ elements appear $\lambda(v-1)$ times each;

(iv) each of the $v(v-1)$ ordered pairs of distinct elements appears λ times among the $\lambda v(v-1)$ pairs of consecutive elements;

(v) aside from the first element, each element appears precisely once in each of the $\lambda(v-1)$ successive sets of v elements.

We denote such a sequence by $SBS2(v, \lambda)$.

Thus, where a row complete Latin square uses v subjects on v occasions to compare v treatments, an $SBS2(v, 1)$ uses one subject on $v(v-1)+1$ occasions to compare v treatments. In each case, each ordered pair of distinct treatments appears exactly once.

Example 24. Various sequences of types 1 and 2 are given in Table 14.9. (There are no type 1 sequences with $\lambda = 1$ for $v = 3, 4$ or 5.) The 'successive sets of v elements' are marked off by semicolons. The $SBS2(5, 1)$ is said to be *completely reversible* since the sequence can be read from left to right or from right to left (in which case the first 'successive set' (or block) is 2 4 1 3 5). □

Table 14.9

Sequence	Design
1; 1 2; 2 1	$SBS1(2, 1)$
1; 2 1	$SBS2(2, 1)$
1; 2 3 1; 3 2 1	$SBS2(3, 1)$
1; 2 3 4 1; 3 2 1 4; 2 4 3 1	$SBS2(4, 1)$
5; 1 2 3 4 5; 2 4 1 3 5; 4 3 2 1 5; 3 1 4 2 5	$SBS2(5, 1)$
1; 1 2 3; 3 1 2; 2 3 1; 1 3 2; 2 1 3; 3 2 1	$SBS1(3, 2)$
1; 2 1 3; 1 2 3; 1 3 2; 3 2 1	$SBS2(3, 2)$

Lemma 25. *If there exists an* $SBSi(v, \lambda)$, *then there is an* $SBSi(v, m\lambda)$ *for every integer* $m \geq 1$, $i = 1, 2$.

Proof. The second sequence is obtained from the first by concatenating m copies of the first sequence. □

Note, however, that the $SBS1(3, 2)$ and the $SBS2(3, 2)$ given in Table 14.9 are not obtained from smaller sequences in this manner.

Theorem 26. *There exists an* $SBS2(v, 1)$ *for all values of* $v \geq 4$.

Proof. The sequences are obtained by developing a starter block of size v, concatenating the blocks in the natural order and adding the final treatment to the beginning of the sequence. Thus, if the starter block is $(b_1, b_2, \ldots, b_v)$ we need to check that the sequence of forward differences $F = \{b_2 - b_1, b_3 - b_2, \ldots, b_v - b_{v-1}, b_1 + 1 - b_v\}$ contains each of the non-zero differences once. The starter blocks are:

$1, 2, 2m, 3, 2m-1, \ldots, m, m+2, m+1, \infty,$ if $v-1 = 2m$;
$1, \infty, 2, 4m+1, 3, 4m, 4, \ldots,$
$m, 3m+3, m+1, m+2, 3m+2, m+3, 3m+1, \ldots,$
$2m+1, 2m+3, 2m+2,$ if $v-1 = 4m+1$;
$1, \infty, 2, 4m+3, 3, 4m+2, 4, \ldots,$
$m+1, 3m+4, m+2, m+3, 3m+3, m+4, 3m+2, \ldots,$
$2m+2, 2m+4, 2m+3,$ if $v-1 = 4m+3$;

where the blocks are developed modulo $v-1$ and, as usual, $\infty + i = \infty$ for all i.

We have already checked the block given for $v-1 = 2m$ in Theorem 15. The checks for the other two are similar and are left as exercises. □

Example 27. In Table 14.10 we give $SBS2(5, 1)$, $SBS2(6, 1)$ and $SBS2(8, 1)$ constructed using the method of Theorem 26. □

Table 14.10

∞; 1 2 4 3 ∞; 2 3 1 4 ∞; 3 4 2 1 ∞; 4 1 3 2 ∞

$SBS2(5, 1)$

3; 1 ∞ 2 3 5 4; 2 ∞ 3 4 1 5; 3 ∞ 4 5 2 1; 4 ∞ 5 1 3 2; 5 ∞ 1 2 4 3

$SBS2(6, 1)$

4; 1 ∞ 2 7 3 4 6 5; 2 ∞ 3 1 4 5 7 6; 3 ∞ 4 2 5 6 1 7;
4 ∞ 5 3 6 7 2 1; 5 ∞ 6 4 7 1 3 2; 6 ∞ 7 5 1 2 4 3;
7 ∞ 1 6 2 3 5 4

$SBS2(8, 1)$

When $v-1 = 2m$ the sequences constructed in Theorem 26 are completely reversible.

No general constructions are known for $SBS1(v, 1)$ but a general construction is known for $SBS1(v, 2)$.

Theorem 28. *An $SBS1(v, 2)$ exists for all $v \geq 4$.*

Proof. If $v = 2m$ let M be the complete Latin square constructed in Theorem 15. Let N be the Latin square obtained from M by applying the permutation $\pi = (1\,2\,3 \ldots m)$ to the elements of M. The sequence is obtained by writing down the first row of M, then the row of N beginning with element $m+1$, followed by the row of M beginning with the element at the end of the row of N, and so on, until m rows of both M and N have been used. The

$(2m+1)$st row to be used is taken from N and then the alternation continues until all rows of both squares have been used. Since both M and N are row-complete, all ordered pairs of distinct elements appear twice. If we can show that each row of M and N is used precisely once, then pairs of the form (i, i) will appear twice in the final sequence. Let us index the rows in M and N by their first elements. In M, if the first element in a row is i, then the last element is $i+m \pmod{2m}$ whereas in N, it is

$$\pi^{-1}(i)+m \pmod{2m}, \quad \text{if } 1 \le i \le m,$$

or

$$\pi(i+m), \quad \text{if } m+1 \le i \le 2m.$$

Hence the rows used from M and N are

$$\begin{array}{llllllllll} M\colon 1 & 2 & 3 & \ldots & m & 2m & 2m-1 & & \ldots & m+1 \\ N\colon & m+1 & m+2 & m+3 & \ldots & 2m & 1 & m & m-1 & \ldots\ 2 \end{array}$$

and we see each row is used once, as required.

If $v = 2m+1$, let M be the quasi-complete Latin square constructed in Theorem 16 and N be the quasi-complete Latin square obtained from M by adding m to each element $\pmod{2m+1}$ and reversing each row. The construction is similar to that for v even, except that now all the rows in M are used and then all the rows in N. Verification is left as an exercise. □

Example 29. In Table 14.11 we give $SBS1(4, 2)$ and $SBS1(5, 2)$ constructed using the method of Theorem 28. □

Table 14.11

1; 1 2 4 3; 3 4 1 2; 2 3 1 4; 4 2 3 1; 1 3 2 4; 4 1 3 2; 2 1 4 3; 3 4 2 1

$SBS1(4, 2)$

1; 1 2 5 3 4; 4 5 3 1 2; 2 3 1 4 5; 5 1 4 2 3; 3 4 2 5 1; 1 5 2 4 3;
3 2 4 1 5; 5 4 1 3 2; 2 1 3 5 4; 4 3 5 2 1

$SBS1(5, 2)$

We finish this section by noting that if at least two subjects, but fewer than v, are available, then we can break the SBS up into smaller sequences and use one of these to determine which treatments to apply to each subject. Note that the treatment immediately to the left of a 'break-point' must also be applied to retain balance.

Example 30. The sequences 1; 2 3 1 and 1; 3 2 1 could be used to test three treatments using two subjects when pairs of the form (i, i) are not required. □

14.6. Linearly ordered block designs

An *equi-neighboured balanced incomplete block design* (EBIBD) is a BIBD in which any two varieties appear adjacent equally often. (The blocks are not circular so two varieties in each block have only one neighbour.) We write the parameters of the design as $EB[k, \lambda, \lambda'; v]$ where λ is the number of times that two varieties appear in the same block and λ' is the number of times that two varieties appear in adjacent positions in blocks.

Lemma 31. *In an $EB[k, \lambda, \lambda'; v]$ design we have:*
(i) $vr = bk$;
(ii) $\lambda(v-1) = r(k-1)$;
(iii) $\lambda' = 2\lambda/k$.

Proof. Both (i) and (ii) follow from the fact that the design is a BIBD. For (iii) let N_{ij} be the number of times that varieties i and j are adjacent in blocks and let e_i be the number of times that variety i appears in the end position in a block. Then

$$\sum_{j \neq i} N_{ij} = 2r - e_i$$

and

$$\sum_i e_i = 2b.$$

Hence

$$\sum_i \sum_{j \neq i} N_{ij} = 2vr - 2b = 2b(k-1) = 2b(v-1)\lambda/r = 2v(v-1)\lambda/k.$$

In an EBIBD $N_{ij} = \lambda'$ for all i, j with $i \neq j$. Thus

$$v(v-1)\lambda' = 2v(v-1)\lambda/k$$

and the result follows. □

Thus a necessary condition for a BIBD to be an EBIBD is that $k|2\lambda$.

Example 32. The design obtained by developing the starter blocks $(0, 1, 2)$, $(0, 2, 4)$ and $(0, 3, 6)$ mod 7 is an $EB[3, 3, 2; 7]$, as is the design obtained by developing the starter blocks $(1, 2, 4)$, $(2, 4, 1)$ and $(4, 1, 2)$ mod 7. The design obtained by developing the starter block $(0, 1, 4, 2)$ mod 7 is an $EB[4, 2, 1; 7]$. □

Theorem 33. *If there exists a $B[k, \lambda; v]$ design which is also an*

$$N[k, 2\lambda/(k-1); v]$$

design then there is an $EB[k-1, (k-2)\lambda, 2(k-2)\lambda/(k-1); v]$ design.

Proof. From each block $(a_1, a_2, \ldots, a_k)$ of the neighbour design construct the k blocks

$$(a_1, a_2, \ldots, a_{k-1}), (a_2, a_3, \ldots, a_k), (a_3, a_4, \ldots, a_k, a_1), \ldots, (a_k, a_1, \ldots, a_{k-2}).$$

These are the blocks of the EBIBD; checking the parameters is left as an exercise. □

Example 34. The (11, 5, 2) cyclic difference set $\{1, 3, 4, 5, 9\}$ can be written as (1, 4, 5, 9, 3), which is the starter block for an $N[5, 1; 11]$ design. Hence there is an $EB[4, 6, 3; 11]$ design, which can be obtained by developing the five starter blocks

$$(1, 4, 5, 9), (4, 5, 9, 3), (5, 9, 3, 1), (9, 3, 1, 4) \text{ and } (3, 1, 4, 5) \bmod 11. \qquad \square$$

The rows of a quasi-complete Latin square of order v may be used as the blocks of an $EB[v, v, 2; v]$ design. If v is even, the first $v/2$ rows of the complete Latin square constructed in Theorem 15 may be used as the blocks of an $EB[v, v, 1; v]$ design.

Theorem 35. *If there exists a $B[k, \lambda; v]$ design, then there exists an $EB[k, k\lambda, 2\lambda; v]$ design. If k is even then there also exists an $EB[k, k\lambda/2, \lambda; v]$ design.*

Proof. Use the elements of each block to construct an $EB[k, k, 2; k]$ ($EB[k, k, 1; k]$) design respectively. □

Example 36. The design in Table 14.12 is an $EB[4, 4, 2; 7]$ design. (The blocks of the $B[4, 2; 7]$ design are in the first line of the table.)

Table 14.12. $EB[4, 4, 2; 7]$ design

1234	1256	1357	1467	2367	2457	3456
2413	2615	3715	4716	3726	4725	4635

Let $D = \{d_1, d_2, \ldots, d_k\}$ be a (v, k, λ) difference set. We say that the difference set D is an *equi-neighboured difference set* if the set $\{\pm(d_{i+1} - d_i) | i = 1, 2, \ldots, k-1\}$ contains each of the non-zero elements equally often, say λ' times. Then D is a starter block for an $EB[k, \lambda, \lambda'; v]$ design. Equi-neighboured supplementary difference sets are defined similarly.

The designs in Examples 32 and 34 illustrate these concepts. In general ordering a difference set may be difficult, if not impossible. We give one general result.

Theorem 37. *If v is a prime and $v \equiv 3 \pmod 4$, then*

$$D = \left\{0, 1^2, 2^2, \ldots, \left(\frac{v-1}{2}\right)^2\right\} \pmod v$$

is an equi-neighboured difference set.

Proof. D is the set of quadratic residues together with 0 and so is a difference set; see Exercise 14.16. Also

$$\begin{aligned} d_{i+1} - d_i &= i^2 - (i-1)^2 \\ &= 2i - 1, \quad i = 1, 2, \ldots, (v-1)/2, \end{aligned}$$

showing that $\{\pm(d_{i+1} - d_i) | i = 1, 2, \ldots, (v-1)/2\} = GF(v)\backslash\{0\}$. Thus $\lambda' = 1$. □

Example 38. Let $v = 11$. Then $(0, 1, 4, 9, 5, 3)$ is a starter block for an $EB[6, 3, 1; 11]$ design. □

14.7. References and comments

Neighbour designs were first described by Rees (1967), in connection with the gel diffusion test. The proof that the necessary conditions for the existence of a neighbour design are sufficient is due to Hwang and Lin (1974, 1976, 1977, 1978). The results in Section 14.2 are from these papers and from Hwang (1973). The underlying unordered design of a neighbour design may have repeated treatments in a block; such designs are related to balanced n-ary designs. Billington (1984) gives a survey of balanced n-ary designs.

Bugelski (1949) and E. J. Williams (1949), both quoted in Denes and Keedwell (1974), appear to have been the first to construct complete Latin squares and Theorems 15 and 16 are due to E. J. Williams (1949). The square in Table 14.3(d) is due to Archdeacon, Dinitz, Stinson, and Tillson (1980). In view of Exercise 14.7, this is the smallest row-complete Latin square of odd order. Bailey (1984) looked at aspects of the construction of quasi-complete Latin squares. The definitions of directional and non-directional balance are from Freeman (1979), as are the arrays in Table 14.4(a) and (b) and in Table 14.6. Examples 17 and 19 and Theorem 18 are due to D. J. Street (1985).

For results related to Sections 14.1, 14.2, and 14.3, see also the survey paper by A. P. Street (1982) and references there cited. See Lindner and Stinson (1984) for related work.

Theorem 22 is given in Freeman (1967). Some *ad hoc* constructions for non-directional seed orchard designs are given in Freeman (1969).

Serially balanced sequences were introduced by Finney and Outhwaite (1955, 1956); Theorems 26 and 28 are due to Sampford (1957). R. M. Williams (1952) considered sequences in which balance was required only for unordered pairs of treatments. Nair (1967) looked at serially balanced sequences in which, in addition, every ordered pair of treatments appears equally often in plots separated by one plot. Dyke and Shelley (1976) looked at the construction of sequences in which triples of treatments appear with equal frequency.

The concepts of an equi-neighboured block design and of an equi-neighboured difference set are due to Kiefer and Wynn (1981) as are Lemma 31, Theorem 37, and the $EB[v, v, 2; v]$ and $EB[v, v, 1; v]$ designs. Theorems 33 and 35 are due to Cheng (1983). Dawson (1985) and D. J. Street and A. P. Street (1985) give other results on EBIBDs. Russell (1985) has discussed the construction of incomplete block designs when not just neighbouring plots are correlated. His methods are algorithmic, not combinatorial, however.

Sometimes allowance is made for any treatment which has been applied previously (in the same experiment) to an experimental unit. Hence we must consider the $\binom{k}{2}$ ordered pairs in a block of size k. The existence of such designs has been considered by Hung and Mendelsohn (1973), Seberry and Skillicorn (1980), D. J. Street and Seberry (1980), and D. J. Street and W. H. Wilson (1980) (see Exercise 14.17).

Exercises

14.1. Prove Lemma 6.

14.2. Construct an $N_H[5, 1; 11]$ design. Hence construct an $N_H[10, 2; 11]$ design.

14.3. Construct an $N_H[6, 1; 25]$ design. What designs can be obtained from this one using Lemma 7?

14.4. Show that a $B[3, \lambda; v]$ design is an $N[3, \lambda; v]$ design.

14.5. Complete the proofs of Theorems 10 and 12.

14.6. Construct a complete Latin square of order 8 and a quasi-complete Latin square of order 7.

14.7. Show that no row-complete Latin square of order 3, 5, or 7 can exist.

14.8. Show that the Latin square with first row and column

$$1 \;\; 2m+1 \;\; 2 \;\; 2m \;\; \ldots \;\; m+4 \;\; m-1 \;\; m+3 \;\; m \;\; m+2 \;\; m+1$$

and with $l_{ij} \equiv l_{i1} + l_{1j} - 1 \pmod{2m+1}$, for $i > 1$, $j > 1$, is quasi-complete.

Consider also the quasi-complete Latin square constructed in Theorem 16. Show that altogether in these two quasi-complete Latin squares, every ordered pair appears exactly twice in rows and exactly twice in columns.

14.9. Let C be the complete Latin square of order 4 with first row and column equal to (1423). Then the design of Table 14.4(c) may be written as $C \; C+1 \; C+1 \; C$. Verify that $C \; C+1 \; C+1 \; C \; C+1 \; C+1 \; C$ is an array with $v = r = 4$, $c = 28$ and with directional balance.

14.10. Let $v = 4$. Construct a 12×12 array with non-directional balance using Theorem 18.

14.11. Use Theorem 22 to construct a directional seed orchard design for seven treatments. Is it the same as (or isomorphic to) the one of Example 21?

14.12. Construct an $SBS2(7, 1)$, an $SBS2(10, 1)$, an $SBS1(6, 2)$, and an $SBS1(7, 2)$.

14.13. Complete the proof of Theorem 28.

14.14. Show that the m-supplementary difference sets of Exercise 3.15 are also the starter blocks for a neighbour design. Use Theorem 33 to construct a family of neighbour designs.

14.15. Construct an $EB[4, 6, 3; 8]$ design.

14.16. Apply Theorem 3.4.7 to show that if v is a prime, $v \equiv 3 \pmod 4$, then the set of quadratic residues, together with 0, is a difference set (mod v).

14.17. Consider the $B[4, 2; 19]$ design constructed from the starter blocks $(0, 3, 12, 1)$, $(13, 1, 5, 0)$ and $(4, 9, 6, 0)$. If the block (a, b, c, d) is said to *cover* the ordered pairs (a, b), (a, c), (a, d), (b, c), (b, d), and (c, d), show that each ordered pair is covered by precisely one block of the design.

15 Competition designs

15.1. Motivation

In this chapter we look at some of the designs which have been proposed to investigate aspects of plant competition. We consider only designs for either one or two varieties (or species) of plants.

If one variety is involved, then the factors of interest may be density of planting, distance between neighbouring plants, number of neighbours at a given distance, or the effect of lighting in a greenhouse experiment. In this last example, a lamp, which may be turned on or off, is positioned above each plant. Interest centres on the intensity of light falling on each plant, which in turn depends on whether the plant's own lamp is on or off, and on how many of its neighbours' lamps are on or off.

If the experiment involves two varieties then the factors of interest may be density of planting, and the effect on a plant of the number and proximity of neighbours of the same, or the other, variety.

Within this context we look at three general types of designs.

Designs of the first type are those laid out on a regular grid (triangular, square or hexagonal); measurements are made on each plant. These designs can be used to investigate competition between two varieties or the effect of crowding on one variety.

Those of the second type are designed to investigate the effect of spacing on one variety.

Those of the third type are designs suitable for intercropping experiments. By *intercropping* we will mean 'the mixture of more than one crop simultaneously for some, if not all, of their lifecycles and irrespective of their spatial arrangement' (Mead and Riley 1981). So far all such designs have allowed for mixtures of only two crops.

15.2. The regular grids

The design consists of a regular grid. Each region defined by a square, triangular or hexagonal grid is occupied by one of two treatments, chosen from a set consisting of two varieties, or of one variety and an empty space. The design is a planar array and the treatments will be represented by blank and shaded regions or by zeros and ones. We call such a design a *planar design*.

For each of the regular grids each plant has $n(=3, 4, 6)$ neighbours and can lie at the centre of 2^n possible neighbourhoods. In Fig. 15.1 we give the eight neighbourhoods of the triangular array, divided into four equivalence classes, the sixteen neighbourhoods of the square array, divided into five equivalence classes, and representatives of each of the seven equivalence classes for the hexagonal grid. Note that the configurations are labelled according to the number of neighbours which match the central element.

A planar design is said to be *balanced* if the number of zeros and ones in the design is equal and if there are equal numbers of each symbol in each configuration. A design is said to have *nested balance* if it is balanced both between and within equivalence classes.

Example 1. The design in Fig. 15.2(a) is balanced; of course, only plants with four neighbours are considered part of the design. Treatment 1 appears thirty-five times in the design, as does treatment 0. There are seven α configurations with a 0 at the centre and seven with a 1 at the centre, seven β configurations with a 0 at the centre and so on. The design of Fig. 15.2(b) has nested balance. There are one hundred and twenty plants in the design; sixty are treatment 0 and sixty are treatment 1. There are twelve α configurations with a 0 at the centre, twelve β configurations with a 0 at the centre and so on. In addition, however, of the twelve β configurations with a 0 at the centre, three have the 1 to the left of the 0, three have the 1 above the 0, three have the 1 to the right of the 0, and three have the 1 below the 0. Hence there is balance within the equivalence class. A similar property holds for the other equivalence classes for both 0s and 1s. □

Several approaches to the construction of balanced grids have been tried. These are:

(i) the construction of a tiling of the plane, of finite period, and with appropriate balance properties;

(ii) the construction of a periodic array which, when considered together with its complement, is balanced;

(iii) the construction of finite blocks which are balanced.

The designs in Fig. 15.2 are examples of (iii). The next example illustrates (i).

Example 2. In Fig. 15.3(a) we give a 13×14 array. If the plants in the border rows and columns, and those in the row marked with the arrow, are ignored, the array is balanced. The equivalence class of each 1 of the array is given in Fig. 15.3(b). Further, the internal positions of every internal row are occupied by the same binary sequence of period 12, namely 1 0 1 0 0 0 0 1 0 1 1 1. Figure 15.3(c) shows how the array can be used to tile the plane. The plants in the border rows and columns, and those in the rows marked with arrows, are not measured. □

In fact, the design given in Example 2 has nested balance.

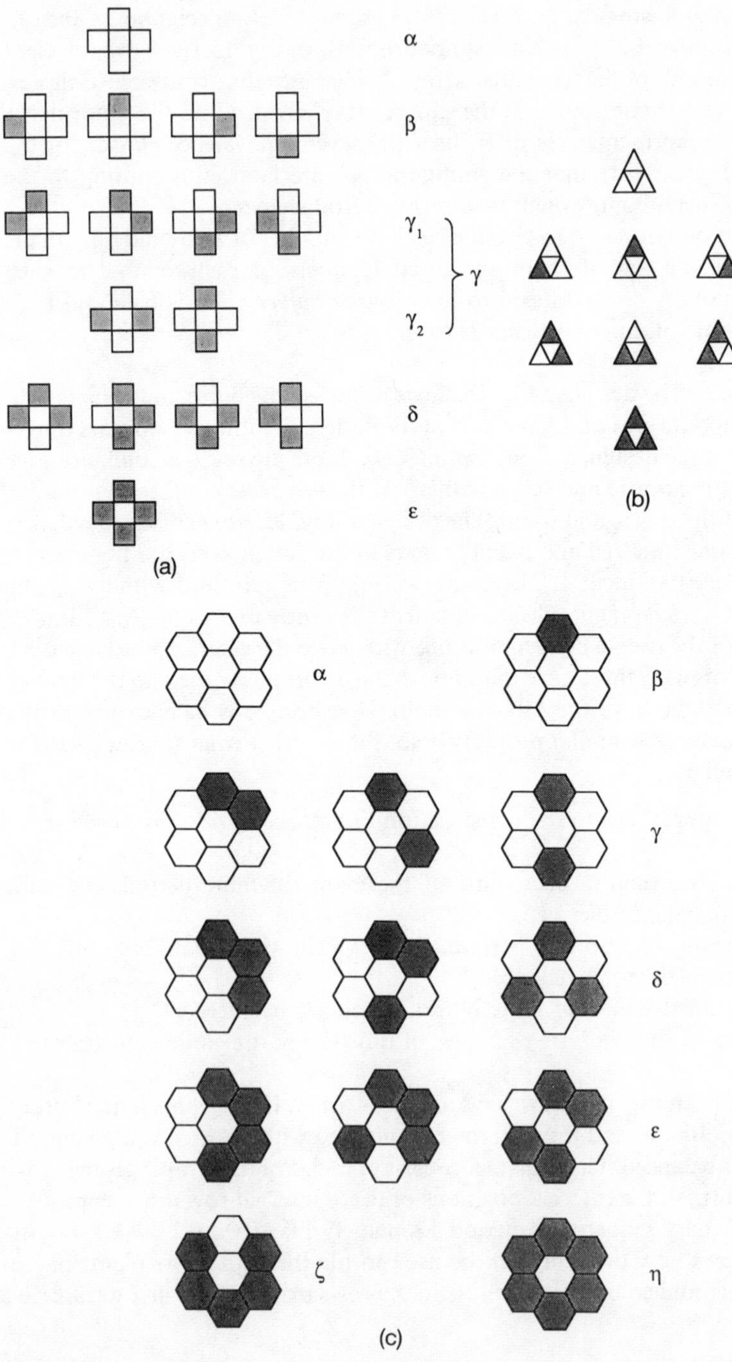

Fig. 15.1

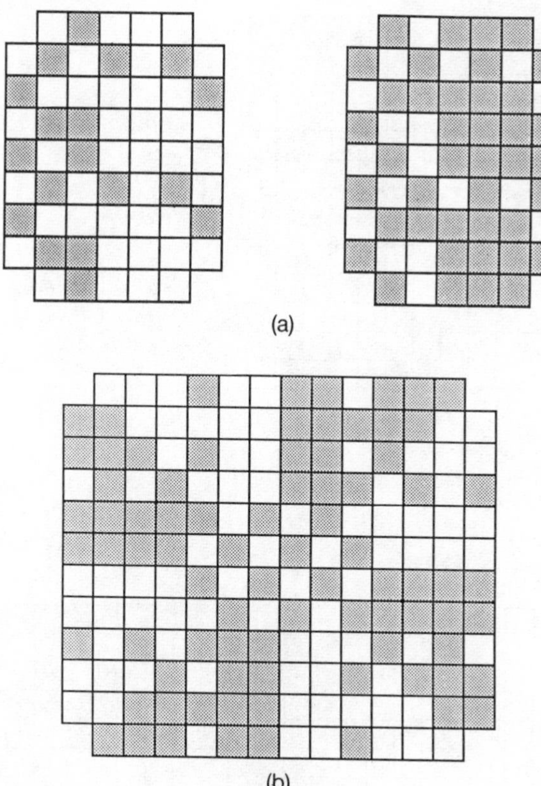

(a)

(b)

Fig. 15.2

Because of the large number of plants used in these experiments, further simplifications in the construction of the layout are desirable. For the periodic tilings of the plane, two approaches have been tried. The first is to require that every row along the lattice directions of the grid be occupied by the same binary sequence; such an array is said to be *sequential*. The second is to choose *planting rows* in the array, which need not be parallel with the lattice directions, and to require that each planting row should be occupied by 0000 . . . , 1111 . . . , 0101 . . . or 1010 . . . , starting from some given point. For finite blocks, symmetry has been used to simplify the layout.

Example 3. The array in Fig. 15.4 is a sequential balanced array of period 20. The same sequence, 0 0 0 0 0 1 0 1 0 0 1 1 1 1 1 0 1 0 1 1, appears in both the rows and columns of the array and is reversed in alternate lines. (The rows and columns of a square grid are the lattice directions of the square grid.)

The design in Fig. 15.5 is a balanced array obtained using columns as planting rows. □

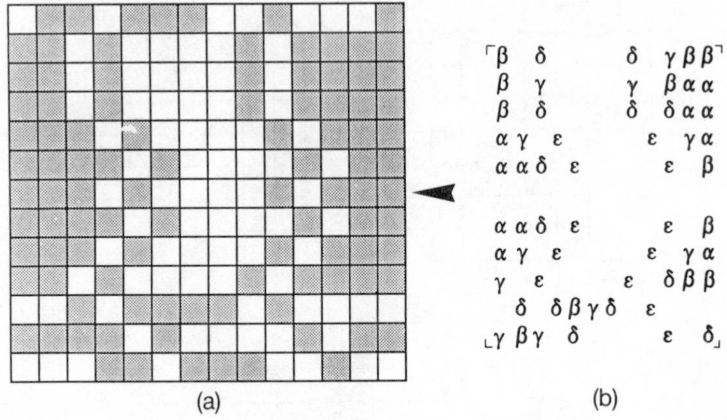
β δ δ γ β β
β γ γ β α α
β δ δ δ α α
α γ ε ε γ α
α α δ ε ε β
α α δ ε ε β
α γ ε ε γ α
γ ε ε δ β β
δ δ β γ δ ε
γ β γ δ ε δ

(a) (b)

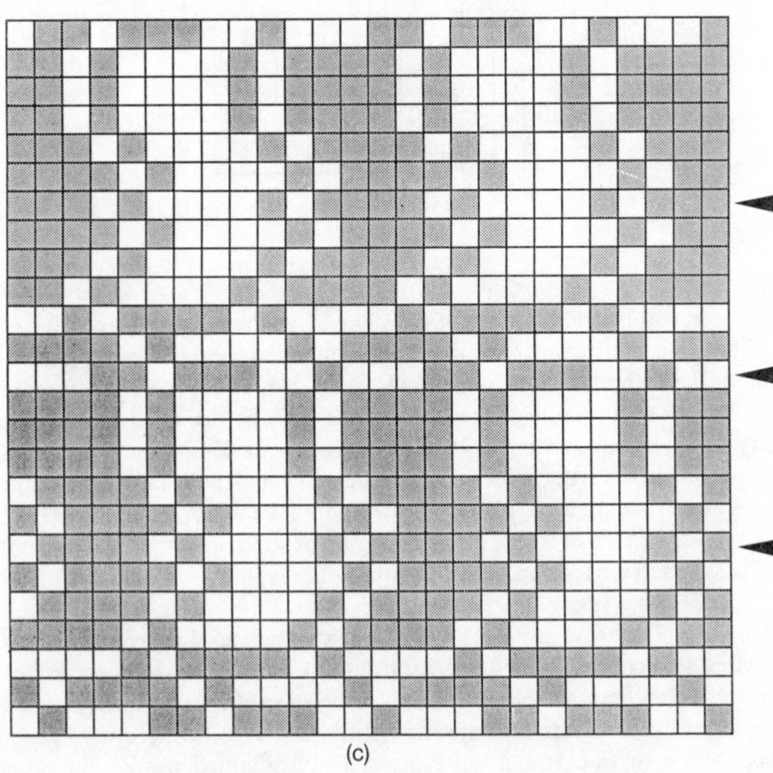

(c)

Fig. 15.3

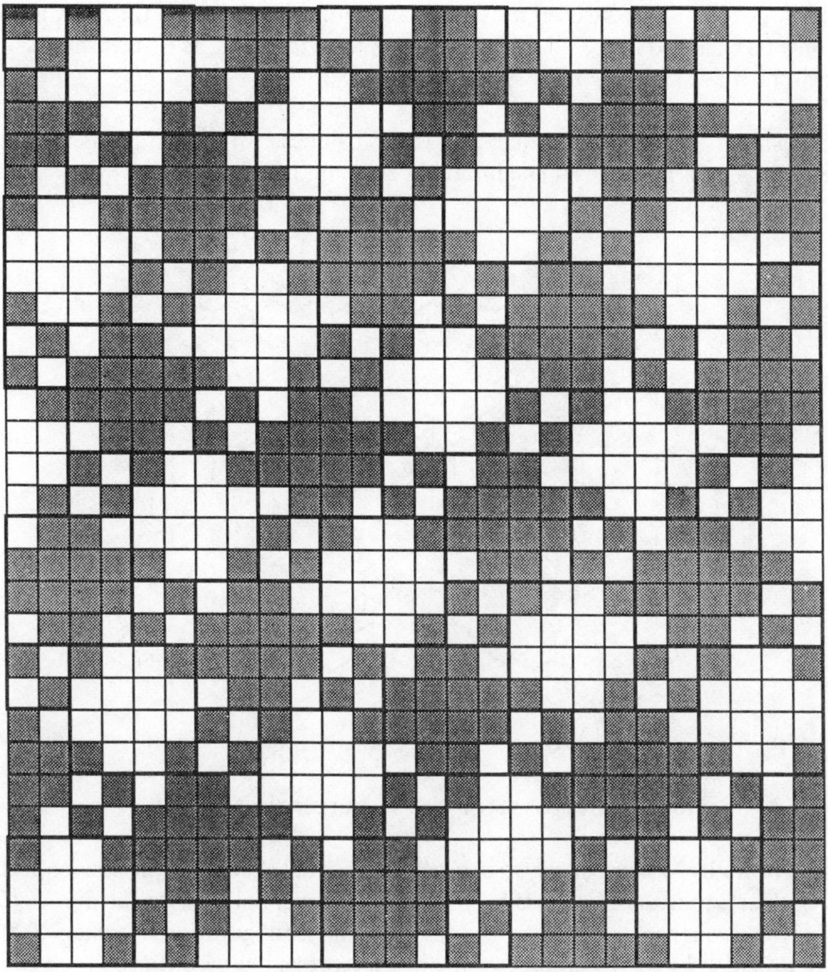

Fig. 15.4. Doubly-uniform sequential balanced array of period 20, unique up to reflection, with sequence 00000101001111101011. (The heavy lines show how the array can also be viewed as a uniform tiling not made up of edge-to-edge tiles.)

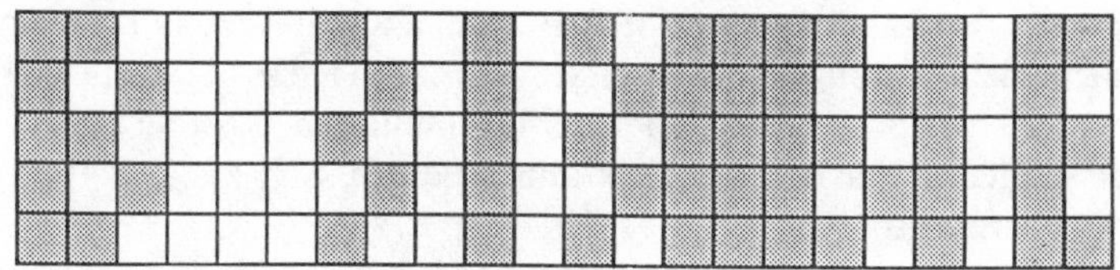

Fig. 15.5

If the design is to be used to test the effects of crowding, then the array need only be balanced in the symbol which represents the plant, as no measurement will be made on the empty space anyway.

Example 4. The array given in Fig. 15.6 is balanced for shaded hexagons but not for blank hexagons. (The internal plants are indicated by the heavy line.) □

Fig. 15.6

15.3. Sequential arrays for the square grid

We begin by looking at the structure of sequential arrays for the square grid.

What is the minimum possible period for a sequential balanced array? Any 0 in the sequence is in one of four possible configurations; namely 000, represented by M (for middle), 001 or 100, represented by E (for end), or 101, represented by I (for isolate). The average contribution of each of the neighbourhoods of 0 to a sequence of types of 0 is shown in Table 15.1. A sequence of five 0s, one each of α, β, γ_1, δ, and ε per period, would have (on average) $1\frac{1}{2}$ type M, 2 type E and $1\frac{1}{2}$ type I. To have whole numbers we must double this, giving a sequence with ten 0s and $3M$, $4E$, and $3I$. We will say such a sequence is of type (3, 4, 3). Similarly, if we had one each of α, β, γ_2, δ, and ε, the sequence would be of type (2, 1, 2). But we cannot have only $1E$ in a period; so we again have to double and have a sequence with ten 0s of type (4, 2, 4). This argument also applies to the 1s. Thus the minimum possible period is 20. Longer sequences have period $20n$ and type $a(3, 4, 3)+b(4, 2, 4)$ in either symbol, where $a+b=n$.

Consider a sequence of period 20 of type (4, 2, 4) in 0. It must contain four isolated 0s and one 6-string of 0s. A sequence of period 20 of type (3, 4, 3) in 0

Table 15.1

	M	E	I
α	1	0	0
β	$\frac{1}{2}$	$\frac{1}{2}$	0
γ_1	0	1	0
γ_2	$\frac{1}{2}$	0	$\frac{1}{2}$
δ	0	$\frac{1}{2}$	$\frac{1}{2}$
ε	0	0	1

must contain three isolated 0s and either a 5-string and a 2-string, or a 4-string and a 3-string of 0s. The possible configurations for these 0s are shown in Fig. 15.7.

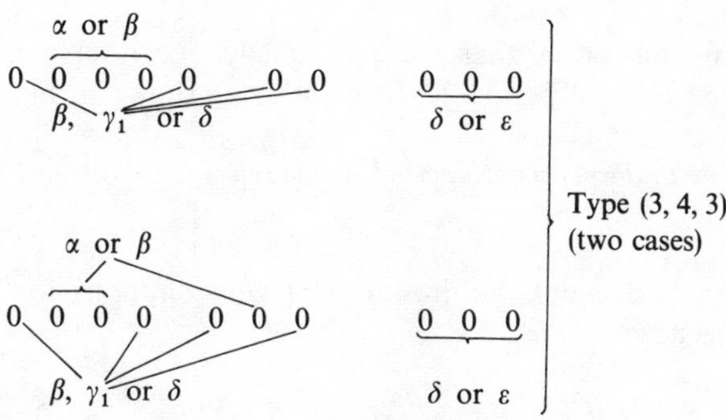

Fig. 15.7

To restrict further the number of arrays which have to be considered and to concentrate on arrays with easy planting properties, the following generalization of a cyclic array is considered. An array of minimum period is said to be *row-uniform* if the symbols of corresponding strings of each row are in the same configurations, and similarly for *column-uniformity*. If an array is both row-uniform and column-uniform and the configurations are the same for rows and columns, the array is said to be *doubly-uniform*.

The array in Fig. 15.4 is doubly-uniform.

In a row-uniform array each row must contain two symbols in each configuration. This restricts the possibilities shown in Fig. 15.7 to those shown in Fig. 15.8.

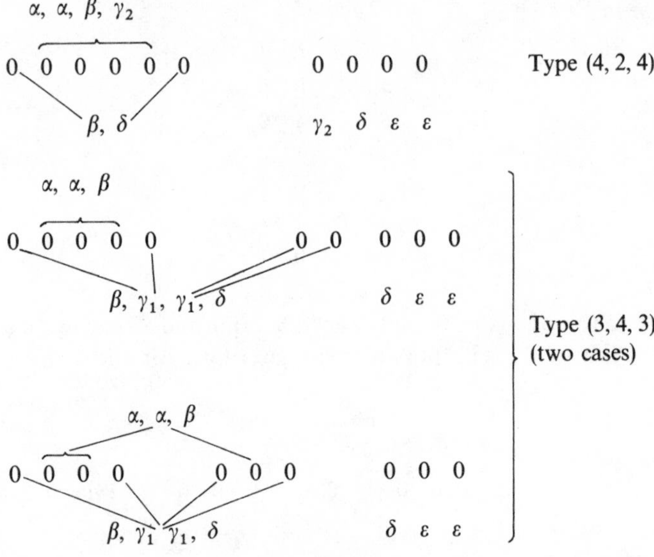

Fig. 15.8

We now outline the proof that there is exactly one doubly uniform sequential balanced array of period 20. (It is the array shown in Fig. 15.4.)

Lemma 5. *No row-uniform array of period* 20 *and type* (4, 2, 4) *in one* (*or both*) *symbols can exist.*

Proof. There are twelve possible inequivalent configurations for the 6-string in the sequence:

(1) $\beta\ \alpha\ \alpha\ \beta\ \gamma_2\ \delta$; (7) $\beta\ \beta\ \alpha\ \alpha\ \gamma_2\ \delta$;
(2) $\beta\ \alpha\ \alpha\ \gamma_2\ \beta\ \delta$; (8) $\beta\ \gamma_2\ \alpha\ \alpha\ \beta\ \delta$;
(3) $\beta\ \alpha\ \beta\ \alpha\ \gamma_2\ \delta$; (9) $\beta\ \beta\ \alpha\ \gamma_2\ \alpha\ \delta$;
(4) $\beta\ \alpha\ \gamma_2\ \alpha\ \beta\ \delta$; (10) $\beta\ \gamma_2\ \alpha\ \beta\ \alpha\ \delta$;
(5) $\beta\ \alpha\ \beta\ \gamma_2\ \alpha\ \delta$; (11) $\beta\ \beta\ \gamma_2\ \alpha\ \alpha\ \delta$;
(6) $\beta\ \alpha\ \gamma_2\ \beta\ \alpha\ \delta$; (12) $\beta\ \gamma_2\ \beta\ \alpha\ \alpha\ \delta$.

Seven of these possibilities can be ruled out immediately, since they imply the existence of 2- or 3-strings of 0s, which are not possible in this sequence. These are shown in Fig. 15.9, where the short strings are indicated by a curly bracket.

We have five remaining configurations to consider. The proof that each of them is impossible is shown in Fig. 15.10. The original configuration is shown in roman print (0, 1) in the diagram. The sequences implied by it are shown in italics (*0*, *1*); these usually involve additional strings of 0s whose neighbour-

(4)
```
  0 0 1 0 0 1
1 0 0 0 0 0 0 1
  0 0 1 0 1 1
```

(6)
```
  0 0 1 0 0 1
1 0 0 0 0 0 0 1
  0 0 1 1 0 1
```

(7)
```
  0 0 0 0 1 1
1 0 0 0 0 0 0 1
  0 1 0 0 1 1
```

(8)
```
  0 1 0 0 0 1
1 0 0 0 0 0 0 1
  0 1 0 0 1 1
```

(10)
```
  0 1 0 0 0 1
1 0 0 0 0 0 0 1
  0 1 0 1 0 1
```

(11)
```
  0 0 1 0 0 1
1 0 0 0 0 0 0 1
  0 1 1 0 0 1
```

(12)
```
  0 1 0 0 0 1
1 0 0 0 0 0 0 1
  0 1 1 0 0 1
```

Fig. 15.9. Forbidden configurations for the 6-string in type (4, 2, 4) sequence of period 20.

hoods are determined by the uniformity condition and shown in italics also. The third stage (if needed) is shown in bold print (**0**, **1**).

In (1), the column marked with an arrow contains a 7-string of 0s. Case (2) divides into two subcases: (a) leads to a 3-string of 0s in a column; (b) leads to two strings of 0s within one period of the sequence, also shown in a column. The two adjacent 1s indicated in (3) make it impossible for the 6-string below them to be in the required configuration for row-uniformity. Again in (5), the 101 subsequence indicated makes it impossible for the 6-string below it to be in the required configuration. Finally (9) leads to a 2-string of 0s in a column.

This completes the proof of the lemma. Note that column-uniformity was not needed in the proof. □

The next result also shows the non-existence of certain arrays.

Lemma 6. *No row-uniform and column-uniform array of period* 20, *having a* 4-*string, a* 3-*string, and three isolates of one* (*or both*) *symbols, can exist.*

Proof. There are eleven possible inequivalent configurations for the 3- and 4-strings in the sequence:

(1) $\beta\alpha\alpha\delta$ with $\gamma_1\beta\gamma_1$;
(2) $\beta\alpha\alpha\gamma_1$ with $\gamma_1\beta\delta$;
(3) $\gamma_1\alpha\alpha\delta$ with $\beta\beta\gamma_1$;
(4) $\gamma_1\alpha\alpha\gamma_1$ with $\beta\beta\delta$;
(5) $\beta\alpha\beta\delta$ with $\gamma_1\alpha\gamma_1$;
(6) $\delta\alpha\beta\beta$ with $\gamma_1\alpha\gamma_1$;
(7) $\beta\alpha\beta\gamma_1$ with $\gamma_1\alpha\delta$;
(8) $\gamma_1\alpha\beta\beta$ with $\gamma_1\alpha\delta$;
(9) $\gamma_1\alpha\beta\delta$ with $\beta\alpha\gamma_1$;
(10) $\delta\alpha\beta\gamma_1$ with $\beta\alpha\gamma_1$;
(11) $\gamma_1\alpha\beta\gamma_1$ with $\beta\alpha\delta$.

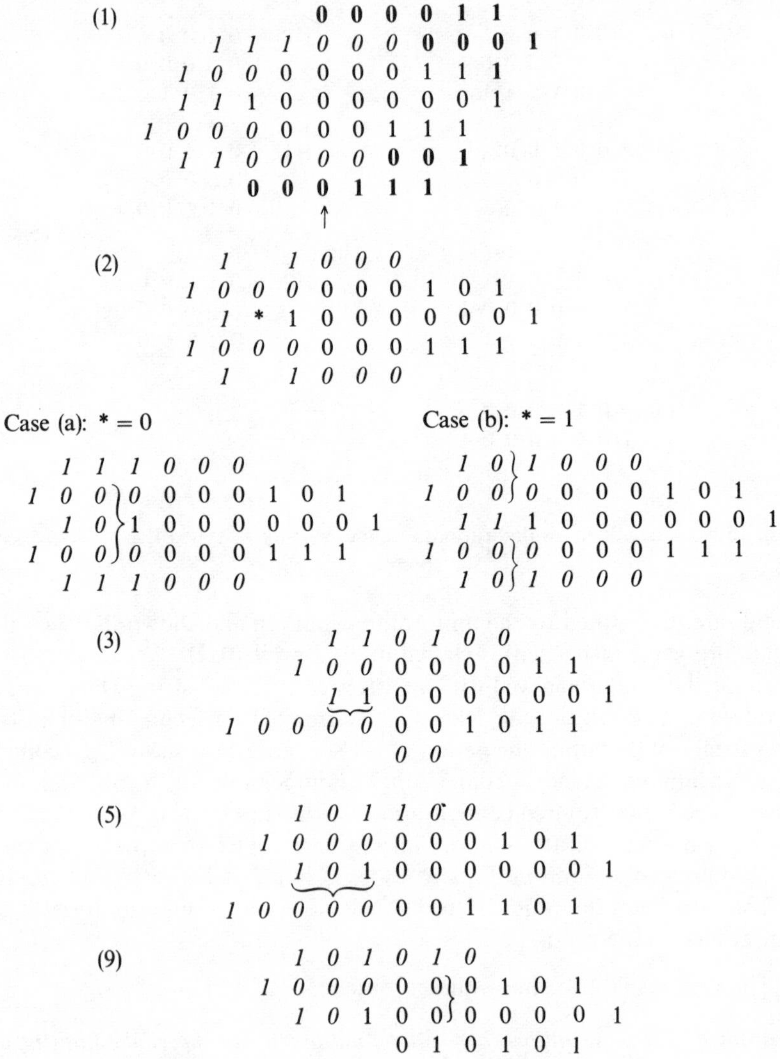

```
(1)
          0 0 0 0 1 1
    1 1 1 0 0 0 0 0 0 1
  1 0 0 0 0 0 0 1 1 1
  1 1 1 0 0 0 0 0 0 1
1 0 0 0 0 0 0 1 1 1
  1 1 0 0 0 0 0 0 1
      0 0 0 1 1 1
          ↑

(2)
    1   1 0 0 0
  1 0 0 0 0 0 0 1 0 1
    1 * 1 0 0 0 0 0 0 1
  1 0 0 0 0 0 0 1 1 1
    1   1 0 0 0

Case (a): * = 0

  1 1 1 0 0 0
1 0 0 0 0 0 0 1 0 1
  1 0 1 0 0 0 0 0 0 1
1 0 0 0 0 0 0 1 1 1
  1 1 1 0 0 0

Case (b): * = 1

  1 0 1 0 0 0
1 0 0 0 0 0 0 1 0 1
  1 1 1 0 0 0 0 0 0 1
1 0 0 0 0 0 0 1 1 1
  1 0 1 0 0 0

(3)
      1 1 0 1 0 0
    1 0 0 0 0 0 0 1 1
      1 1 0 0 0 0 0 0 1
1 0 0 0 0 0 0 1 0 1 1
          0 0

(5)
    1 0 1 1 0 0
  1 0 0 0 0 0 0 1 0 1
    1 0 1 0 0 0 0 0 0 1
1 0 0 0 0 0 0 1 1 0 1

(9)
    1 0 1 0 1 0
  1 0 0 0 0 0 0 1 0 1
    1 0 1 0 0 0 0 0 0 1
          0 1 0 1 0 1
```

Fig. 15.10

Again, some cases can be ruled out immediately, though because γ_1 is not symmetrical with respect to the string, more subcases arise than in the previous lemma. The configurations are shown in Fig. 15.11. The cases (3), (4a), (9b), (10b), and (11d) all imply the existence of 2-strings of 0s and are clearly impossible.

The other cases are more awkward but are no more difficult. We only look at three of them, (1), (2), and (4b). The others are left as a (long) exercise.

```
          0 0 0 1              0 0 0             0 0 1             1 0 1
(1)     1 0 0 0 0 1  with  (a) 1 0 0 0 1   or  (b) 1 0 0 0 1  or  (c) 1 0 0 0 1
          0 0 0 1              1 1 1             1 1 0             0 1 0

          0 0 0 0              0 0 1             1 0 1
(2)     1 0 0 0 0 1  with  (a) 1 0 0 0 1   or  (b) 1 0 0 0 1
          0 0 0 1              1 1 1             0 1 1

          0 0 0 1              0 0 0             0 0 1
(3)     1 0 0 0 0 1  with  (a) 1 0 0 0 1   or  (b) 1 0 0 0 1
          1 0 0 1              0 1 1             0 1 0

          0 0 0 0              0 0 0 1             0 0 1
(4) (a) 1 0 0 0 0 1  or   (b) 1 0 0 0 0 1  with  1 0 0 0 1
          1 0 0 1              1 0 0 0             0 1 1

          0 0 0 1              0 0 0             0 0 1
(5)     1 0 0 0 0 1  with  (a) 1 0 0 0 1   or  (b) 1 0 0 0 1
          0 0 1 1              1 0 1             1 0 0

          1 0 0 0              0 0 0             0 0 1
(6)     1 0 0 0 0 1  with  (a) 1 0 0 0 1   or  (b) 1 0 0 0 1
          1 0 1 0              1 0 1             1 0 0

          0 0 0 0              0 0 0 1             0 0 1
(7) (a) 1 0 0 0 0 1  or   (b) 1 0 0 0 0 1  with  1 0 0 0 1
          0 0 1 1              0 0 1 0             1 0 1

          0 0 0 0              1 0 0 0             0 0 1
(8) (a) 1 0 0 0 0 1  or   (b) 1 0 0 0 0 1  with  1 0 0 0 1
          1 0 1 0              0 0 1 0             1 0 1

          0 0 0 1              1 0 0 1             0 0 0
(9) (a) 1 0 0 0 0 1  or   (b) 1 0 0 0 0 1  with  1 0 0 0 1
          1 0 1 1              0 0 1 1             0 0 1

           1 0 0 0              1 0 0 1             0 0 0
(10) (a) 1 0 0 0 0 1  or   (b) 1 0 0 0 0 1  with  1 0 0 0 1
           1 0 1 1              1 0 1 0             0 0 1

           0 0 0 0              0 0 0 1              1 0 0 0
(11) (a) 1 0 0 0 0 1  or   (b) 1 0 0 0 0 1  or  (c) 1 0 0 0 0 1
           1 0 1 1              1 0 1 0              0 0 1 1

           1 0 0 1              0 0 1
or   (d) 1 0 0 0 0 1  with   1 0 0 0 1
           0 0 1 0              0 0 1
```

Fig. 15.11

Case (1): If the 4-string is bordered by two 3-strings, they must both be of type (a), leading to Fig. 15.12(i). This is impossible because the 3-strings in the columns are not uniform. If the 4-string is bordered by a 3-string (which must be of type (a)) and a 4-string, we have Fig. 15.12(ii), which is impossible because the column 4-strings are not uniform. If the 4-string is bordered by two 4-strings, we have Fig. 15.12(iii), which is impossible because we have 5-strings of 0s in the columns. Hence case (1) does not occur.

(i)	(ii)	(iii)	(iv)
1 1 1	1 1 1	1 0 0 0	1 0 0 0
1 0 0 0 1	1 0 0 0 1	1 0 0 0 0 1	1 0 0 0 0 1
1 0 0 0 0 1	1 0 0 0 0 1	1 0 0 0 0 1	1 0 0 0 0 1
1 0 0 0 1	1 0 0 0 0 1	1 0 0 0 0 1	1 0 0 0 0 1
1 1 1	1 0 0 0 1	1 0 0 0	1 0 0 0 0 1
	1 1 1	↑ ↑ ↑	0 0 0 1
			↑ ↑ ↑

Fig. 15.12

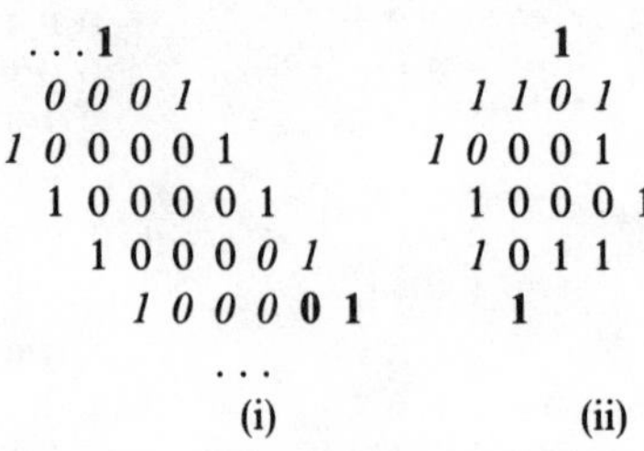

Fig. 15.13

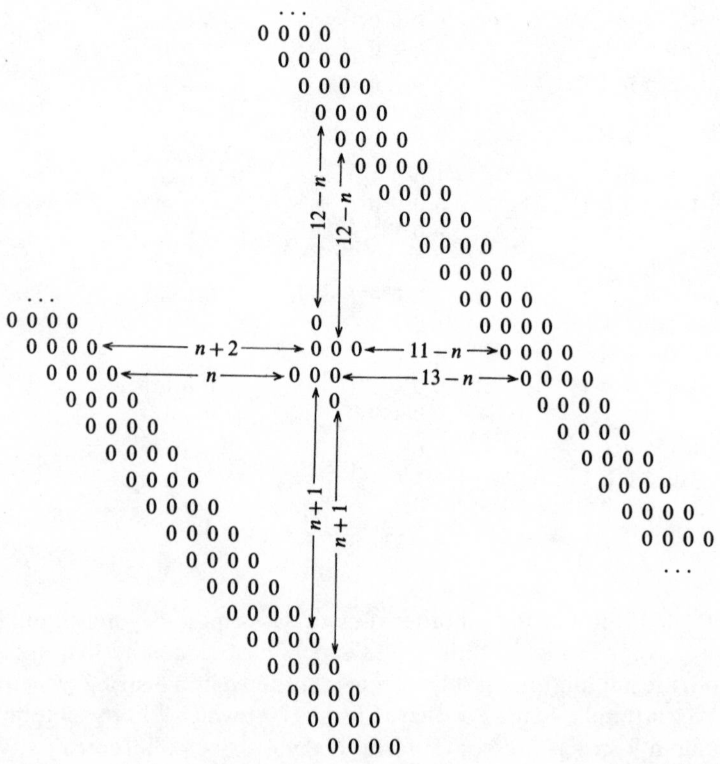

Case (2): Comparison of neighbourhoods shows that in neither (a) nor (b) can a 3-string border a 4-string; so we have Fig. 15.12(iv). But this is impossible, because we have 5-strings of 0s in the marked columns. So case (2) does not occur.

Case (4b): Comparison of neighbourhoods shows that the strings bordering the 4-string must themselves be 4-strings, and similarly that the string above the 3-string must itself be a 3-string. This leads to Fig. 15.13(i), from which we see that the 4-strings must form an infinite diagonal stripe and that the 3-strings must form connected blocks containing eight 0s each. The bold **1s** in Fig. 15.13(ii) are adjoined to prevent the 0s adjacent to them from being in the γ_2 configuration.

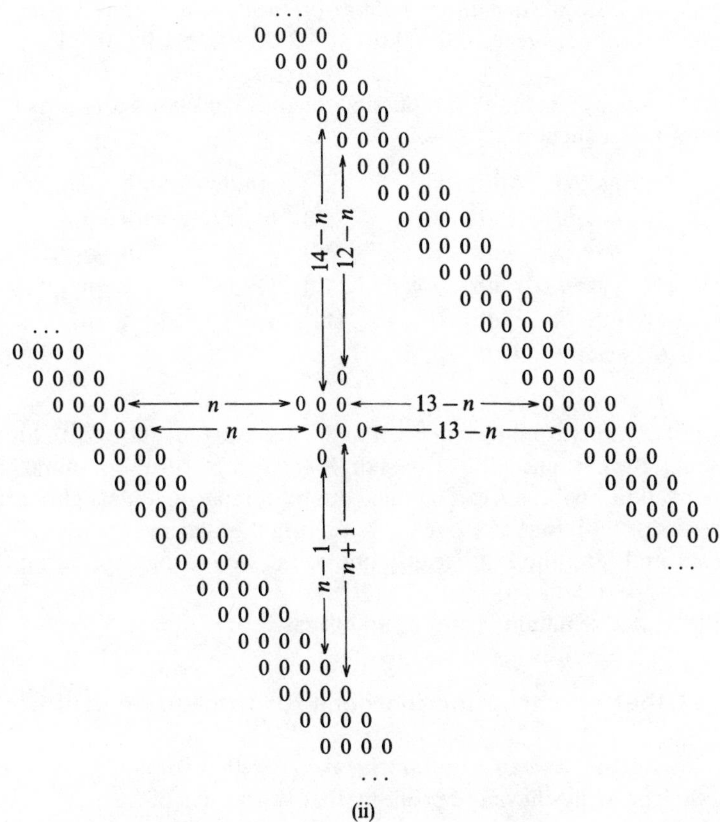

(ii)

Fig. 15.14. Spacing of an array based on configuration of case 4(b).

Suppose that there exists a minimal balanced row-uniform array based on this configuration. It must have period 20; it must also be doubly-uniform, as is obvious from Fig. 15.13. The block of 3-strings can have two essentially different positions relative to the stripe of 4-strings, as shown in Fig. 15.14.

The sequence must contain thirteen symbols besides the 3- and 4-strings of 0s; hence, if the strings of 0s are separated by n symbols as shown in Fig. 15.14(i), they must be separated by $13-n$ symbols in the remaining part of the sequence. The position shown in Fig. 15.14(i) leads to the distances between the strings taking the values $n, n+2$, $11-n$ and $13-n$ for the rows, and $n+1$, $12-n$, for the columns. But among these values, more than two must be distinct modulo 20; so the array is not sequential. Similarly, the distances between the strings in Fig. 15.14(ii) are n and $13-n$ for the rows, and $n-1$, $n+1$, $12-n$, and $14-n$ for the columns, and again more than two of these are distinct modulo 20. This rules out case (4b). □

Theorem 7. *There exists exactly one doubly uniform array of period* 20, *having a 5-string, a 2-string, and three isolates of each symbol. It is unique up to reflection and has sequence* 0 0 0 0 0 1 0 1 0 0 1 1 1 1 1 0 1 0 1 1.

Proof. There are again eleven possible inequivalent configurations for the strings of the sequence:

(1)	$\beta\alpha\alpha\beta\delta$	with	$\gamma_1\gamma_1$;	(7)	$\gamma_1\alpha\alpha\beta\gamma_1$	with	$\beta\delta$;
(2)	$\delta\alpha\alpha\beta\beta$	with	$\gamma_1\gamma_1$;	(8)	$\beta\alpha\beta\alpha\gamma_1$	with	$\gamma_1\delta$;
(3)	$\beta\alpha\alpha\beta\gamma_1$	with	$\gamma_1\delta$;	(9)	$\beta\alpha\beta\alpha\delta$	with	$\gamma_1\gamma_1$;
(4)	$\gamma_1\alpha\alpha\beta\beta$	with	$\gamma_1\delta$;	(10)	$\gamma_1\alpha\beta\alpha\delta$	with	$\beta\gamma_1$;
(5)	$\gamma_1\alpha\alpha\beta\delta$	with	$\beta\gamma_1$;	(11)	$\gamma_1\alpha\beta\alpha\gamma_1$	with	$\beta\delta$.
(6)	$\delta\alpha\alpha\beta\gamma_1$	with	$\beta\gamma_1$;				

The configurations are shown in Fig. 15.15. The cases (5b), (6b), (7d), (10b), and (11c) imply the existence of 3-strings of 0s and can be ruled out immediately.

Several additional cases can be ruled out by comparing the neighbourhood of the 2-string with that of a 2-string bordering the 5-string, namely (2b), (4a), (7a), (7b), and (8b), the last because one of the two strings bordering the 5-string must be a 2-string.

We leave the remaining cases as an exercise. □

15.4. Other construction methods for the square grid

A *building design* is a small array which, taken together with its complement, is balanced. The arrays have the property that two arrays of the same type can interlock and form a larger array which, again taken together with its complement, is balanced.

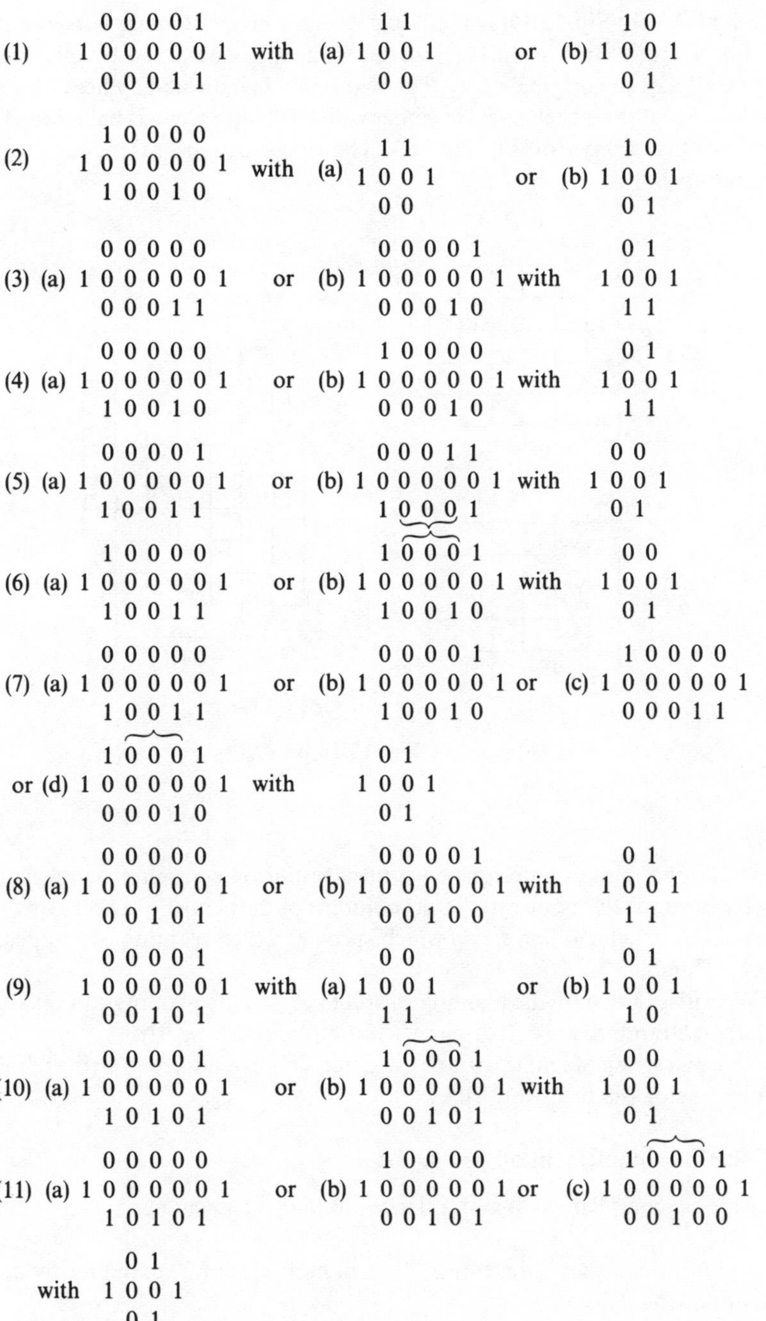

Fig. 15.15. Configurations of the 5- and 2-strings in type (3, 4, 3) sequence of period 20.

Example 8. In Fig. 15.16(a) a building design is given. Its complement is given in Fig. 15.16(b). In Fig. 15.16(c) we show the array obtained by interlocking two copies of the original array. The two symbols indicated by heavy lines are added so that the larger array, together with its complement, is balanced. These are the two arrays of Fig. 15.2(a). The array can be extended vertically indefinitely. □

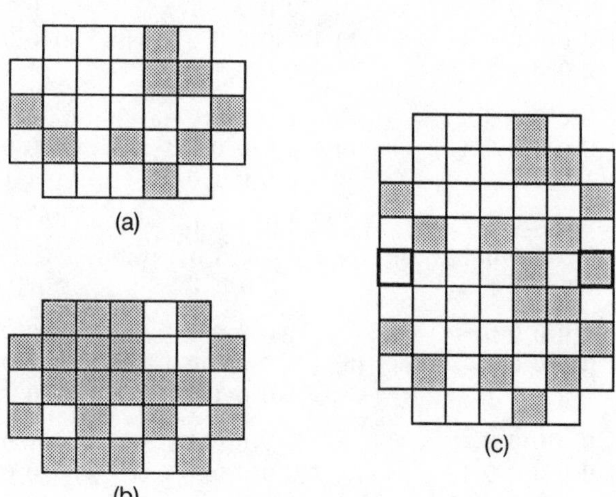

Fig. 15.16

All known building designs have been found by computer searches.

The second technique is to let the columns of the array be the planting rows and to construct a *design string* which specifies which planting row appears in each column.

We will let **1** denote the planting row 1111 . . . , **0** the planting row 0000 . . . , *A***1** the planting row 1010 . . . and *A***0** the planting row 0101 . . . .

A design string of length 10 is constructed. It is extended to a string of length 20 by one of the following rules:

Rule A: complement all 0s and 1s;

Rule B: complement **0** and **1**, leave *A***0** and *A***1** unchanged.

The complete design is developed now, in multiples of 20 columns, by one of the following rules:

Rule 1: extend each planting row to the same arbitrary length;

Rule 2: extend each planting row to the same even length.

To obtain balance the final column must be repeated to the left of the array and the first column to the right of the array.

A design string is said to be of *type* 1*A* if it gives rise to a balanced design by applying rules 1 and *A*, and so on. The twenty-eight non-isomorphic design strings of length 10, found by computer search, are given in Table 15.2.

Example 9. We construct a design from the type 1*B* design string

(***A*1, *A*0, 0, 0, 0, *A*1, *A*0, 0, 1, 0**).

First we extend it to length 20 using rule B; this gives

***A*1, *A*0, 0, 0, 0, *A*1, *A*0, 0, 1, 0, *A*1, *A*0, 1, 1, 1, *A*1, *A*0, 1, 0, 1.**

We now use Rule 1 to extend the design. Suppose we want the design to have five rows. Then the layout, including the border columns, is that given in Fig. 15.5. □

Finally, balanced arrays may be constructed by tiling the planes with polyominoes. A *polyomino* is a plane figure obtained by joining unit squares in a rook-wise fashion; that is, so that a chess rook could travel through all the squares in the piece. The design in Fig. 15.4, for instance, can be tiled using the 16-omino in Fig. 15.17(a), in both 0 and 1, and single squares. The polyominoes

Table 15.2

A1 0 0 0 A0 0 1 A0 A1 A0; A1 0 0 0 A0 A1 A0 1 0 A0;
A1 0 0 0 1 A0 A0 0 A0 A1; A1 0 0 A0 0 A1 0 A1 1 A1;
A1 0 0 A0 0 A1 1 A1 A0 1; A1 0 0 A0 0 A0 A1 0 A0 1;
A1 0 0 A0 A1 0 A1 0 A0 0; A1 0 0 0 A0 0 1 A0 A0 A1;
A1 0 0 0 1 A0 A1 A0 0 A0; A1 0 0 A0 0 A1 0 A1 A0 0;
A1 0 0 A0 0 A1 A0 0 A0 1; A1 0 0 A0 0 A0 A1 0 A1 0;
A1 0 0 A0 0 A0 1 A0 1 A1.

Type 1A

A1 A0 0 0 0 A1 A0 0 1 0; A1 A0 0 0 0 1 0 A0 A1 0;
A1 A0 0 A0 0 A1 1 0 1 A0; A1 A0 0 A0 A1 0 0 0 1 0.

Type 1B

A1 A1 0 0 0 A0 0 A1 A0 A1; A1 A1 0 0 A1 A0 A1 1 1 A1;
A1 A1 0 0 A0 A1 A0 1 1 A1; A1 A0 0 0 0 A1 A1 A1 1 A0;
A1 A0 0 0 A1 A1 A1 1 1 A0; A0 A0 0 A0 1 1 1 0 1 A0;
A1 A1 0 0 0 A0 A1 A0 1 A1; A1 A1 0 0 A0 0 0 A1 A0 A1;
A1 A1 0 A0 0 0 0 A1 A0 A1; A1 A0 0 0 0 1 0 A0 1 A0;
A1 A0 0 0 A0 A0 A0 1 1 A0.

Type 2B

in Fig. 15.17(b), in both 0 and 1, together with single squares, can be used to tile the plane in at least two different ways, given in Figs. 15.17(c) and 15.17(d). Again, the problem of constructing balanced arrays in this way has not been explored systematically.

15.5. Periodic balanced binary arrays for the hexagonal grid

The main result here is

Theorem 10. *For the hexagonal grid, no periodic balanced binary array can exist.*

Proof. The proof is a counting argument in which it is shown it is not possible for such an array to contain enough ζs and ηs.

Consider the neighbourhoods given in Fig. 15.1. We will label each hexagon by its type (shaded or blank) and equivalence class. A blank α cannot border a shaded η or a shaded or blank ζ. A blank β can border a shaded ζ or shaded η but not a blank ζ and so on. Thus, with each blank hexagon in the array we will associate a vector (a, b, c), where a is the number of blank ζs, b the number of shaded ζs and c the number of shaded ηs bordered by the original blank hexagon.

We introduce an ordering on these vectors, saying that (a_1, b_1, c_1) *covers* (a_2, b_2, c_2) if and only if $a_1 \geq a_2, b_1 \geq b_2$ and $c_1 \geq c_2$. For each configuration there is a *covering family of vectors*; that is, a family of vectors such that any vector associated with a configuration is covered by a member of the covering family. In Table 15.3, we list the covering family for each configuration and in Fig. 15.18 we give the associated neighbourhoods. Within each of the equivalence classes γ, δ and ε, the subclasses are labelled 1, 2, and 3 from left to right in Fig. 15.1(c).

Suppose a periodic balanced binary array exists. Then, for some integer n, it contains a balanced block of $14n$ hexagons, half shaded, half blank, such that there are n of each colour in each configuration. Let f_i, g_i and h_i denote the number of blank hexagons in the balanced block which border i shaded ζs, i blank ζs and i shaded ηs respectively. Similarly, let t_i denote the number which border a total of i shaded ζs, blank ζs and shaded ηs.

We now do some counting. Each shaded ζ has five of its six edges adjacent to blank hexagons and from Table 15.3 we know that $f_5 = f_6 = 0$; so

$$f_1 + 2f_2 + 3f_3 + 4f_4 = 5n. \tag{15.1}$$

Similarly, counting edges of blank ζs, we have

$$g_1 + 2g_2 + 3g_3 = n; \tag{15.2}$$

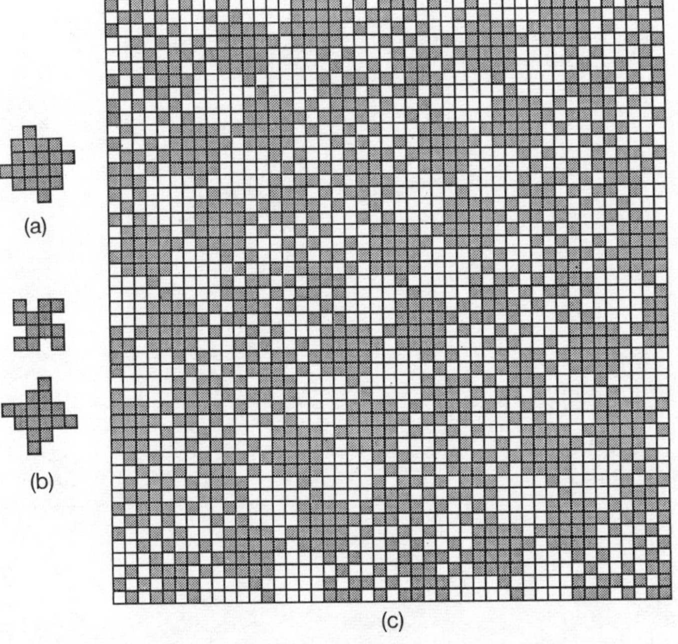

(a)

(b)

(c)

(d)

Fig. 15.17

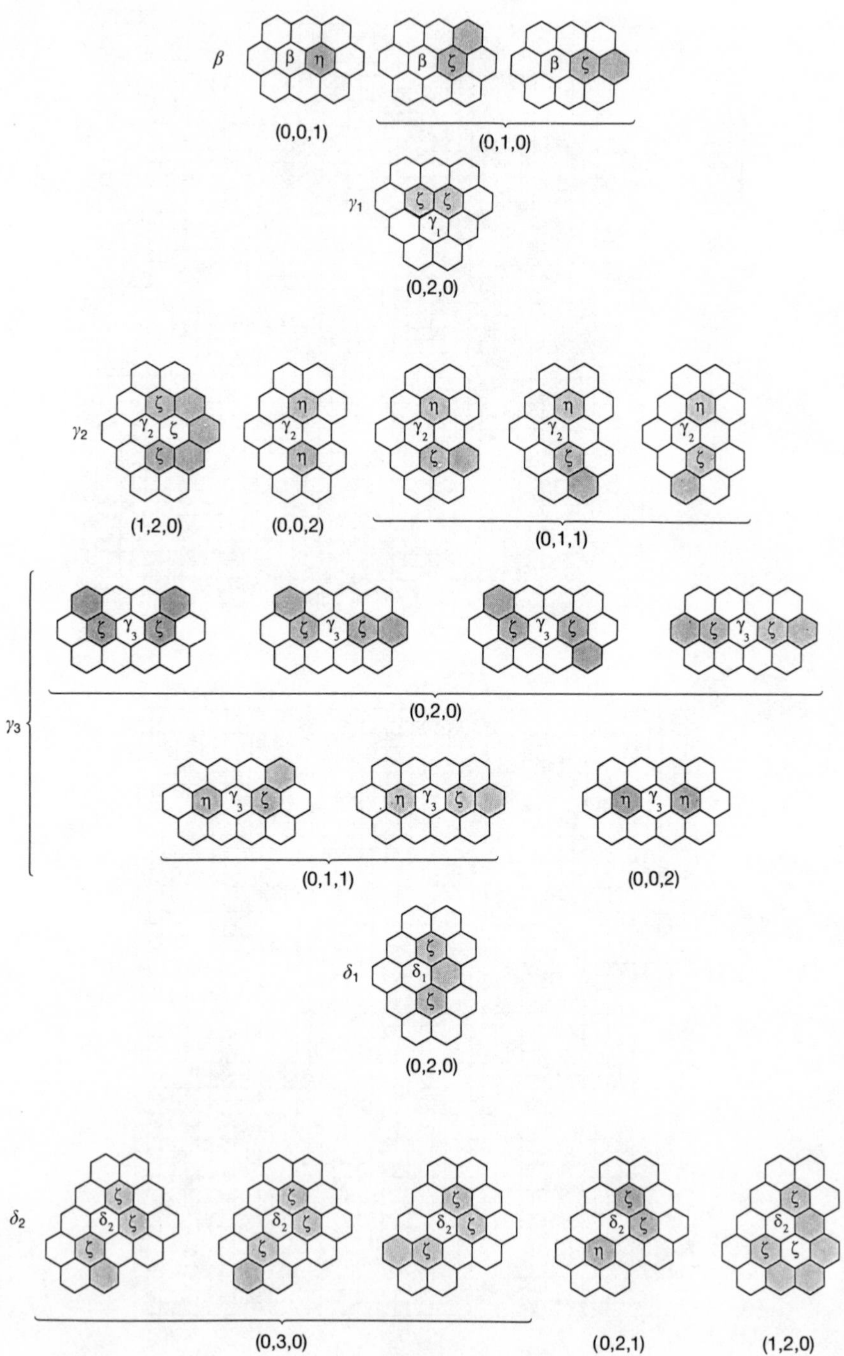

β
β η
β ζ
β ζ
(0,0,1)
(0,1,0)
γ_1
ζ ζ
γ_1
(0,2,0)
γ_2
(1,2,0)
(0,0,2)
(0,1,1)
γ_3
(0,2,0)
(0,1,1)
(0,0,2)
δ_1
(0,2,0)
δ_2
(0,3,0)
(0,2,1)
(1,2,0)

Fig. 15.18

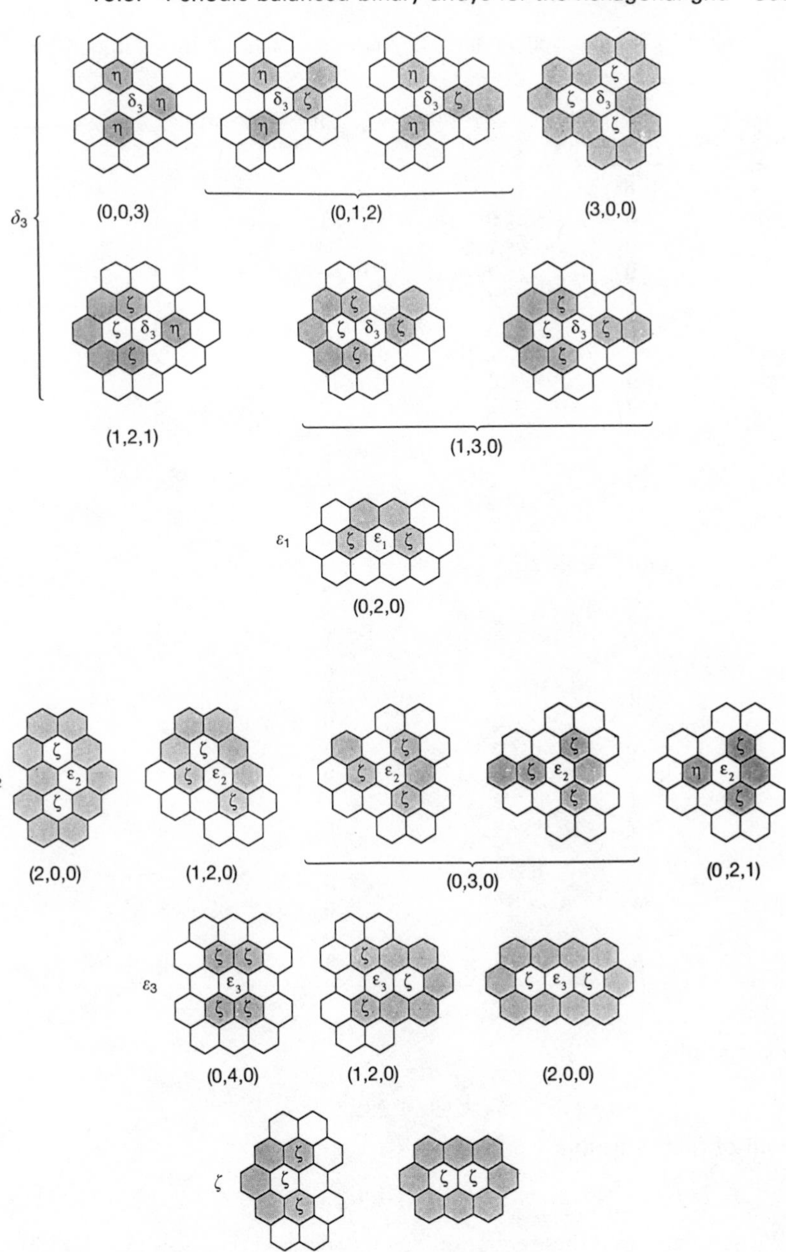
δ_3
(0,0,3)
(0,1,2)
(3,0,0)
(1,2,1)
(1,3,0)
ε_1
(0,2,0)
ε_2
(2,0,0)
(1,2,0)
(0,3,0)
(0,2,1)
ε_3
(0,4,0)
(1,2,0)
(2,0,0)
ζ
(0,2,0)
(1,0,0)

Fig. 15.18

Table 15.3. Covering families of vectors and sums of their components

Blank hexagon	Bordering a blank ζs	Bordering b shaded ζs	Bordering c shaded ηs	$a+b+c$
α	0	0	0	0
β	0	0	1	1
	0	1	0	1
γ_1	0	2	0	2
γ_2	1	2	0	3
	0	0	2	2
	0	1	1	2
γ_3	0	2	0	2
	0	1	1	2
	0	0	2	2
δ_1	0	2	0	2
δ_2	0	3	0	3
	0	2	1	3
	1	2	0	3
δ_3	0	0	3	3
	0	1	2	3
	3	0	0	3
	1	2	1	4
	1	3	0	4
ε_1	0	2	0	2
ε_2	2	0	0	2
	1	2	0	3
	0	3	0	3
	0	2	1	3
ε_3	0	4	0	4
	1	2	0	3
	2	0	0	2
ζ	0	2	0	2
	1	0	0	1
η	0	0	0	0

for the shaded ηs

$$h_1 + 2h_2 + 3h_3 = 6n; \tag{15.3}$$

for all of them together

$$t_1 + 2t_2 + 3t_3 + 4t_4 = 12n. \tag{15.4}$$

We now show that these equations are inconsistent. Suppose that of the n ε_3s in the array, $n\lambda$ of them contribute to t_4; that is, are of type $(0, 4, 0)$. Suppose $n\mu$ δ_3s also contribute to t_4 (and so are of type $(1, 2, 1)$ or $(1, 3, 0)$), and that $n\nu$ γ_2s contribute to t_3 (and so are of type $(1, 2, 0)$), where $0 \leq \lambda, \mu, \nu \leq 1$.

Now (see Fig. 15.18), each blank ε_3 of type $(0, 4, 0)$ has two blank neighbours and these neighbours are either γ_2 or δ_3. Furthermore, distinct ε_3 of type

(0, 4, 0) do not share any common blank neighbours. The γ_2s are of type (0, 2, 0) (covered by (1, 2, 0)) and the δ_3s are of type (0, 2, 1), (0, 3, 0) or (0, 2, 0) (covered by (1, 2, 1), (1, 3, 0) or either, respectively). These γ_2s contribute to none of the h_i and the δ_3s to at most h_1. Let there be $n\lambda_1$ such γ_2s and $n\lambda_2$ such δ_3s. Then

$$\lambda_1 + \lambda_2 = 2\lambda. \tag{15.5}$$

Also, the δ_3s contributing to t_4 are of types (1, 2, 1) or (1, 3, 0) and so contribute to at most h_1 and the γ_2s contributing to t_3 are of type (1, 2, 0) and contribute to no h_i.

Consider the blank hexagons bordering the shaded ηs. A contribution to h_3 can only come from δ_3s, and these must not contribute to t_4 nor border ε_3s of type (0, 4, 0). A contribution to h_2 could arise from γ_3s or γ_2s not contributing to t_3 nor bordering ε_3s of type (0, 4, 0) (or from the δ_3s just mentioned but we are interested in finding the maximum possible total contribution). A contribution to h_1 can arise from βs, from ε_2s, and from δ_3s contributing to t_4 or bordering ε_3s of type (0, 4, 0). Thus, from eqn (15.3),

$$\begin{aligned} 6n &= h_1 + 2h_2 + 3h_3 \\ &\le n(1 + (1 - \lambda) + \lambda_2 + \mu) + 2n(1 - \lambda_1 - \nu) + 3n(1 - \lambda_2 - \mu) \\ &= n(7 - \lambda - 2(\lambda_1 + \lambda_2) - 2\mu - 2\nu) \\ &= n(7 - (5\lambda + 2\mu + 2\nu)), \text{ using (15.5)}. \end{aligned}$$

Hence

$$5\lambda + 2\mu + 2\nu \le 1. \tag{15.6}$$

We treat eqn (15.4) in the same way. Contributions to t_4 come from the $n\lambda$ ε_3s and $n\mu$ δ_3s only, to t_3 from the remaining δs and εs and from the $n\nu$ γ_2s, to t_2 from the ζs and the remaining γs, and to t_1 from the βs. Thus we have

$$\begin{aligned} 12n &= t_1 + 2t_2 + 3t_3 + 4t_4 \\ &\le n + 2n(1 + 1 - \nu) + 3n((1 - \lambda) + (1 - \mu) + \nu) + 4n(\lambda + \mu) \\ &= n(11 + \lambda + \mu + \nu) \\ &\le 11\tfrac{1}{2}n \end{aligned}$$

by eqn (15.6), and this is a contradiction. □

15.6. Beehive designs

A *sixfold-symmetric beehive design* is a pair of complementary hexagonal blocks each containing $3r(r + 1) + 1$ plots, r being the *radius* of the design. The blocks have six-fold symmetry about an *axial* plot; so the whole design is determined by a *basic triangle* which may be assumed to have 0 at its apex.

Example 11. Let $r = 4$ and consider the basic triangle given in Fig. 15.19(a). The design obtained from it is given in Fig. 15.19(b). The type of each internal

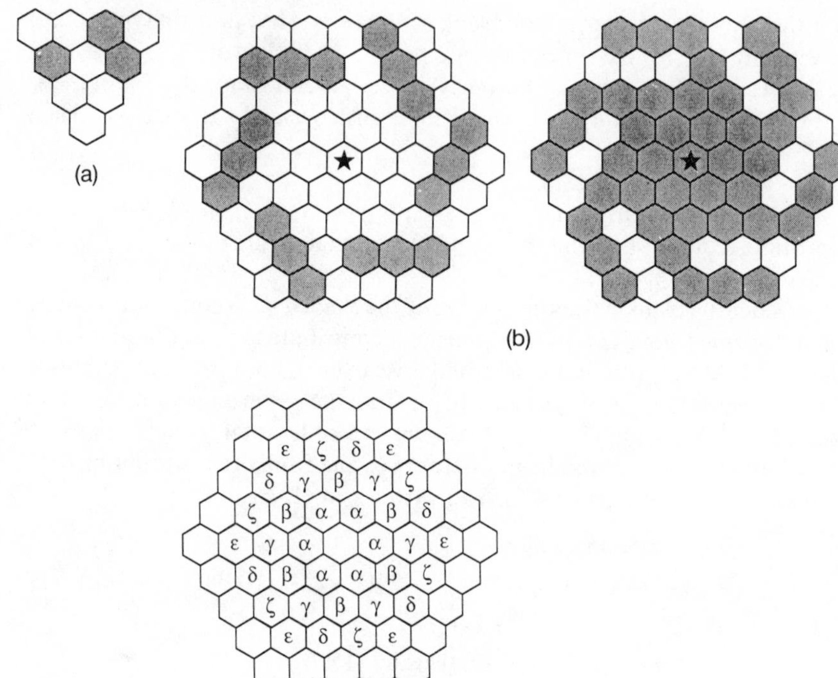

Fig. 15.19

plot of each block, except the axial plot which is marked with an asterisk, is given in Fig. 15.19(c). □

In general, we will represent a basic triangle by a sequence of octal numbers, one for each row, starting with the apex and separated by commas. Thus the triangle given in Example 11 is represented by [0, 0, 5, 2].

Let n_i be the number of plots in the design containing a 0 and adjacent to i neighbours of variety 1. So n_0 is the number of plots in the design containing a 0 and of type α, n_1 is the number containing a 0 and of type β, and so on. (Because of complementarity, n_i is also the number of plots in the design containing a 1 and adjacent to i neighbours of variety 0.) Then

$$\boldsymbol{n} = (n_0, n_1, \ldots, n_6) = 6\boldsymbol{k} = 6(k_0, k_1, \ldots, k_6)$$

is called the *profile* of the design. The profile of the design given in Example 11 is (6, 6, 6, 6, 6, 6, 0).

Clearly

$$\sum_{i=0}^{6} k_i = r(r-1)/2 \tag{15.7}$$

and so a sixfold-symmetric beehive design could only be balanced for $r \equiv 0$ or 1 (mod 7). In fact, a counting argument similar to that of Theorem 10 can be used to show

Theorem 12. *There are no balanced sixfold-symmetric beehive designs.* □

Because of this, we attempt to construct designs in which

$$S = \sum_i k_i^2$$

is minimized subject to eqn (15.7), and to the requirement that the k_i must be non-negative integers. We will denote the minimum value for given r by $S^*(r)$. If a design can be constructed with S equal to $S^*(r)$ we say that the design is *levelled*. In cases where no levelled designs exist we will say that a design with minimum realizable S is *optimal*.

Again, the results have been obtained by complete enumeration on a computer. Some of the optimal designs found are given in Table 15.4. (P is the number of designs with the given profile – one example is given for each profile for radius 5 and 6.)

Table 15.4

Radius	n	P	S	Basic triangle (octal representation)
4	(6, 6, 6, 6, 6, 6, 0)	4	6	[0, 0, 3, 4]
	,,			[0, 0, 3, 10]
	,,			[0, 0, 5, 2]
	,,			[0, 0, 5, 4]
5	(12, 6, 12, 12, 6, 6, 6)	5	16	[0, 0, 1, 4, 26]
	(12, 6, 12, 6, 12, 6, 6)	5		[0, 0, 1, 4, 27]
	(6, 12, 12, 12, 6, 6, 6)	12		[0, 0, 2, 5, 10]
	(6, 12, 6, 12, 12, 6, 6)	10		[0, 0, 2, 11, 6]
	(6, 12, 12, 6, 12, 6, 6)	2		[0, 0, 2, 12, 4]
	(6, 6, 12, 12, 12, 6, 6)	2		[0, 0, 3, 5, 37]
6	(18, 18, 12, 12, 12, 12, 6)	5	35	[0, 0, 0, 5, 22, 1]
	(18, 12, 18, 12, 12, 12, 6)	10		[0, 0, 0, 6, 22, 10]
	(18, 12, 12, 18, 12, 12, 6)	20		[0, 0, 0, 6, 22, 11]
	(18, 12, 18, 12, 12, 6, 12)	2		[0, 0, 0, 12, 5, 4]
	(12, 18, 18, 12, 12, 12, 6)	8		[0, 0, 1, 0, 17, 0]
	(12, 18, 12, 18, 12, 12, 6)	3		[0, 0, 1, 0, 36, 0]
	(12, 12, 18, 18, 12, 12, 6)	22		[0, 0, 1, 2, 11, 34]
	(12, 18, 18, 12, 12, 6, 12)	2		[0, 0, 1, 4, 24, 10]
	(12, 6, 18, 18, 12, 12, 12)	5		[0, 0, 1, 4, 26, 0]
	(6, 12, 18, 18, 12, 12, 12)	4		[0, 0, 2, 12, 5, 0]
	(12, 12, 18, 12, 18, 12, 6)	1		[0, 2, 1, 17, 23, 77]

We can relax the requirement of sixfold symmetry to twofold or threefold symmetry. The basic triangle is replaced by a *basic array* which is repeated three or two times.

Example 13. A threefold symmetric beehive design is given in Fig. 15.20. The basic array is indicated by heavy lines. □

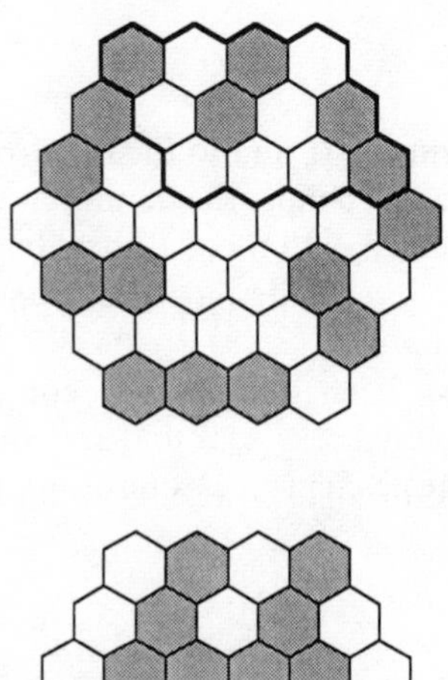

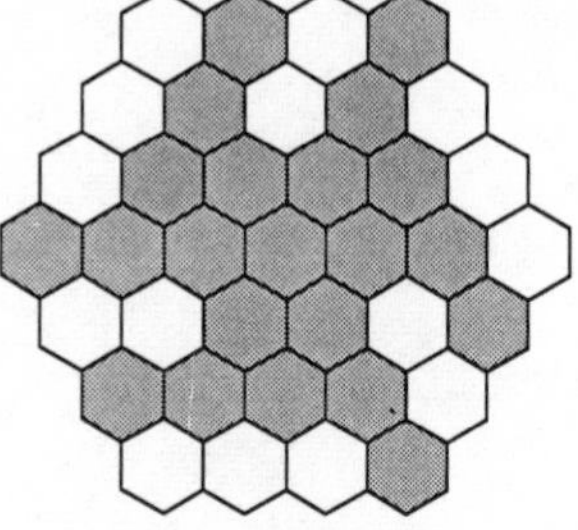

Fig. 15.20

Again, we can represent each basic array in octal, recording each triangle separately. Thus the basic array of the design in Example 13 is 0, 1, 5; 0, 0, 3. Some examples of optimal two- and threefold symmetric beehive designs are given in Table 15.5.

15.7. Other designs for the hexagonal grid

First we look at building designs for the hexagonal grid.

Example 14. In Fig. 15.21(a) a building design is given. In Fig. 15.21(b) we show the array obtained by interlocking two copies of this array. The heavy hexagon, the *augmenting variety*, is added so that the larger array, together with its complement, would still be balanced. □

Table 15.5

m	r	Profile	Basic array
3	3	(3, 3, 6, 3, 0, 3, 0)	[0, 1, 5; 0, 0, 3]
		(0, 0, 3, 3, 6, 3, 3)	[1, 1, 3; 0, 1, 4]
	4	(3, 6, 6, 6, 6, 6, 3)	[1, 2, 3, 10; 0, 3, 7, 17]
		(6, 6, 6, 6, 6, 3, 3)	[1, 0, 4, 10; 0, 1, 0, 2]
	5	(9, 12, 9, 9, 9, 6, 6)	[0, 0, 0, 5, 20; 0, 0, 5, 6, 2]
		(9, 6, 12, 9, 9, 9, 6)	[0, 0, 2, 12, 13; 0, 0, 4, 12, 31]
2	3	(2, 4, 2, 4, 2, 2, 2)	[0, 1, 0; 0, 0, 0; 1, 1, 2]
		(2, 2, 4, 4, 2, 2, 2)	[0, 2, 7; 1, 2, 7; 1, 3, 7]
	4	(6,6,6,6,4,4,4)	[0, 0, 0, 0; 1, 0, 2, 1; 1, 0, 5, 14]

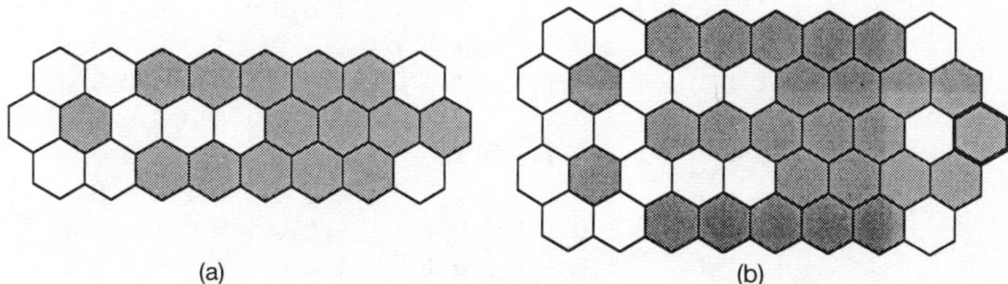

Fig. 15.21

We need only record the first two rows of a building design and we can represent them in octal. Thus the design above is [76; 217]. There are twelve building designs. We list them, together with augmenting varieties, in Table 15.6.

Table 15.6

Two-row octal representation	Augmenting variety
[10; 207]	1
[36; 357]	1
[13; 204]	0
[41; 320]	0
[16; 203]	1
[76; 217]	1
[15; 204]	0
[41; 260]	0
[40; 207]	1
[36; 373]	1
[57; 274]	0
[75; 364]	0

Finally, we mention some designs in which balance is obtained only for the 1s. These designs are appropriate in experiments to examine the effects of crowding, as one is hardly interested in the effect of neighbours on an unoccupied site. As with the building designs, these strings were found by a complete search. Each element in the string corresponds to one column of the design which is to be planted exclusively with that variety. Each column contains the same number of internal plants. The nineteen strings are given in Table 15.7. The design derived from the first string is that of Fig. 15.6. Figure 15.22 shows how the planting row (A, B, C, . . . , X, Y) is placed on the grid.

Table 15.7

Number	String
1	0 1 1 1 1 1 1 0 1 1 0 1 0 1 0 1 0 0 1 0 0 1 0 0 1
2	0 1 1 1 1 1 1 0 1 1 0 1 0 1 0 0 1 0 0 1 0 0 1 0 1
3	0 1 1 1 1 1 1 0 1 1 0 1 0 0 1 1 0 0 1 0 0 1 0 0 1
4	0 1 1 1 1 1 1 0 1 1 0 1 0 0 1 0 0 1 1 0 0 1 0 0 1
5	0 1 1 1 1 1 1 0 1 1 0 1 0 0 1 0 0 1 0 0 1 1 0 0 1
6	0 1 1 1 1 1 1 0 1 1 0 1 0 0 1 0 0 1 0 0 1 0 1 0 1
7	0 1 1 1 1 1 1 0 1 1 0 0 1 1 0 1 0 0 1 0 0 1 0 0 1
8	0 1 1 1 1 1 1 0 1 1 0 0 1 0 1 1 0 0 1 0 0 1 0 0 1
9	0 1 1 1 1 1 1 0 1 1 0 0 1 0 1 0 0 1 0 0 1 0 0 1 1
10	0 1 1 1 1 1 1 0 1 1 0 0 1 0 0 1 1 0 1 0 0 1 0 0 1
11	0 1 1 1 1 1 1 0 1 1 0 0 1 0 0 1 0 1 1 0 0 1 0 0 1
12	0 1 1 1 1 1 1 0 1 1 0 0 1 0 0 1 0 1 0 0 1 0 0 1 1
13	0 1 1 1 1 1 1 0 1 1 0 0 1 0 0 1 0 0 1 1 0 1 0 0 1
14	0 1 1 1 1 1 1 0 1 1 0 0 1 0 0 1 0 0 1 0 1 1 0 0 1
15	0 1 1 1 1 1 1 0 1 0 1 1 0 1 0 1 0 0 1 0 0 1 0 0 1
16	0 1 1 1 1 1 1 0 1 0 1 1 0 1 0 0 1 0 0 1 0 0 1 0 1
17	0 1 1 1 1 1 1 0 1 0 1 0 1 1 0 1 0 0 1 0 0 1 0 0 1
18	0 1 1 1 1 1 1 0 1 0 0 1 1 0 1 1 0 0 1 0 0 1 0 0 1
19	0 1 1 1 1 1 1 0 1 0 0 1 0 0 1 1 0 1 1 0 0 1 0 0 1

15.8. Spacing experiments

In spacing experiments involving only plants of one species (monocultures), the aim is to compare the effect of different spacings on the growth of the plants. The plants are situated on a regular grid and the factors of interest are the area around each plant and the *rectangularity* (defined as breadth/length) of that area.

If the plots are to be of a constant size then the number of plants per plot will vary. This is appropriate if the interest is in yield per unit area but can involve either an excessive number of plants at the close spacings or too few at the larger spacings.

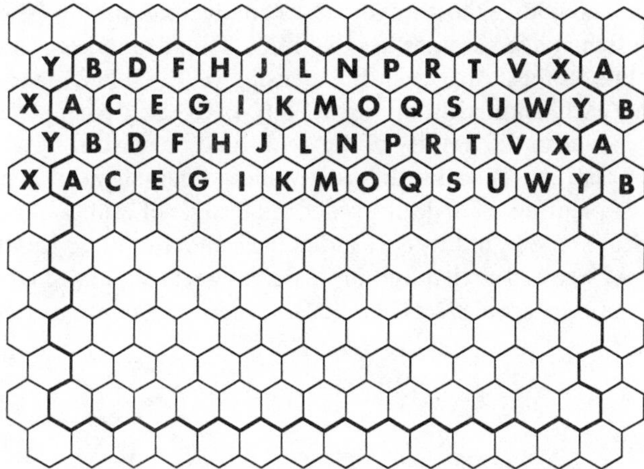

Fig. 15.22

Two sorts of designs for spacing experiments have been proposed. One is a design in which spacings are assigned at random to the rows and columns of each plot; see Fig. 15.23. The plots are of unequal sizes but have a constant number of plants in them. Plants are planted where a row and a column intersect. Plants in the rows and the columns indicated by a heavy line are guard plants and are not harvested.

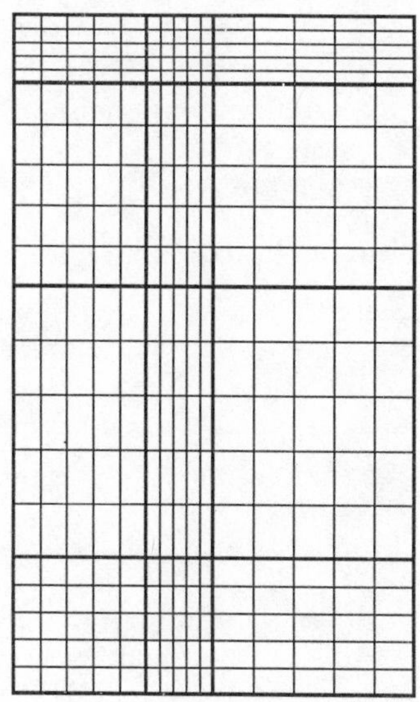

Fig. 15.23

The other, and to date most usual, way of conducting spacing experiments is to use the systematic fan designs; see Fig. 15.24. The advantages of a systematic design in this context are that the density of the plants and the area surrounding each plant can vary gradually over the design. Of course, a systematic change in the fertility of the field could be masked by a systematic design but orienting replicates differently should help avoid this problem.

For ease of planting, a fan design is defined by a set of grid points where two sets of curves intersect, in this case radial lines and arcs of concentric circles. The grid is constructed so that the curves defining sets of plots of equal area or equal rectangularity are also sets of radial lines or arcs of concentric circles. The

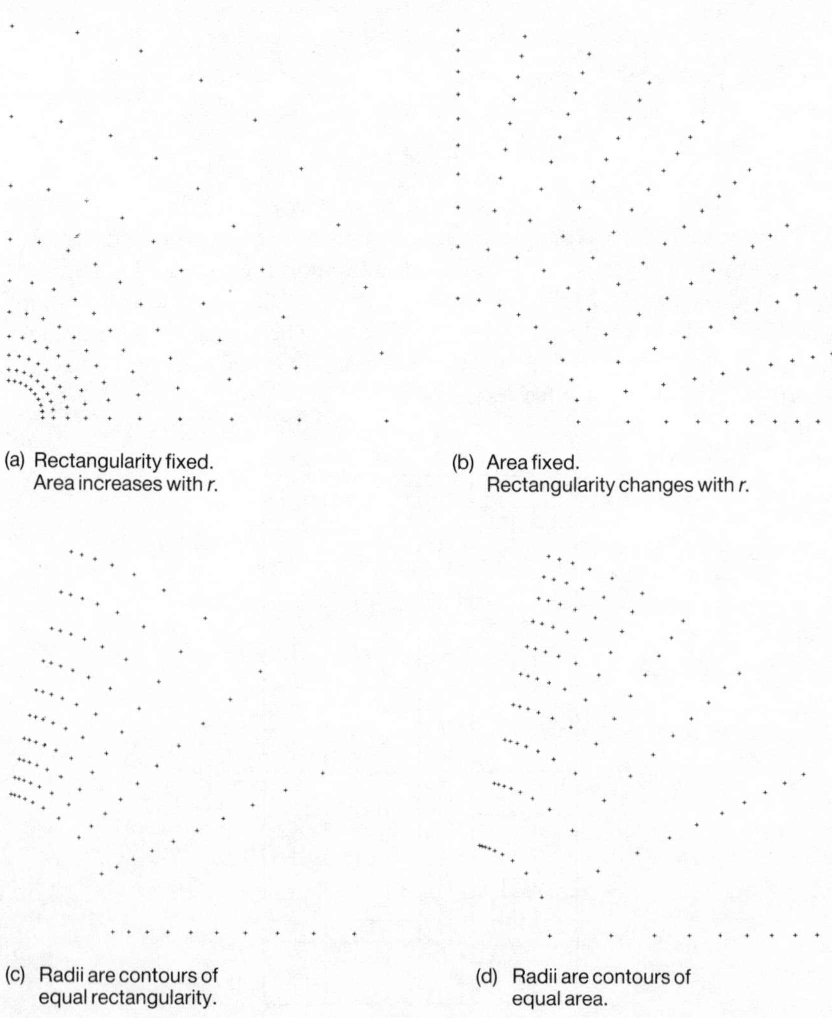

(a) Rectangularity fixed. Area increases with r.

(b) Area fixed. Rectangularity changes with r.

(c) Radii are contours of equal rectangularity.

(d) Radii are contours of equal area.

Fig. 15.24

sets of plots of equal area or equal rectangularity are distinct so that the effects of rectangularity and area are not confounded.

Formally, a *fan design* will be defined by an n-sequence of radii $(r_1, r_2, \ldots, r_n)$ and an m-sequence of angles $(\theta_1, \theta_2, \ldots, \theta_m)$ where, from a given point on a base line, n arcs of radii $r_1, r_2, \ldots, r_n$, and m lines at angles $\theta_1, \theta_2, \ldots, \theta_m$, are drawn. The grid given by the intersection of these sets of lines is the set of points where plants are to be planted. In Fig. 15.25 we illustrate the plot associated with a given plant. The rectangularity of such a plot is defined as the ratio of the length of the arc which contains the point to the length of the radial line through the point.

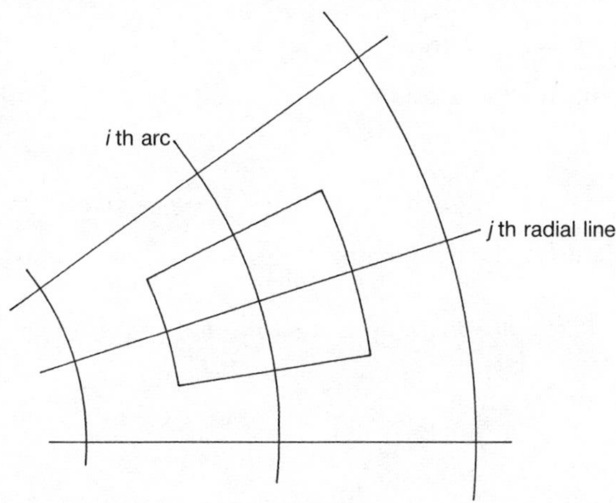

Fig. 15.25

Theorem 15. (a) *A fan design in which the rectangularity is constant throughout and the area is constant along a given arc is given by* $r_i = r_0\alpha^i$, $\theta_j = j\theta$, $i \geq 1$, $j \geq 1$.

(b) *A fan design in which the area is constant throughout and the rectangularity is constant along a given arc is given by* $r_i = (ai+b)^{\frac{1}{2}}, \theta_j = j\theta, i \geq 1, j \geq 1$.

(c) *A fan design in which the rectangularity is constant along a given radial line is given by* $r_i = r_0\alpha^i, \theta_j = \arctan((j-1)\lambda)$, $i \geq 1$, $j \geq 1$.

(d) *A fan design in which the area is constant along a given radial line is given by* $r_i = (ai+b)^{\frac{1}{2}}$, $\theta_j = \arctan((j-1)\lambda)$, $i \geq 1$, $j \geq 1$.

Proof. Consider the plant at the intersection of the ith arc and the jth radial line.

The area of this plot is

$$\left(\frac{\theta_{j+\frac{1}{2}}}{2\pi}\pi r^2_{i+\frac{1}{2}} - \frac{\theta_{j-\frac{1}{2}}}{2\pi}\pi r^2_{i+\frac{1}{2}}\right) - \left(\frac{\theta_{j+\frac{1}{2}}}{2\pi}\pi r^2_{i-\frac{1}{2}} - \frac{\theta_{j-\frac{1}{2}}}{2\pi}\pi r^2_{i-\frac{1}{2}}\right)$$

$$= \tfrac{1}{2}(\theta_{j+\frac{1}{2}} - \theta_{j-\frac{1}{2}})(r^2_{i+\frac{1}{2}} - r^2_{i-\frac{1}{2}}) = A, \text{ say,}$$

and the rectangularity is

$$\left(\frac{\theta_{j+\frac{1}{2}}}{2\pi}2\pi r_i - \frac{\theta_{j-\frac{1}{2}}}{2\pi}2\pi r_i\right)\bigg/ (r_{i+\frac{1}{2}} - r_{i-\frac{1}{2}}) = r_i(\theta_{j+\frac{1}{2}} - \theta_{j-\frac{1}{2}})/(r_{i+\frac{1}{2}} - r_{i-\frac{1}{2}}) = R, \text{ say.}$$

We now calculate these values for each of the designs in turn.

(a) Here

$$\begin{aligned} A &= \tfrac{1}{2}((j+\tfrac{1}{2})\theta - (j-\tfrac{1}{2})\theta)\,(r_0^2 a^{2i+1} - r_0^2 \alpha^{2i-1}) \\ &= \tfrac{1}{2}\theta r_0^2 \alpha^{2i}(\alpha - \alpha^{-1}), \end{aligned}$$

which is constant for given i, and

$$R = r_0 \alpha^i \theta/(r_0\alpha^{i+\frac{1}{2}} - r_0\alpha^{i-\frac{1}{2}}) = \theta/(\alpha^{\frac{1}{2}} - \alpha^{-\frac{1}{2}}),$$

which is constant for the whole design.

(b) Here

$$A = \tfrac{1}{2}\theta(a(i+\tfrac{1}{2}) + b - a(i-\tfrac{1}{2}) - b) = \tfrac{1}{2}\theta a,$$

which is constant for the design, and

$$R = \theta(ai+b)^{\frac{1}{2}}/((a(i+\tfrac{1}{2})+b)^{\frac{1}{2}} - (a(i-\tfrac{1}{2})+b)^{\frac{1}{2}}),$$

which is constant for given i.

(c) Here

$$\begin{aligned} R &= r_0\alpha^i\,(\arctan((j-\tfrac{1}{2})\lambda) - \arctan((j-3/2)\lambda))/(r_0\alpha^{i+\frac{1}{2}} - r_0\alpha^{i-\frac{1}{2}}) \\ &= \arctan(\lambda/(1+\lambda^2(j^2-2j+3/4)))/(\alpha^{\frac{1}{2}} - \alpha^{-\frac{1}{2}}), \end{aligned}$$

which is constant for given j.

(d) Here

$$\begin{aligned} A &= \tfrac{1}{2}(\arctan((j-\tfrac{1}{2})\lambda) - \arctan((j-3/2)\lambda))\,(a(i+\tfrac{1}{2}) + b - a(i-\tfrac{1}{2}) - b) \\ &= \tfrac{1}{2}a(\arctan((j-\tfrac{1}{2})\lambda) - \arctan((j-3/2)\lambda)) \\ &= a(\arctan(\lambda/2(1+\lambda^2(j^2-2j+3/4)))), \end{aligned}$$

which is constant for given j. □

15.9. Designs for intercropping experiments

We look at three designs which have been proposed to study two-variety intercrops. Even after the two varieties have been decided, there are still five other factors of interest; namely, the area and rectangularity of each of the varieties and the intimacy, or relative arrangement, of the two varieties.

Example 16. The designs in Fig. 15.26 have the rectangularity of each variety and the intimacy varying separately. The number of plants (of a particular type) per unit area is kept constant throughout, however. The intimacy is greater for (a), (c), (e), and (g) than for (b), (d), (f), and (h). The rectangularity is close to 1 for both crops in the (a) and (b) configurations and is more extreme in the (c) and (d) configurations. In the remaining four configurations one crop is much 'squarer' than the other. □

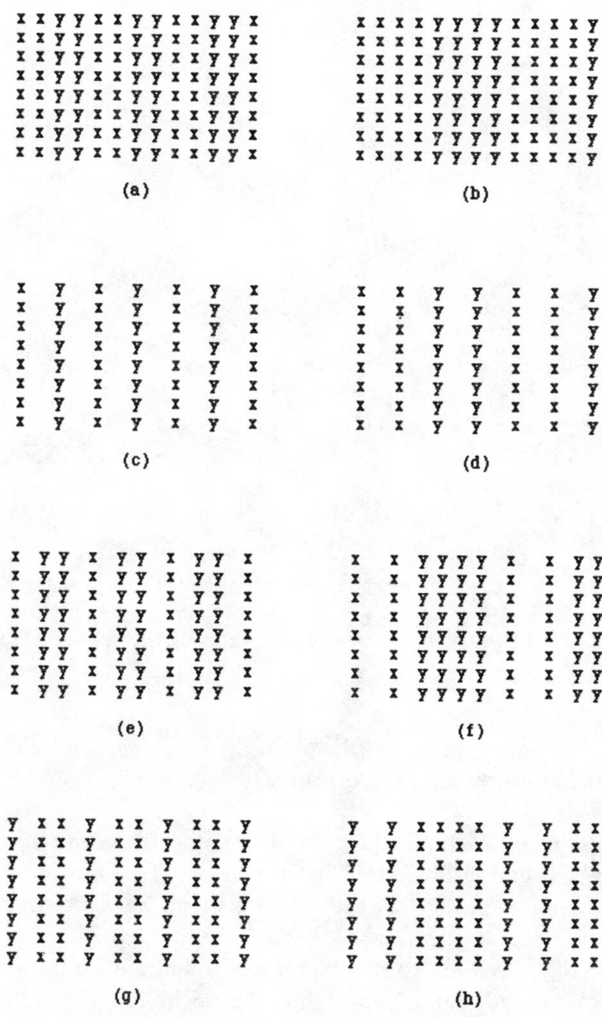

Fig. 15.26

Example 17. The fan design can be modified to have one variety planted on 'even' spokes and the other variety planted on the 'odd' spokes. One such design, which includes a segment planted just with one variety, appears in Fig. 15.27. □

The final example, given in Fig. 15.28 is a two-way systematic design for two crops in which the densities vary in the perpendicular directions.

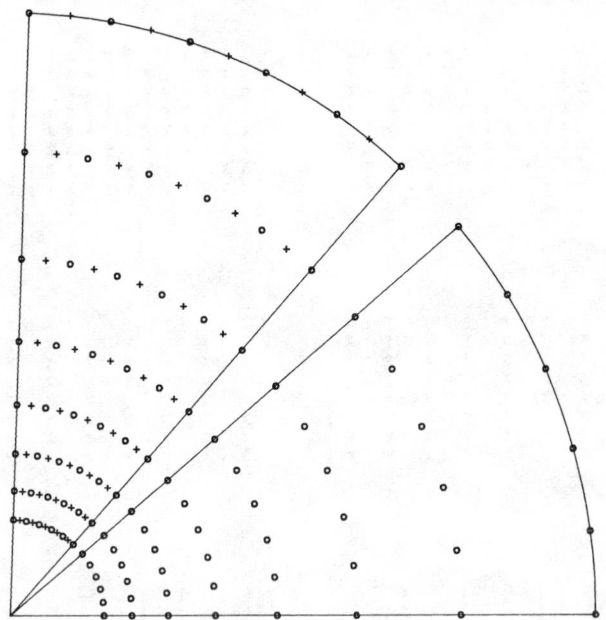

Fig. 15.27. The left-hand segment contains two varieties, the right-hand segment only one.

15.10. References and comments

A general discussion of some aspects of plant competition and of competition designs can be found in Besag (1974), Cormack (1979), Donald (1963), Mead (1979), and Mead and Riley (1981). A. P. Street (1982) gives a survey of design constructions.

Figure 15.2(a) is the first known example of a building design and is due to Veevers, Boffey, and Zafar-Yab (1980). Those in Exercise 15.6 are from D. J. Street and W. H. Wilson (1985), who have also shown that the design of Fig. 15.2(b) is unique and that there are two hundred and forty-four

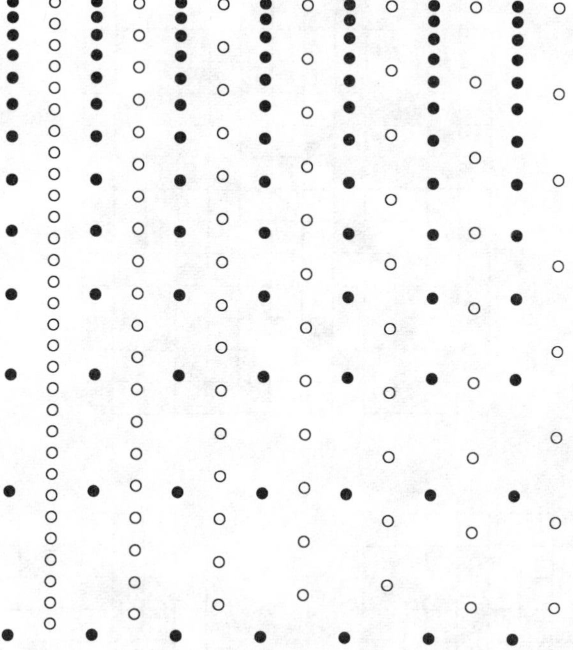

Fig. 15.28

inequivalent designs of the type given in Exercise 15.1. Both of these designs originally appeared in E. R. Williams and Bailey (1981). The design in Example 2 is due to Gates (1980).

The designs in Figs 15.4 and 15.17, and the results in Section 15.3, are due to A. P. Street and Macdonald (1979). Exercise 15.8 is from P. J. Robinson (1986). Macdonald and A. P. Street (1979) also considered the triangular grid.

The design in Fig. 15.5 and the design strings discussed in Section 15.4 and tabulated in Table 15.2 are due to Veevers and Zafar-Yab (1980, 1982). The design in Fig. 15.6 and the strings in Table 15.7 are due to Veevers (1982). The results in Section 15.5, as well as Exercise 15.9, are due to Oates-Williams and A. P. Street (1979).

Beehive designs were introduced by Martin (1973) and studied extensively by Veevers and Boffey: Theorem 12 is due to Boffey and Veevers (1979), Table 15.4 and Exercise 15.10 are from Veevers and Boffey (1975) and Tables 15.5 and 15.6 and Fig. 15.21 are from Veevers and Boffey (1979). Other designs based on a hexagonal grid are given by Fasoulas (1981); see also Bos (1983).

The design of Fig. 15.23 is due to Lin and Morse (1975). Fan designs were introduced by Nelder (1962) who also proved Theorem 15. Figures 15.26 and 15.28 are based on those of Mead and Stern (1980). Figure 15.27 is a modification of Nelder's fan design due to Wahua and Miller (1978).

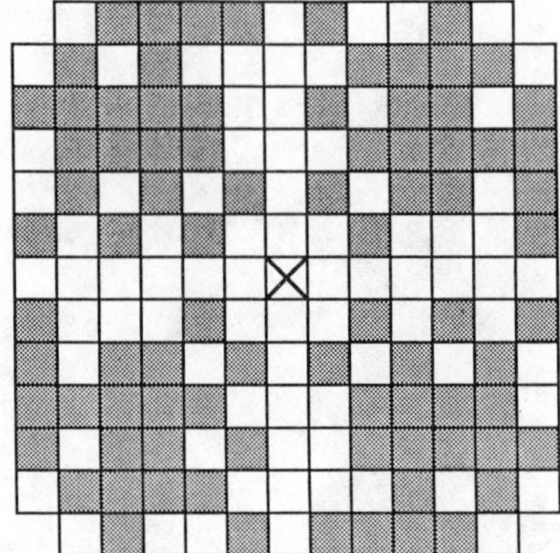

Fig. 15.29

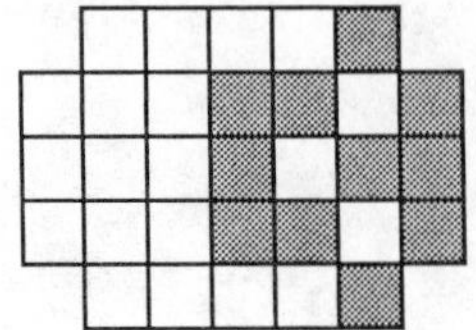

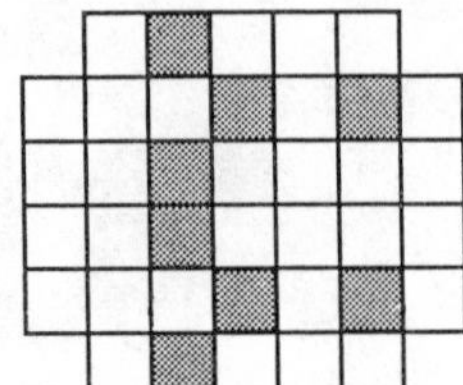

Fig. 15.30

(a)

(b)

Fig. 15.31

Exercises

15.1. Verify that, if the square marked with X is ignored, the array in Fig. 15.29 has nested balance.
15.2. Check that the design in Fig. 15.3 has nested balance.
15.3. Check that the design in Fig. 15.6 is balanced for 1s but not for 0s.
15.4. Complete the proof of Lemma 6. (Cases (5) and (6) are fairly short.)
15.5. Complete the proof of Theorem 7.
15.6. Verify that the two designs in Fig. 15.30 are building designs.
15.7. Check that (some of) the design strings in Table 15.2 do give balanced designs.
15.8. Show that the array with first row

$$1\ 0\ 1\ 1\ 1\ 1\ 1\ 1\ 0\ 1\ 0\ 1\ 0\ 0\ 0\ 0\ 0\ 0\ 1\ 0,$$

and subsequent rows obtained by cycling three steps to the right, is balanced.
15.9. Another family of sequential arrays with sixfold symmetry are the snowflake arrays, two of which appear in Fig. 15.31. Let n denote the number of small shaded hexagons above the central blank hexagon. Show that the period of these arrays is given by $g(n) = 3(3n^2 + 3n + 1)$.
15.10. Construct the beehive design with radius 7 and basic triangle [0, 0, 0, 5, 2, 51, 6]. Calculate its profile.
15.11. Construct the threefold symmetric beehive design with radius 4 and basic array [1, 2, 3, 10; 0, 3, 7, 17]. Calculate its profile.
15.12. Construct an hexagonal building design with octal representation [75; 364]. What is the augmenting variety?

References

The numbers in square brackets which follow each entry indicate the chapters in which that entry is referred to in the text.

Archdeacon, D.S., Dinitz, J.H., Stinson, D.R., and Tillson, T.W. (1980). Some new row-complete Latin squares. *Journal of Combinatorial Theory* A **29**, 395–8. [14]

Baartmans, A., Danhof, K., and Tan, Soon-teck (1980). Quasi-residual quasi-symmetric designs. *Discrete Mathematics* **30**, 69–81. [12]

Bailey, R.A. (1977). Patterns of confounding in factorial designs. *Biometrika* **64**, 597–603. [10]

—— (1984). Quasi-complete Latin squares: Construction and randomization. *Journal of the Royal Statistical Society* B **46**, 323–34. [14]

——, Gilchrist, F.H.L., and Patterson, H.D. (1977). Identification of effects and confounding patterns in factorial designs. *Biometrika* **64**, 347–54. [10]

Baker, Ronald D. (1983). Resolvable BIBD and SOLS. *Discrete Mathematics* **44**, 13–29. [8]

Baranyai, Z. (1975). On the factorization of the complete uniform hypergraph. In *Infinite and finite sets*, Proceedings of the Colloquia Mathematica Societatis Janos Bolyai 10 (eds. A. Hajnal, R. Rado, and Vera T. Sos) pp. 91–108. North-Holland, Amsterdam. [2]

Batten, Lynn Margaret (1986). *Combinatorics of finite geometries.* Cambridge University Press, Cambridge. [8]

Baumert, Leonard D. (1971). *Cyclic difference sets*, Lecture Notes in Mathematics 182. Springer-Verlag, Heidelberg. [3, 4]

Besag, Julian (1974). Spatial interaction and the statistical analysis of lattice systems (with discussion). *Journal of the Royal Statistical Society* **B 36**, 192–225. [15]

Beth, T. (1983). Eine Bemerkung zur Abschätzung der Anzahl orthogonaler lateinscher Quadrat mittels Siebverfahren. *Abhandlungen aus dem Mathematischen Seminar der Universität Hamburg* **53**, 284–8. [7]

——, Jungnickel, D., and Lenz, H. (1985). *Design theory.* Bibliographisches Institut, Mannheim. [2]

Betten, D. (1983). Zum Satz von Euler–Tarry. *Der Mathematische und Naturwissenschaftliche Unterricht* **36**, 449–53. [6]

Beutelspacher, Albrecht (1986). 21 − 6 = 15: A connection between two distinguished geometries. *American Mathematical Monthly* **93**, 29–41. [8]

Bhat-Nayak, V.N. and Kane, V.D. (1984). Construction and irreducibility of quasi-multiple designs. *Ars Combinatoria* **18**, 113–29. [4]

—— and —— (1986). Two families of simple and irreducible BIBDs. *Discrete Mathematics* (to appear). [3, 4]

Bhattacharya, K.N. (1943). A note on two-fold triple systems. *Sankhya* **6**, 313–14. [3]

—— (1944). A new balanced incomplete block design. *Science and Culture* **9**, 508. [12]

Biggs, Norman (1974). *Algebraic graph theory*. Cambridge Tracts in Mathematics 67. Cambridge University Press, Cambridge. [11]

—— and White, Arthur T. (1979). *Permutation groups and combinatorial structures*, London Mathematical Society Lecture Note Series 33. Cambridge University Press, Cambridge. [4]

Billington, Elizabeth J. (1982). Further constructions of irreducible designs. *Congressus Numerantium* **35**, 77–89. [4]

—— (1984). Balanced *n*-ary designs: A combinatorial survey and some new results. *Ars Combinatoria* **17** A, 37–72. [14]

Boffey, T.B. and Veevers, A. (1979). On the non-existence of balanced designs for two-variety competition experiments. *Utilitas Mathematica* **16**, 131–43. [15]

Bos, I. (1983). Some remarks on honeycomb selection. *Euphytica* **32**, 329–35. [15]

Bose, R. C. (1939). On the construction of balanced incomplete block designs. *Annals of Eugenics* **9**, 353–99. [1, 3, 5]

—— (1942). A note on the resolvability of balanced incomplete block designs. *Sankhya* 6, 105–10. [8]

—— (1947). Mathematical theory of the symmetrical factorial design. *Sankhya* **8**, 107–66. [9]

—— (1949). A note on Fisher's inequality for balanced incomplete block designs. *Annals of Mathematical Statistics* **20**, 619–20. [2]

—— (1980). Combinatorial problems of experimental design II: Factorial designs. In *Combinatorial mathematics, optimal designs and their applications* (ed. J. Srivastava), *Annals of Discrete Mathematics* **6**, 7–18. North-Holland, Amsterdam. [9]

—— and Connor, W.S. (1952). Combinatorial properties of group divisible incomplete block designs. *Annals of Mathematical Statistics* **23**, 367–83. [11]

—— and Kishen, K. (1940). On the problem of confounding in the general symmetrical factorial design. *Sankhya* **5**, 21–36. [9]

—— and Nair, K.R. (1939). Partially balanced incomplete block designs. *Sankhya* **4**, 337–72. [11]

—— and —— (1962). Resolvable incomplete block designs with two replications. *Sankhya* A **24**, 9–24. [8]

—— and Shrikhande, S.S. (1960). On the composition of balanced incomplete block designs. *Canadian Journal of Mathematics* **12**, 177–88. [8]

——, ——, and Bhattacharya, K.N. (1953). On the construction of group divisible incomplete block designs. *Annals of Mathematical Statistics* **24**, 167–95. [11]

——, ——, and Parker, E.T. (1960). Further results on the construction of mutually orthogonal Latin squares and the falsity of Euler's conjecture. *Canadian Journal of Mathematics* **12**, 189–203. [1, 7]

——, ——, and Singhi, N.M. (1976). Edge-regular multigraphs and partial geometric designs with an application to the embedding of quasi-residual designs. *Atti del convegni Lincei* **17**, 49–81. [12]

Breach, D.R. and Street, Anne Penfold (1985). Irreducible designs from supplementary difference sets. *Bulletin of the Australian Mathematical Society* **31**, 105–15. [4]

Brouwer, A.E. (1979). Wilson's theory. In *Packing and covering in combinatorics* (ed. A. Schrijver), Mathematical Centre Tracts 106, pp.77–88, Amsterdam. [13]

Bruck, R.H. (1955). Difference sets in a finite group. *Transactions of the American Mathematical Society* **78**, 464–81. [3]

—— and Ryser, H.J. (1949). The non-existence of certain finite projective planes. *Canadian Journal of Mathematics* **1**, 88–93. [12]

Bugelski, B.R. (1949). A note on Grant's discussion of the Latin square principle in the design of experiments. *Physiological Bulletin* **46**, 49–50. [14]

Burnside, W. (1911). *Theory of groups of finite order*, 2nd edn. Cambridge University Press; reprinted (1955) Dover Publications. [4]

Bush, K.A. (1952). Orthogonal arrays of index unity. *Annals of Mathematical Statistics* **23**, 426–34. [9]

Butson, A.T. (1963). Relations among generalized Hadamard matrices, relative difference sets, and maximal length linear recurring sequences. *Canadian Journal of Mathematics* **15**, 42–8. [3]

Cameron, Peter J. (1976). *Parallelisms of complete designs*, London Mathematical Society Lecture Note Series 23. Cambridge University Press, Cambridge. [2]

Carmichael, Robert D. (1937). *Introduction to the theory of groups of finite order*. Dover Publications, New York; reprinted 1956. [8]

Chakrabarti, M.C. (1964). *Design of experiments*, Calcutta Research and Training School, Indian Statistical Institute (Summer Course (advanced) for Statisticians), Lecture Notes No. 1. [1]

Chang, L.-C. (1960). Association schemes of partially balanced designs with parameters $v = 28$, $n_1 = 12$, $n_2 = 15$ and $p_{11}^2 = 4$, *Science Record* (New Series) **4**, 12–18. [11]

——, Liu, C.-W., and Liu, W.-R. (1965). Incomplete block designs with triangular parameters for which $k \leq 10$ and $r \leq 10$. *Scientia Sinica* **4**, 329–38. [11]

Cheng, Ching-Shui (1983). Construction of optimal balanced incomplete block designs for correlated observations. *Annals of Statistics* **11**, 240–6. [14]

Chowla, S. (1949). A property of biquadratic residues. *Proceedings of the National Academy of Science of India* A **14**, 45–6. [3].

——, Erdös, P., and Straus, E.G. (1960). On the maximal number of pairwise orthogonal Latin squares of a given order. *Canadian Journal of Mathematics* **12**, 204–8. [7].

—— and Ryser, H.J. (1950). Combinatorial problems. *Canadian Journal of Mathematics* **2**, 93–9. [12]

Clatworthy, Willard H. (1973). *Tables of two-associate-class partially balanced designs*, National Bureau of Standards (US) Applied Mathematics Series No.63. [11]

Cochran, William G. and Cox, Gertrude M. (1957). *Experimental designs*, 2nd edn. John Wiley, New York. [6, 9].

Colbourn, Charles J., Manson, Karen J., and Wallis, W. D. (1984). Frames for two-fold triple systems. *Ars Combinatoria* **17**, 69–78. [6]

Colbourn, Marlene J. and Colbourn, Charles J. (1984). Recursive constructions for cyclic block designs. *Journal of Statistical Planning and Inference* **10**, 97–103. [3]

—— and Mathon, Rudolf A. (1980). On cyclic Steiner 2-designs. In *Topics on Steiner systems* (eds. C.C. Lindner and A. Rosa), *Annals of Discrete Mathematics* **7**, 215–53. North-Holland, Amsterdam. [3].

Collings, Bruce Jay (1984). Generating the intrablock and interblock subgroups for confounding in general factorial experiments. *Annals of Statistics* **12**, 1500–9. [10]

Connor, W.S., Jr. (1952). On the structure of balanced incomplete block designs.

Annals of Mathematical Statistics **23**, 57–71. [11]

—— and Clatworthy, W.H. (1954). Some theorems for partially balanced designs. *Annals of Mathematical Statistics* **25**, 100–12. [11].

Cormack, R.M. (1979). Spatial aspects of competition between individuals. In *Spatial and temporal analysis in ecology* (eds. R.M. Cormack and J.K. Ord), Statistical Ecology Series 8 (1979) pp.151–212, Satellite Program in Statistical Ecology, International Statistical Ecology Program, International Publishing House, Fairfield, Maryland, USA. [15]

Cotter, Sarah C. (1974). A general method of confounding for symmetrical factorial experiments. *Journal of the Royal Statistical Society* B **36**, 267–76. [10]

——, John, J.A., and Smith, T.M.F. (1973). Multi-factor experiments in non-orthogonal designs. *Journal of the Royal Statistical Society* B **35**, 361–7. [10]

Cox, David R. (1958). *Planning of experiments*. John Wiley, New York. [6]

David, H.A. (1967). Resolvable cyclic designs. *Sankhya* A **29**, 191–8. [8].

Dawson, Jeremy E. (1985). Equineighboured designs of block size four. *Ars Combinatoria* **19A**, 295–301. [14]

De Palluel, Cretté (1788). Sur les avantages and l'économie que procurent les racines employées a l'engrais des moutons a l'étable. *Mémoires d'Agriculture*, trimestre d'été, pp.17–23. English translation (1790): *Annals of Agriculture* **14**, 133–9. [1]

De Vries, H.L. (1984). Historical notes on Steiner systems. *Discrete Mathematics* **52**, 293–7. [1]

Dean, A.M. and John, J.A. (1975). Single replicate factorial designs in generalized cyclic designs: II. Asymmetrical arrangements. *Journal of the Royal Statistical Society* B **37**, 72–6. [10]

—— and Lewis, S.M. (1980). A unified theory for generalized cyclic designs. *Journal of Statistical Planning and Inference* **4**, 13–23. [10]

Dembowski, P. (1967). *Finite geometries*. Springer-Verlag, New York. [8]

Dénes, J. and Keedwell, A.D. (1974). *Latin squares and their applications*. English Universities Press, London. [1, 5, 6, 14]

Donald, C.M. (1963). Competition among crop and pasture plants. *Advances in Agronomy* **15**, 1–118. [15]

Doyen, J. (1972). Constructions of disjoint Steiner triple systems. *Proceedings of the American Mathematical Society* **32**, 409–16. [5]

—— and Rosa, Alexander (1980). An updated bibliography and survey of Steiner systems. In *Topics on Steiner systems* (eds. C.C. Lindner and A. Rosa), *Annals of Discrete Mathematics* **7**, 317–49. North-Holland, Amsterdam. [1]

Dudeney, H.E. (1917). *Amusements in mathematics*. Nelson, Edinburgh. [1]

Dyke, G.V., and Shelley, Christine F. (1976). Serial designs balanced for effects of neighbours on both sides. *Journal of Agricultural Science, Cambridge* **87**, 303–5. [14]

Euler, L. (1782). Recherches sur une nouvelle espèce de quarres magiques. *Verhandlingen Zeeuwach Genootschap Wetenschappen Vlissengen* **9**, 85–239; see also *Leonardi Euleri Opera Omnia*, Série 1, **7** (1923) 291–392. [1]

Evans, Trevor (1969). Universal algebra and Euler's officer problem. *American Mathematical Monthly* **86**, 466–73. [1]

Fasoulas, A. (1981). *Principles and methods of plant breeding*. Publication No.11, Department of Genetics and Plant Breeding, Aristotalian University of Thessaloniki, Greece. [15]

Federer, Walter T. (1980). Some recent results in experiment design with a bibliography. *International Statistical Review* **48**, 357–68; **49**, 95–109 and 185–97. [1]

Finney, D.J. and Outhwaite, Anne D. (1955). Serially balanced sequences, *Nature* **176**, 748. [14]

—— and —— (1956). Serially balanced sequences in bioassay. *Proceedings of the Royal Society* B **145**, 493–507. [14]

Fisher, R.A. (1926). The arrangement of field experiments. *Journal of Ministry of Agriculture* **33**, 503–13. [1, 9]

—— (1935–66). *The design of experiments*. Oliver & Boyd, Edinburgh. [1]

—— (1940). An examination of the different possible solutions of a problem in incomplete blocks. *Annals of Eugenics* **10**, 52–75. [2]

—— and Yates, F. (1938–63). *Statistical tables for biological, agricultural and medical research*. Oliver & Boyd, Edinburgh. [1]

Foody, W. and Hedayat, A. (1977). On theory and application of BIB designs with repeated blocks. *Annals of Statistics* **5**, 932–45; Corrigendum 7 (1979), 925. [1]

Franklin, M.F. (1984*a*). Cyclic generation of self-orthogonal Latin squares. *Utilitas Mathematica* **25**, 135–46. [6, 7]

—— (1984*b*). Cyclic generation of orthogonal Latin squares. *Ars Combinatoria* **17**, 129–39. [7]

—— (1984*c*). Triples of almost orthogonal 10×10 Latin squares useful in experimental design. *Ars Combinatoria* **17**, 141–6. [7]

Freeman, G.H. (1967). The use of cyclic balanced incomplete block designs for directional seed orchards. *Biometrics* **23**, 761–78. [14]

—— (1969). The use of cyclic balanced incomplete block designs for non-directional seed orchards. *Biometrics* **25**, 561–71. [14]

—— (1979). Some two-dimensional designs balanced for nearest neighbours. *Journal of the Royal Statistical Society* B **41**, 88–95. [14]

Ganley, Michael J. and Spence, Edward (1975). Relative difference sets and quasiregular collineation groups. *Journal of Combinatorial Theory* A **19**, 134–53. [3]

Gates, D.J. (1980). Competition between two types of plants with specified neighbour configurations. *Mathematical Biosciences* **48**, 195–209. [15]

Geramita, A.V. and Seberry, Jennifer (1979). *Orthogonal designs*. Marcel Dekker, New York. [2, 12]

Gilbert, William J. (1976). *Modern algebra with applications*. John Wiley, New York. [3]

Glynn, David J. (1978). Finite projective planes and related combinatorial structures. Ph.D. thesis, University of Adelaide. [3]

Graybill, Franklin A. (1961). *An introduction to linear statistical models*. McGraw-Hill, New York. [1]

Griffin, Harriet (1954). *Elementary theory of numbers*. McGraw-Hill, New York. [12]

Gupta, S.C. (1983) Some new methods for constructing block designs having orthogonal factorial structure. *Journal of the Royal Statistical Society* B **45**, 297–307. [10]

Hadamard, Jacques (1893). Résolution d'une question relative aux déterminants. *Bulletin des Sciences Mathématiques* **17**, 240–6. [12]

Hall, Marshall, Jr. (1956). A survey of difference sets. *Proceedings of the American Mathematical Society* **7**, 975–86. [3]

—— (1967). *Combinatorial theory*. Blaisdell Publishing Company, Waltham, Mass. [2, 3, 4, 8]

—— and Connor, W.S. (1953). An embedding theorem for balanced incomplete block designs. *Canadian Journal of Mathematics* **6**, 35–41. [11]

—— and Ryser, H.J. (1951). Cyclic incidence matrices. *Canadian Journal of Mathematics* **3**, 495–502. [4, 12]

Hanani, Haim (1961). The existence and construction of balanced incomplete block designs. *Annals of Mathematical Statistics* **32**, 371–86. [5, 13]

—— (1963). On some tactical configurations. *Canadian Journal of Mathematics* **15**, 702–22. [5]

—— (1975). Balanced incomplete block designs and related designs. *Discrete Mathematics* **11**, 255–369. [6, 13]

——, Ray-Chaudhuri, D.K., and Wilson, Richard M. (1972). On resolvable designs. *Discrete Mathematics* **3**, 343–57. [13]

Harshbarger, B. (1947). *Rectangular lattices*, Virginia Agricultural Experimental Station Memoir 1. [8]

—— (1950). Triple rectangular lattices. *Biometrics* **5**, 1–13. [8]

Hedayat, A. and Li Shuo-Yen, R. (1979). The trade-off method in the construction of BIB designs with repeated blocks. *Annals of Statistics* **7**, 1277–87. [1]

—— and —— (1980). Combinatorial topology and the trade-off method in BIB designs. In *Combinatorial mathematics, optimal designs and their applications* (ed. J. Srivastava), *Annals of Discrete Mathematics* **6**, 189–200. North-Holland, Amsterdam. [1]

—— and Shrikhande, S.S. (1971). Experimental designs and combinatorial systems associated with Latin squares and sets of mutually orthogonal Latin squares. *Sankhya* A **33**, 423–32. [6]

Heinrich, Katherine (1977). Approximation to a self-orthogonal Latin square of order 6. *Ars Combinatoria* **4**, 17–24. [6]

Hirschfeld, J.W.P. (1979). *Projective geometries over finite fields*. Clarendon Press, Oxford. [8]

Hoffman, A.J. (1960). On the uniqueness of the triangular association scheme. *Annals of Mathematical Statistics* **31**, 492–7. [11]

Hoffman, D. G. (1984). Private communication. [4]

——, Schellenberg, P., and Vanstone, S.A. (1976). A starter-adder approach to equidistant permutation arrays and generalised permutation arrays. *Ars Combinatoria* **1**, 307–19. [1]

Hohn, Franz E. (1973). *Elementary matrix algebra*, 3rd edn. Collier-Macmillan, London. [8, 11]

Honsberger, Ross (1976). *Mathematical gems II*, Dolciani Mathematical Expositions. Mathematical Association of America (No.2). [12]

Horton, J.D. (1974). Sub-Latin squares and incomplete orthogonal arrays. *Journal of Combinatorial Theory* A **16**, 23–33. [6]

Hughes, D.R. (1957). Collineations and generalized incidence matrices. *Transactions of the American Mathematical Society* **86**, 284–96. [12]

—— and Piper, F.C. (1973). *Projective planes*. Springer-Verlag, New York. [8]

—— and —— (1985). *Design theory*. Cambridge University Press, Cambridge. [1]

Hung, Stephen H.Y. and Mendelsohn, N.S. (1973). Directed triple systems. *Journal of Combinatorial Theory* A **14**, 310–18. [14]

Husain, Q.M. (1945). On the totality of the solutions for the symmetrical incomplete block designs: $\lambda = 2$, $k = 5$ or 6. *Sankhya* **7**, 204–8. [2]

Hwang, F.K. (1973). Constructions for some classes of neighbour designs. *Annals of Statistics* **1**, 786–90. [14]

—— and Lin, S. (1974). A direct method to construct triple systems. *Journal of Combinatorial Theory*. A **17**, 84–94. [3, 14]

—— and —— (1976). Construction of 2-balanced (n, k, λ) arrays. *Pacific Journal of Mathematics* **64**, 437–53. [14]

—— and —— (1977). Neighbour designs. *Journal of Combinatorial Theory* A **23**, 303–13. [14]

—— and —— (1978). Distributions of integers into k-tuples with prescribed conditions. *Journal of Combinatorial Theory* A **25**, 105–16. [14]

Jeffcott, Barbara and Spears, William T. (1977). Decompositions of difference sets: A generalization of balanced weighing matrices. *Proceedings of the Eighth Southeastern Conference on Combinatorics, Graph Theory and Computing*, (eds. F. Hoffman *et al.*) pp. 385–414. Utilitas Mathematica, Winnipeg. [3]

John, J.A. (1966). Cyclic incomplete block designs. *Journal of the Royal Statistical Society* **B 28**, 345–60. [8]

—— (1981). Factorial-balance and the analysis of designs with factorial structure, *Journal of Statistical Planning and Inference* **5**, 99–105. [10]

—— and Dean, A. M. (1975). Single replicate factorial experiments in generalized cyclic designs: I. Symmetrical arrangements. *Journal of the Royal Statistical Society* **B 37**, 63–71. [10]

—— and Williams, E.R. (1982). Conjectures for optimal block designs. *Journal of the Royal Statistical Society* **B 44**, 221–5. [1]

——, Wolock, F.W., and David, H.A. (1972). *Cyclic designs*, National Bureau of Standards (US) Applied Mathematics Series No.62. [8]

Johnson, D.M., Dulmage, A.L., and Mendelsohn, N.S. (1961). Orthomorphisms of groups and orthogonal Latin squares 1. *Canadian Journal of Mathematics* **13**, 356–72. [6, 7]

Kageyama, Sanpei (1974). Reduction of associate classes for block designs and related combinatorial arrangements, *Hiroshima Mathematical Journal* 4, 527–618. [11]

—— (1976). Some observations on affine μ-resolvable incomplete block designs. *Bulletin of the Faculty of Education, Hiroshima University* 25, 9–14. [8]

—— and Tsuji, Takumi (1977). Characterization of certain incomplete block designs. *Journal of Statistical Planning and Inference* **1**, 151–61. [8]

Kelly, Graham (1982). On the uniqueness of embedding of residual designs. *Discrete Mathematics* **39**, 153–60. [12]

Kempthorne, Oscar (1952). *The design and analysis of experiments*. John Wiley, New York. [9]

Kiefer, J. (1980). Optimal design theory in relation to combinatorial design. In *Combinatorial mathematics, optimal designs and their applications* (ed. J. Srivastava), *Annals of Discrete Mathematics* **6**, 225–41. North-Holland, Amsterdam. [1]

—— (1981). The interplay of optimality and combinatorics in experimental design. *Canadian Journal of Statistics* **9**, 1–10. [1]

—— and Wynn, H.P. (1981). Optimum balanced block and Latin square designs for correlated observations. *Annals of Statistics* **9**, 737–57. [14]

Kirkman, T.P. (1847). On a problem in combinations. *Cambridge and Dublin Mathematics Journal* **2**, 191–204. [1]

—— (1850*a*). Note on an unanswered prize question. *Cambridge and Dublin Mathematics Journal* **5**, 255–62. [1]

—— (1850*b*), Query VI on "Fifteen young ladies . . . ", *Lady's and Gentleman's Diary* No.147, 48. [8]

König, D. (1950). *Theorie der endlichen und unendlichen Graphen.* Chelsea Publishing Company, New York. [2]

Kramer, Earl S. (1974). Indecomposable triple systems. *Discrete Mathematics* **8**, 173–80. [4]

Kurkjian, B. and Zelen, M. (1962). A calculus for factorial arrangements. *Annals of Mathematical Statistics* **33**, 600–19. [10]

—— and —— (1963). Applications of the calculus of factorial arrangements I. Block and direct product designs. *Biometrika* **50**, 63–73. [10]

Lander, Eric S. (1983). *Symmetric designs: an algebraic approach*, London Mathematical Society Lecture Note Series 74. Cambridge University Press, Cambridge. [4]

Lawless, J.F. (1970). Block intersections in quasi-residual designs. *Aequationes Mathematicae* **5**, 40–6. [12]

Lehmer, Emma (1953). On residue difference sets. *Canadian Journal of Mathematics* **5**, 425–32. [3]

Leonard, Philip A. (1986). Cyclic relative difference sets. *American Mathematical Monthly* **93**, 106–11. [3]

Lewis, S.M. (1982). Generators for asymmetrical factorial experiments. *Journal of Statistical Planning and Inference* **6**, 59–64. [10]

—— and Dean, A.M. (1980). Factorial experiments in resolvable generalized cyclic designs. *BIAS* **7**, 159–67. [10]

Lidl, Rudolf and Niederreiter, Harald (1983). Finite fields. *Encyclopedia of mathematics and its applications*, Vol. 20. Addison-Wesley, Reading, Mass. [3]

Lin, Chuang-Sheng and Morse, Pamela M. (1975). A compact design for spacing experiments. *Biometrics* **31**, 661–71. [15]

Lindner, C.C. (1980). A survey of embedding theorems for Steiner systems. In *Topics on Steiner systems* (eds. C.C. Lindner and A. Rosa), *Annals of Discrete Mathematics* **7**, 175–202. North-Holland, Amsterdam. [5]

—— and Stinson, D. R. (1984). Steiner pentagon systems. *Discrete Mathematics* **52**, 67–74. [14]

—— and Street, Anne Penfold (1986). Disjoint designs, and irreducible designs without repeated blocks. *Ars Combinatoria* **21A**, 229–36. [4]

Lorimer, Peter (1980). An introduction to projective planes: Some of the properties of a particular plane of order 16. *Mathematical Chronicle* **9**, 53–66. [8]

Lu, Jia-Xu (1983). Large sets of disjoint Steiner triple systems, I, II, III. *Journal of Combinatorial Theory* A **34**, 140–46, 147–55, and 156–82. [2]

—— (1984). Large sets of disjoint Steiner triple systems, IV, V, VI. *Journal of Combinatorial Theory* A **37**, 136–63; 164–88; 189–92. [2]

Macdonald, Sheila Oates and Street, Anne Penfold (1979). Balanced binary arrays II: the triangular grid. *Ars Combinatoria* **8**, 65–84. [15]

Mann, Henry B. (1965). *Addition theorems: The addition theorems of group theory and number theory*. Interscience Publishers/John Wiley, New York. [4]

Mandl, Robert (1985). Orthogonal Latin squares: An application of experiment design to compiler testing. *Communications of the Association for Computing Machinery* **28**, 1054–8. [7]

Martin, Frank B. (1973). Beehive designs for observing variety competition. *Biometrics* **29**, 397–402. [15]

Mathon, R.A., Phelps. K.T., and Rosa, A. (1983). Small Steiner triple systems and their properties. *Ars Combinatoria* **15**, 3–110. [5]

—— and Rosa, A. (1977). A census of Mendelsohn triple systems of order nine. *Ars Combinatoria* **4**, 309–15. [2]

—— and —— (1985). Tables of parameters of BIBDs with $r \leqslant 41$ including existence, enumeration and resolvability results. In *Algorithms in combinatorial design theory* (eds. C.J. Colbourn and M.J. Colbourn), *Annals of Discrete Mathematics* **26**, 275–308. North-Holland, Amsterdam. [2, 8]

Mavron, V.C., Mullin, R.C., and Rosa, A. (1984). Further properties of generalized residual designs. *Journal of Statistical Planning and Inference* **10**, 59–68. [12]

McLean, Robert A. and Anderson, Virgil L. (1984). *Applied factorial and fractional designs*. Marcel Dekker, New York. [9]

Mead, R. (1979). Competition experiments, *Biometrics* **35**, 41–54. [15]

—— and Riley, Janet (1981). A review of statistical ideas relevant to intercropping research. *Journal of the Royal Statistical Society* A **144**, 462–509. [15]

—— and Stern, R.D. (1980). Designing experiments for intercropping research. *Experimental Agriculture* **16**, 329–42. [15]

Mendelsohn, N.S. (1971). Latin squares orthogonal to their transposes. *Journal of Combinatorial Theory* A **11**, 187–9. [5]

Moore, E.H. (1893). Concerning triple systems. *Mathematische Annalen* **43**, 271–85. [5]

Mordell, L.J. (1969). *Diophantine equations*. Academic Press, London. [12]

Morgan, Elizabeth J. (1977). Some small quasi-multiple designs. *Ars Combinatoria* **3**, 233–50. [2]

Mukerjee, R. (1979). Inter-effect orthogonality in factorial experiments. *Calcutta Statistical Association Bulletin* **28**, 83–108. [10]

—— (1981). Construction of effect-wise orthogonal factorial designs. *Journal of Statistical Planning and Inference* **5**, 221–9. [10]

—— and Kageyama, Sanpei (1985). On resolvable and affine resolvable variance-balanced designs. *Biometrika* **72**, 165–72. [8]

Nair, C. Ramankutty (1967). Sequences balanced for pairs of residual effects. *Journal of the American Statistical Association* **62**, 205–25. [14]

Nelder, J.A. (1962). New kinds of systematic designs for spacing experiments. *Biometrics* **18**, 283–307. [15]

—— (1965). The analysis of randomized experiments with orthogonal block structure I: Block structure and the null analysis of variance. *Proceedings of the Royal Society* A **283**, 147–62. [10]

Netto, E. (1893). Zur Theorie der Tripelsysteme. *Mathematische Annalen* **42**, 143–52. [3]

Neumaier, A. (1982). Quasi-residual 2-designs, 1 1/2-designs and strongly regular multigraphs. *Geometriae Dedicata* **12**, 351–66. [12]

Oates-Williams, Sheila and Street, Anne Penfold (1979). Balanced binary arrays III: the hexagonal grid. *Journal of the Australian Mathematical Society* A **28**, 479–98. [15]

Parker, E.T. (1957). On collineations of symmetric designs. *Proceedings of the American Mathematical Society* **8**, 350–1. [4]

Paterson, Lindsay (1983). Circuits and efficiency in incomplete block designs. *Biometrika* **70**, 215–25. [2]

Patterson, H.D. (1965). The factorial combination of treatments in rotation experiments. *Journal of Agricultural Science* **65**, 171–82. [10]

—— (1976). Generation of factorial designs. *Journal of the Royal Statistical Society* B **38**, 175–79. [10]

—— and Bailey, R.A. (1978). Design keys for factorial experiments. *Applied Statistics* **27**, 335–43. [10]

—— and Williams, E.R. (1975). Some theoretical results on general block designs. *Proceedings of the Fifth British Combinatorial Conference* (eds. C. St. J. Nash-Williams and J. Sheehan), *Congressus Numerantium* **15**, 489–96. Utilitas Mathematica, Winnipeg. [8]

—— and —— (1976). A new class of resolvable incomplete block designs. *Biometrika* **63**, 83–92. [8]

Pearce, S.C. (1983). *The agricultural field experiment*. John Wiley, London. [1, 6]

Peltesohn, Rose (1938). Eine Losung der beiden Hefferschen Differenzenprobleme. *Compositio Mathematica* **6**, 251–7. [3]

Plucker, J. (1835). *System der analytischen Geometrie, auf neue Betrachtung-weisen gegrundet, und insbesondere eine ausführliche Theorie der Curven dritter Ordnung enthaltend*. Duncker und Humblot, Berlin. [1]

—— (1839). *Theorie der algebraischen Curven, gegrundet auf eine neue Behandlungweise der analytischen Geometrie*. Marcus, Bonn. [1]

Preece, D.A. (1966). Some balanced incomplete block designs for two sets of treatments. *Biometrika* **53**, 497–506. [6]

—— (1967). Nested balanced incomplete block designs. *Biometrika* **54**, 479–86. [2]

—— (1976). Non-orthogonal Graeco-Latin designs. *Combinatorial Mathematics IV, Proceedings of the Fourth Australian Conference* (eds. Louis R.A. Casse and Walter D. Wallis), Lecture Notes in Mathematics 560, pp. 7–26. Springer-Verlag, Heidelberg [6]

—— (1982). Balance and designs: Another terminological tangle. *Utilitas Mathematica* **21**C, 85–186. [8]

—— (1983). Latin squares, Latin cubes, Latin rectangles etc. In Kotz-Johnson, *Encyclopedia of statistical sciences*, Vol. 4, 504–10. John Wiley, New York. [1]

Puri, P.D. and Nigam, A.K. (1977). Balanced block designs. *Communications in Statistics – Theory and Methods* A **6**, 1171–9. [1].

Raghavarao, Damaraju (1971). *Constructions and combinatorial problems in design of experiments*. John Wiley, New York. [2, 3, 8, 9, 11]

Raj, Des (1971). *Sampling theory*. Tata/McGraw-Hill, New Delhi. [1]

Raktoe, B.L., Hedayat, A., and Federer, W.T. (1981). *Factorial designs*. John Wiley, New York. [9]

Rao, C.R. (1946). Hypercubes of strength ‘d’ leading to confounded designs in factorial experiments. *Bulletin of the Calcutta Mathematical Society* **38**, 67–78. [9]

Ray-Chaudhuri, D.K. and Wilson, Richard M. (1971). Solution of Kirkman's schoolgirl problem. *Proceedings of the Symposium on Mathematics XIX*, American Mathematical Society, Providence, Rhode Island, pp. 187–203. [8, 13]

—— and —— (1973). Existence of resolvable block designs. In *Survey of combinatorial theory*, pp. 361–75. [13]

Rees, D.H. (1967). Some designs of use in serology. *Biometics* **23**, 779–91. [14]

Reiss, M. (1858). Über eine Steinersche combinatorische Aufgabe. *Journal für die reine und angewandte Mathematik* **56**, 273–80. [5]

Robinson, Peter J. (1986). Balanced cyclic binary arrays. *Ars Combinatoria* **21**, 189–99. [15]

Robinson, Philip Wylie (1984). How to construct a nearly Kirkman triple system. M.Sc. thesis, Auburn University, Auburn, Alabama. [13]

Rohde, Charles A. (1966). Some results on generalized inverses. *Society for Industrial and Applied Mathematics Review* **8**, 201–5. [10]

Room, T.G. and Kirkpatrick, P.B. (1971). *Miniquaternion geometry: an introduction to the study of projective planes*, Cambridge Tracts in Mathematics and Mathematical Physics 60. Cambridge University Press, Cambridge. [8]

Russell, K.G. (1986). An algorithm for the construction of optimal block designs when observations from within a block may be correlated. In *Pacific Statistical Congress* (ed. I. S. Francis, B. F. J. Manly, and F. C. Lam), pp. 163–6. Elsevier–North Holland, Amsterdam. [14].

Ryser, Herbert John (1963). *Combinatorial mathematics*. Mathematical Association of America/John Wiley, New York. [2]

Sampford, M.R. (1957). Methods of construction and analysis of serially balanced sequences. *Journal of the Royal Statistical Society* B **19**, 286–304. [14]

Schellenberg, P.J., Van Rees, G.H.J., and Vanstone, S.A. (1978). Four pairwise orthogonal Latin squares of order 15. *Ars Combinatoria* **6**, 141–50. [7]

Schutzenberger, M.P. (1949). A non-existence theorem for an infinite family of symmetrical block designs. *Annals of Eugenics* **14**, 286–7. [12]

Searle, S.R. (1971). *Linear models*. John Wiley, New York. [1, 10]

Seberry, Jennifer (1979). A note on orthogonal Graeco-Latin designs. *Ars Combinatoria* **8**, 85–94. [6]

—— and Skillicorn, David (1980). All directed BIBDs with $k = 3$ exist. *Journal of Combinatorial Theory* A **29**, 244–8. [14]

Shanks, Daniel (1978). *Solved and unsolved problems in number theory*, 2nd edn. Chelsea Publishing Company, New York. [12]

Shrikhande, S.S. (1950). The impossibility of certain symmetrical balanced incomplete block designs. *Annals of Mathematical Statistics* **21**, 106–11. [12]

—— (1959). On a characterisation of the triangular association scheme. *Annals of Mathematical Statistics* **30**, 39–47. [11]

—— (1960). Relations between certain incomplete block designs. In *Contributions to probability and statistics*, pp. 388–95. Stanford University Press, Stanford. [11]

—— (1976). Affine resolvable balanced incomplete block designs: A survey. *Aequationes Mathematicae* **14**, 251–69. [8]

—— and Raghavarao, D. (1963). Affine α-resolvable incomplete block designs. In *Contributions to statistics*, presented to Prof. P.C. Mahalanobis on the occasion of his 70th birthday (ed. C.R. Rao), pp. 471–80. Pergamon Press, Oxford. [8]

Silvey, S.D. (1980). *Optimal design: An introduction to the theory for parameter estimation*. Chapman & Hall, New York. [1]

Singer, James (1938). A theorem in finite projective geometry and some applications to number theory. *Transactions of the American Mathematical Society* **43**, 377–85. [3, 8]

Skolem, T. (1958). Some remarks on the triple systems of Steiner. *Mathematica Scandinavica* **6**, 273–80. [5]

Sprott, D.A. (1954). A note on balanced incomplete block designs. *Canadian Journal of Mathematics* **6**, 341–6. [3]

—— and Stanton, R.G. (1964). Block intersections in balanced incomplete block designs. *Canadian Mathematical Bulletin* **7**, 539–48. [2]

Stanton, R.G. and Goulden, I.P. (1981). Graph factorisations, general triple systems and cyclic triple systems. *Aequationes Mathematicae* **22**, 1–28. [2, 5]

——, Mullin, R.C., and Lawless, J.F. (1969). Quasi-residual designs. *Aequationes Mathematicae* **2**, 274–81. [12]

—— and Sprott, D.A. (1958). A family of difference sets. *Canadian Journal of Mathematics* **10**, 73–7. [3]

Steinberg, David M. and Hunter, William G. (1984). Experimental design: Review and comment. *Technometrics* **26**, 71–130. [1]

Steiner, J. (1853). Combinatorische Aufgabe. *Journal für die reine und angewandte Mathematik* **45**, 181–2. [1]

Stewart, Ian (1973). *Galois theory*. Chapman & Hall, London. [3]

Stinson, D.R. (1984). A short proof of the non-existence of a pair of orthogonal Latin squares of order six. *Journal of Combinatorial Theory* A **36**, 373–6. [6]

—— and Wallis, W.D. (1983*a*). Snappy constructions for triple systems. *Australian Mathematical Society Gazette* **10**, 84–8. [5]

—— and —— (1983*b*). Triple systems without repeated blocks. *Discrete Mathematics* **47**, 125–8. [5]

Storer, Thomas (1967). *Cyclotomy and difference sets*. Markham Publishing Company, Chicago. [3]

Street, Anne Penfold (1982). A survey of neighbour designs. *Congressus Numerantium* **34**, 119–55. [14, 15]

—— (1985). A survey of irreducible balanced incomplete block designs. *Ars Combinatoria* **19** A, 43–60. [4]

—— and Macdonald, Sheila Oates (1979). Balanced binary arrays I: The square grid. *Combinatorial Mathematics VI, Proceedings of the Sixth Australian Conference* (eds. A.F. Horadam and W.D. Wallis), Lecture Notes in Mathematics 748, pp. 165–98. Springer-Verlag, Heidelberg. [15]

Street, Deborah J. (1981). Graeco-Latin and nested row and column designs. *Combinatorial Mathematics VIII, Proceedings of the Eighth Australian Conference* (ed. Kevin L. McAvaney), Lecture Notes in Mathematics 884, pp. 304–13. Springer-Verlag, Heidelberg. [6]

—— (1982). A difference set construction for inversive planes. *Combinatorial Mathematics IX, Proceedings of the Ninth Australian Conference* (eds. Elizabeth J. Billington, Sheila Oates-Williams, and Anne Penfold Street), Lecture Notes in Mathematics 952, pp. 419–22. Springer-Verlag, Heidelberg. [3]

—— (1986). Unbordered two-dimensional nearest neighbour designs. *Ars Combinatoria* (to appear). [14]

—— and Seberry, Jennifer (1980). All DBIBDs with block size four exist. *Utilitas Mathematica* **18**, 27–34. [14]

—— and Street, Anne Penfold (1985). Designs with partial neighbour balance. *Journal of Statistical Planning and Inference* **12**, 47–59. [14]

—— and Wilson, William H. (1980). On directed balanced incomplete block designs with block size five. *Utilitas Mathematica* **18**, 161–74. [14]

—— and Wilson, William H. (1985). Balanced designs for two-variety competition experiments. *Utilitas Mathematica* **28**, 113–20. [15]

Tarry, G. (1900–01). Le probleme des 36 officiers. *Compte rendu Association française pour l'Avancement des Sciences* **1**, 122–3; **2**, 170–203. [1]

Teirlinck, Luc (1977). On making two Steiner triple systems disjoint. *Journal of Combinatorial Theory* A **23**, 343–50. [4]

Todorov, D.T. (1985). Three mutually orthogonal Latin squares of order 14. *Ars Combinatoria* **20**, 45–8. [7]

Van Buggenhaut, J. (1974). On the existence of the 2-designs $S_2(2, 3, v)$ without repeated blocks. *Discrete Mathematics* **8**, 105–9. [5]

Van der Waerden, B.L. (1950). *Modern Algebra*, Vol. II (translated by Theodore J. Benac). Frederick Ungar, New York. [13]

Van Lint, J.H. and Tonchev, V.D. (1984). Nonembeddable quasi-residual designs with large *k*. *Journal of Combinatorial Theory* A **37**, 359–62. [12]

Vanstone, S.A. (1982). Private communication. [13]

Veblen, O. and Bussey, W.H. (1906). Finite projective geometries. *Transactions of the American Mathematical Society* **7**, 241–59. [1, 8]

—— and Wedderburn, J.H.M. (1907). Non-Desarguesian and non-Pascalian geometries. *Transactions of the American Mathematical Society* **8**, 379–88. [1, 8]

Veevers, Alan (1982). Balanced designs for observing intra-variety nearest-neighbour interactions. *Euphytica* **31**, 465–8. [15]

—— and Boffey, T.B. (1975). On the existence of levelled beehive designs. *Biometrics* **31**, 963–7. [15]

—— and —— (1979). Designs for balanced observation of plant competition. *Journal of Statistical Planning and Inference* **3**, 325–31. [15]

——, ——, and Zafar-Yab, M. (1980). Competition experiments for two varieties. In *Proceedings of the Tenth European Meeting of Statisticians*. Zeuven, Belgium. [15]

—— and Zafar-Yab, M. (1980). Balanced designs for two-component competition experiments on a square lattice. *Euphytica* **29**, 459–64. [15]

—— and —— (1982). A two-variety square lattice design balanced under Besag's coding scheme. *Journal of the Royal Statistical Society* **B 44**, 47–8. [15]

Venables, W.N. (1985). Lecture notes on linear models (to appear). Personal communication. [1]

Von Staudt, K.G.C. (1856). *Beitrage zur Geometrie der Lage*, Vol. 1. F. Korn, Nurnberg. [1, 8]

Wahua, T.A.T. and Miller, D.A. (1978). Relative yield totals and yield components of intercropped sorghum and soybeans. *Agronomy Journal* **70**, 287–91. [15]

Wallis, Jennifer Seberry (1976). On the existence of Hadamard matrices. *Journal of Combinatorial Theory* A **21**, 188–95. [12]

Wallis, W.D. (1984). Three orthogonal Latin squares. *Congressus Numerantium* **42**, 69–86. [7]

——, Street, Anne Penfold and Wallis, Jennifer Seberry (1972), *Combinatorics: Room squares, sum-free sets, Hadamard matrices*, Lecture Notes in Mathematics 292. Springer-Verlag, Heidelberg. [2, 12]

Wang, S.P. (1978). On self-orthogonal Latin squares and partial transversals of Latin squares. Ph.D. thesis, Ohio State University. [7]

Whiteman, Albert Leon (1962). A family of difference sets. *Illinois Journal of Mathematics* **6**, 107–21. [3]

Williams, E.J. (1949). Experimental designs balanced for the estimation of residual effects of treatments. *Australian Journal of Scientific Research* A **2**, 149–68. [14]

Williams, E.R. and Bailey, R.A. (1981). A note on designs for neighbor configurations. *Mathematical Biosciences* **56**, 153–4. [15]

Williams, R.M. (1952). Experimental designs for serially correlated observations. *Biometrika* **39**, 151–67. [14]

Wilson, Richard M. (1972*a*). An existence theory for pairwise balanced designs, I, II. *Journal of Combinatorial Theory* A **13**, 220–45 and 246–73. [13]

—— (1972*b*). Cyclotomy and difference families in elementary Abelian groups. *Journal of Number Theory* **4**, 17–47. [3, 13]

—— (1974*a*). Concerning the number of mutually orthogonal Latin squares. *Discrete Mathematics* **9**, 181–98. [7]

—— (1974*b*). A few more squares. *Congressus Numerantium* **10**, 675–80. [7]

—— (1974*c*). Construction and uses of pairwise balanced designs. In *Combinatorial theory* (eds. M. Hall Jr. and J.H. van Lint) pp. 18–41, Mathematical Centre Tracts 55, Amsterdam. [13]

—— (1975). An existence theory for pairwise balanced designs. III. *Journal of Combinatorial Theory* A **18**, 71–9. [13]

Woolhouse, W. S. B. (1844). Prize Question 1733. *Lady's and Gentleman's Diary.* [1]

Yates, F. (1935). Complex experiments (with discussion). *Journal of the Royal Statistical Society*, Suppl. 2, 181–247. [1, 9]

—— (1936). Incomplete randomized blocks. *Annals of Eugenics* **7**, 121–40. [1]

—— (1940*a*). The recovery of inter-block information on balanced incomplete block designs. *Annals of Eugenics* **10**, 317–25. [8]

—— (1940*b*). Lattice squares. *Journal of Agricultural Science* **30**, 672–87. [8]

Some references to experimental work

Many of the designs discussed in this book have been, and are being, used in experimental work. Some of the references already cited describe one, or more, experiments. In addition the following brief (and incomplete) list gives some papers which describe experiments in which designs mentioned in the text have been used, or in which the advantages and disadvantages of various experimental layouts are considered.

Latin squares

Peters, P. A. (1984). Evaluation of log attachment methods by a Latin square design. *Transactions of the American Society of Agricultural Engineers* **27**, 382–4.

Lattice designs

Fielding, W. J. and Killick, R. J. (1983). The use of generalised lattice designs in potato variety trials (some preliminary results). *Journal of Agricultural Science, Cambridge* **101**, 59–62.

Factorial designs

Cloughley, J. B., Grice, W. J., and Ellis, R. T. (1983). Effects of harvesting policy and nitrogen application rates on the production of tea in Central Africa. I. Yield and crop distribution. *Experimental Agriculture* **19**, 33–46.

Kayode, G. O. (1985). Effects of NPK fertilizer on tuber yield, starch content and dry matter accumulation of white guinea yam (Dioscarea rotundata) in a forest alfisol of south western Nigeria. *Experimental Agriculture* **21**, 389–93.

Province, C. A., Squires, E. L., Pickett, B. W., and Amann, R. P. (1985). Cooling rates, storage temperatures and fertility of extended equine spermatozoa. *Theriogenology* **23**, 925–34.

Competition designs

Antonovics, J. and Fowler, N. L. (1985). Analysis of frequency and density effects on growth in mixtures of *Salvia splendens* and *Linium grandiflorum* using hexagonal fan designs. *Journal of Ecology* **73**, 219–34.

Finch, S., Skinner, G., and Freeman, G. H. (1976). The effect of plant density on populations of the cabbage root fly on four cruciferous crops. *Annals of Applied Biology* **83**, 191–7.

Putnam, D. H., Herbert, S. J., and Vargas, A. (1985). Intercropped corn-soyabean density studies I. Yield complementarity. *Experimental Agriculture* **21**, 41–51.

Rees, D. J. (1986). Crop growth, development and yield in semi-arid conditions in Botswana. I. The effects of population density and row spacing on *Sorghum bicolor*. *Experimental Agriculture* **22**, 153–67.

Neighbour designs

Hancock, T. W. and Mayo, O. (1986). Regression analysis of the interstate wheat variety trials- series 15 year 2 (1985). In *Proceedings of the Fifth Assembly, Wheat Breeding Society of Australia* (ed. R. McLean), pp. 161–74. Western Australian Department of Agriculture.

Comparisons of different designs

Melchinger, A. E. (1984). Analysis of incomplete factorial mating designs. *Vortrage für Pflanzenzuchtung* **7**, 131–50.

Panda, S. K. and Singh, H. P. (1982). Application of PBIBD designs(2) in analysing a diallel cross experiment with parents and F_1s. *Journal of the Indian Society of Agricultural Statistics* **34**, 116–17.

Patterson, H. D. and Hunter, E. A. (1983). The efficiency of incomplete block designs in National List and Recommended List cereal variety trials. *Journal of Agricultural Science, Cambridge* **101**, 427–33.

Index